JO ANNE STOLAROFF COTSEN PRIZE
IMPRINT

The Excavation of the Prehistoric Burial Tumulus at Lofkënd, Albania Volume 2: Illustrations

The Excavation of the Prehistoric Burial Tumulus at Lofkënd, Albania
Volume 2: Illustrations

John K. Papadopoulos, Sarah P. Morris,

Lorenc Bejko, and Lynne A. Schepartz

With contributions by
Esmeralda Agolli, Jamie D. Aprile, Brian N. Damiata, J.E. Foss, Shpresa Gjongecaj, Christopher Johanson, Rovena Kurti, Richard MacDonald, John M. Marston, Samantha Martin-McAuliffe, Laura Menez, Vanessa Muros, Yannis Mylonas, Seth Pevnick, David A. Scott, Sarah C. Sherwood, John Southon, Lyssa C. Stapleton, George Theodorou, Evi G.Vardala-Theodorou, Evangelos Velitzelos

Survey, site plans, and drawings by Max Farrar and Samantha Martin-McAuliffe

Object drawings by Ilir Zaloshnja

Photography by Richard MacDonald (2004–2007),
Anna MacDonald (2006–2007),
and Ian Coyle (2008)

Monumenta Archaeologica 34
Cotsen Institute of Archaeology Press
University of California, Los Angeles
2014

The Cotsen Institute of Archaeology Press is the publishing unit of the Cotsen Institute of Archaeology at UCLA. The Cotsen Institute is a premier research organization dedicated to the creation, dissemination, and conservation of archaeological knowledge and heritage. It is home to both the Interdepartmental Archaeology Graduate Program and the UCLA/Getty Master's Program in the Conservation of Archaeological and Ethnographic Materials. The Cotsen Institute provides a forum for innovative faculty research, graduate education, and public programs at UCLA in an effort to impact positively the academic, local and global communities. Established in 1973, the Cotsen Institute is at the forefront of archaeological research, education, conservation, and publication and is an active contributor to interdisciplinary research at UCLA.

The Cotsen Institute Press specializes in producing high-quality academic volumes in several different series, including *Monographs, World Heritage and Monuments, Cotsen Advanced Seminars,* and *Ideas, Debates, and Perspectives.* The Press is committed to making the fruits of archaeological research accessible to professionals, scholars, students, and the general public. We are able to do this through the generosity of Lloyd E. Cotsen, longtime Institute volunteer and benefactor, who has provided an endowment that allows us to subsidize our publishing program and produce superb volumes at an affordable price. Publishing in nine different series, our award-winning archaeological publications receive critical acclaim in both the academic and popular communities.

Edited by Barbara Kohl
Designed by Carol Leyba
Cover design by Randi Danforth
Index by Matthew White

This publication was undertaken with the assistance of the Institute for Aegean Prehistory

Library of Congress Cataloging-in-Publication Data

Papadopoulos, John K., 1958-
The excavation of the prehistoric burial tumulus at Lofkënd, Albania / by John K. Papadopoulos, Sarah P. Morris, Lorenc Bejko, and Lynne A. Schepartz ; with contributions by Esmeralda Agolli, Jamie D. Aprile, Brian N. Damiata, J.E. Foss, Shpresa Gjongecaj, Christopher Johanson, Rovena Kurti, Richard MacDonald, John M. Marston, Samantha Martin-McAuliffe, Laura Menez, Vanessa Muros, Yannis Mylonas, Seth Pevnick, David A. Scott, Sarah C. Sherwood, John Southon, Lyssa C. Stapleton, George Theodorou, Evi G.Vardala-Theodorou, Evangelos Velitzelos ; survey, site plans, and drawings by Max Farrar and Samantha Martin-McAuliffe ; object drawings by Ilir Zaloshnja ; photography by Richard MacDonald (2004-2007), Anna MacDonald (2006-2007), and Ian Coyle (2008).
pages cm — (Monumenta archaeologica ; 34)
Includes bibliographical references.
ISBN 978-1-938770-00-5 (hardback)
1. Illyria—Antiquities. 2. Illyria—Social life and customs. 3. Illyria—Environmental conditions. 4. Lofkënd (Albania)—Antiquities. 5. Excavations (Archaeology)—Albania—Lofkënd. 6. Mounds—Albania—Lofkënd. 7. Bronze age—Albania—Lofkënd. 8. Iron age—Albania—Lofkënd. 9. Social archaeology—Albania—Lofkënd. 10. Antiquities, Prehistoric—Albania—Lofkënd. I. Morris, Sarah P., 1954- II. Bejko, Lorenc. III. Schepartz, Lynne Alison. IV. Agolli, Esmeralda. V. Title.

DR39.5.P36 2014
939'.865—dc23

2014015415

Contents

Unless otherwise noted, all photographs are by Rich MacDonald and all drawings are by Ilir Zaloshnja.

Figures

Chapter 1 Illustrations

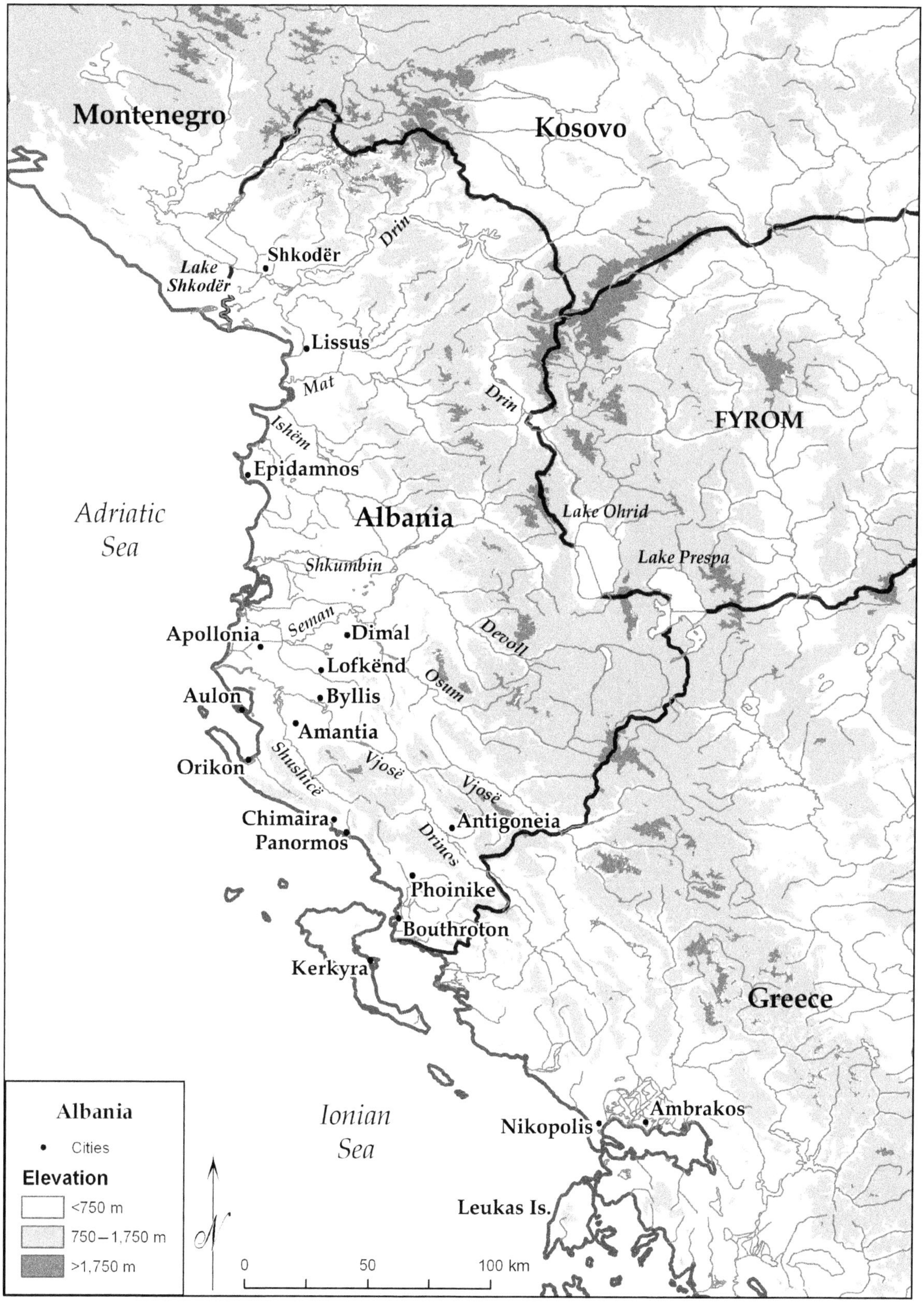

FIGURE 1.1 Map of Albania and neighboring regions showing principal sites in Albania (Stanislav Parfenov)

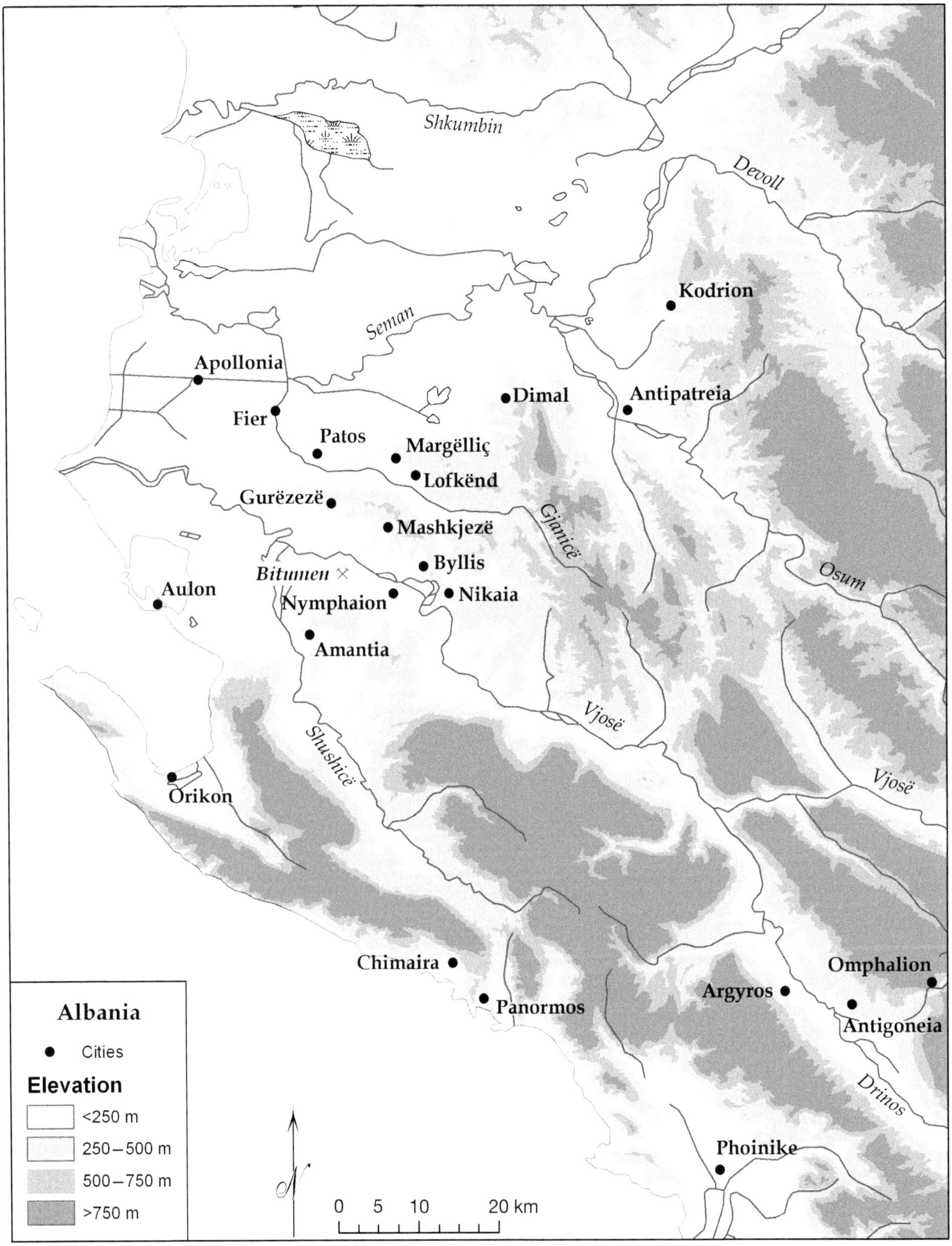

FIGURE 1.2 Map of the immediate region around Lofkënd tumulus (Stanislav Parfenov)

FIGURE 1.3 View of the Lofkënd tumulus from south in 2004

FIGURE 1.4 View of the Lofkënd tumulus from west in 2004

FIGURE 1.5 Margelliç looking west from Lofkënd: (a) from Lofkënd village; (b) close-up view from the Lofkënd tumulus; (c) distant view from the tumulus

a

b

c

FIGURE 1.6 One of the many oil pumps in the Mallakastër Hills, a few kilometers south-southeast of the tumulus, which is just visible in the skyline on the upper left

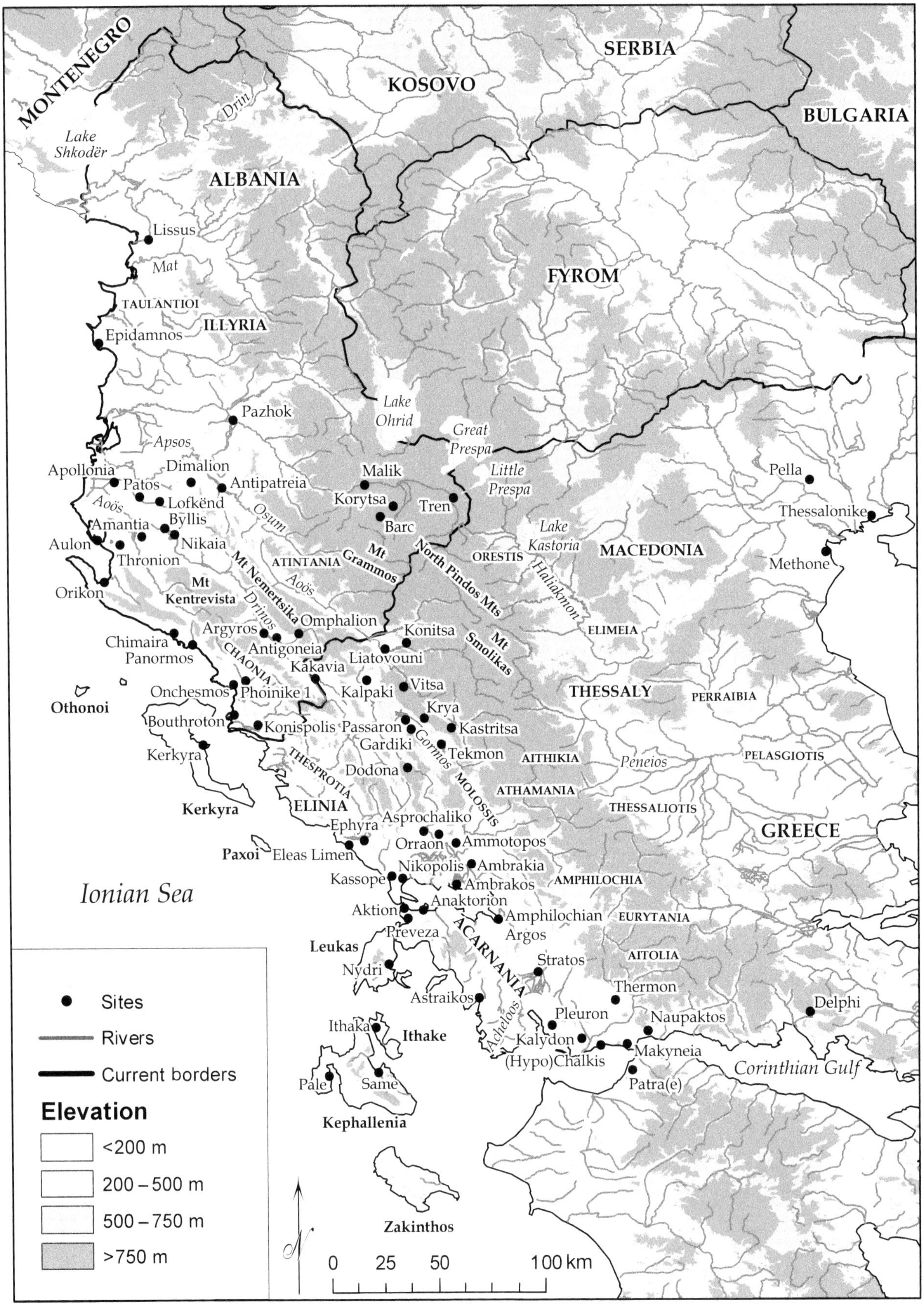

FIGURE 1.7 Map of Albania and northwest Greece showing principal sites and territory of the *ethnē* or tribes (UCLA Experiential Technology Center)

FIGURE 1.8 Exposed human remains at the steep southern scarp of the tumulus in 2003, view from south-west (photo Sarah Morris and John Papadopoulos)

FIGURE 1.9 Exposed bedrock base on which the tumulus was constructed. View from south-southeast at the end of the 2007 season

FIGURE 1.10 The tumulus at Patos in the 1970s (photo courtesy Muzafer Korkuti)

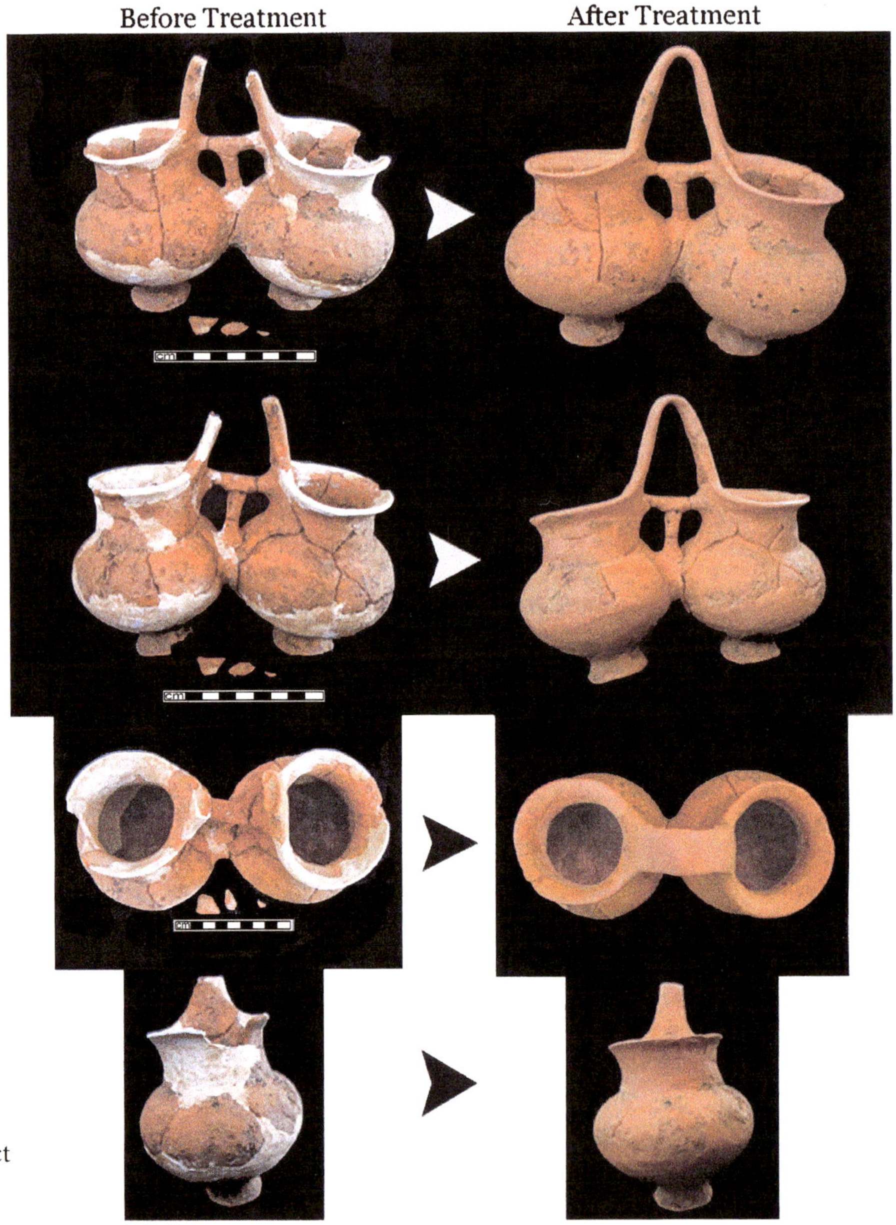

FIGURE 1.11 Double pot from Grave 25 at the tumulus of Patos, now in the Fier Museum, shown before and after conservation by the Lofkënd Archaeological Project (photos Naoko Fukumaru)

FIGURE 1.12 Transportation of troops and ammunition up the Louros River in the Greco-Turkish War of 1912

FIGURE 1.13 Selected pottery from the surface of the Lofkënd tumulus (2003): **9/11**, **9/28–9/29**, **9/88**, **9/279**, prehistoric; **9/330** probably Classical

a

b

FIGURE 1.14 (a) Uninventoried surface pottery from the tumulus collected in 2003 (photo 4068). (b) Uninventoried surface pottery and roof tile from the tumulus collected in 2004 (photo 4077)

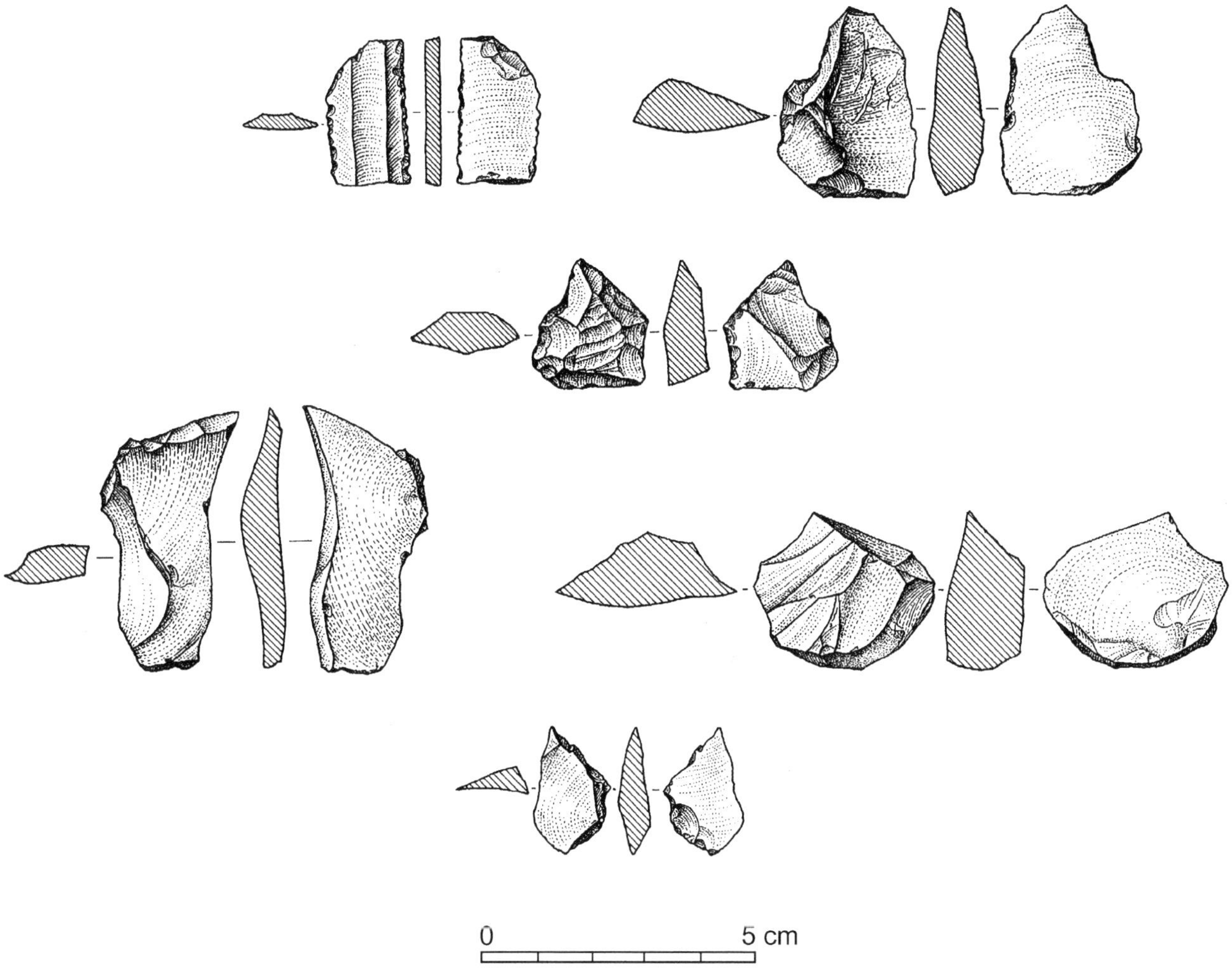

FIGURE 1.15 Chipped stone tools from the surface of the Lofkënd tumulus: *top row*, **13/3** and **13/37**; *second row*, **13/82**; *third row*, **13/89** and **13/59**; *bottom row:* **13/36** (Papadopoulos, Bejko, and Morris 2007:134, fig. 29; see also Chapter 13)

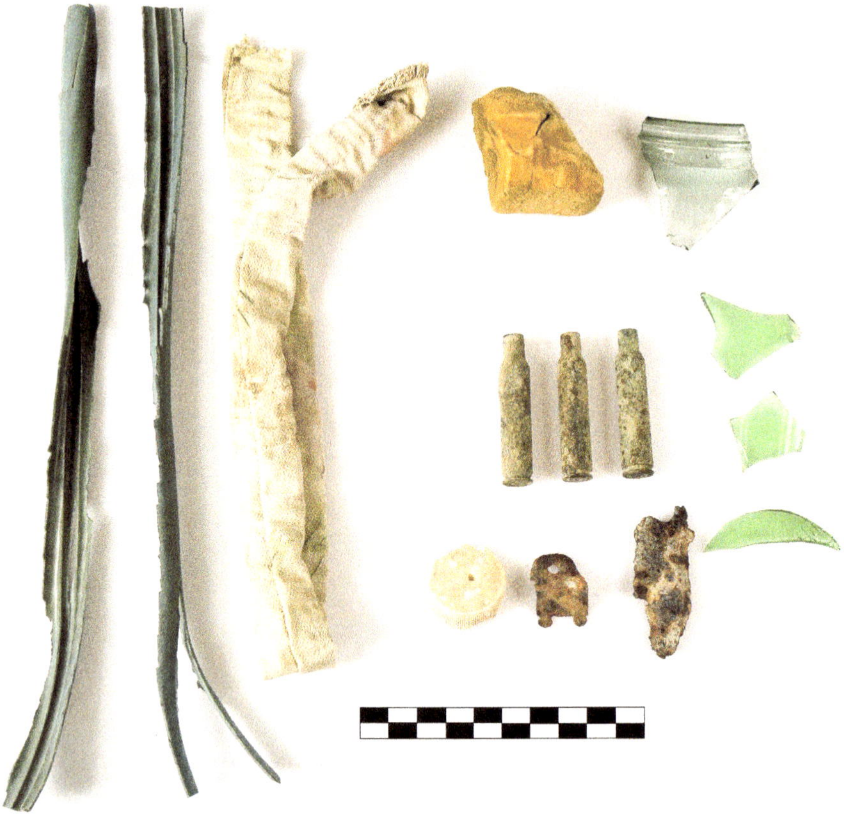

a

Figure 1.16 (a) Surface clearance of tumulus prior to excavation, June 21, 2004: gun cartridges and other modern surface finds (photo 4120); (b) surface clearance of tumulus prior to excavation, June 22, 2004: gun cartridges, glass, and other modern surface finds (photo 4121)

b

Chapter 2 Illustrations

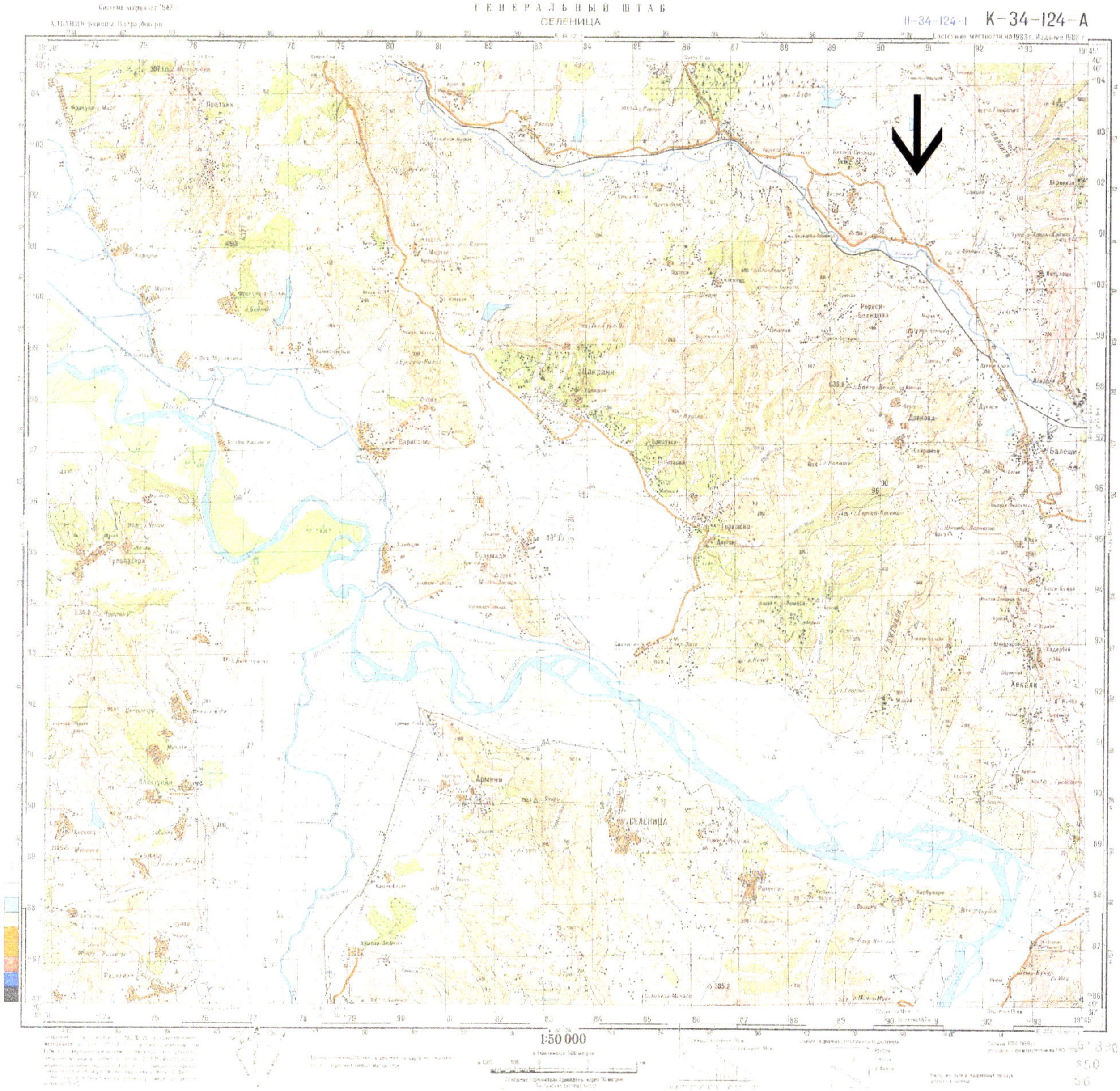

FIGURE 2.1 Albanian Army map (1:50,000) with Vjosë River running diagonally through central portion and region around Lofkënd in top right corner

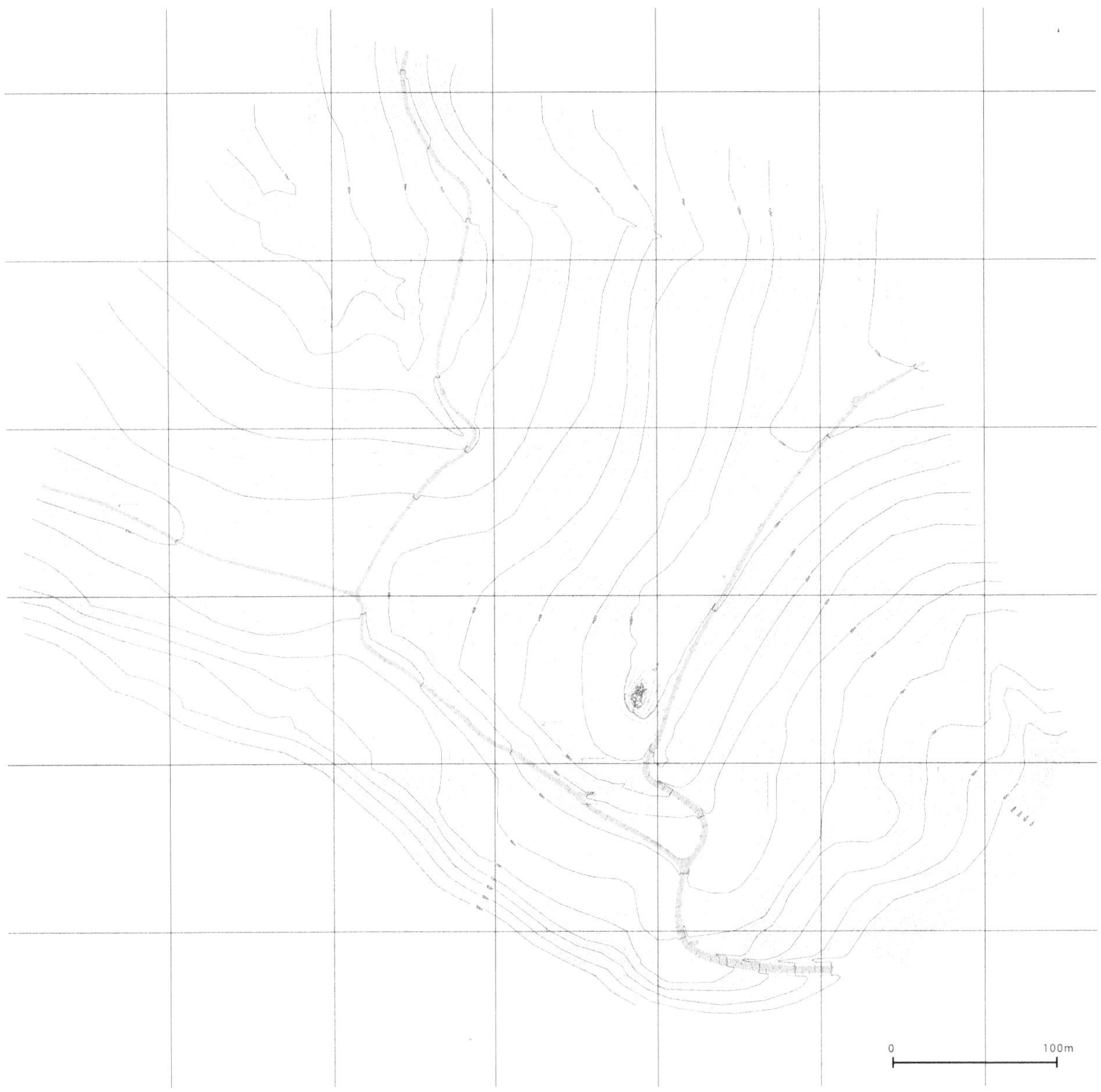

FIGURE 2.2 Location of the Lofkënd tumulus on a contour map of the immediate region (grid 100 × 100 m) (drawing by Max Farrar, formatted by Itay Zaharovitz, in Papadopoulos, Bejko, and Morris 2007:113, fig. 5)

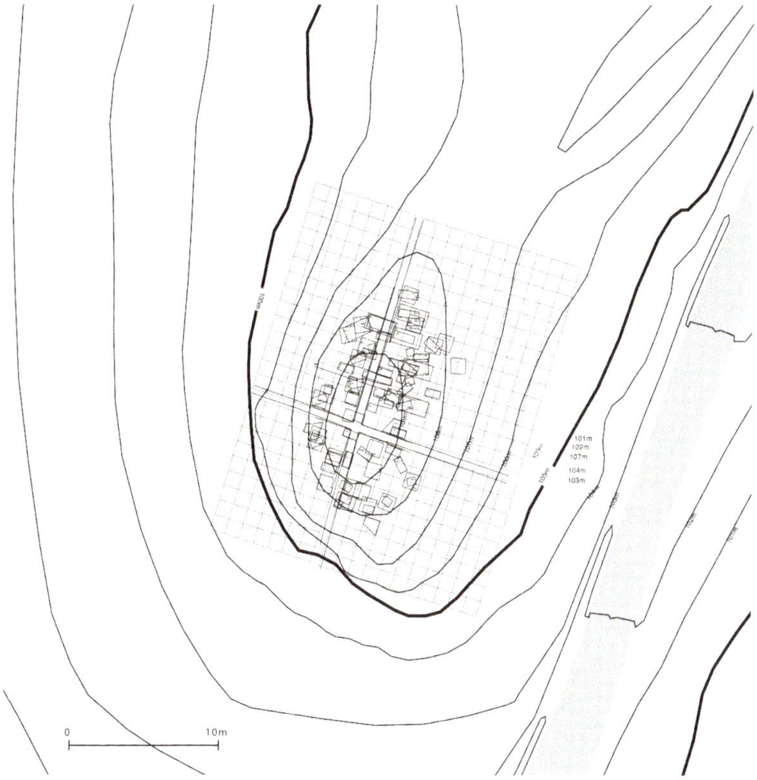

FIGURE 2.3 Plan and 1-m^2 grid of the tumulus (Max Farrar and Itay Zaharovitz, in Papadopoulos, Bejko, and Morris 2007:113, fig. 6)

FIGURE 2.4 Lofkënd 2006: use of wooden tools for the excavation of all graves

a

b

FIGURE 2.5 (a) View from the northeast showing the very beginning of excavation in 2004 (photo 075); (b) view from the north-northwest of the tumulus at end of the 2004 campaign (photo 459)

Figure 2.6 View from the south-southeast of the burial tumulus of Lofkënd at center (at the highest point) with the modern Muslim cemetery of Ngrançija in foreground (photo 085)

Figure 2.7 View from the east showing part of the baulk separating sectors 1 and 2 at the end of the 2006 season (photo 1632)

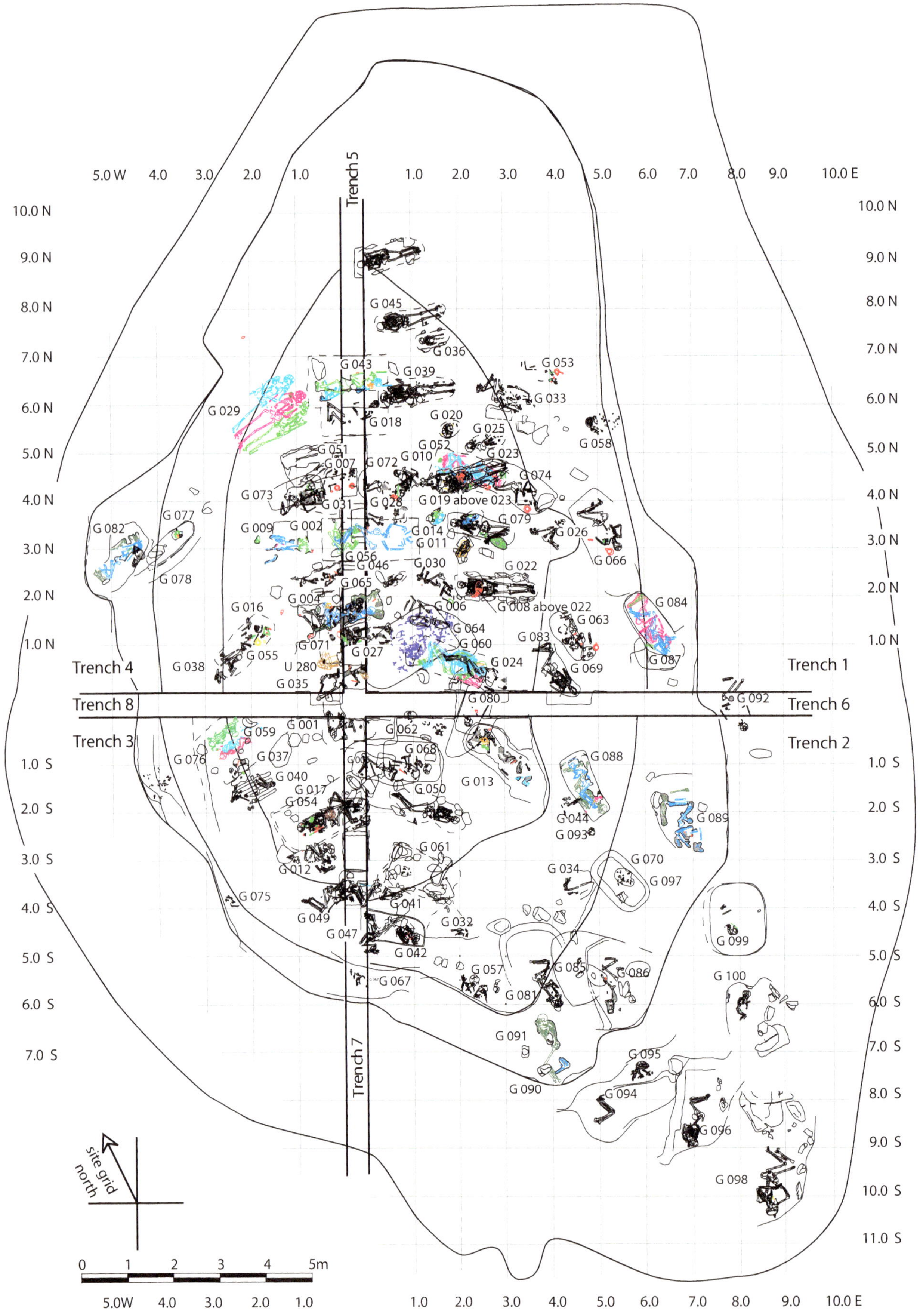

FIGURE 2.8 Final plan of the Lofkënd tumulus at the end of the last season of excavations in 2007, showing all tombs (Max Farrar)

FIGURE 2.9 View of the bedrock platform, from north-northwest, on which the earthen tumulus was heaped; the wooden poles indicate the original height of the mound (photo 1760)

FIGURE 2.10 View from the west of the bedrock platform on which the tumulus was constructed; the wooden poles indicate the original height of the mound (photo 1801)

a

b

Figure 2.11 Two views of the 10×4-m trench laid out and excavated at the southeast of the tumulus to ensure that no tombs were overlooked: (a) from north; (b) from south-southeast (photos 1776–1778)

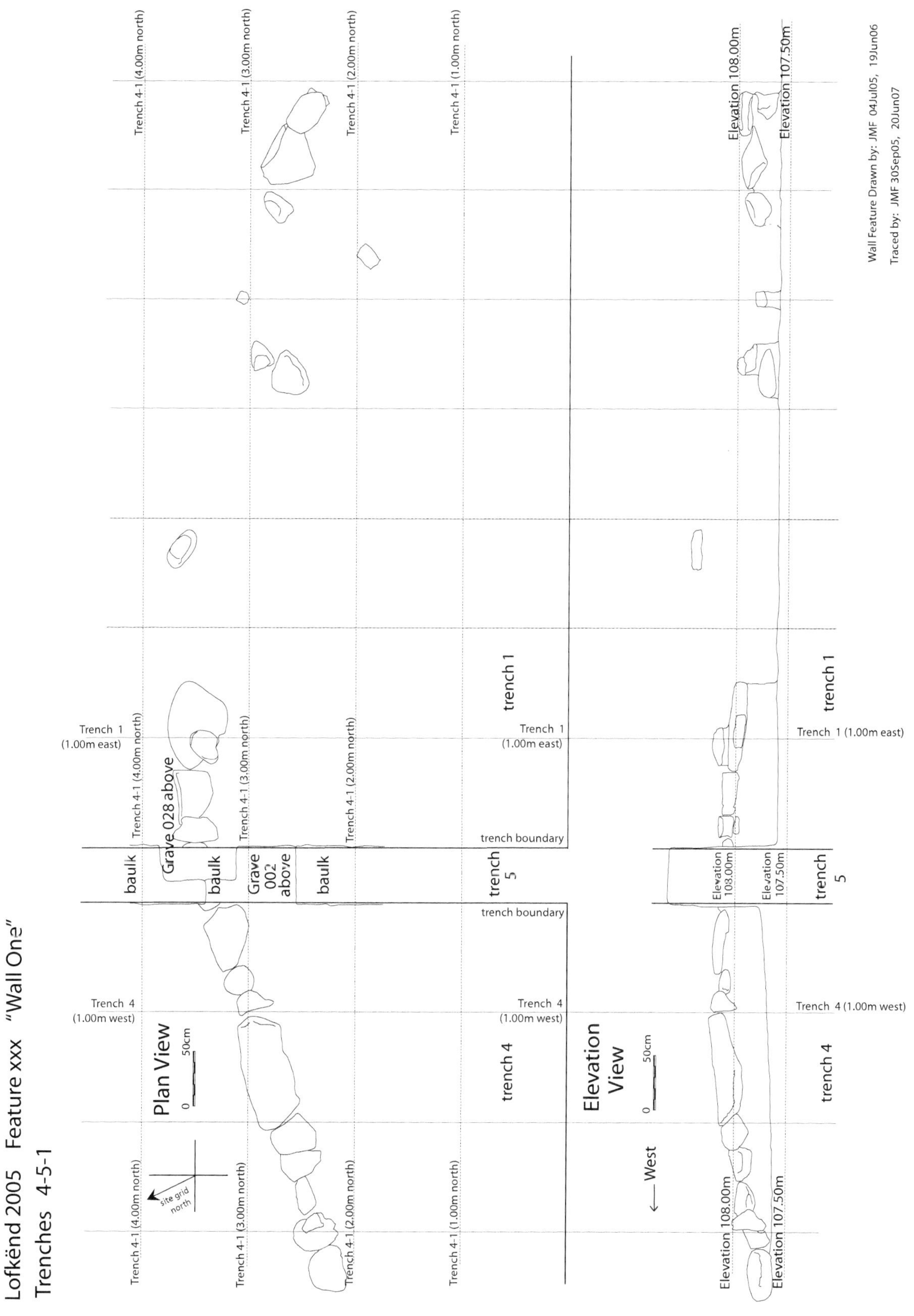

FIGURE 2.12 (a) Plan and section of the feature referred to as "Wall 1" (Max Farrar);
(b) *on next page*

FIGURE 2.12 (*Continued from previous page*). (b) Plan and section of "Wall 1" together with the "gulley" immediately to the north

a b

Figure 2.13 Two views of a large stone, at various stages of excavation, standing upright as encountered, in the fill of the tumulus in the baulk Sector/Trench 8: (a) from south (photo 202; (b) from west-northwest (photo 257)

Figure 2.14 View from south of the stones from Wall 1 and from the fill of the tumulus used as part of the backfill of the mound following the reconstruction of the baulks with sun-dried mud bricks (photo 7835)

FIGURE 2.15 View of the gulley, from the east (photo 1812)

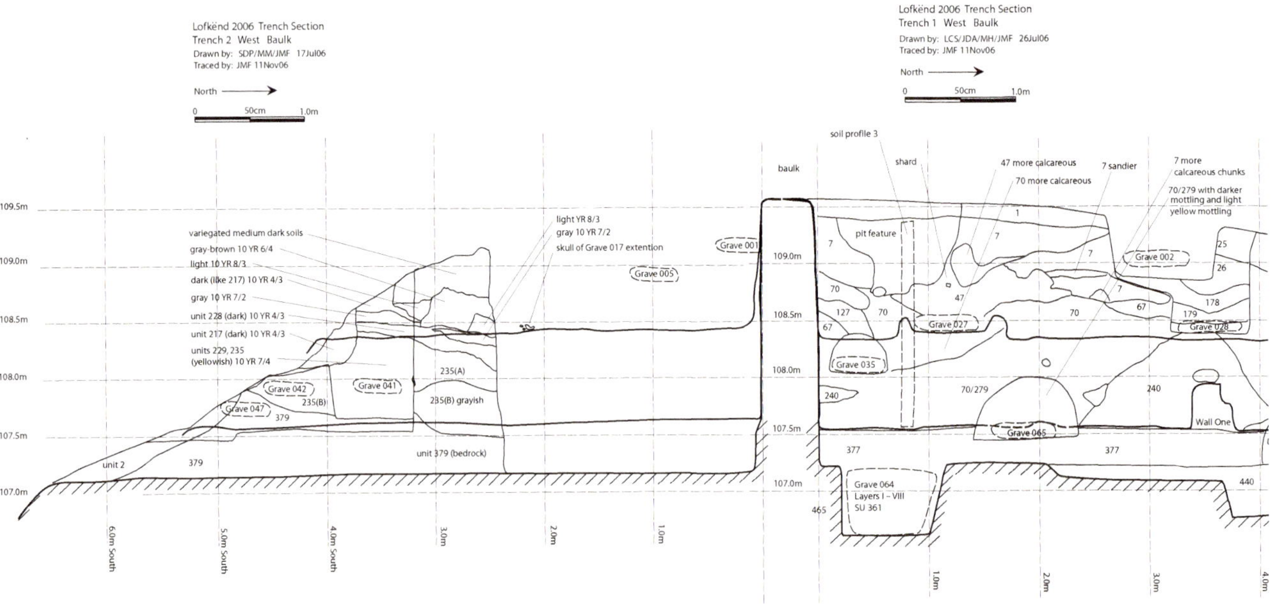

FIGURE 2.16 *(Facing pages)*. North-south section through Sectors/Trenches 1 and 2 (Max Farrar)

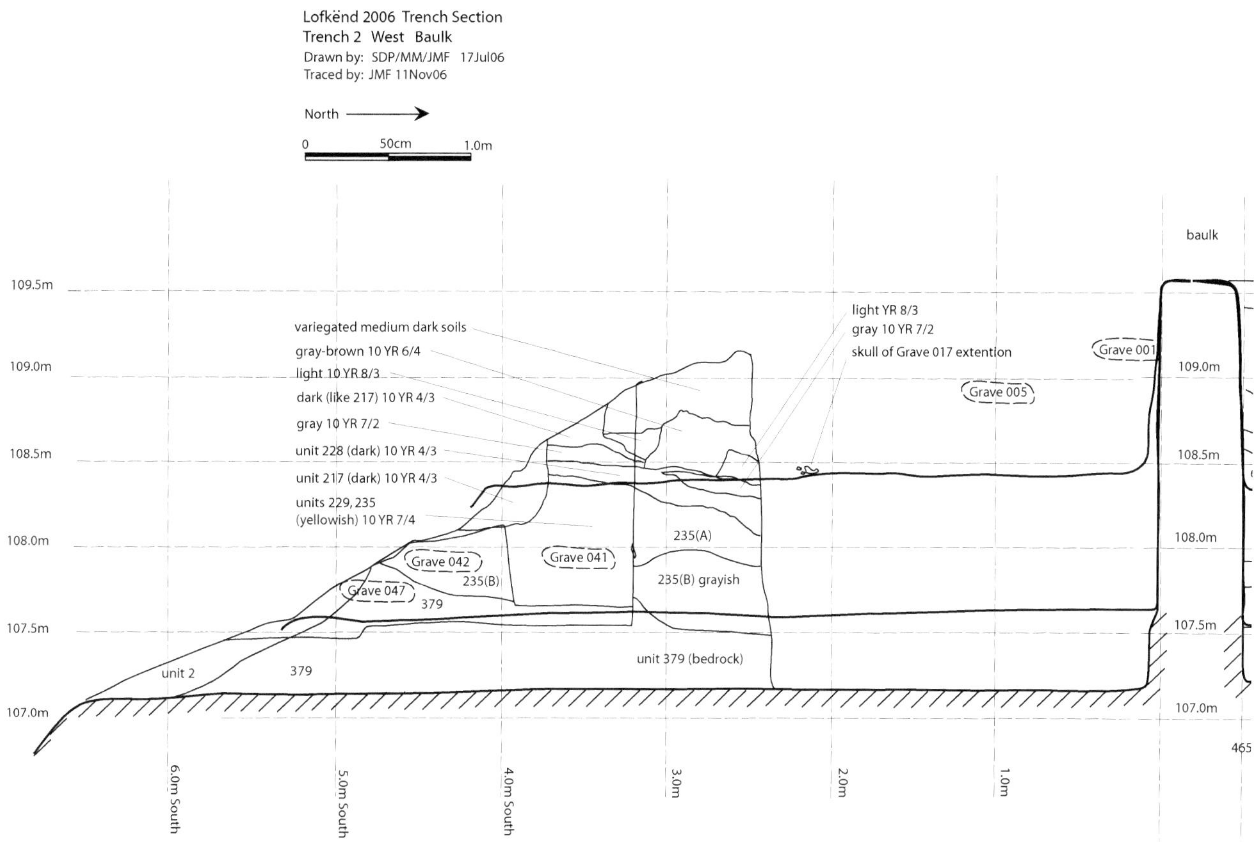

FIGURE 2.16A Detail view of FIGURE 2.16, left side (*see also FIGURES 16B–C on following pages*)

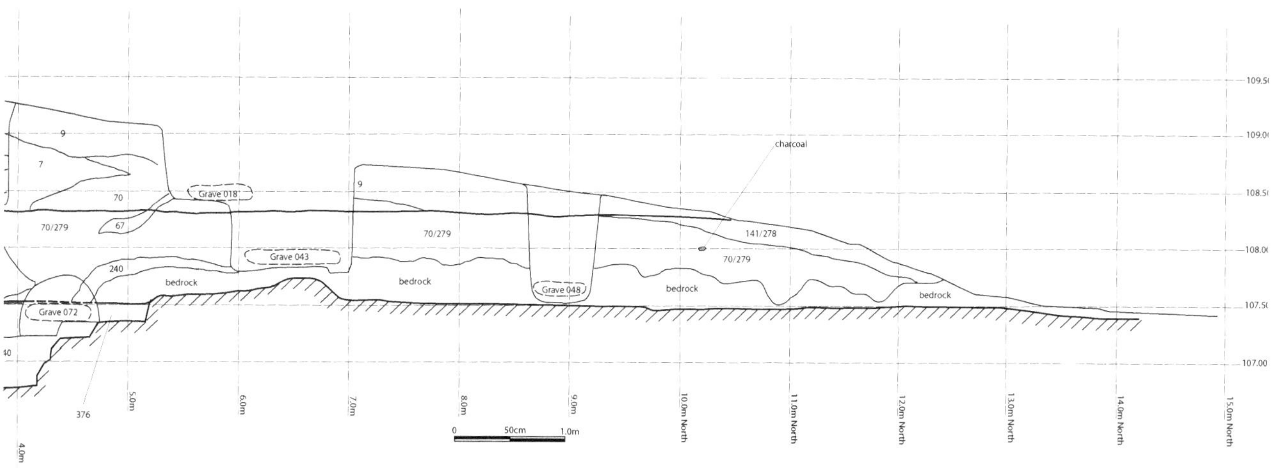

FIGURE 2.16 *(Continued from facing page)*. North-south section through Sectors/Trenches 1 and 2 (Max Farrar); see detail views in FIGURES 2.16A, B, C

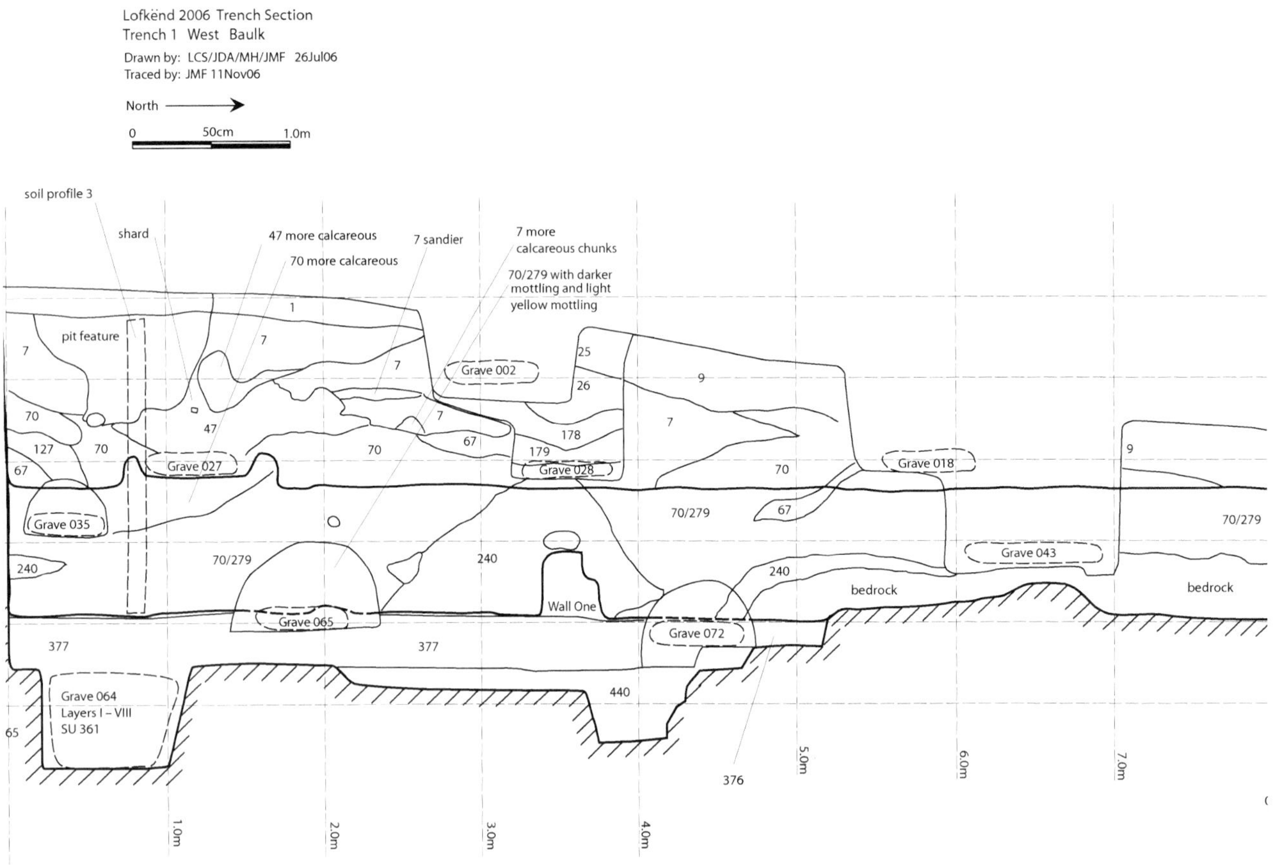

FIGURE 2.16B Detail view of FIGURE 2.16, middle

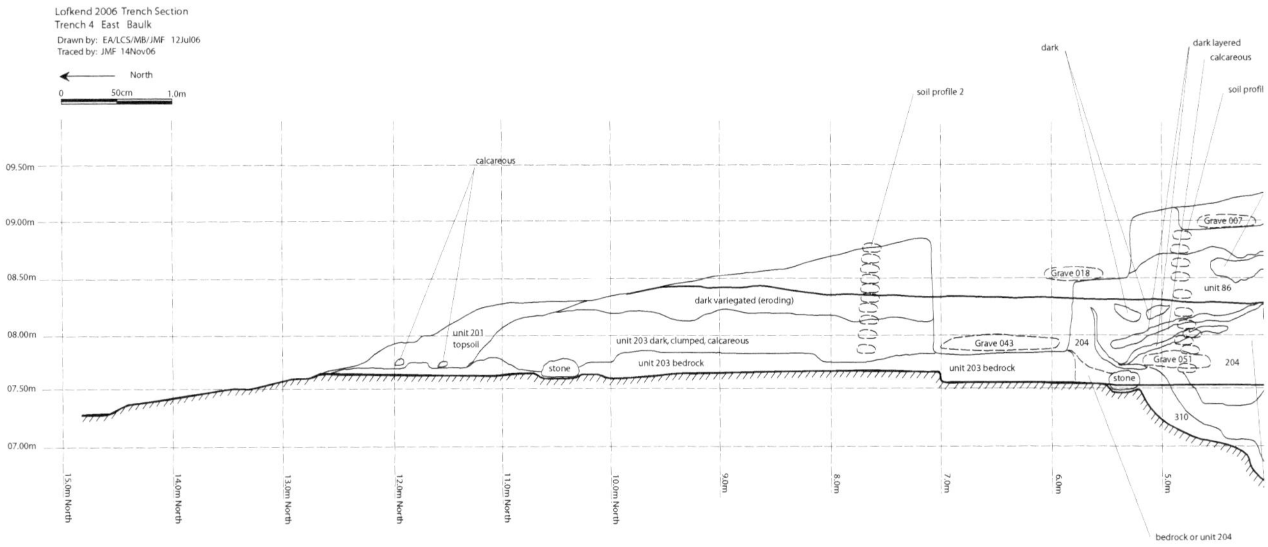

FIGURE 2.17 *(Facing pages)*. North-south section through Sectors/Trenches 3 and 4 (Max Farrar)

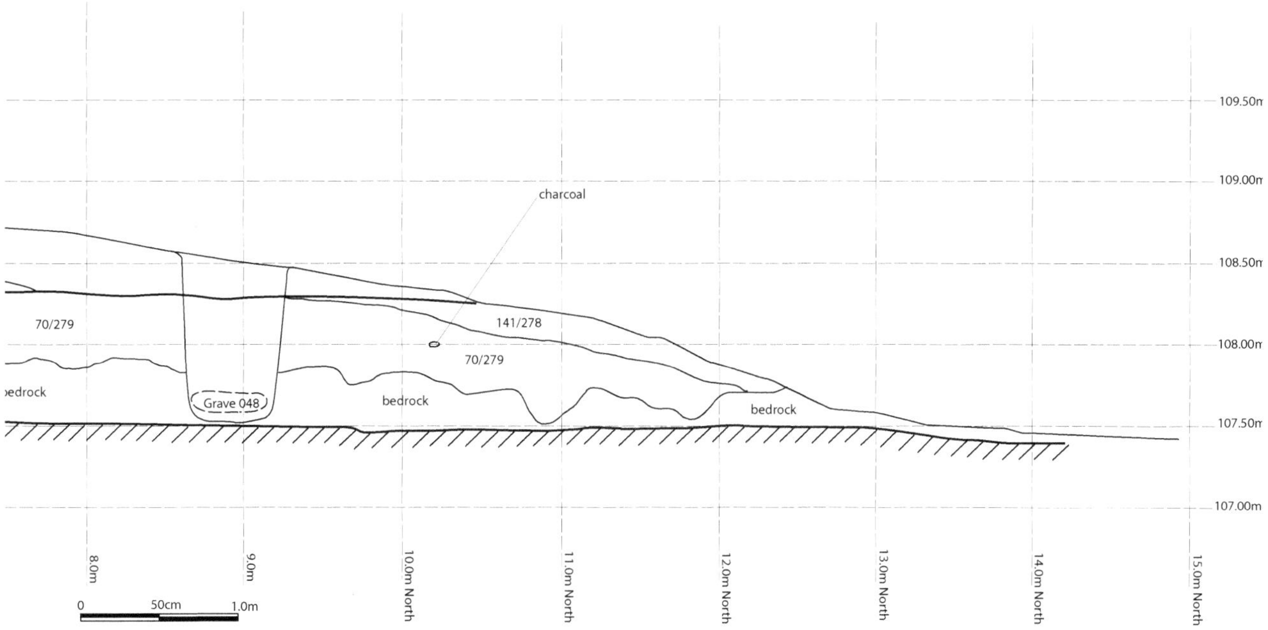

FIGURE 2.16C Detail view of FIGURE 2.16, right side

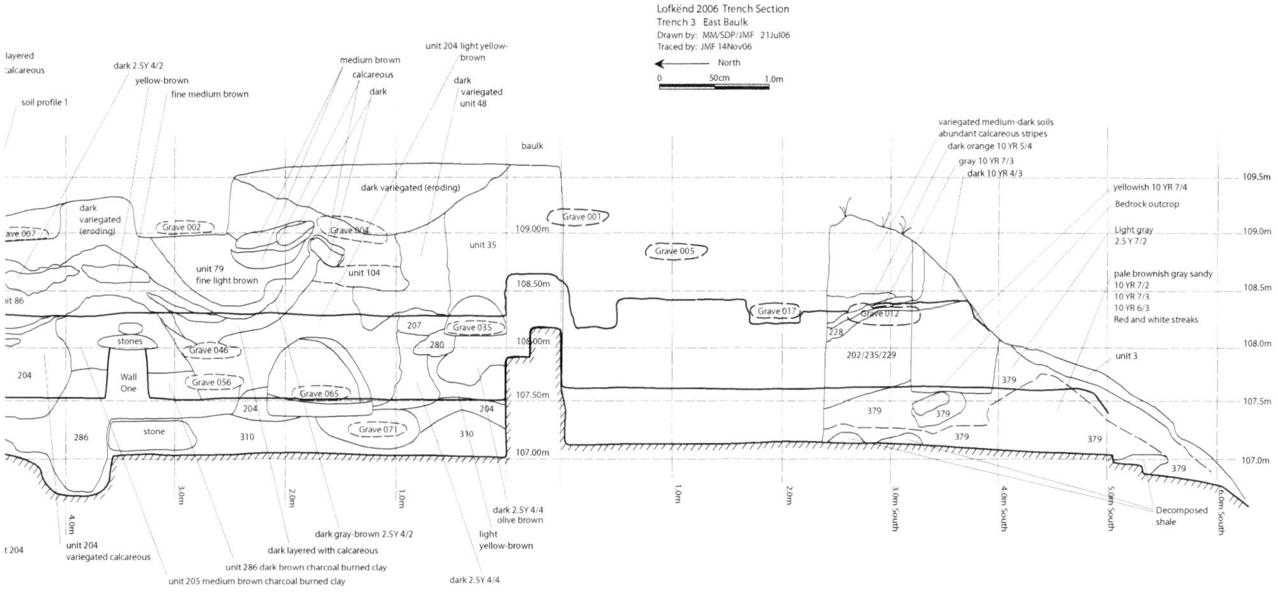

FIGURE 2.17 *(Continued from facing page)*. North-south section through Sectors/Trenches 3 and 4 (Max Farrar); see detail views in FIGURES 2.17A, B, C (*on following pages*)

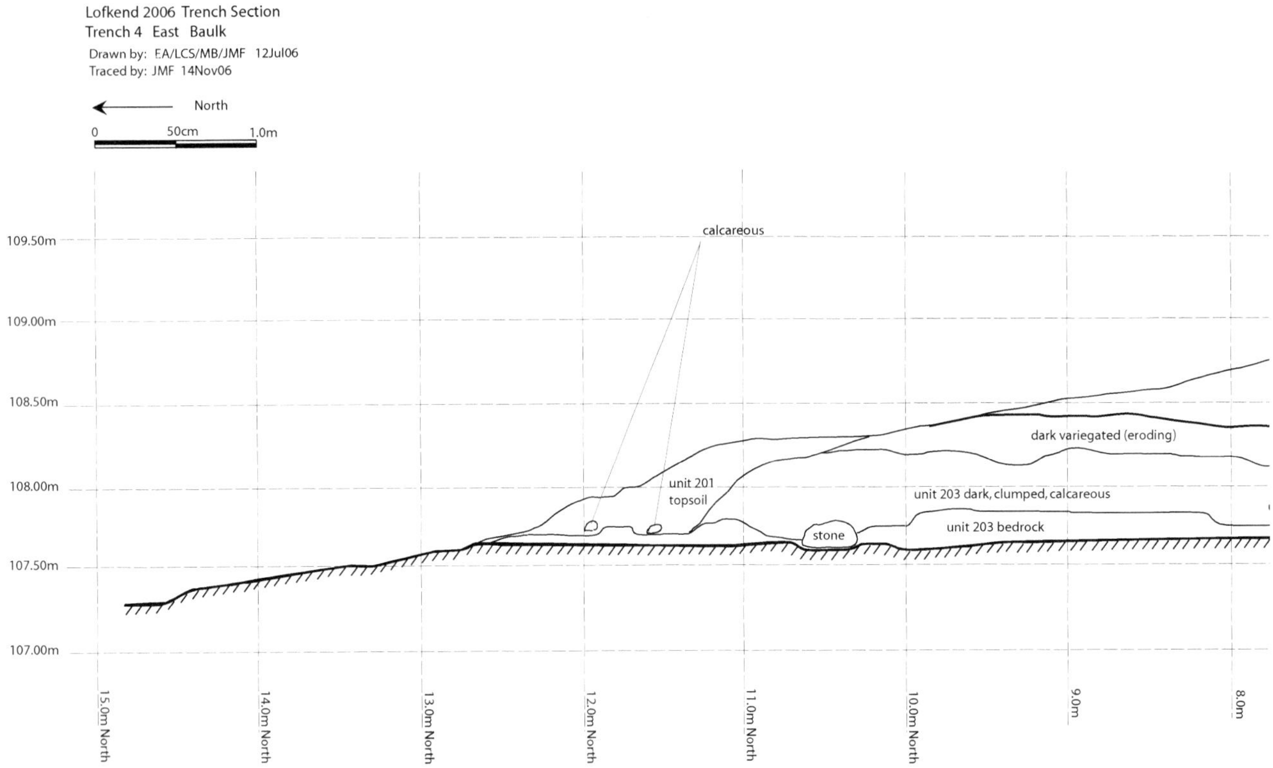

FIGURE 2.17A Detail view of FIGURE 2.17, left side

Facing page: FIGURE 2.17B (*Top*) Detail view of FIGURE 2.17, middle; FIGURE 2.17C (*Bottom*) Detail view of FIGURE 217, right side

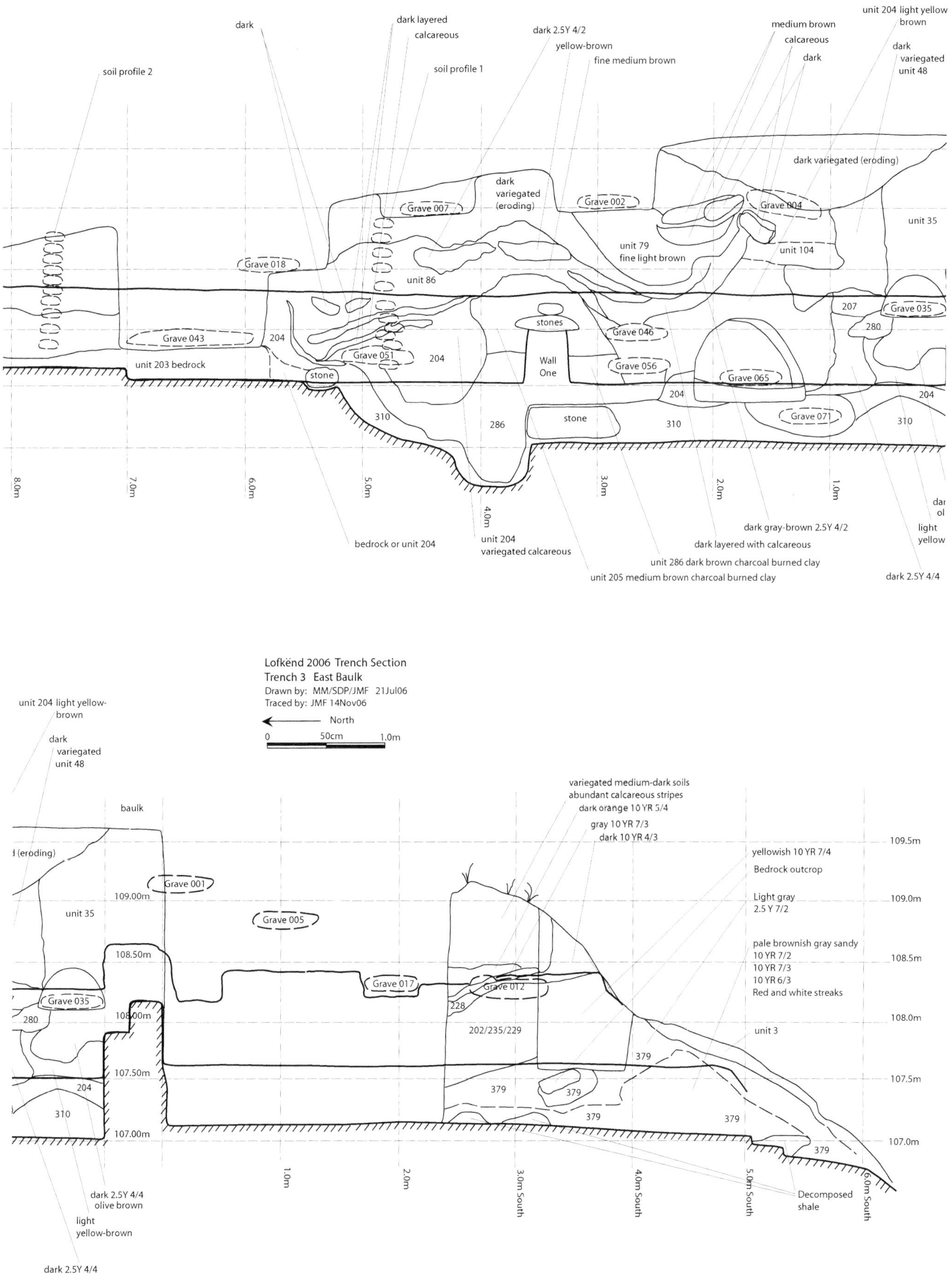

Lofkënd 2006 Trench Section
Trench 3 East Baulk
Drawn by: MM/SDP/JMF 21Jul06
Traced by: JMF 14Nov06
North
0 50cm 1.0m
soil profile 2
dark
dark layered
calcareous
soil profile 1
dark 2.5Y 4/2
yellow-brown
fine medium brown
medium brown
calcareous
dark
unit 204 light yellow brown
dark variegated unit 48
dark variegated (eroding)
unit 35
dark variegated (eroding)
Grave 007
Grave 002
Grave 004
Grave 018
unit 79 fine light brown
unit 104
unit 86
207
Grave 035
280
Grave 043
204
stones
Grave 046
Grave 051
204
Wall One
Grave 056
unit 203 bedrock
stone
Grave 065
204
310
286
stone
310
Grave 071
310
8.0m
7.0m
6.0m
5.0m
4.0m
3.0m
2.0m
1.0m
bedrock or unit 204
unit 204 variegated calcareous
dark gray-brown 2.5Y 4/2
dark layered with calcareous
unit 286 dark brown charcoal burned clay
unit 205 medium brown charcoal burned clay
light yellow
dark 2.5Y 4/4
unit 204 light yellow-brown
dark variegated unit 48
baulk
(eroding)
Grave 001
109.00m
unit 35
Grave 005
108.50m
Grave 017
Grave 012
228
Grave 035
280
108.00m
202/235/229
379
107.50m
204
379
379
310
379
379
107.00m
379
variegated medium-dark soils
abundant calcareous stripes
dark orange 10 YR 5/4
gray 10 YR 7/3
dark 10 YR 4/3
yellowish 10 YR 7/4
Bedrock outcrop
Light gray 2.5 Y 7/2
pale brownish gray sandy
10 YR 7/2
10 YR 7/3
10 YR 6/3
Red and white streaks
unit 3
109.5m
109.0m
108.5m
108.0m
107.5m
107.0m
1.0m
2.0m
3.0m South
4.0m South
5.0m South
6.0m South
Decomposed shale
dark 2.5Y 4/4 olive brown
light yellow-brown
dark 2.5Y 4/4

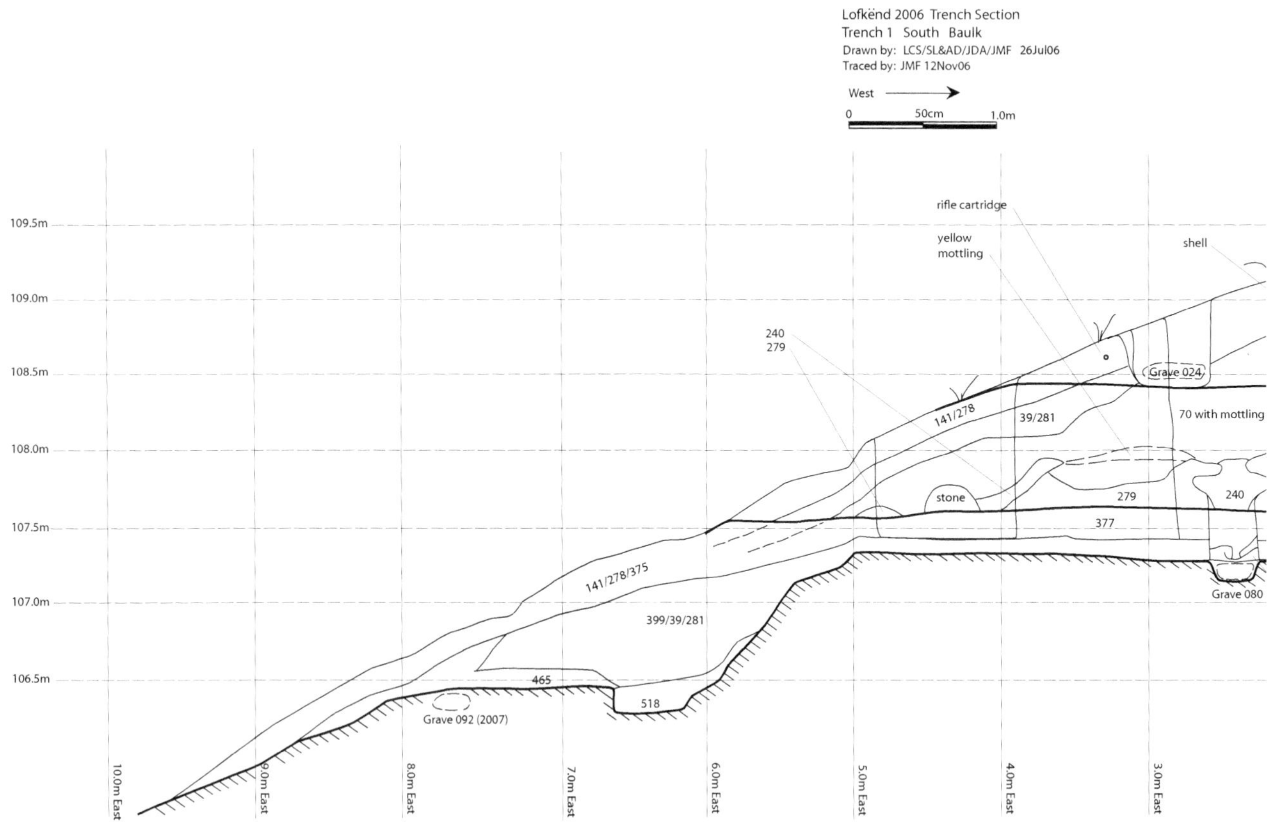

FIGURE 2.18 *(Facing pages).* East-west section through Sectors/Trenches 1 and 4 (Max Farrar)

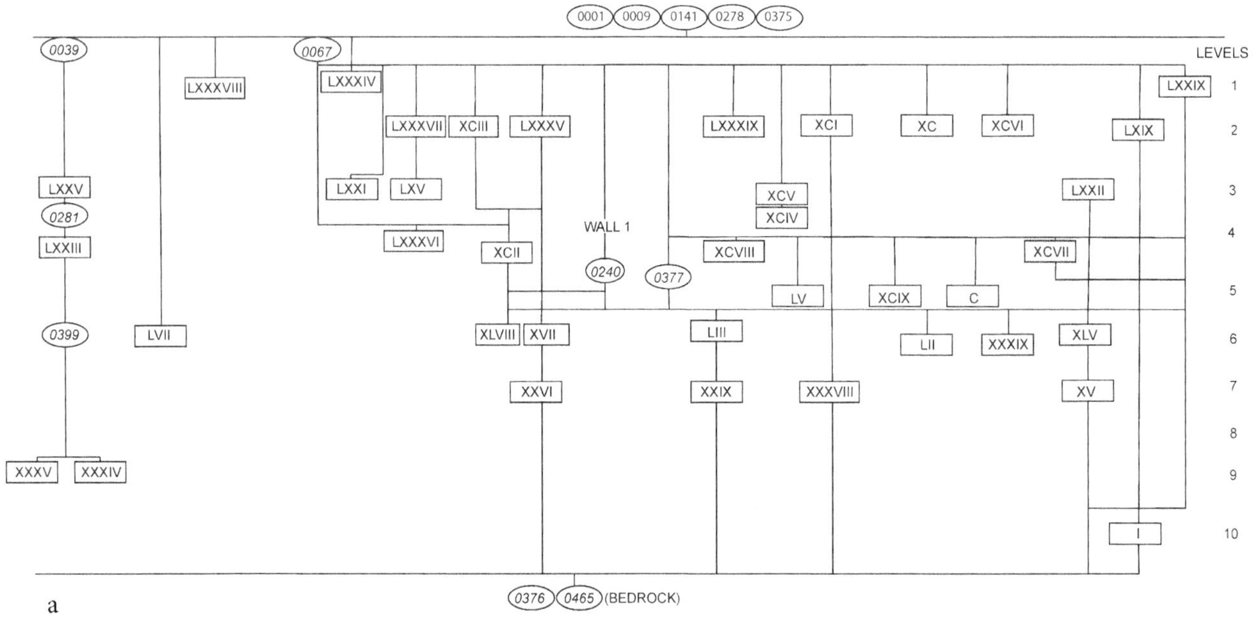

a

FIGURE 2.19 Simplified Harris matrices of (a) Sector/Trench 1 (*above*); (b, c) *on following pages* (prepared by Lyssa Stapleton, based on the original hand-drawn versions of the excavators, Esmeralda Agolli, Jamie Aprile, Seth Pevnick, and Lyssa Stapleton; amended by John Papadopoulos)

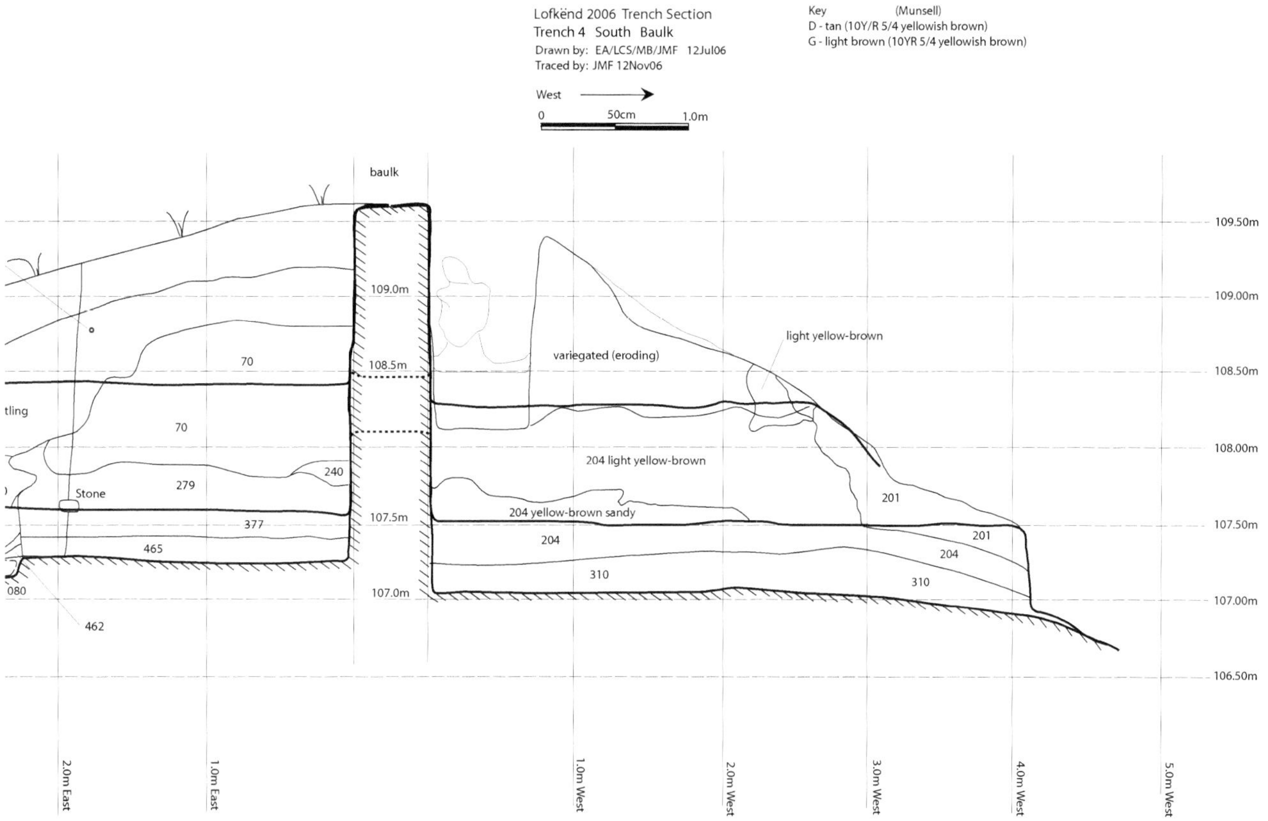

FIGURE 2.18 *(Continued from facing page)*. East-west section through Sectors/Trenches 1 and 4 (Max Farrar)

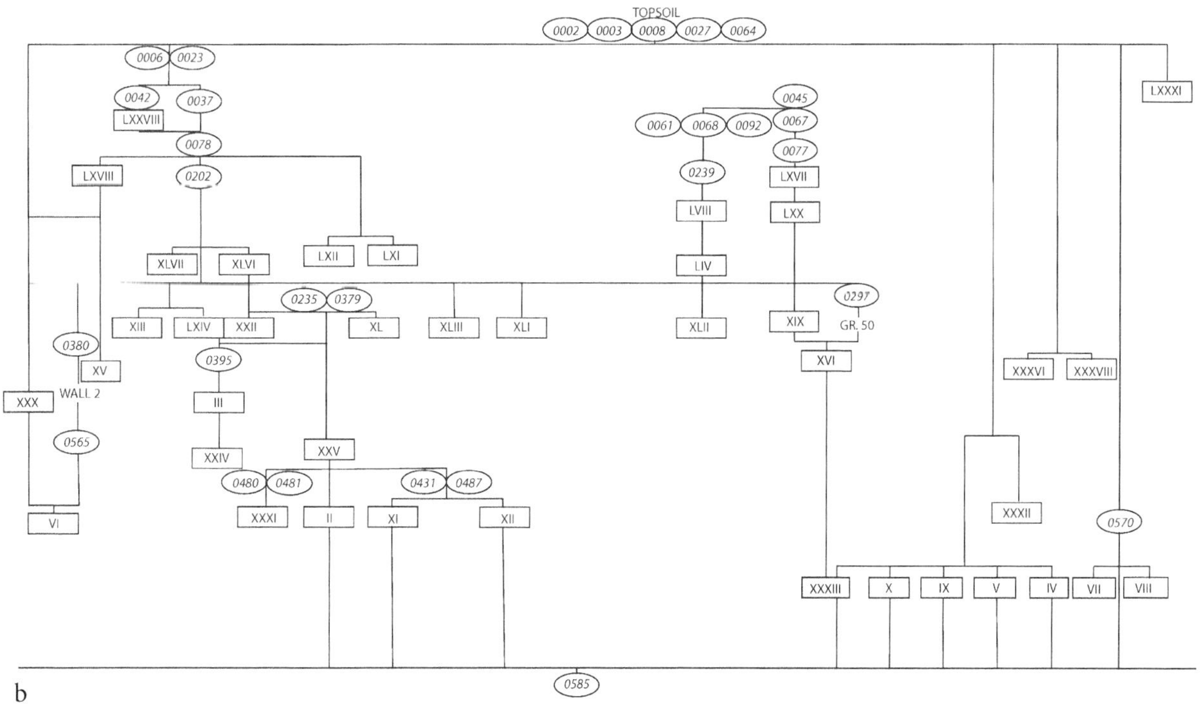

b

FIGURE 2.19 (*continued from facing page*). Simplified Harris matrices of (b) Sector/Trench 2/3; (c) *on following page* (prepared by Lyssa Stapleton, based on the original hand-drawn versions of the excavators, Esmeralda Agolli, Jamie Aprile, Seth Pevnick, and Lyssa Stapleton; amended by John Papadopoulos)

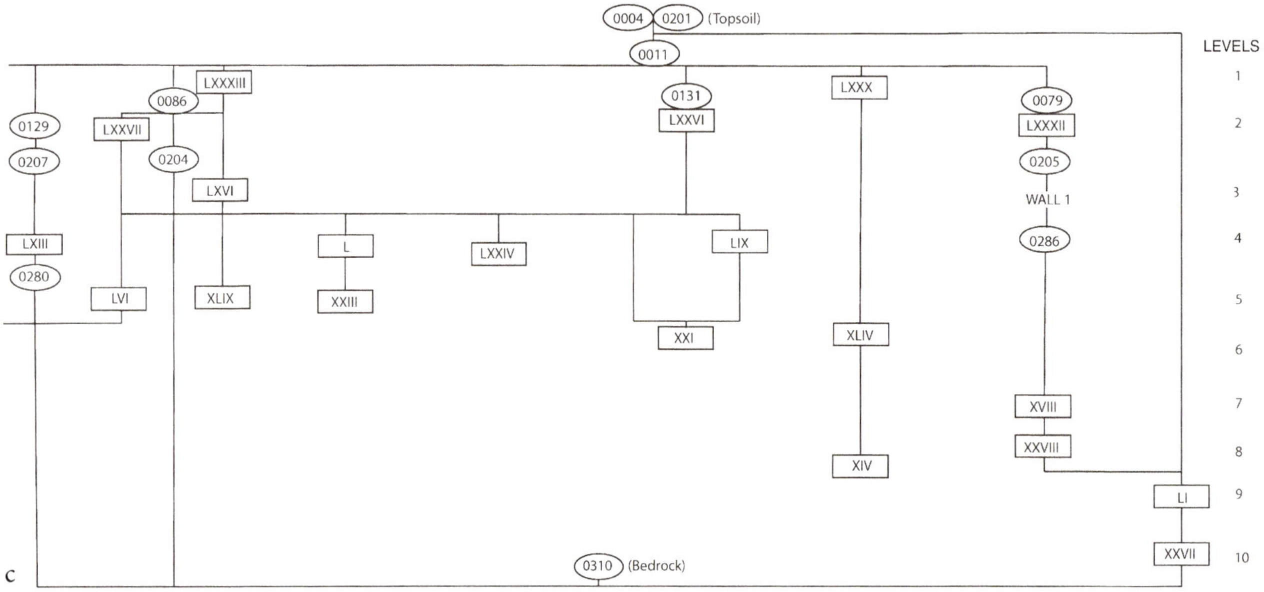

FIGURE 2.19 (*continued from previous page*). Simplified Harris matrices of (c) Sector/Trench 4 (prepared by Lyssa Stapleton, based on the original hand-drawn versions of the excavators, Esmeralda Agolli, Jamie Aprile, Seth Pevnick, and Lyssa Stapleton; amended by John Papadopoulos)

FIGURE 2.20 View of the tumulus at the conclusion of the 2005 season from north-northwest, showing aerial photography of the site by Alket Islami using a paramotor (photo John Papadopoulos)

FIGURE 2.21 Aerial view of the tumulus at the conclusion of the 2005 season, from north-northeast (photo Alket Islami)

FIGURE 2.22 Aerial view of the tumulus at the conclusion of the 2005 season, from above west (photo Alket Islami)

FIGURE 2.23 Long-distance aerial view of the Lofkënd tumulus (in center of photo) and its surrounding landscape from the southeast (photo Alket Islami)

FIGURE 2.24 Aerial view of the tumulus, surrounding fields, and the mountains to the south, from north (photo Alket Islami)

FIGURE 2.25 Two minutes in the life of Lofkënd, a panorama of the site by Rich MacDonald, July 4, 2006. This panorama was shot with the camera placed directly above the highest point of the tumulus. By positioning the camera at the location where the baulks intersect, an equal view into all four trenches was achieved. Many of the everyday excavation tasks are represented in this final image, which is itself composed from around a dozen photos taken over two-minute time span.

Chapter 3 Illustrations

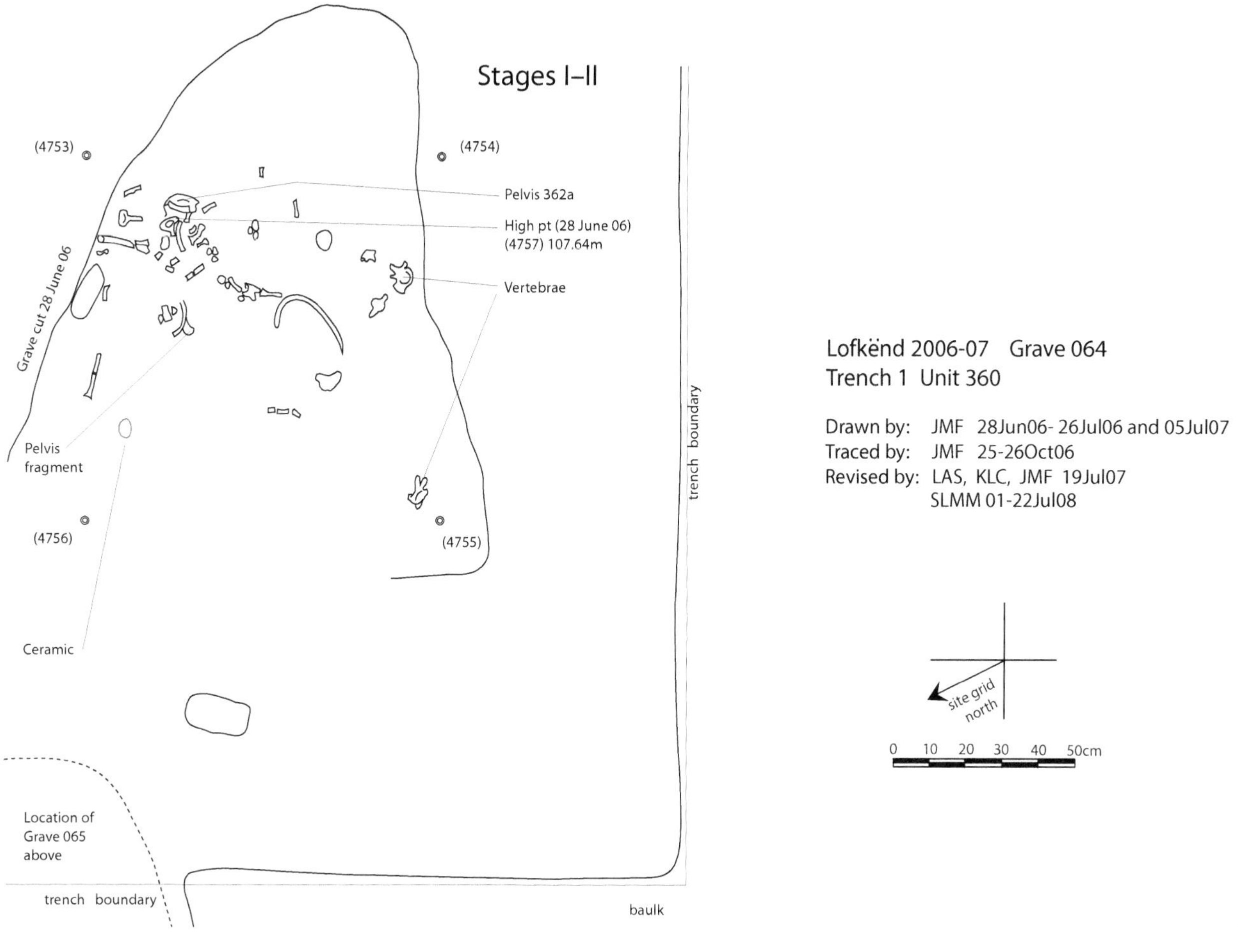

a

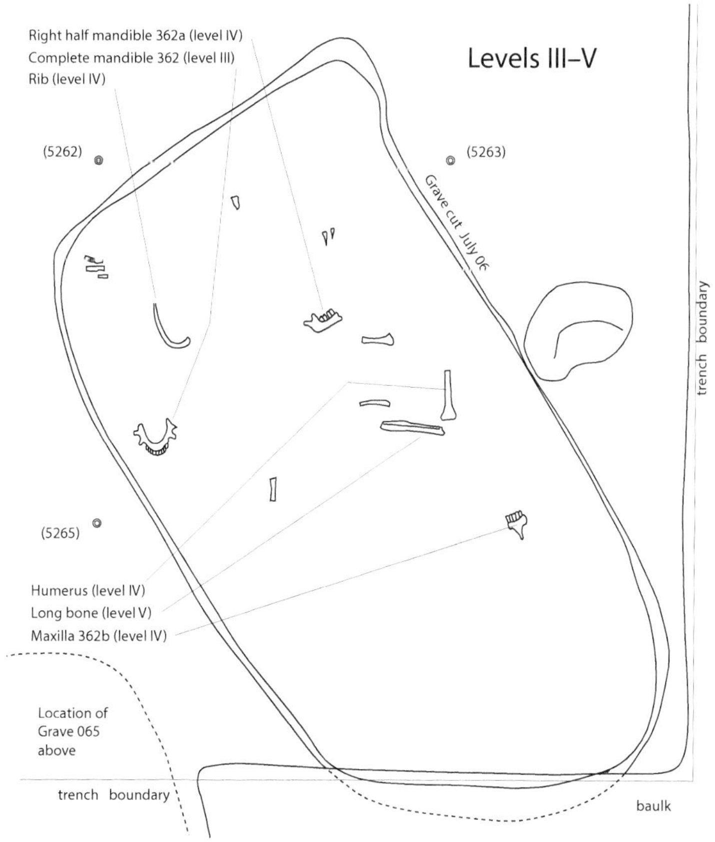

b

FIGURE 3.1 Plans of Tomb I: (a) Stages I–II; (b) Levels III–V; (c–f) *on following pages*

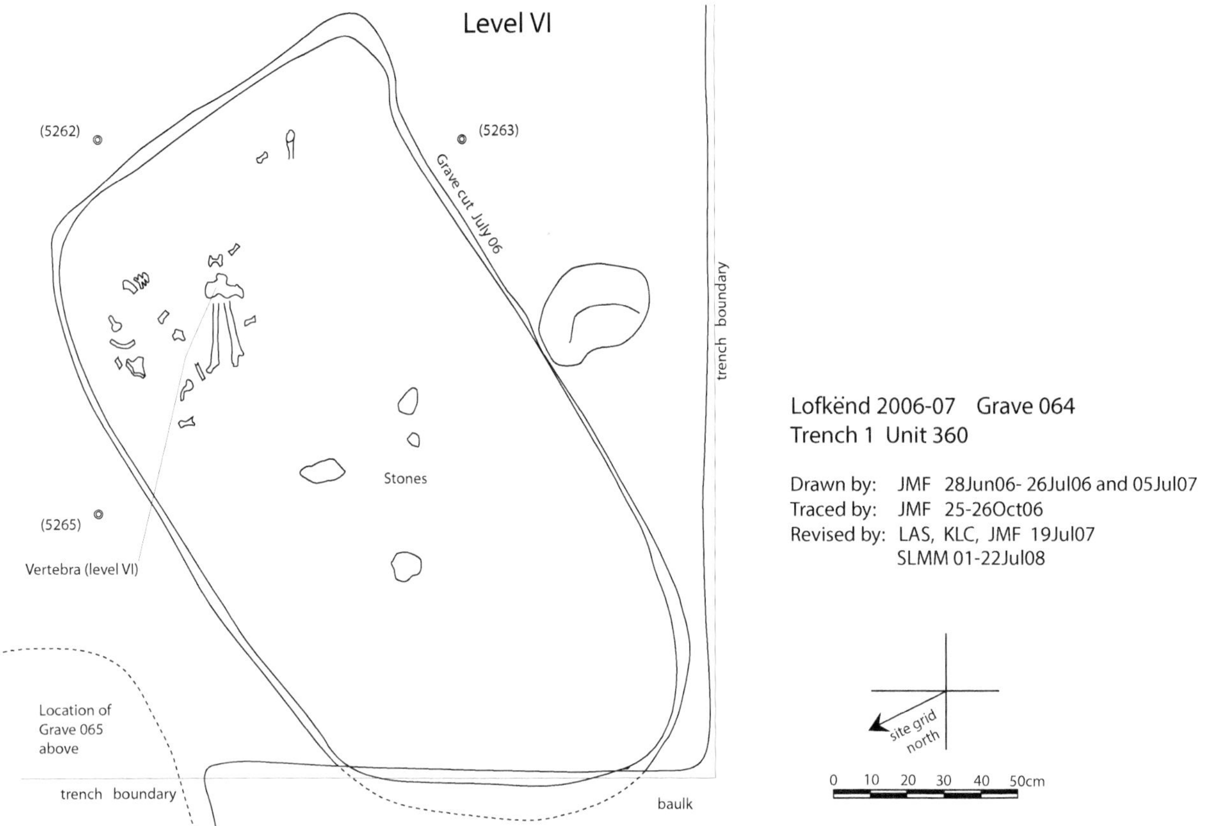

c

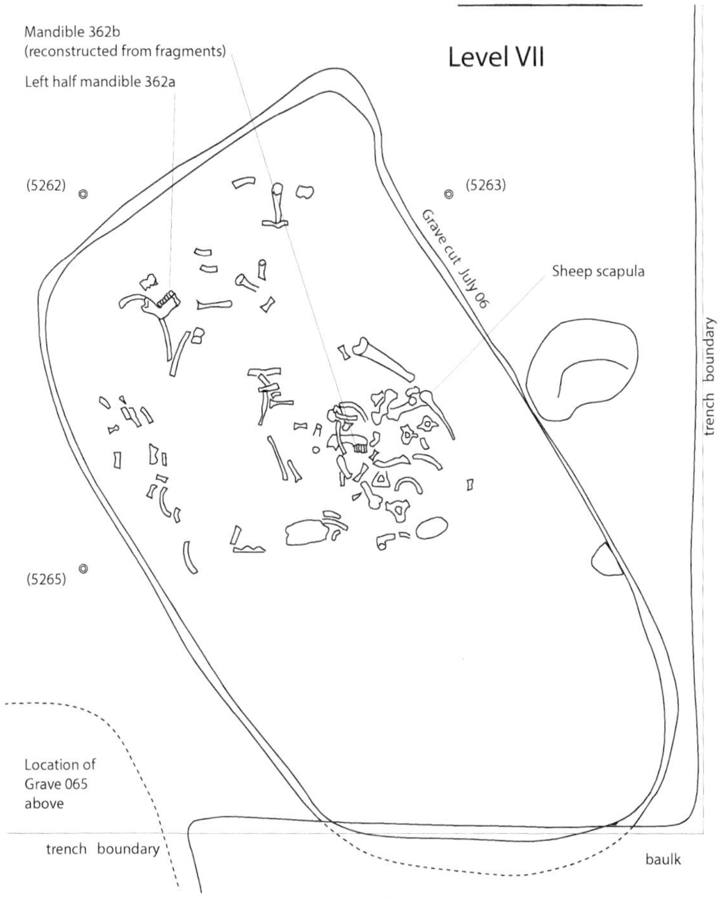

d

FIGURE 3.1 (*continued from previous page*). Plans of Tomb I: (c) Level VI; (d) Level VII; (e, f) *on following page*

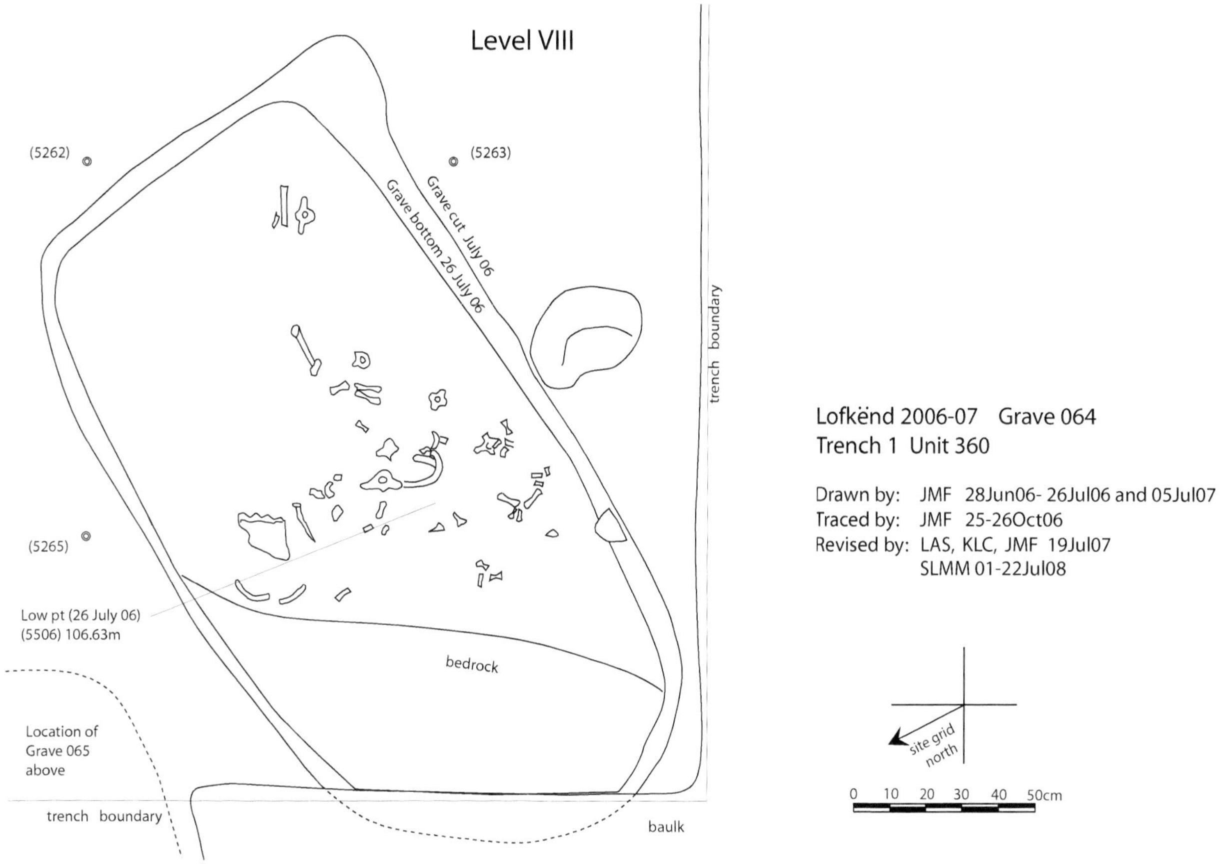

e

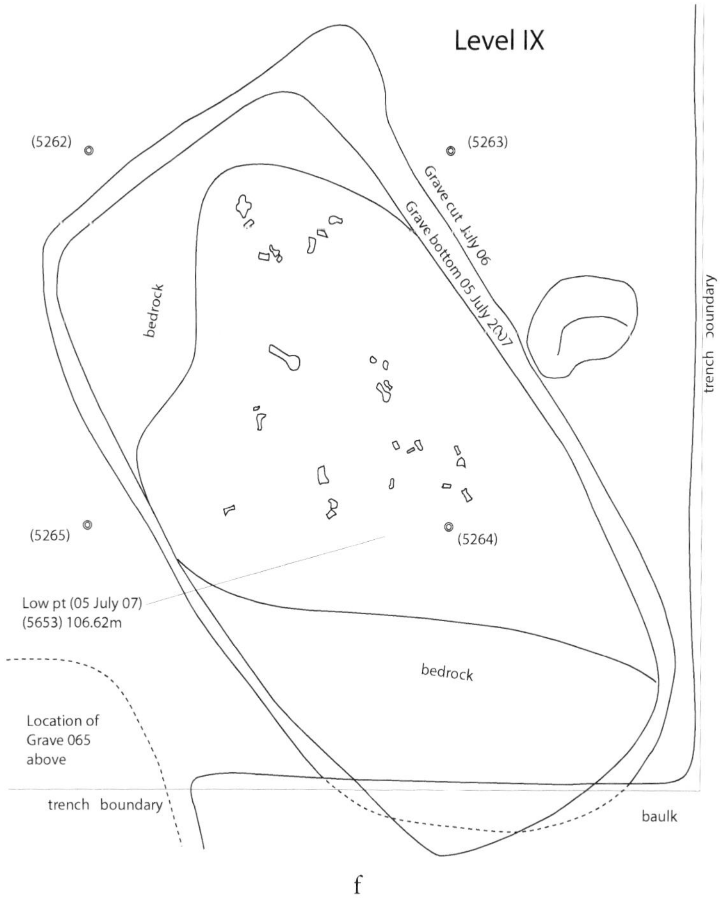

f

FIGURE 3.1 (*continued*). Plans of Tomb I: (e) Level VIII; (f) Level IX

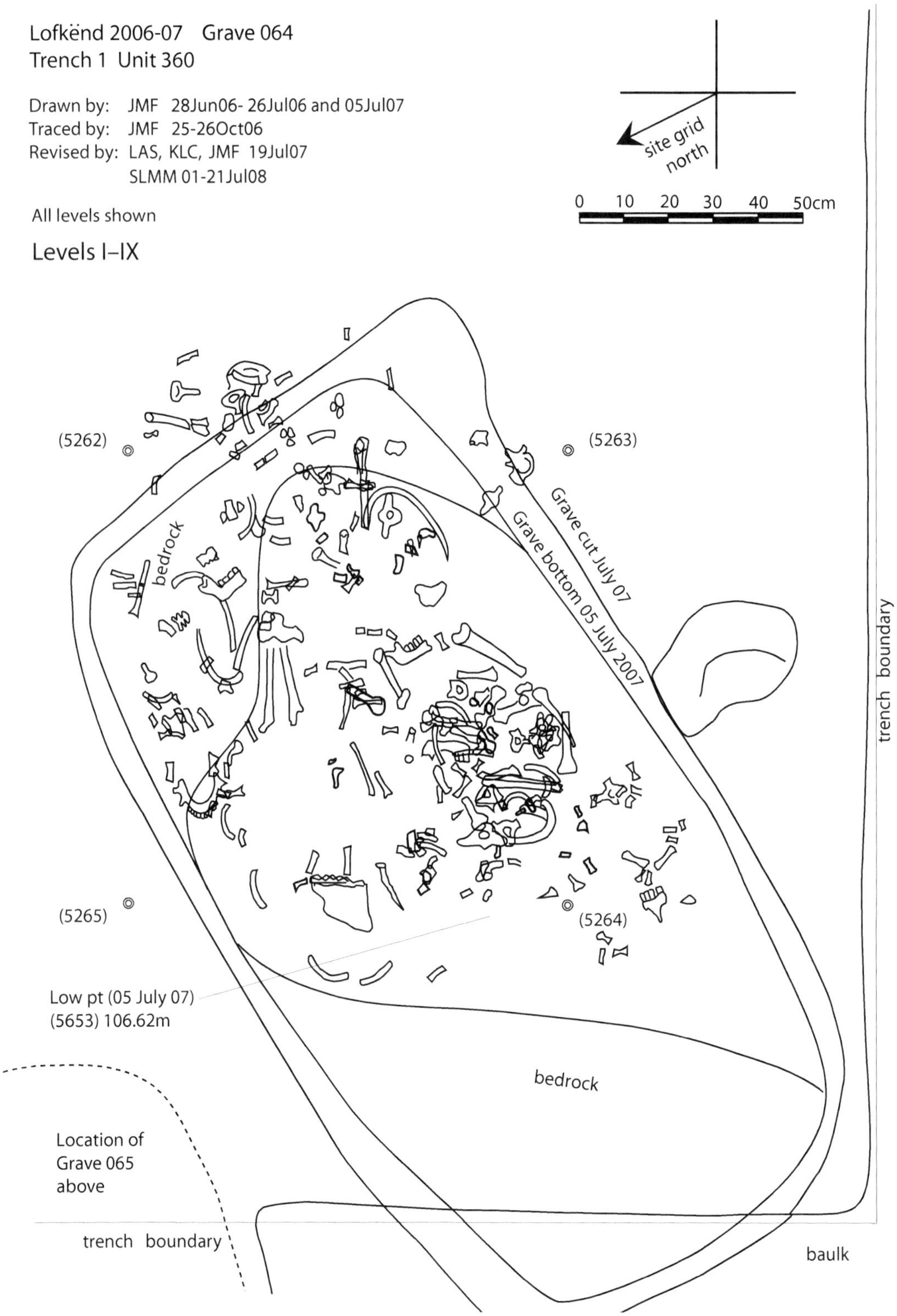

FIGURE 3.2 Plan of Tomb I, all levels, I–IX

a

FIGURE 3.3 Tomb I: (a) (photo 1332) from east; (b) (photo 1344) from northeast; (c–h) *on following pages*

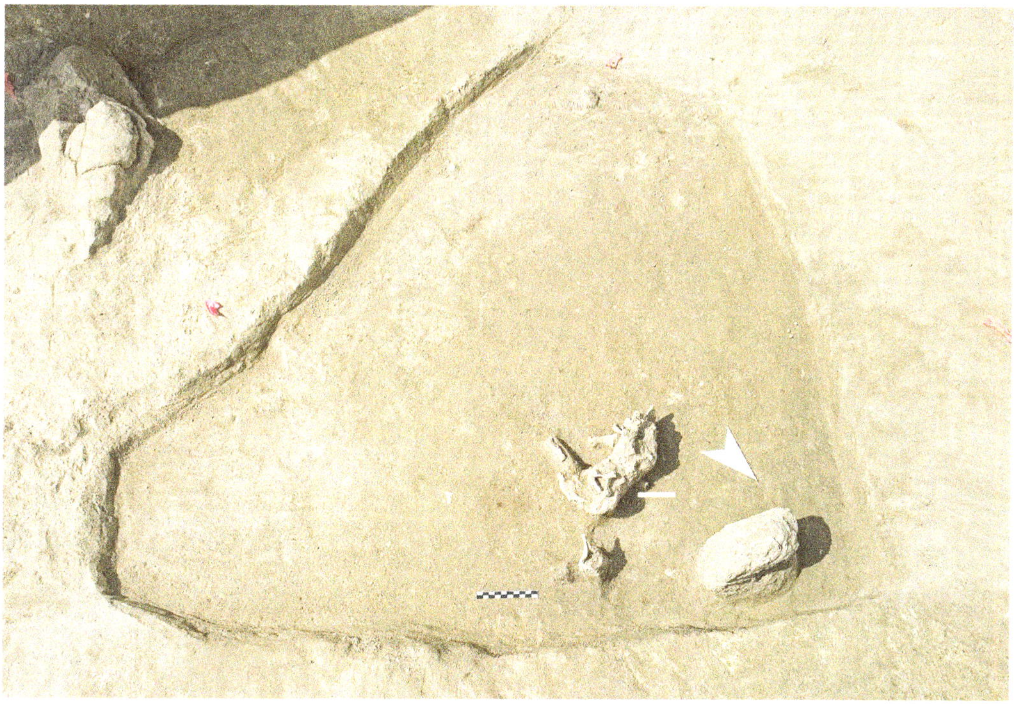

b

c

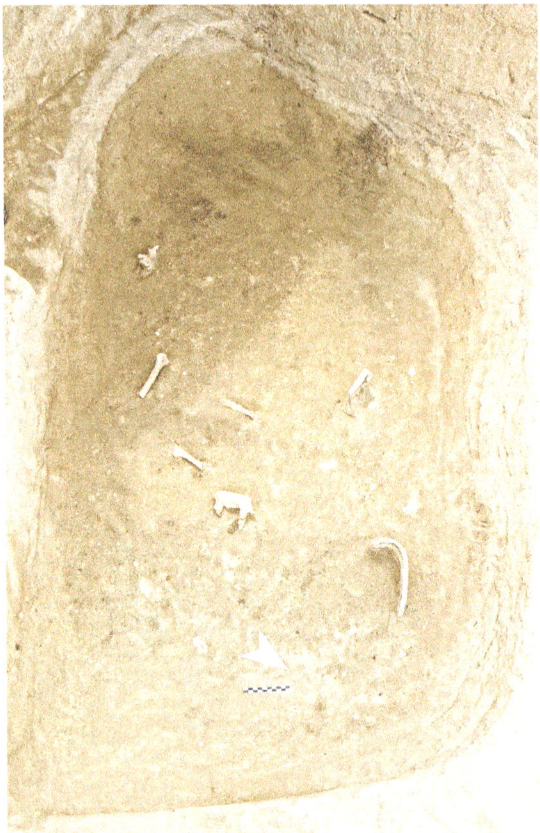

d

e

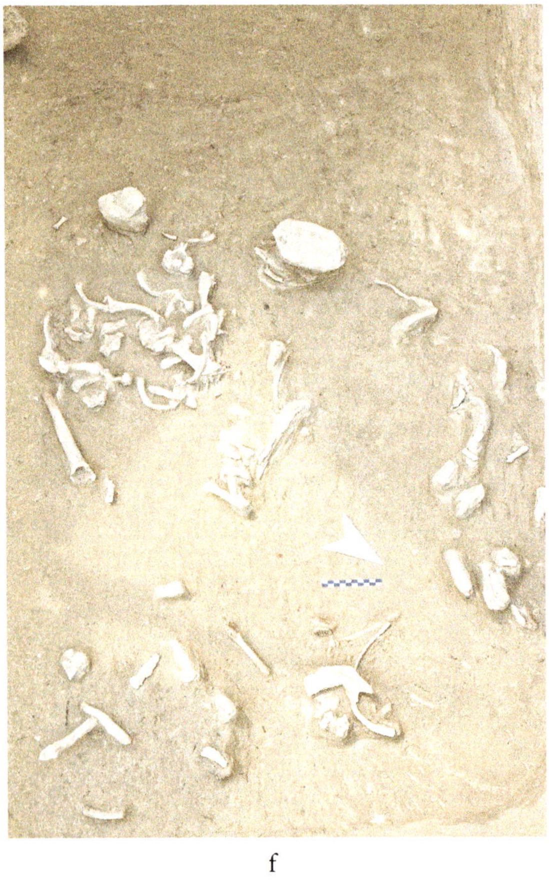

f

FIGURE 3.3 (*continued*). Tomb I: (c) (photo 1479) from northeast; (d) (photo 1493) from northeast; (e) (photo 1577) from northeast; (f) (photo 1607) from east-northeast; (g, h) *on following page*

g

h

FIGURE 3.3 (*continued*). Tomb I: (g) (photo 1577) from northeast; (h) (photo 1743) bedrock, from west-southwest

FIGURE 3.4 Bedrock at the end of the 2007 season, with Tomb I in center foreground (photo 1822), from south

Figure 3.5 Tomb I: **TI-1**

Figure 3.6 Tomb I: **TI-2**

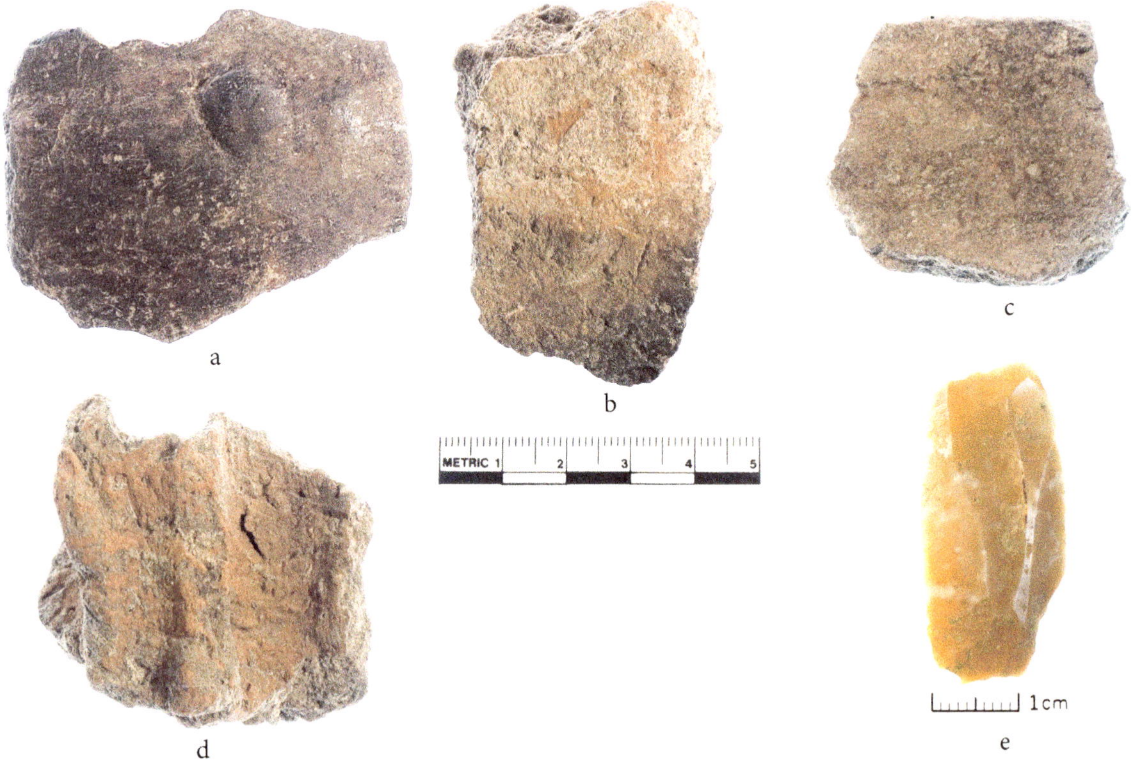

Figure 3.7 Tomb I: (a) P316 (photo 2953); (b) P354 (photo 2959); (c) P366 (photo 3298); (d) SF 387 (photo 323); (e) SF 396 (photo 3267)

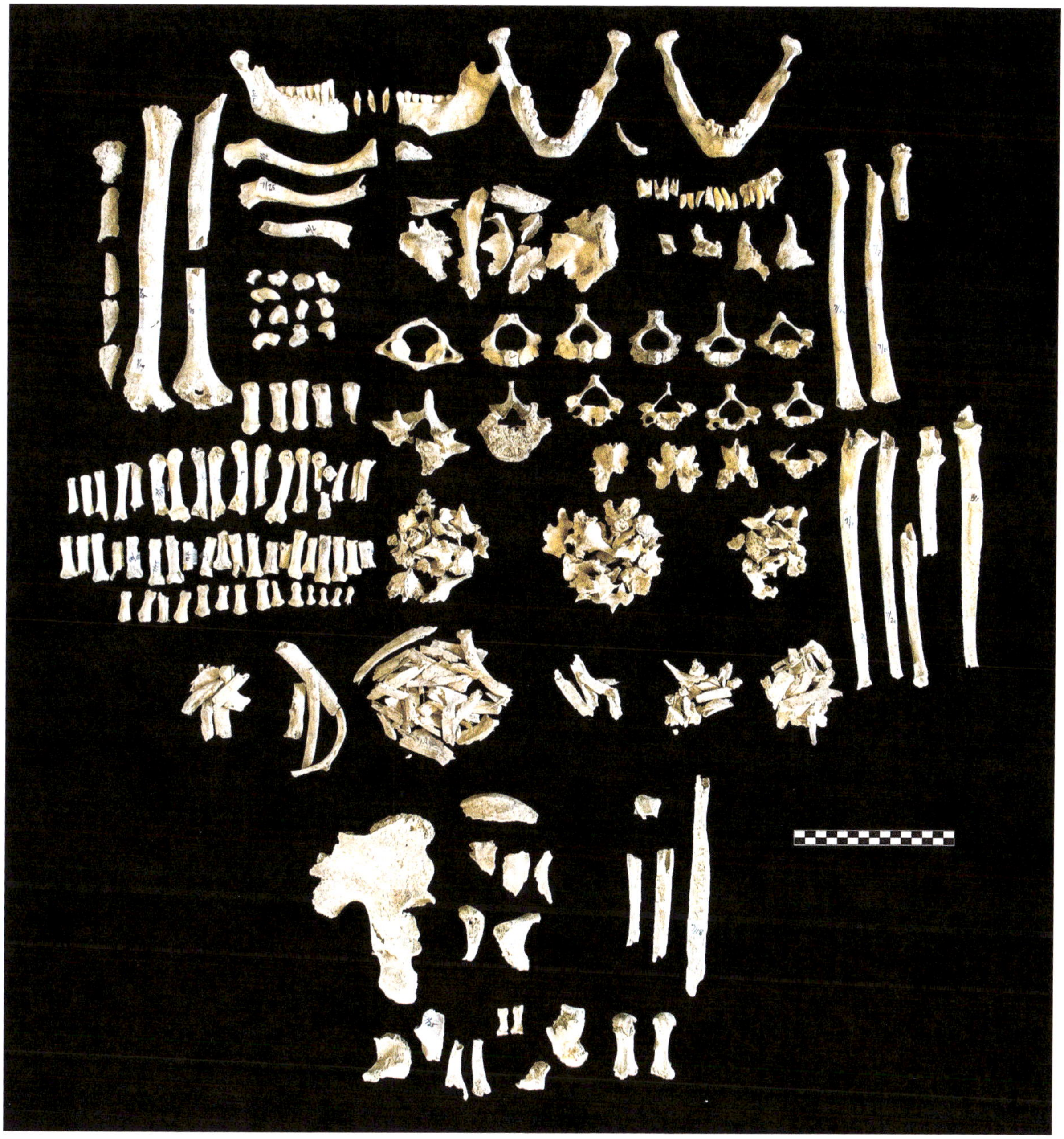

FIGURE 3.8 Tomb I: (photo 3543) all bones from the burial deposit

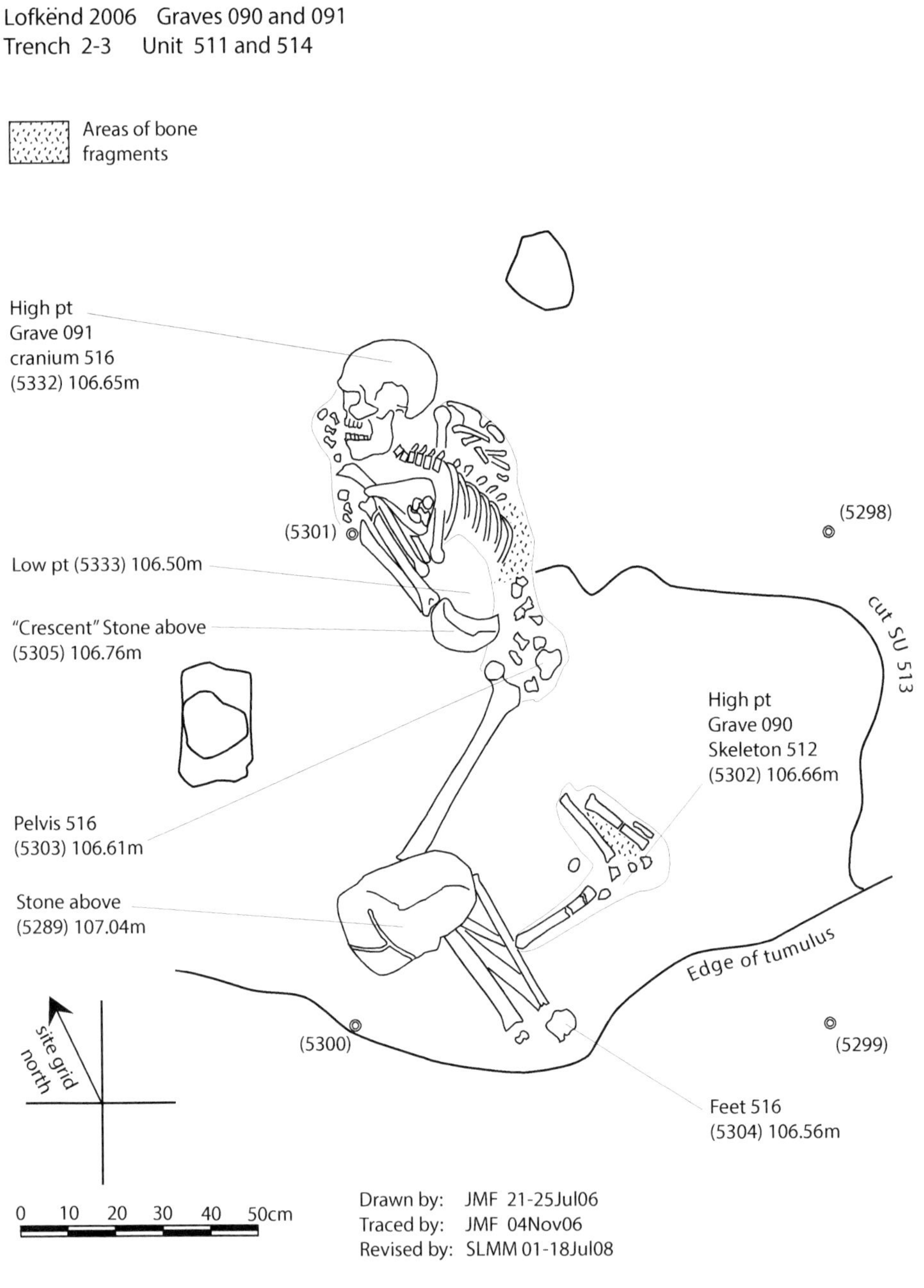

FIGURE 3.9 Plan of Tomb II (and Tomb XXV)

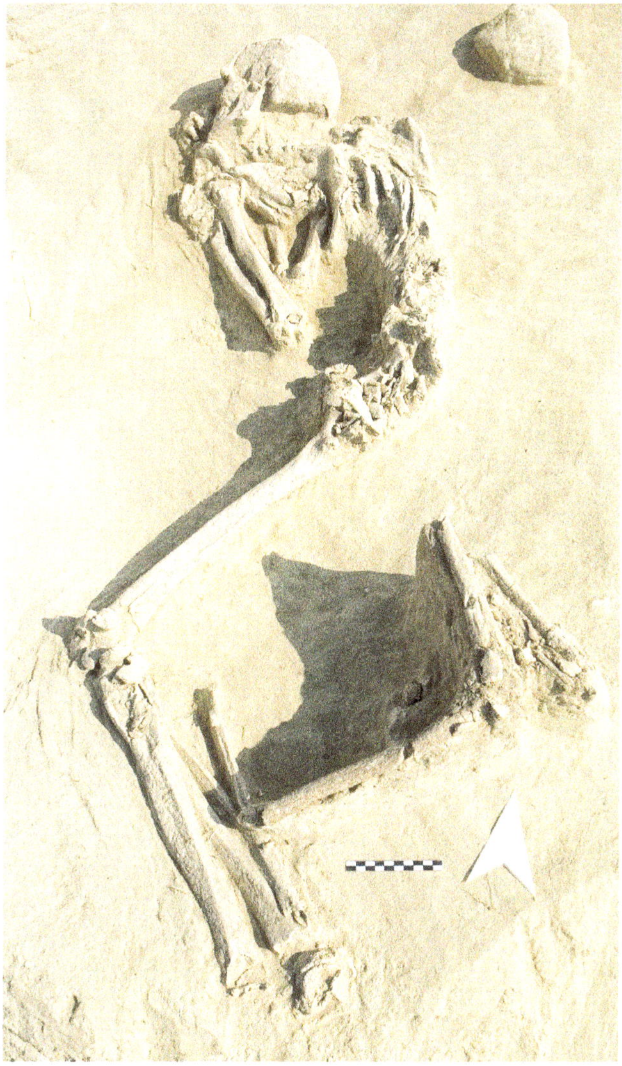

a

b

FIGURE 3.10 Tomb II: (a) Tomb II (and Tomb XXV) (photo 1614) from south; (b) detail of cranium (photo 1618), from southwest

FIGURE 3.11 Tomb II (and Tomb XXV) (photo 1615) from above west-southwest

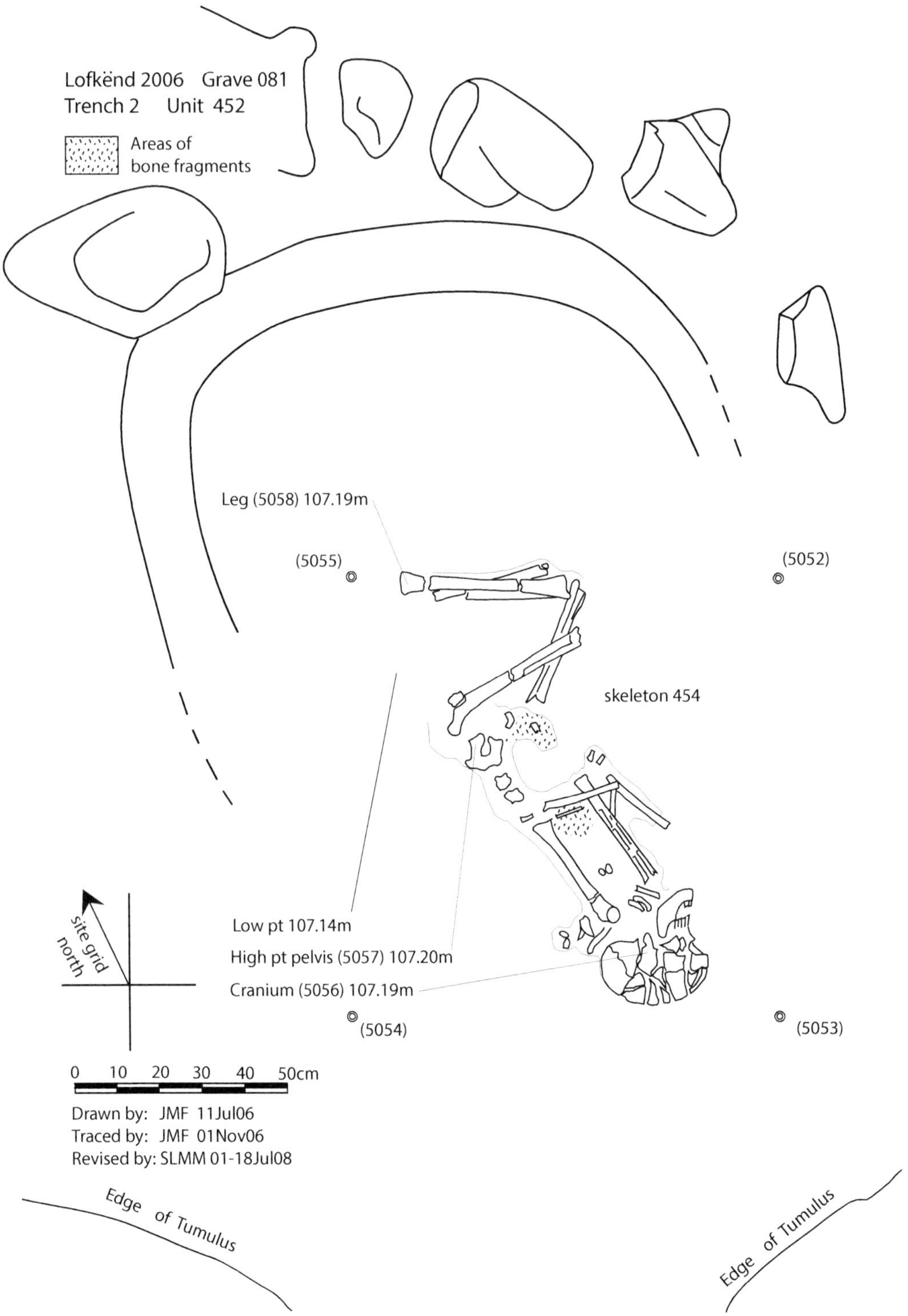

FIGURE 3.12 Plan of Tomb III

a

b

Figure 3.13 Tomb III: (a) (photo 1470) from north-northwest;
(b) (photo 1476) and surrounding area with stones to the east of the tomb, from south

Figure 3.14 General view of stone feature and Tomb III (photo 1482) from northwest

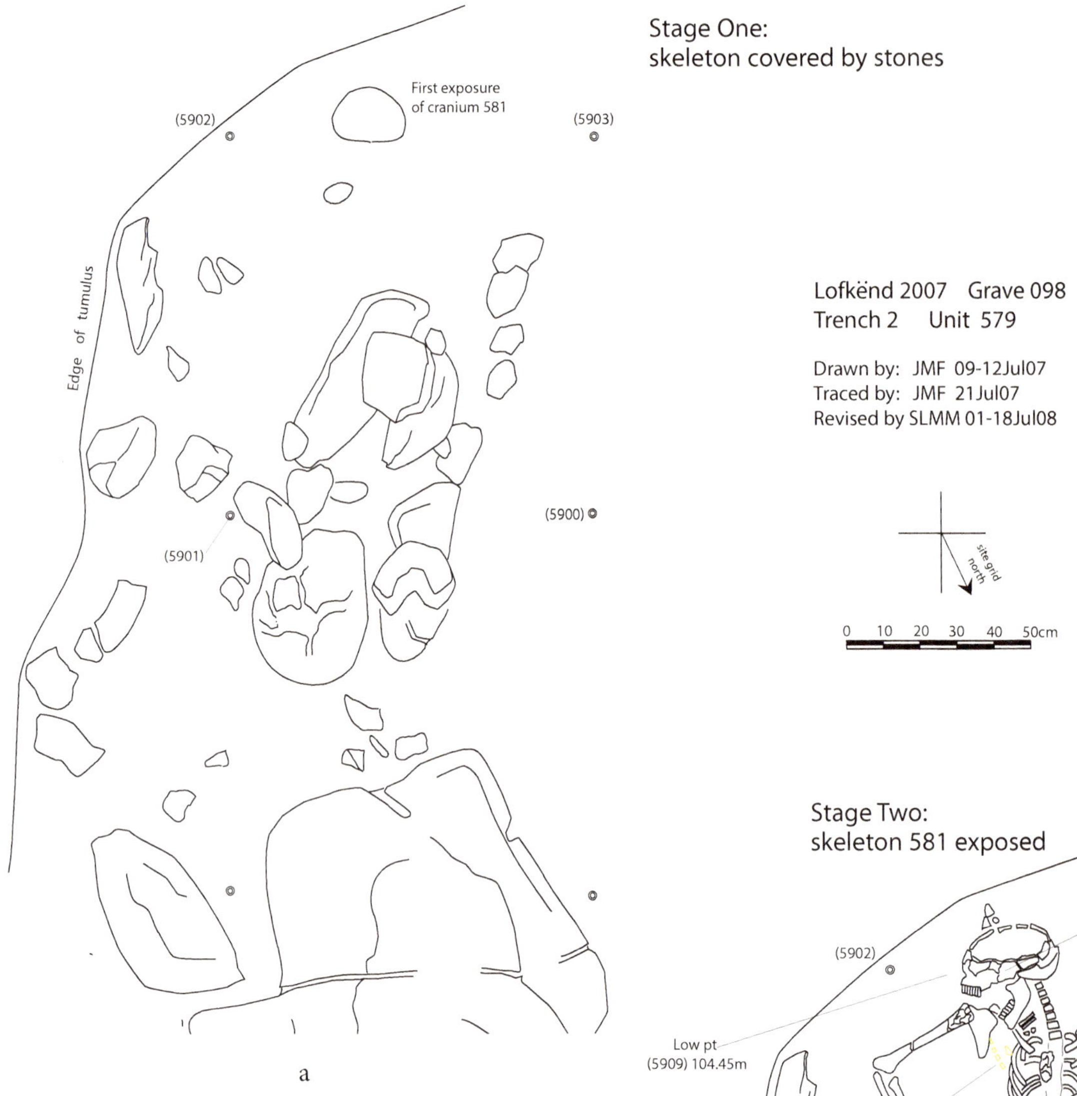

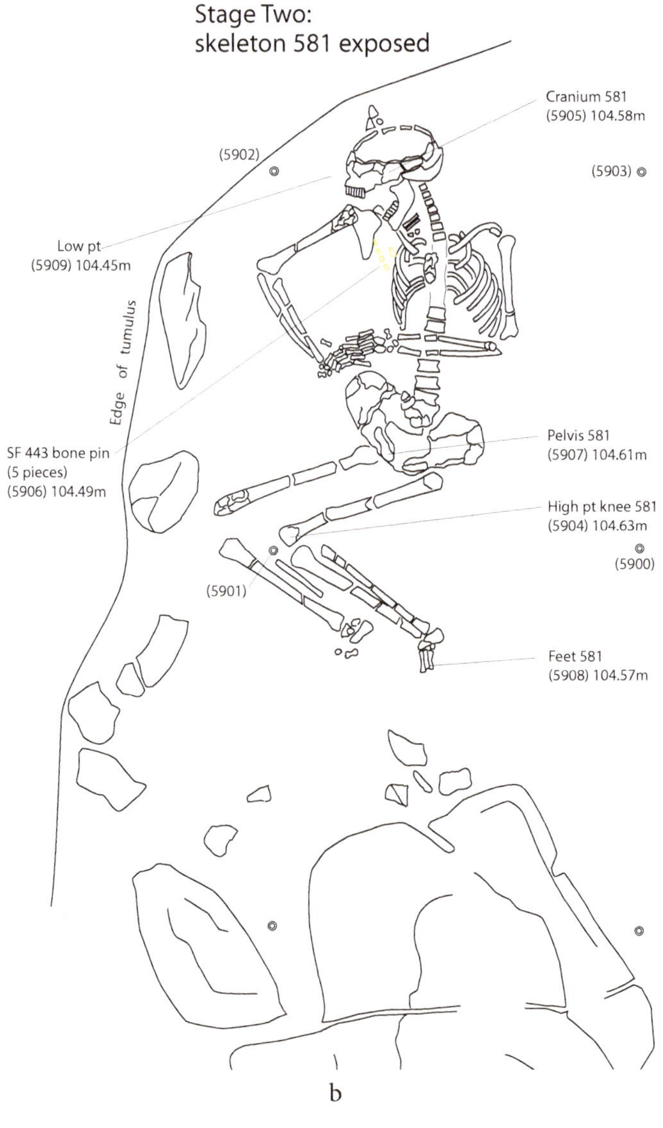

FIGURE 3.15 Plans of Tomb IV: (a) Stage 1; (b) Stage 2

FIGURE 3.16 Tomb IV (photo 1771) showing stone cover above tomb, from south

FIGURE 3.17 Tomb IV: (a) (photo 1780) during excavation, stone cover partially removed, from south; (b) (photo 1781), as previous, from north; (c) (photo 1787) fully exposed, from north; (d) (photo 1792), detail of **TIV-1** in situ, from north

FIGURE 3.18 Tomb IV: **TIV-1**

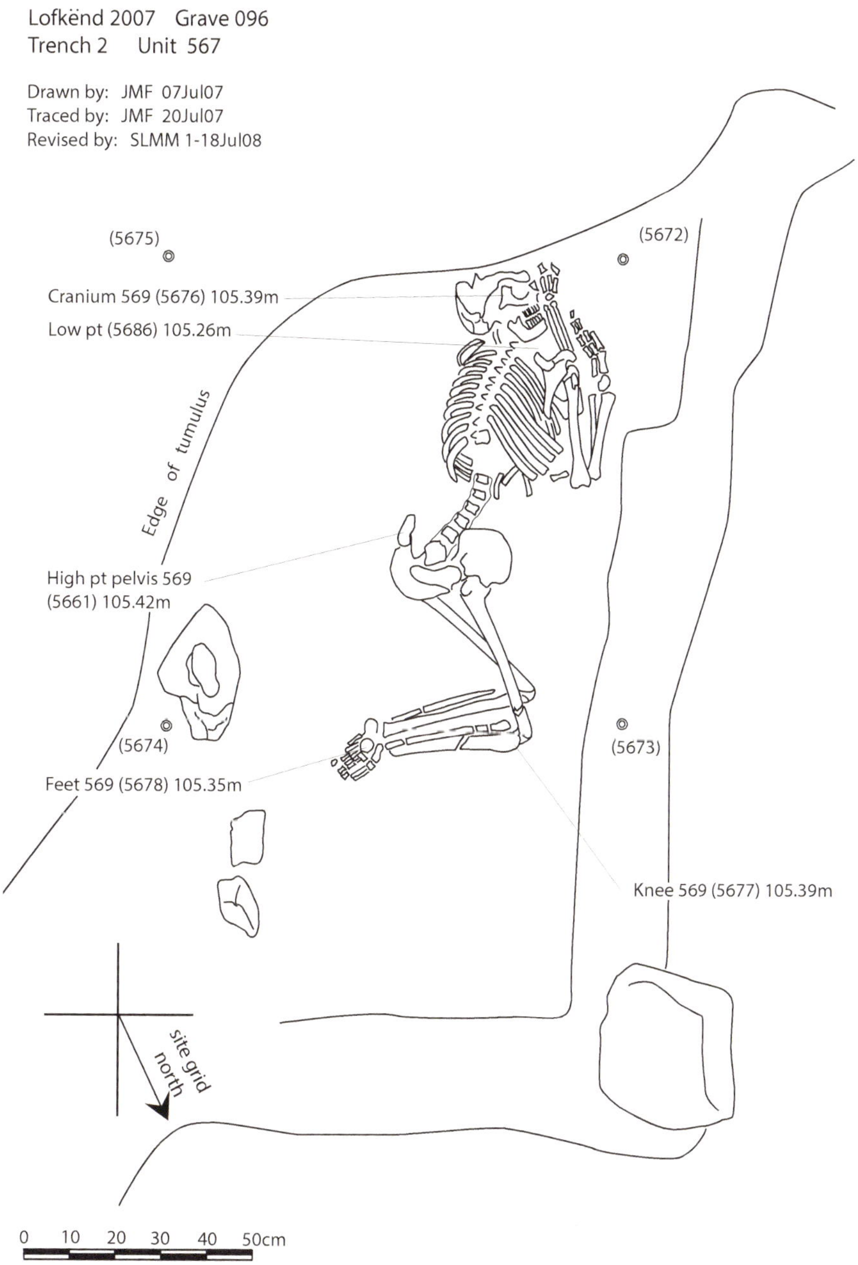

FIGURE 3.19 Plan of Tomb V

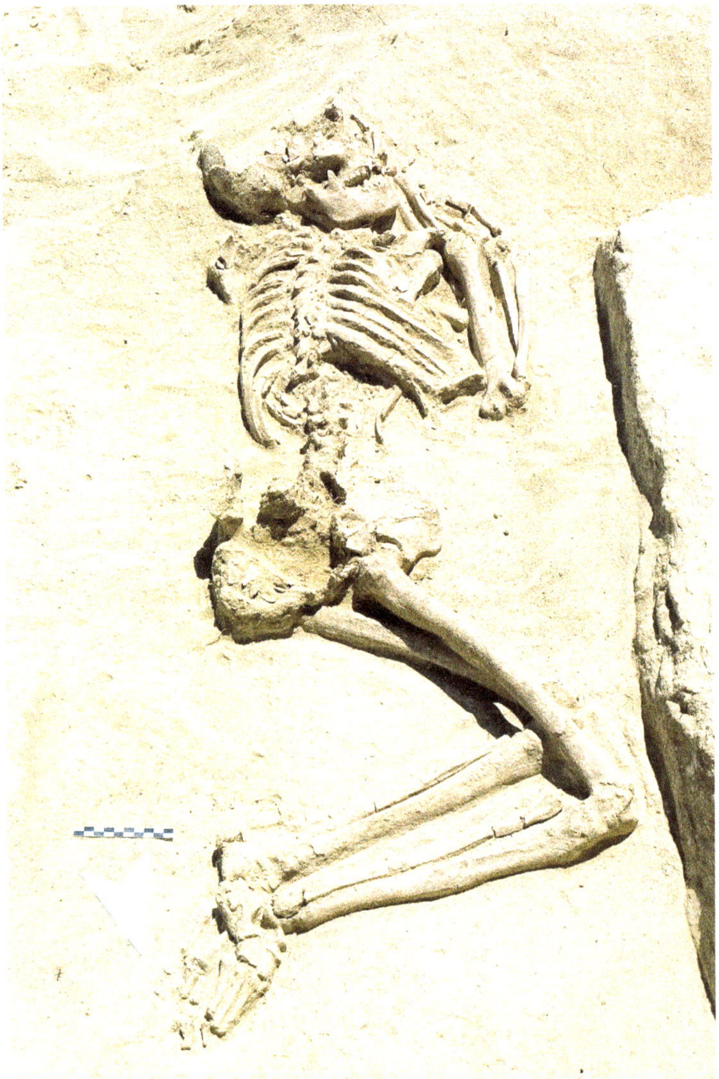

a

b

Figure 3.20 Tomb V: (a) (photo 1749) from north; (b) (photo 1752) from above west

FIGURE 3.21 Tomb V: **TV-1**

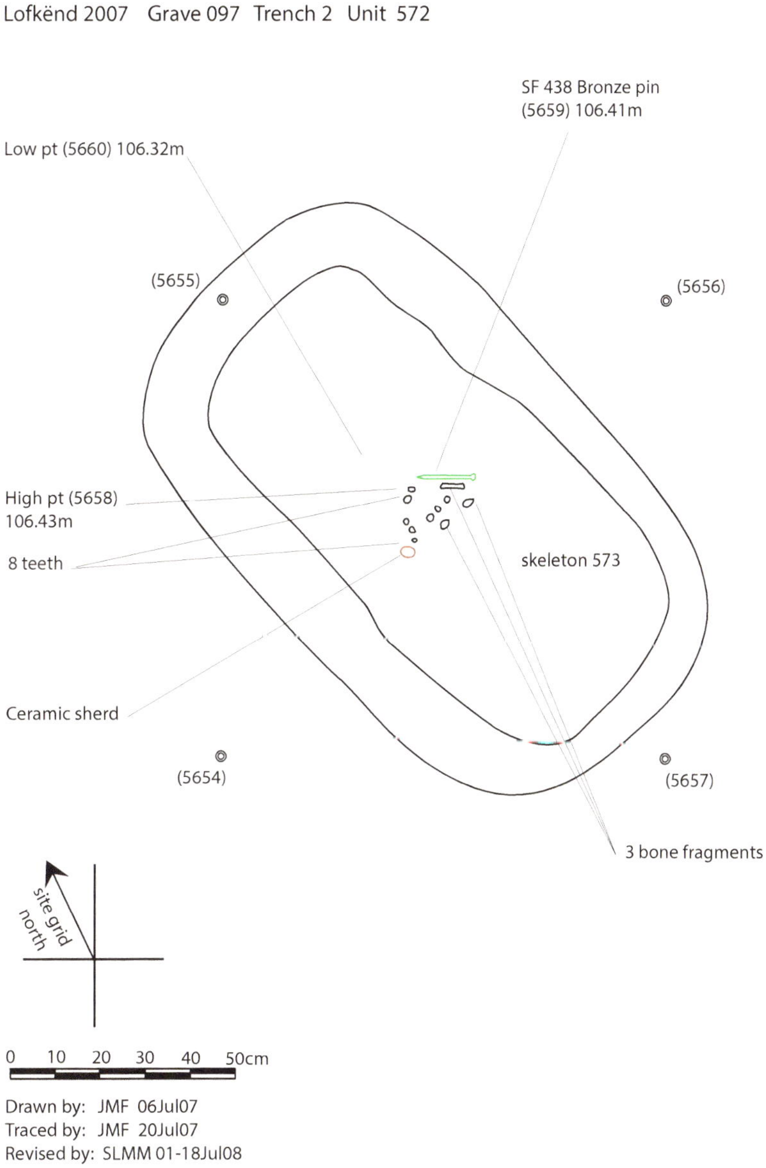

FIGURE 3.22 Plan of Tomb VI

a

b

FIGURE 3.23 Tomb VI: (a) (photo 1747) from northeast;
(b) (photo 1748) detail of skeletal remains and **TVI-1**, from above southeast

FIGURE 3.24 TOMB VI: **TVI-1**

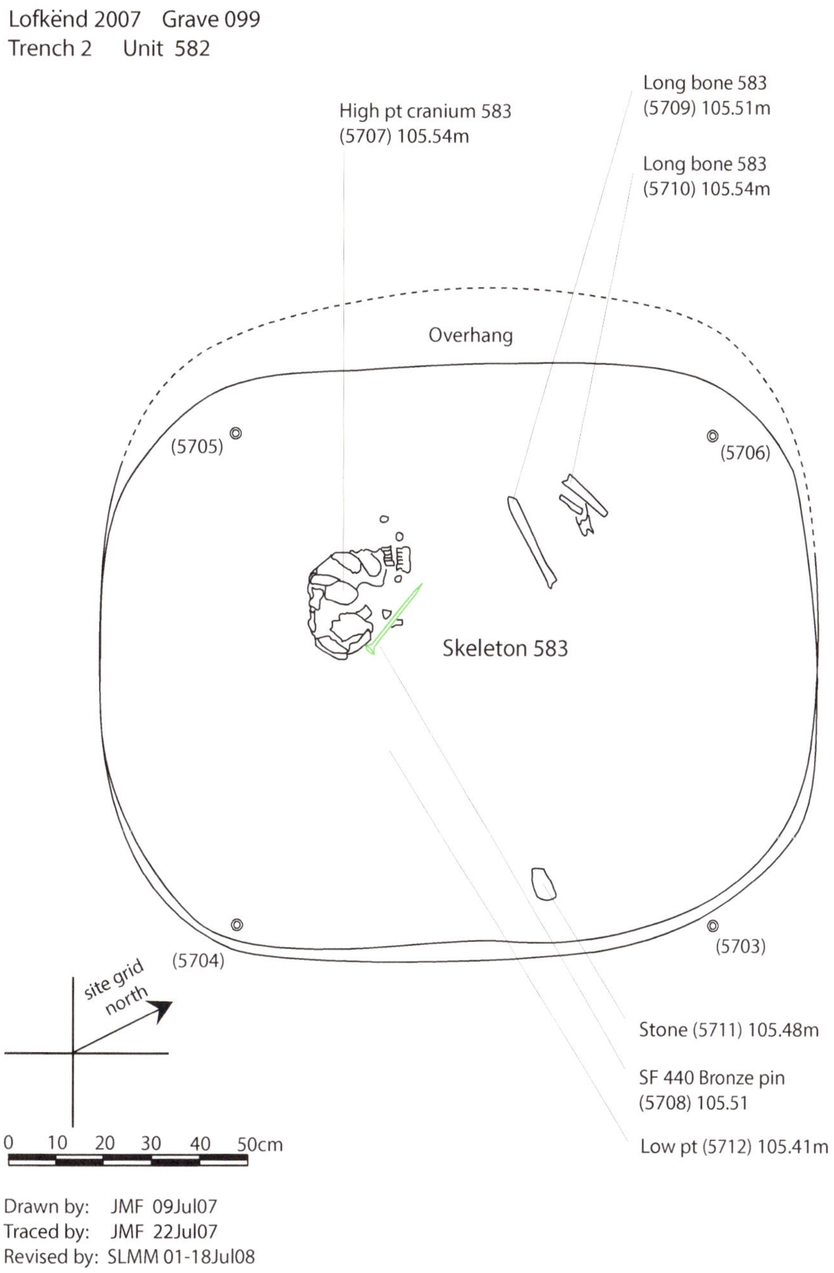

FIGURE 3.25 Plan of Tomb VII

a

c

b

Figure 3.26 Tomb VII: (a) (photo 1763) from south; (b) (photo 1765) from west; (c) (photo 1766) detail of cranium and **TVII-1,** from above north

Figure 3.27 Tomb VII: **TVII-1**

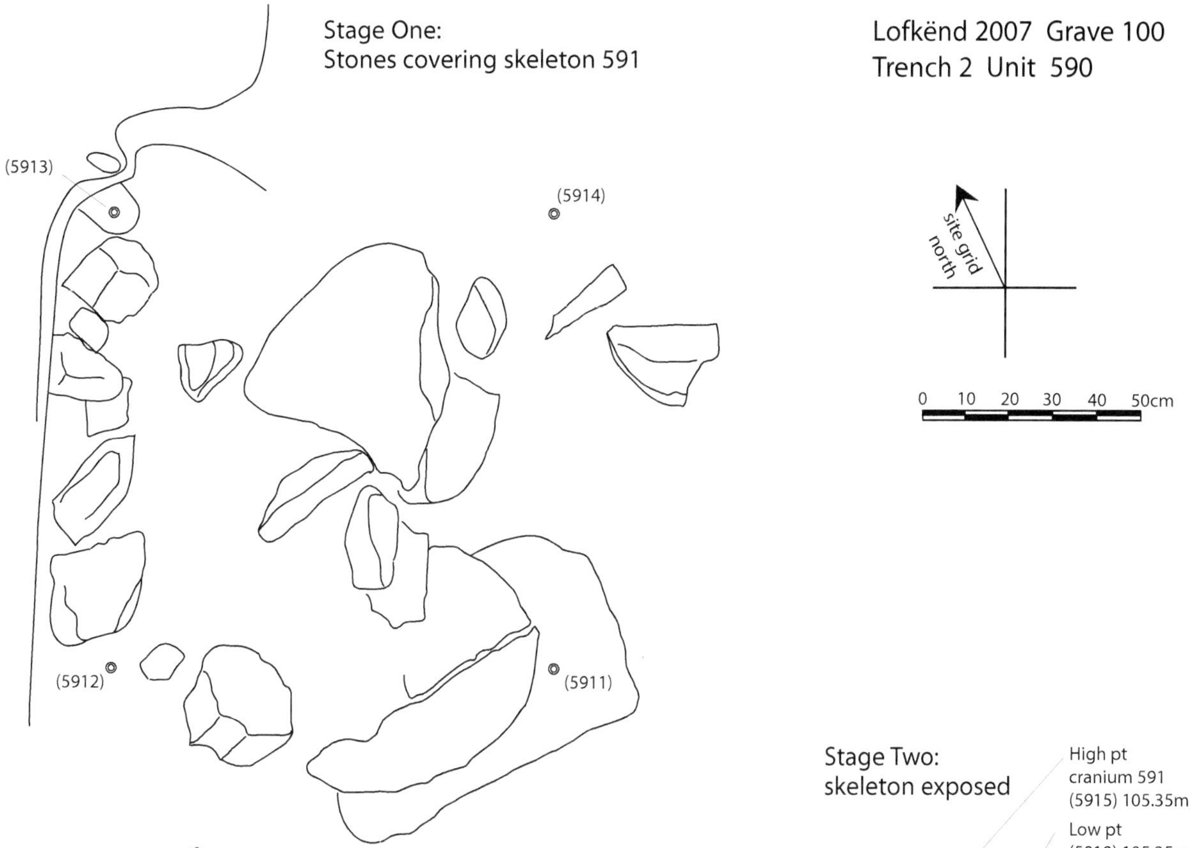

a

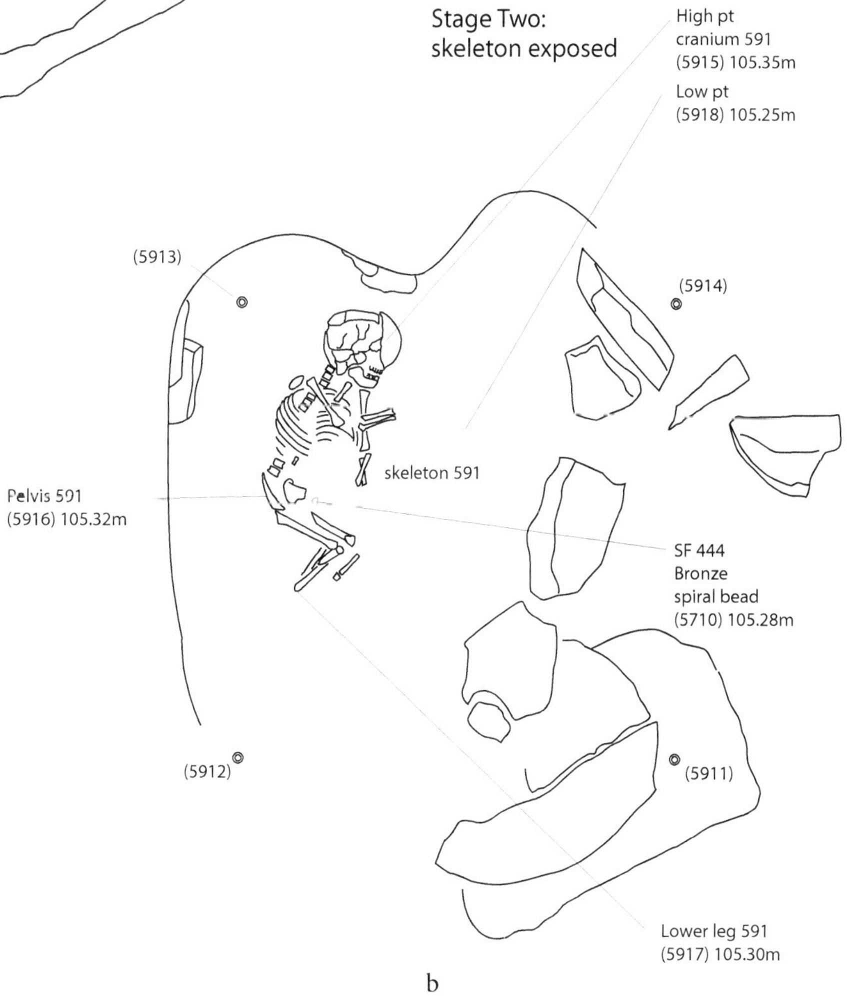

b

FIGURE 3.28 Plans of Tomb VIII: (a) Stage 1, covered with stone; (b) Stage 2, uncovered

a

b

FIGURE 3.29 Tomb VIII: (a) (photo 1796) from east; (b) (photo 1794) detail from south

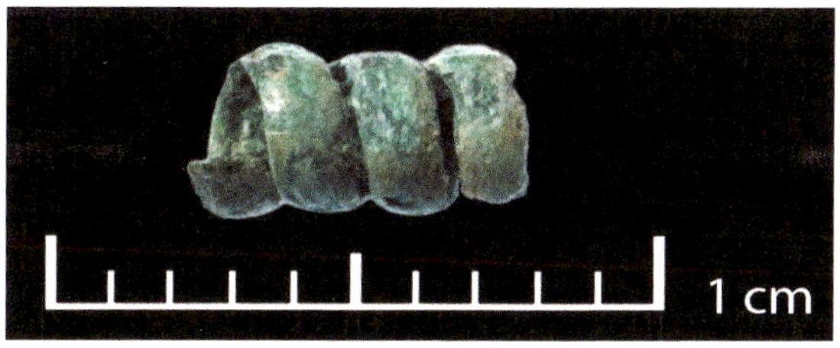

FIGURE 3.30 Tomb VIII: **TVIII-1**

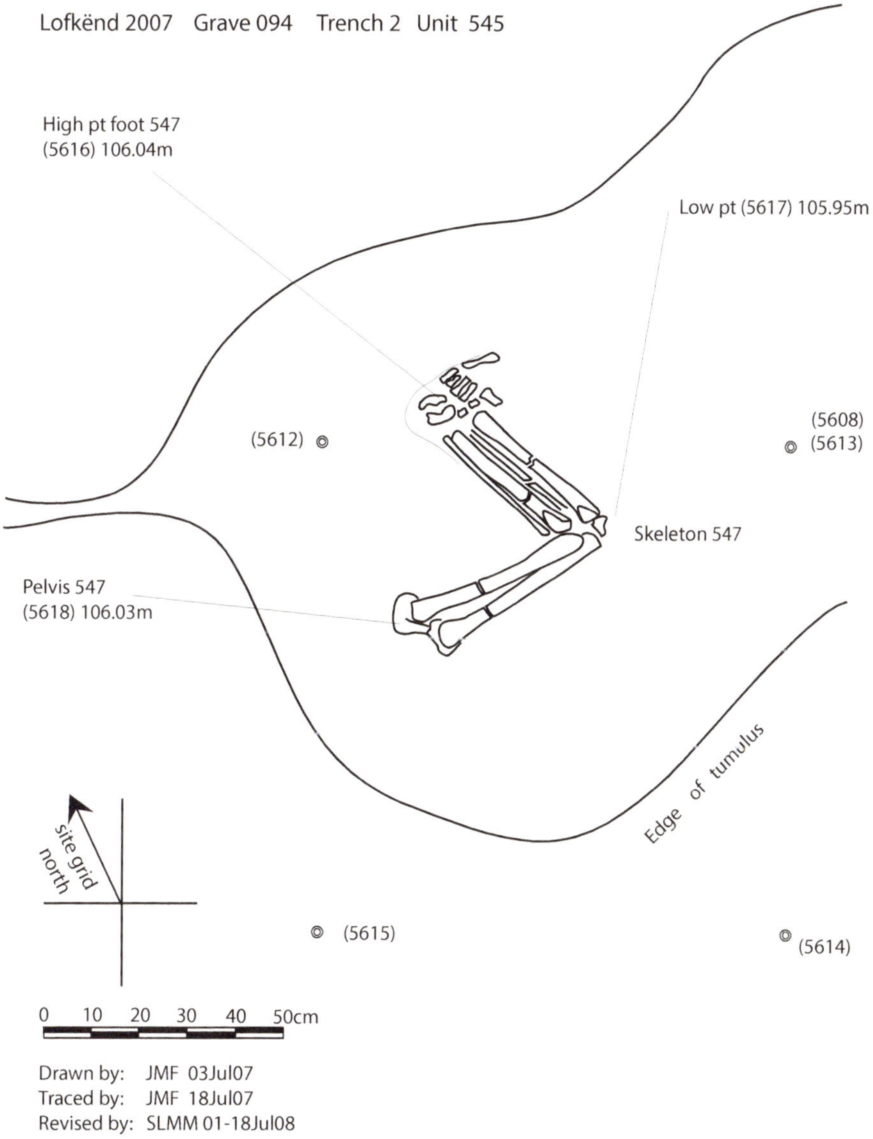

FIGURE 3.31 Plan of Tomb IX

a

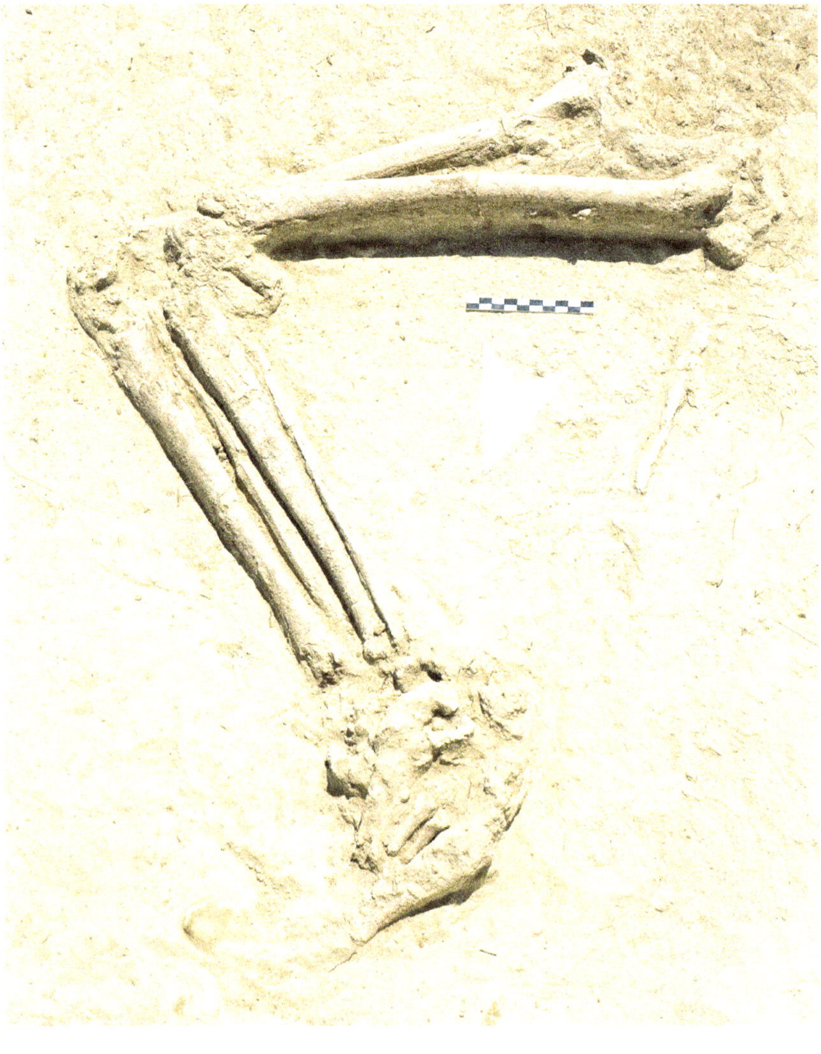

b

FIGURE 3.32 Tomb IX: (a) Tomb IX (left) and Tomb X (photo 1710) from east-southeast; (b) Tomb IX (photo 1717) from north-northwest

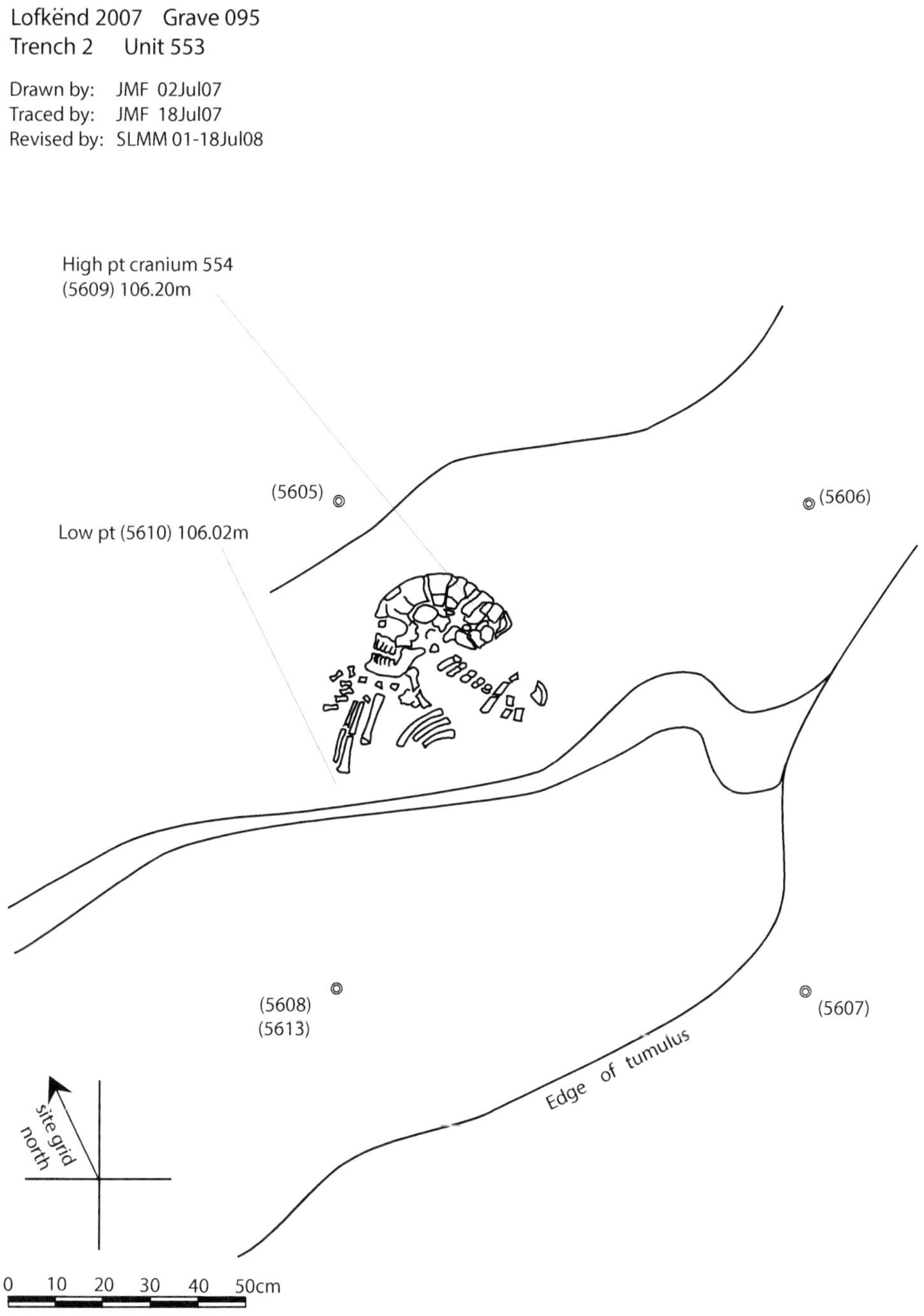

FIGURE 3.33 Plan of Tomb X

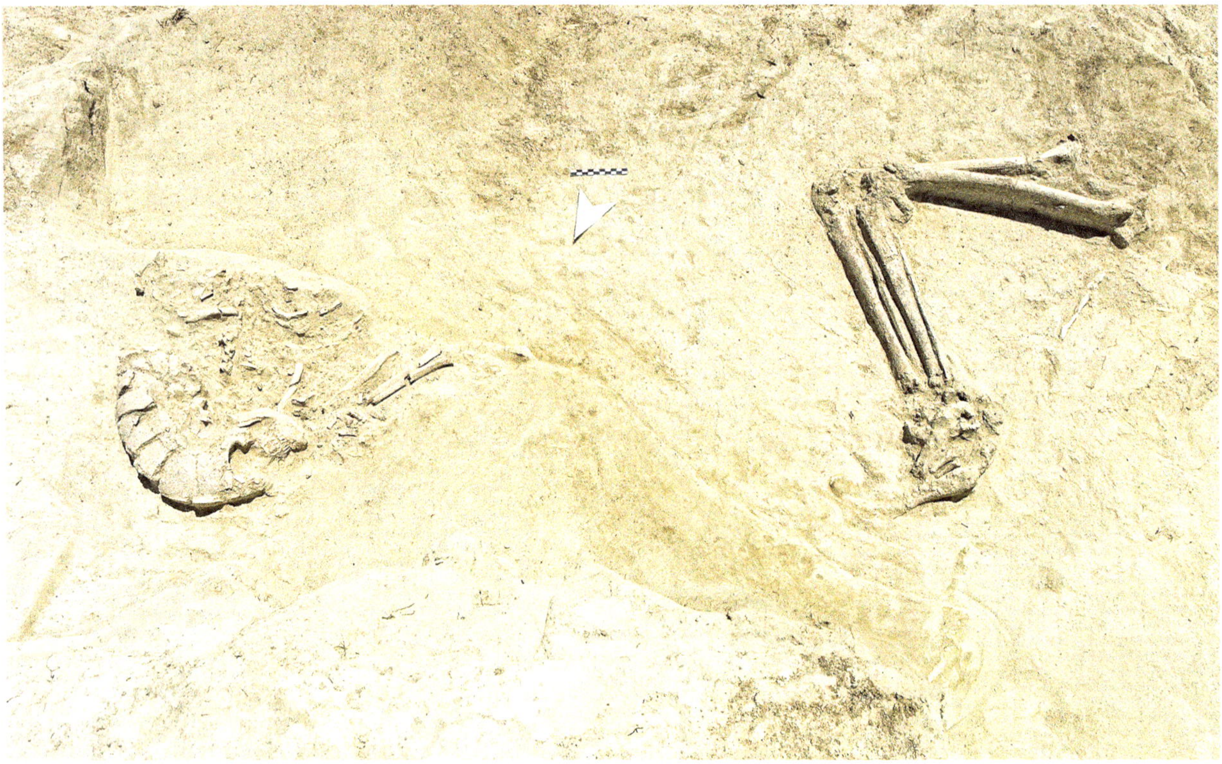

a

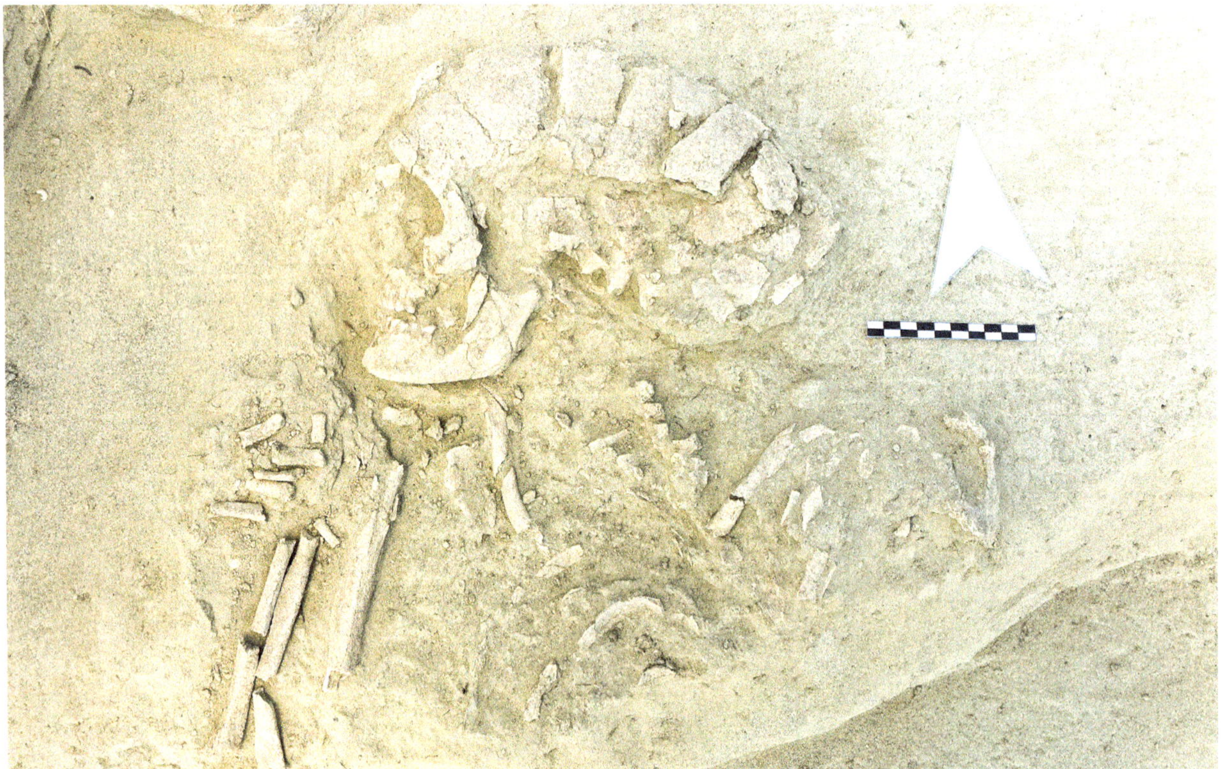

b

FIGURE 3.34 Tomb X: (a) Tomb IX (right) and X (photo 1708) from northwest; (b) (photo 1713), detail of cranium and upper body as preserved, from above south

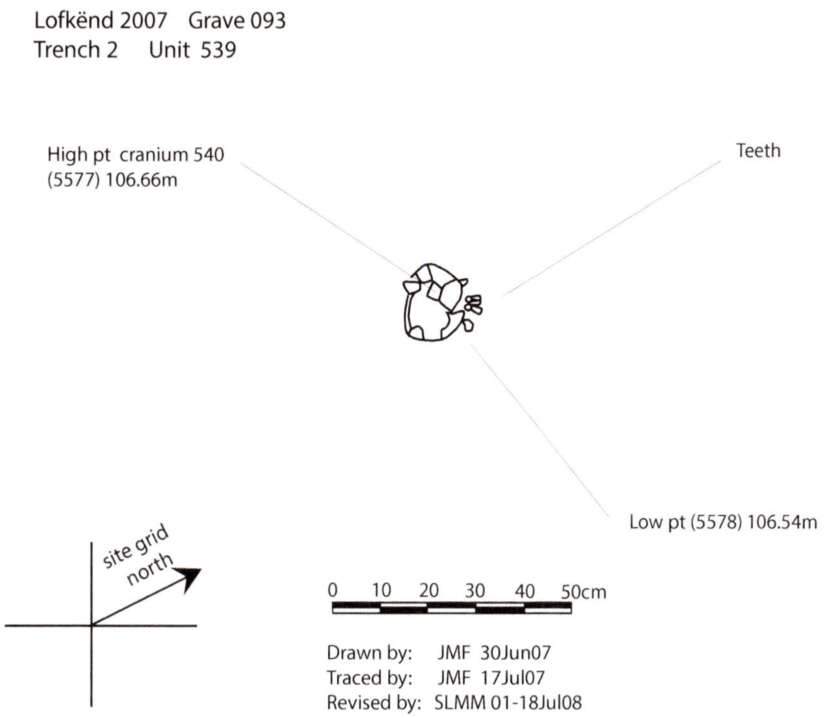

FIGURE 3.35 Plan of Tomb XI as preserved

FIGURE 3.36 Tomb XI (photo 1692) from above south

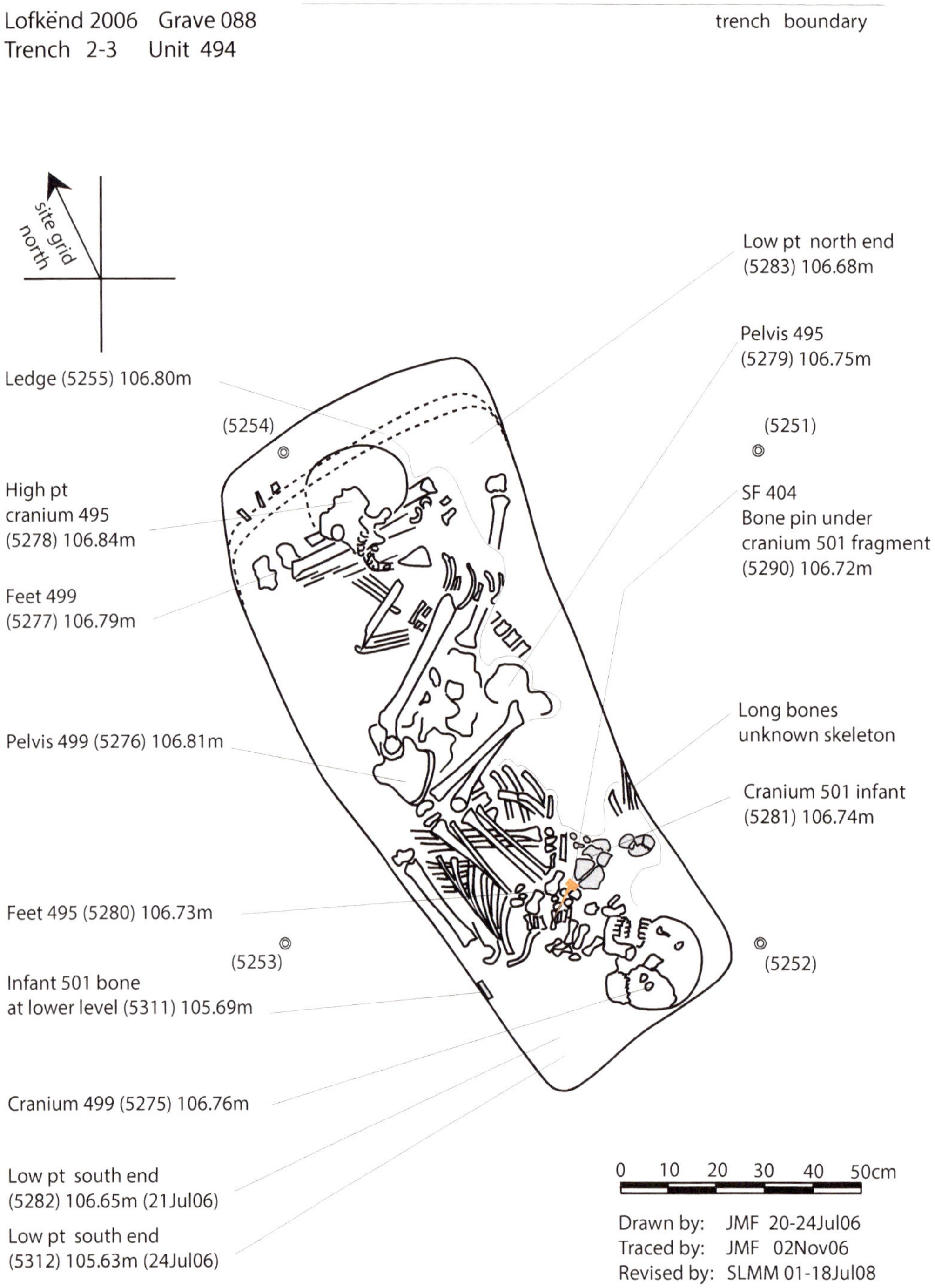

FIGURE 3.37 Plan of Tomb XII

a

b

c

FIGURE 3.38 Tomb XII: (a) (photo 1576) from northeast; (b) (photo 1590) from northwest; (c) (photo 1600) showing cranium of SU 499 and infant 501, together with **TXII-1**

FIGURE 3.39 Tomb XII: **TXII-1**

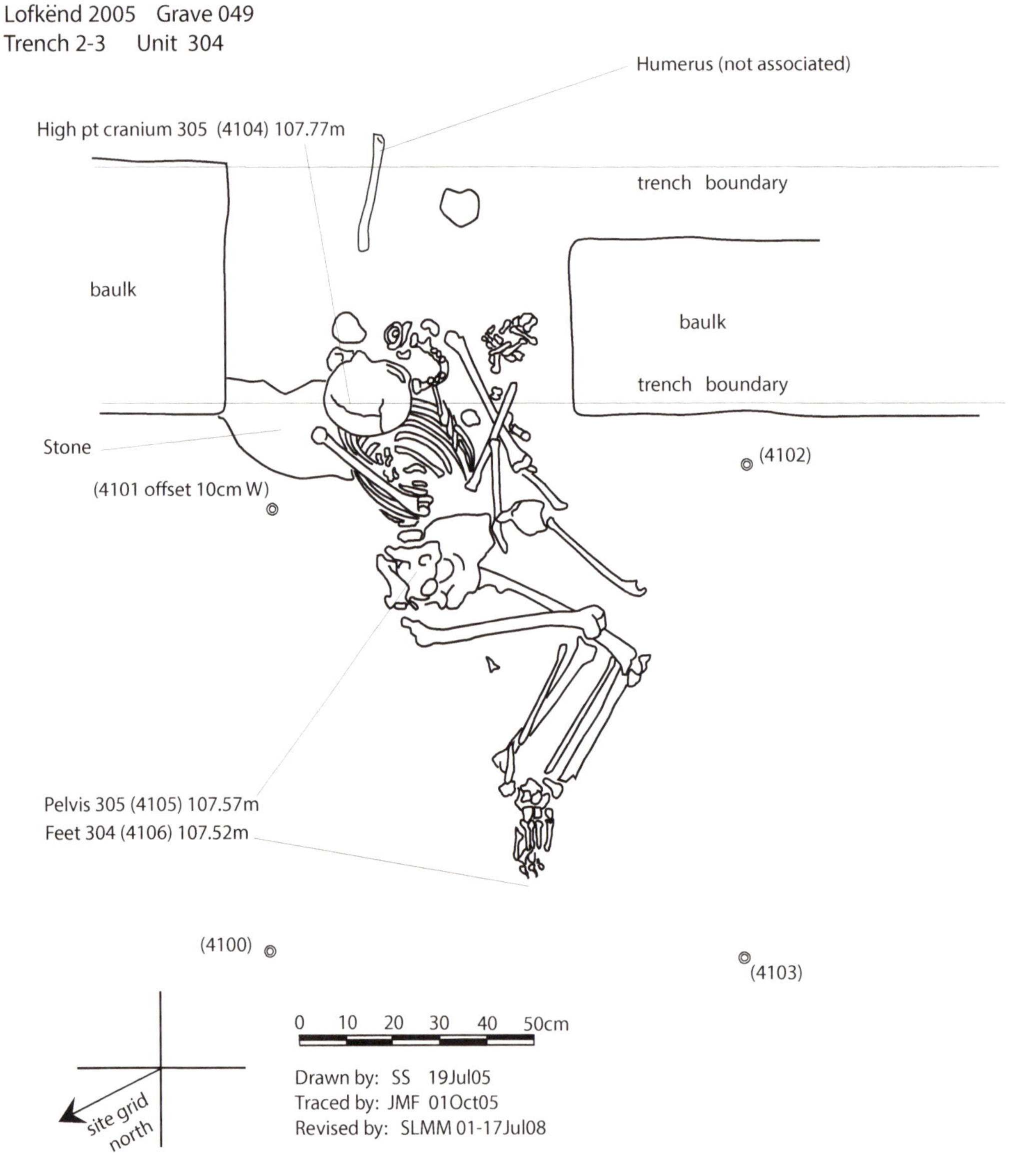

FIGURE 3.40 Plan of Tomb XIII

FIGURE 3.41 Tomb XIII: (a) (photo 1139) from above west-southwest; (b) (photo 1141) from south-southwest

a

b

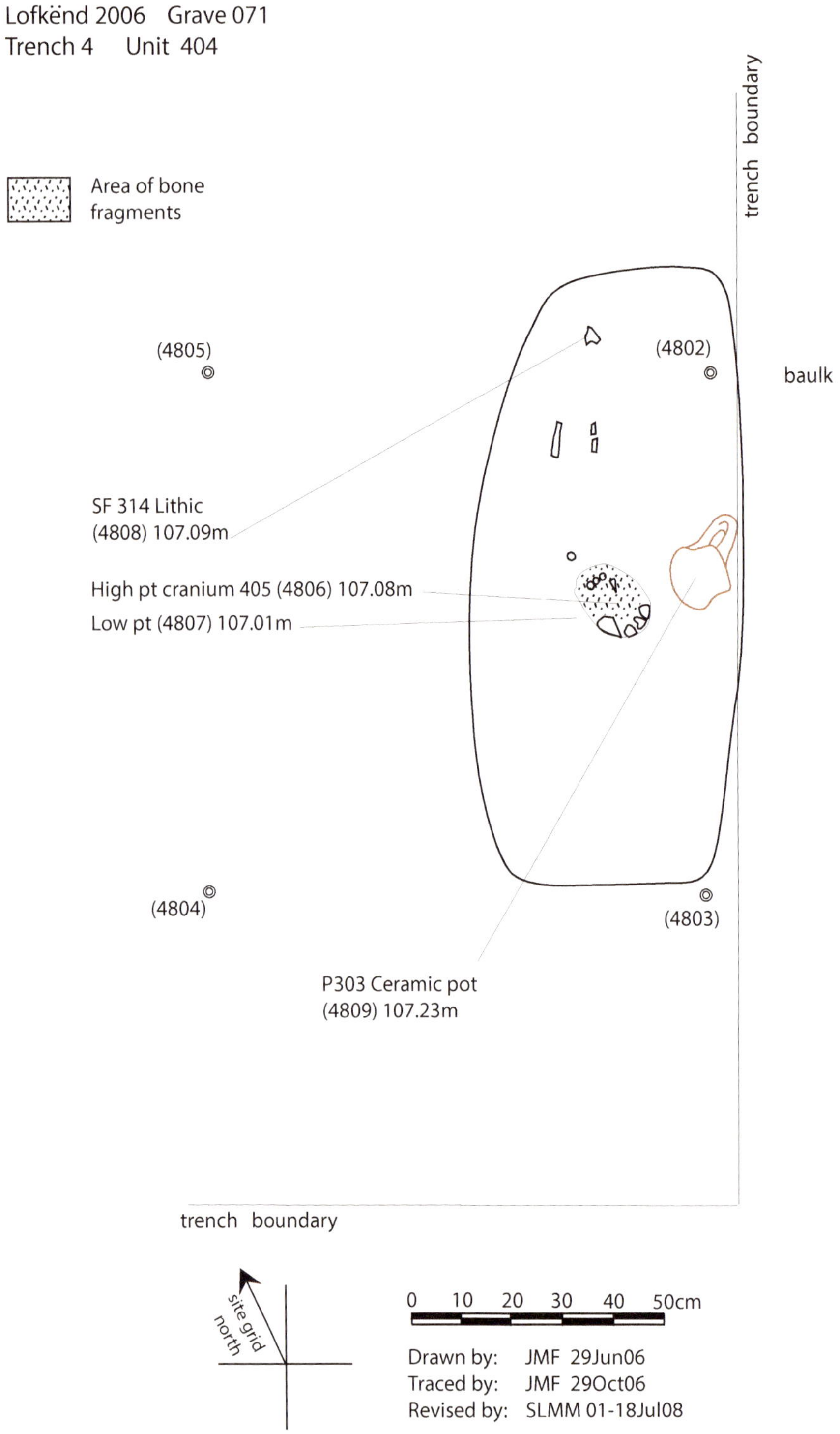

FIGURE 3.42 Plan of Tomb XIV

a

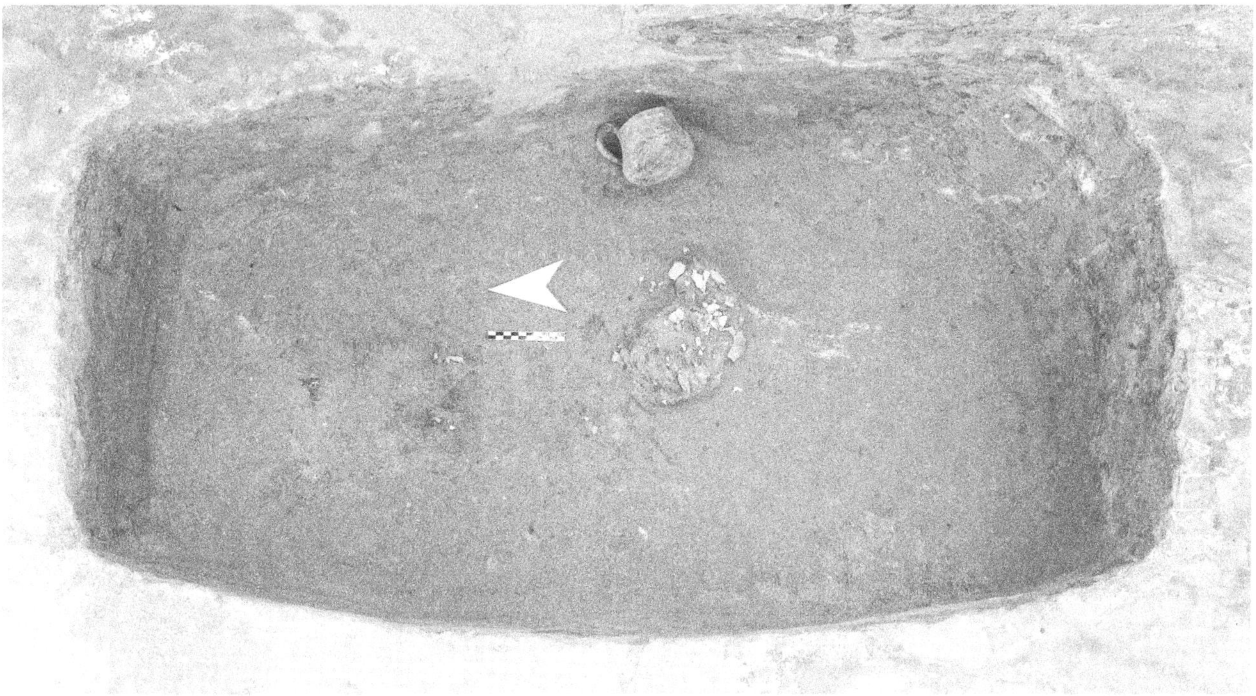

b

FIGURE 3.43 Tomb XIV: (a) The grave pit as first encountered (photo 1336) from above west; (b) the tomb as excavated (photo 1346) from above west

a

b

FIGURE 3.44 Tomb XIV: (a) (photo 1349) detail of **TXIV-1**, from above northwest; (b) **TXIV-1**

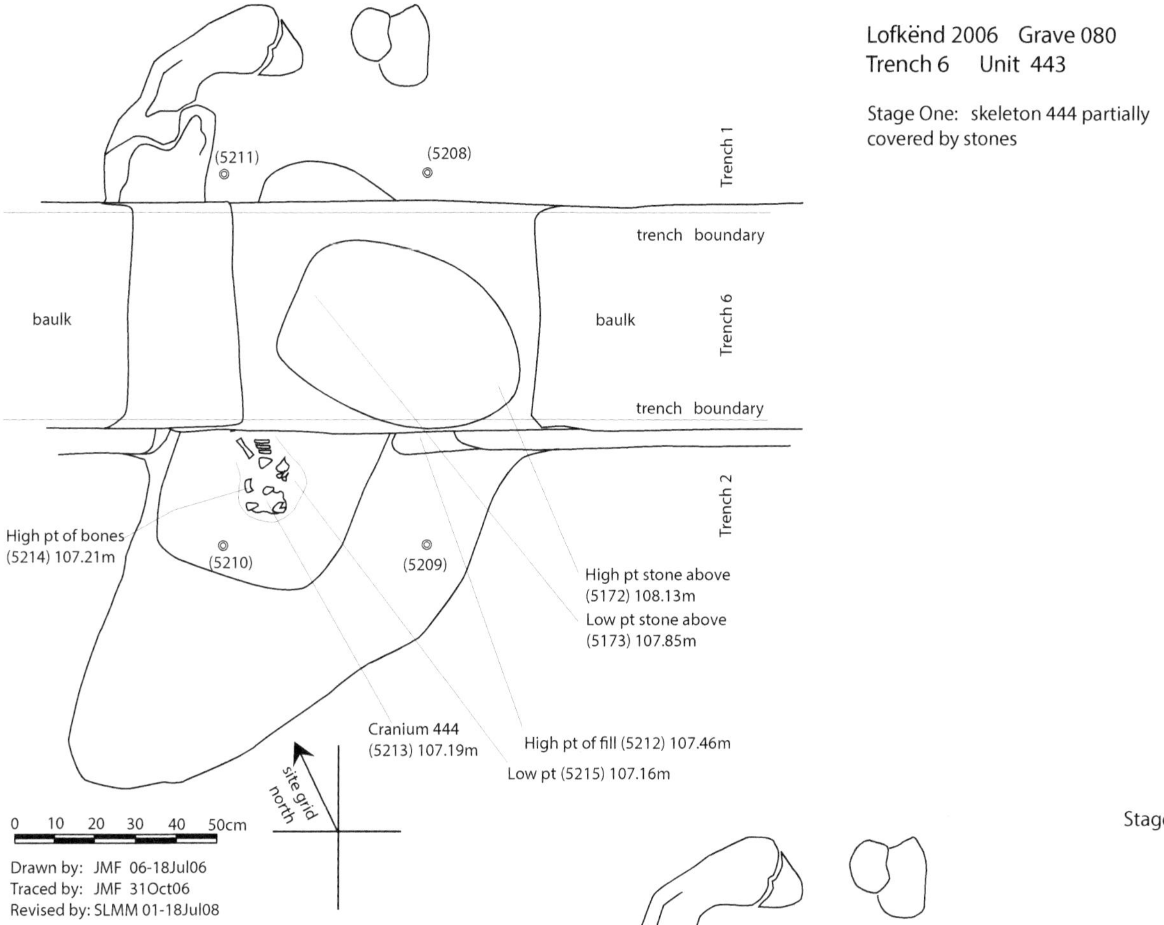

a

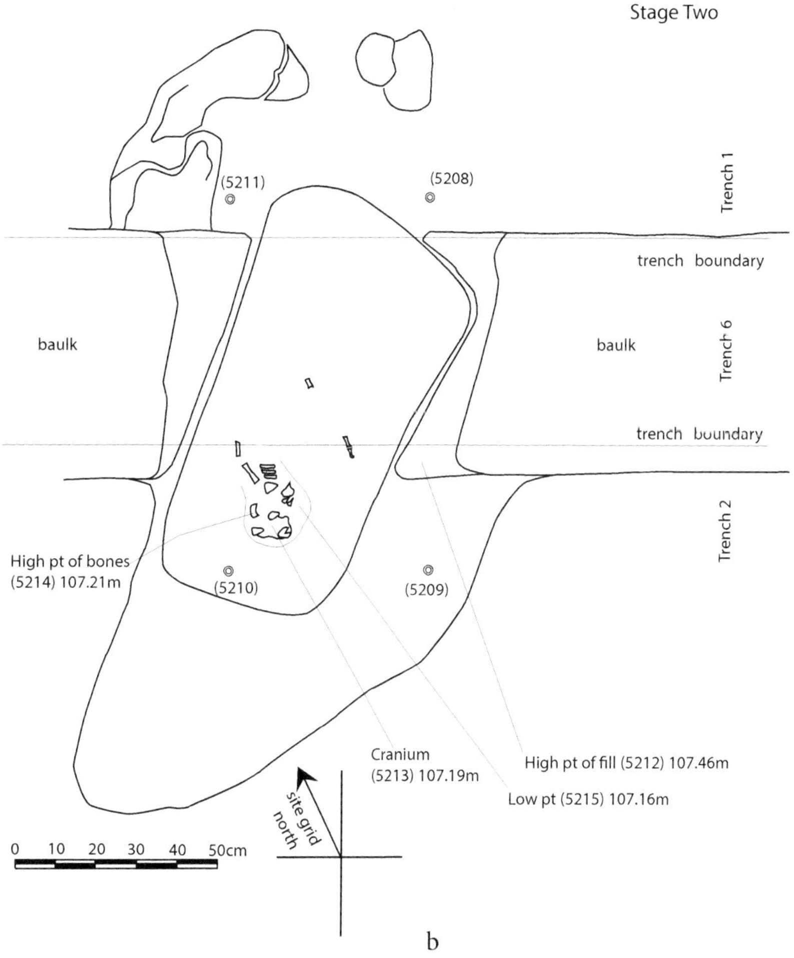

b

FIGURE 3.45 TXV: (a) Plan, Stage 1; (b) plan, Stage 2; (c) *on next page*

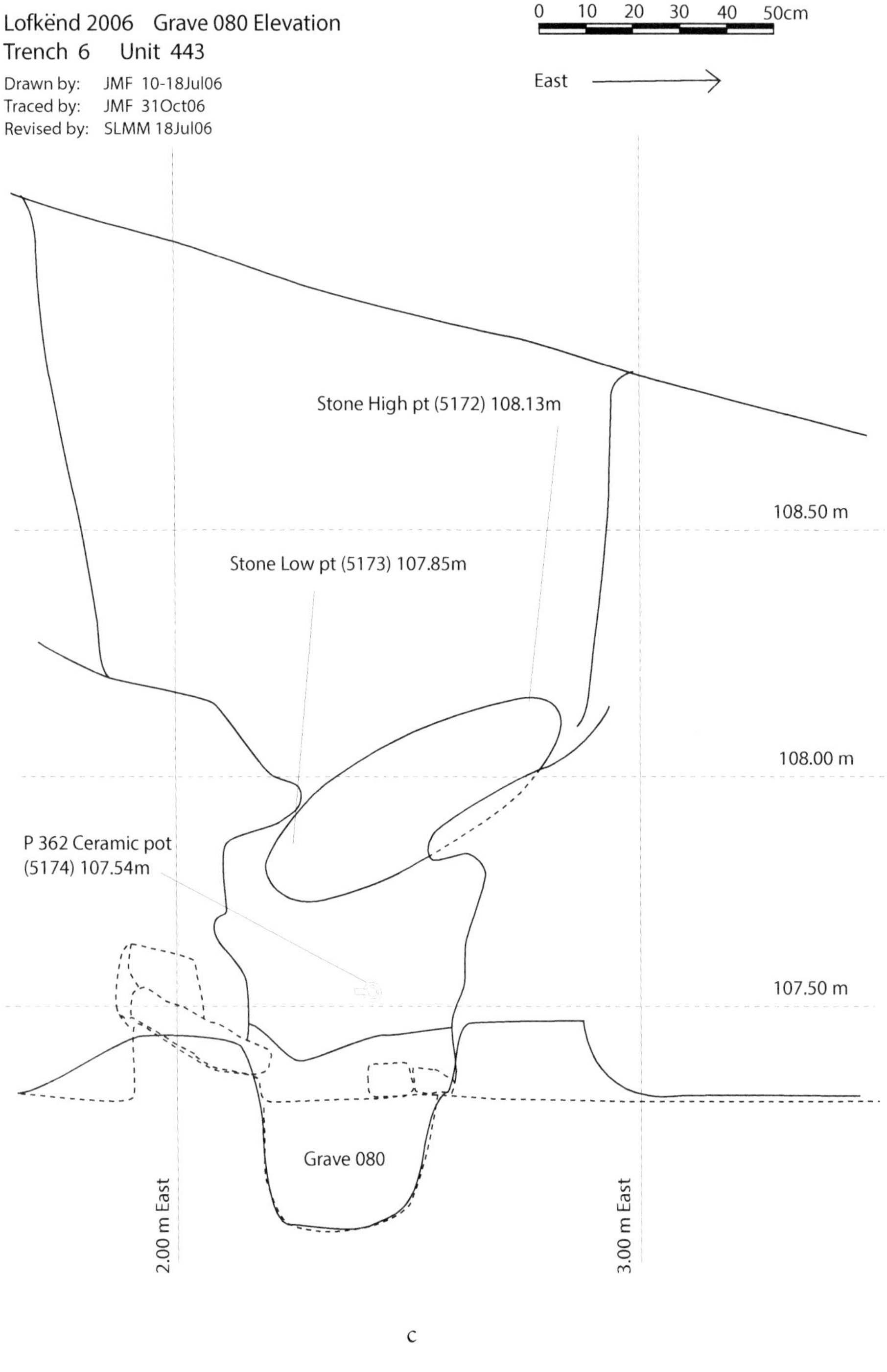

c

FIGURE 3.45 (*continued from previous page*). TXV: (c) elevation

a

c

b

(*Above and left*): FIGURE 3.46 TXV: (a) as first encountered in Sector 1, extending into the baulk, Sector 6 (photo 1463), from north; (b) large stone directly above Tomb XV, in baulk, Sector 6 (photo 1529), from west; (c) tomb (photo 1553) as fully exposed, from north-northeast

FIGURE 3.47 Tomb XV: **TXV-1**

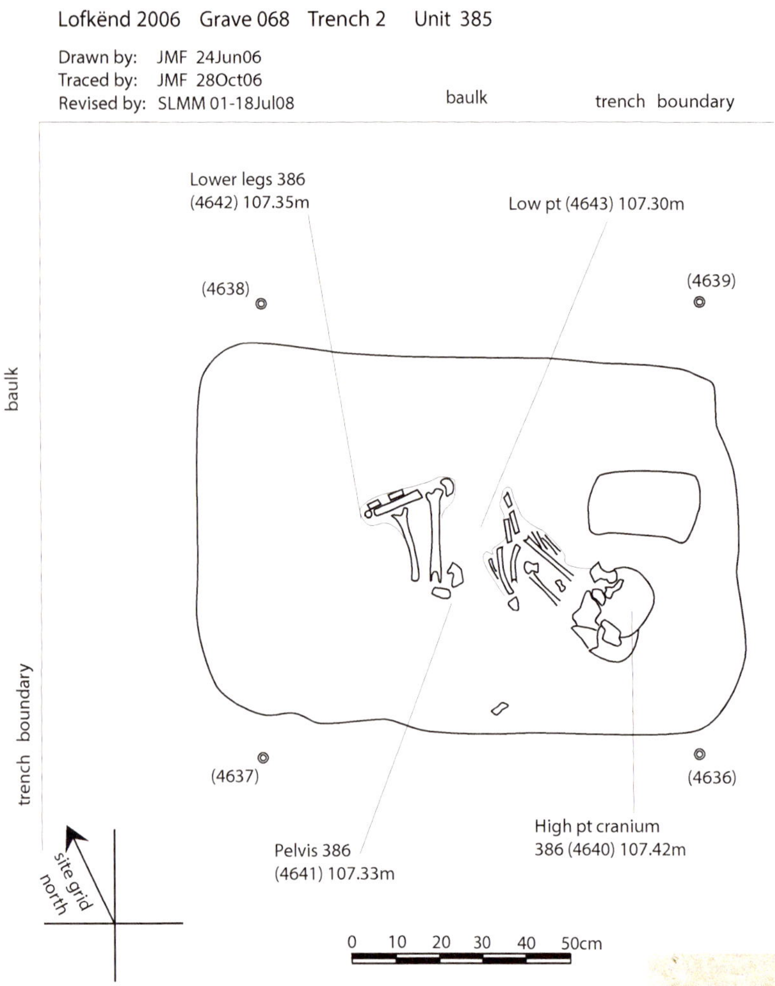

FIGURE 3.48 Plan of Tomb XVI

FIGURE 3.49 Tomb XVI: (photo 1312) from west

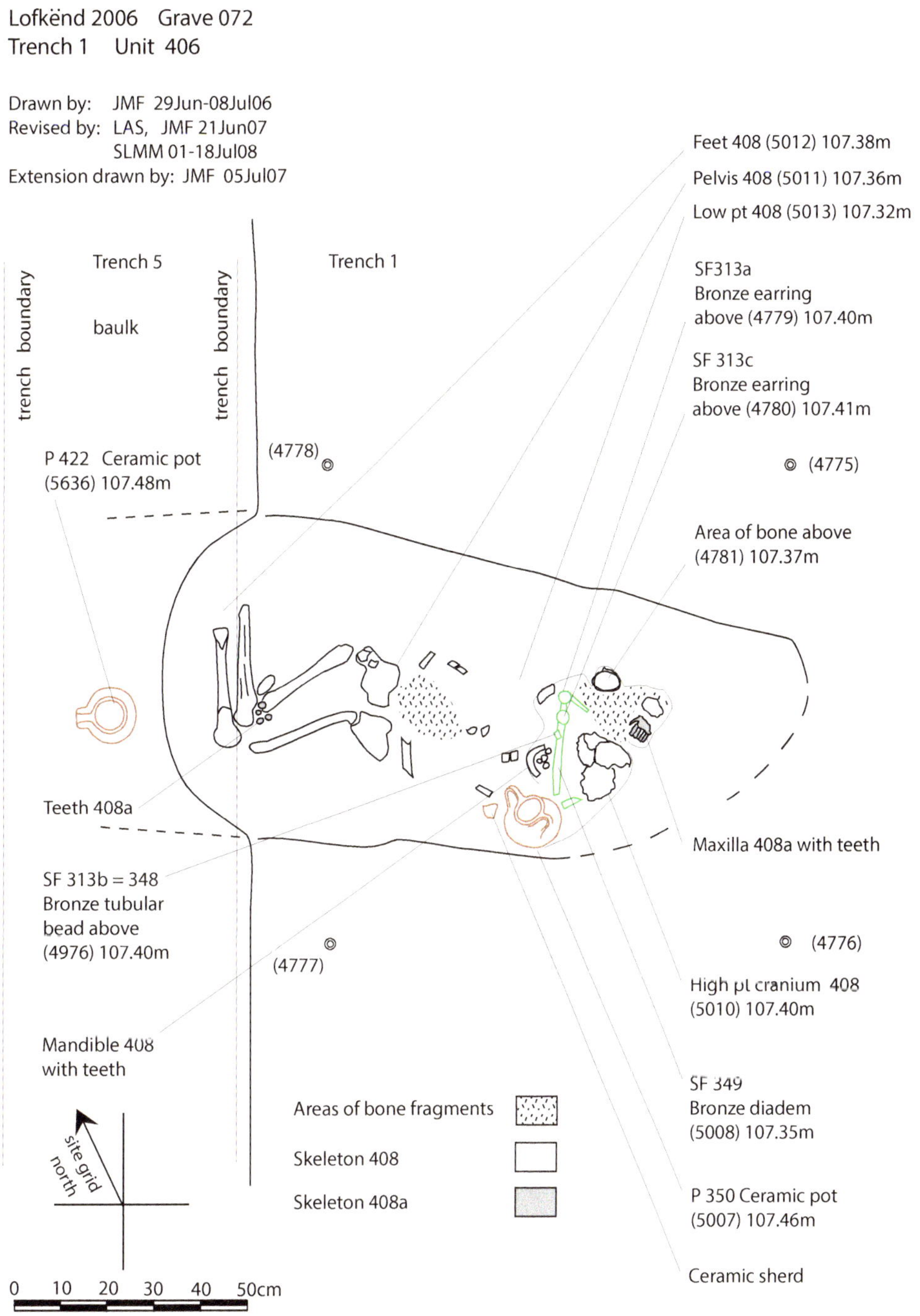

FIGURE 3.50 Plan of Tomb XVII

a

FIGURE 3.51 Tomb XVII: (a) (photo 1426) as first exposed, from east; (b) (photo 1429) detail of kantharos, **TXVII-1**, and headband, **TXVII-2,** from above south; (c, d) *on facing page*

b

c

d

FIGURE 3.51 *(continued)* Tomb XVII: (c) (photo 1434) tomb fully exposed, from above south; (d) (photo 1728) photo of **TXVII-7** found in 2007 in the baulk (Sector 5), from east (the yellow label indicates the lowest level of Tomb 72 as excavated in 2006)

Figure 3.52 Tomb XVII: (a) **TXVII-1**; (b) **TXVII-7**

Figure 3.53 Tomb XVII: **TXVII-2**

Figure 3.54 Tomb XVII: **TXVII-4** and **TXVII-6** (photo 3055)

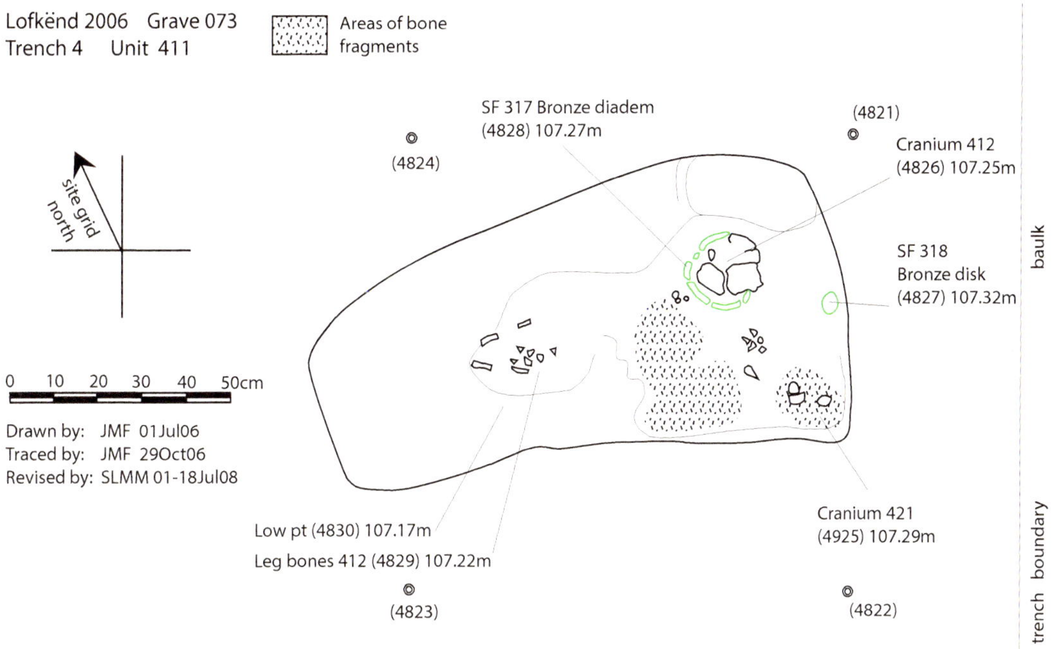

FIGURE 3.55 Plan of Tomb XVIII

a

b

(*Left and above*) FIGURE 3.56 Tomb XVIII: (a) (photo 1360), from west; (b) (photo 1369) **TXVIII-1** in situ after cleaning, from above west

(*Right*): FIGURE 3.57 Tomb XVIII: **TXVIII-1**

FIGURE 3.58 Tomb XVIII: **TXVIII-2**

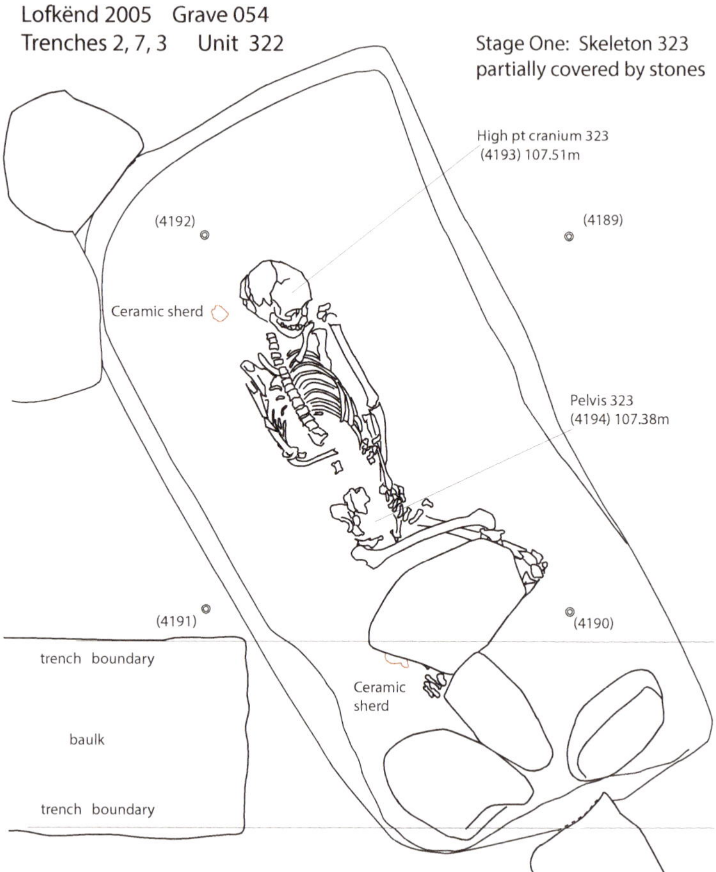

a

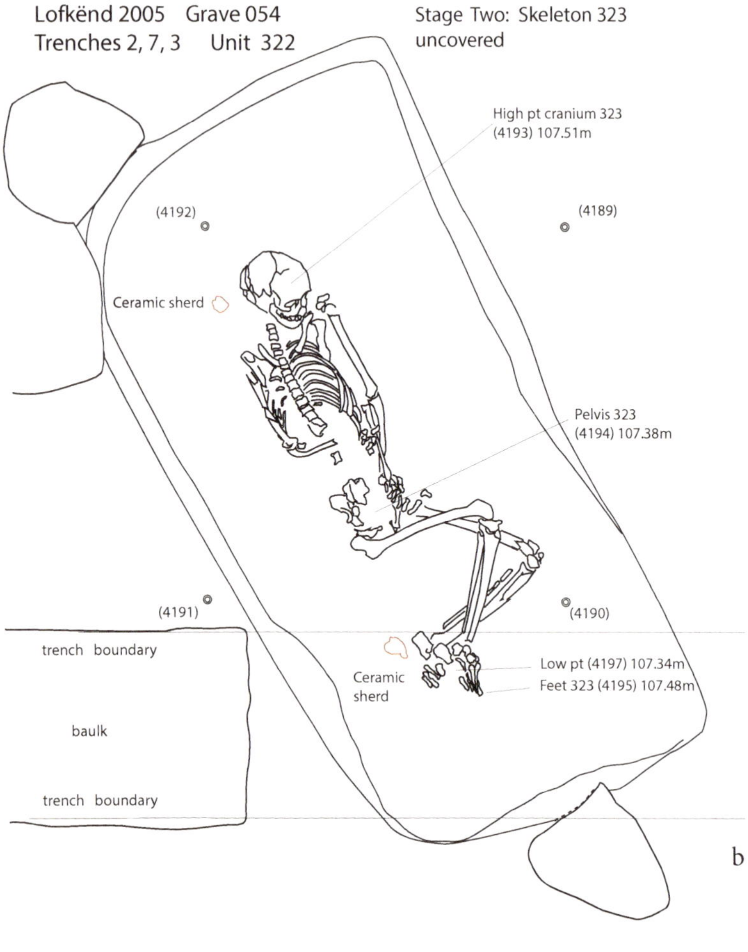

b

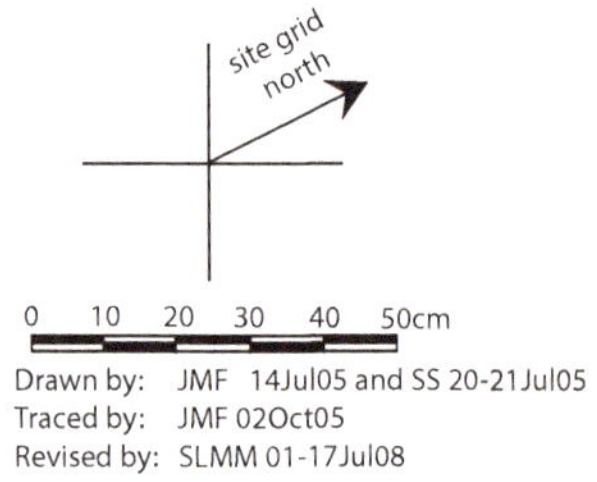

FIGURE 3.59 Tomb XIX:
(a) plan, Stage 1;
(b) plan, Stage 2

a b c d

Figure 3.60 Tomb XIX: (a) the pit for Tomb XIX as first encountered (photo 1138), from southwest; (b) (photo 1153b), during excavation, showing dark stripe to south of skeleton, from northeast; (c) (photo 1171) as covered, from northeast; (d) (photo 1188), detail of skeleton exposed, from north

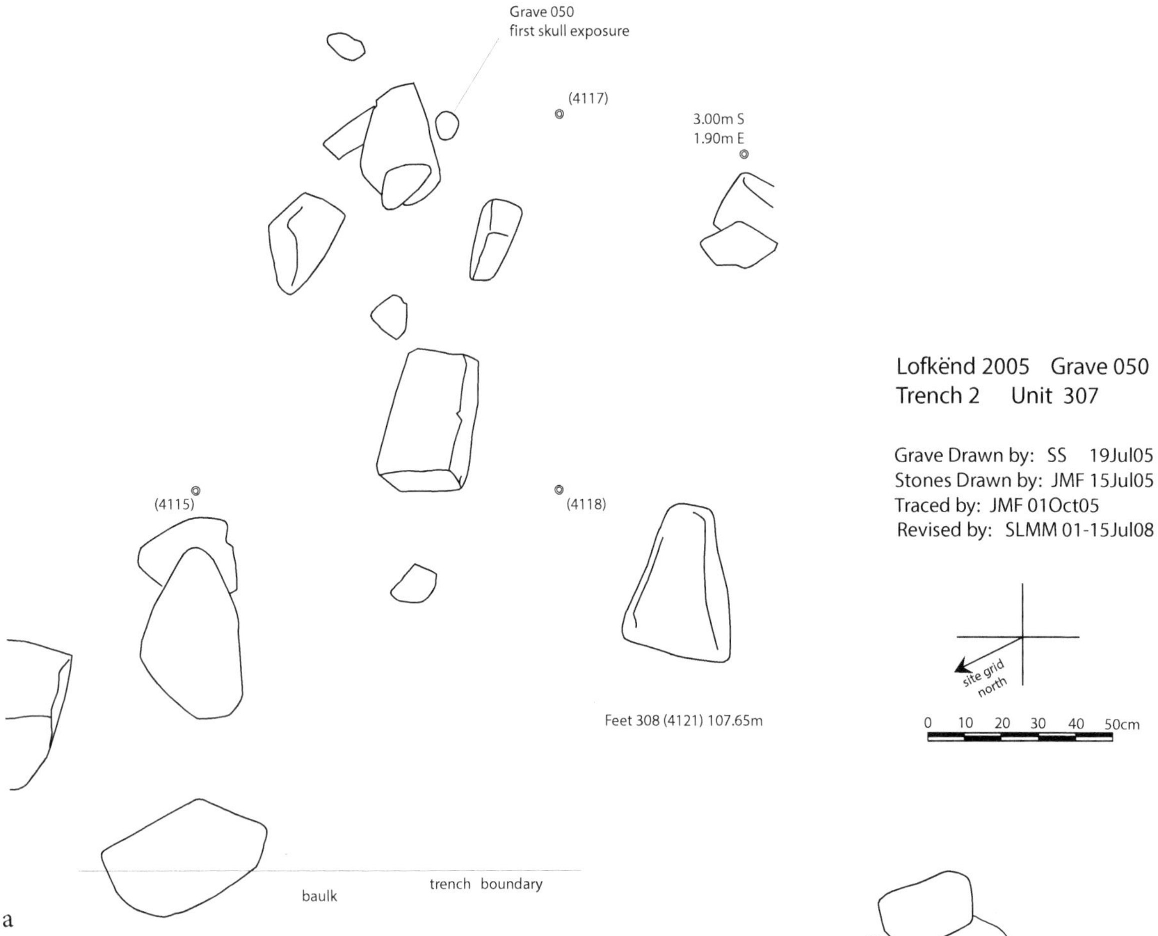

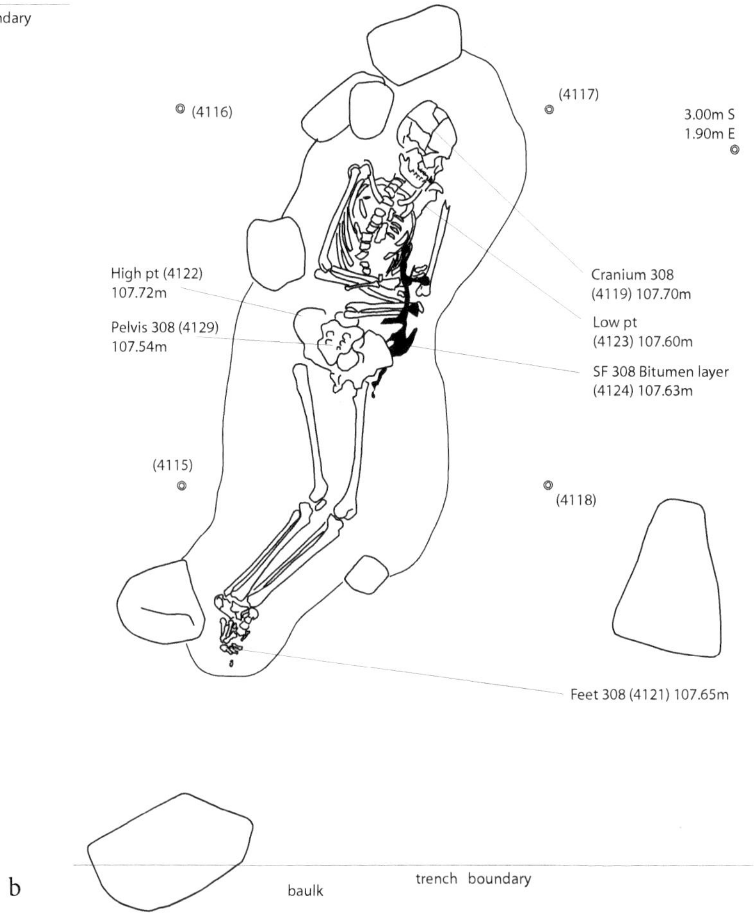

FIGURE 3.61 Tomb XX:
(a) plan, Stage 1; (b) plan, Stage 2

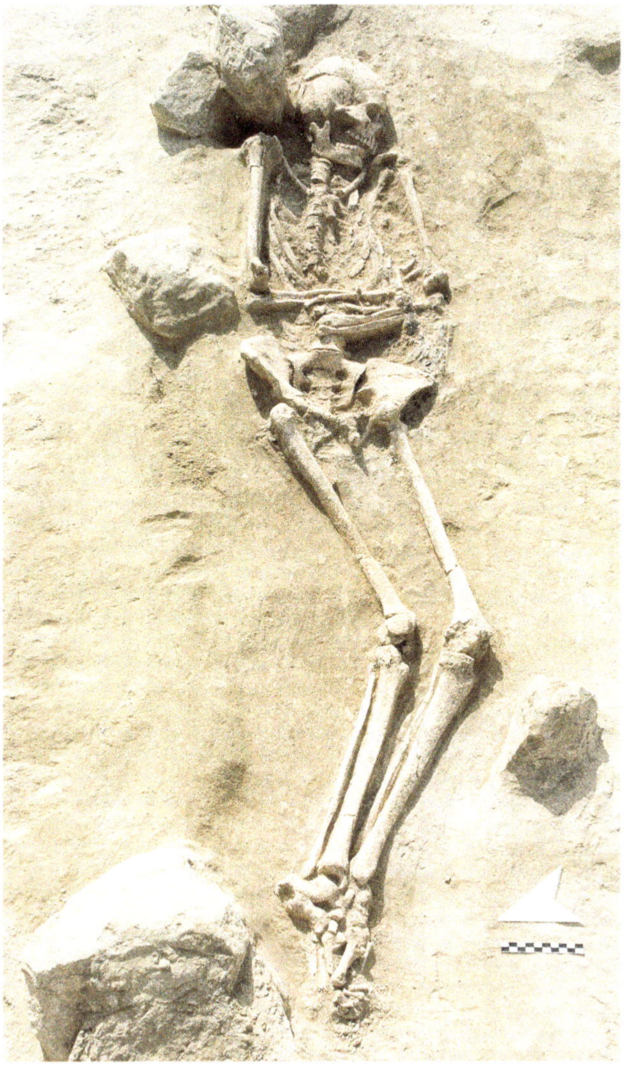

a

FIGURE 3.62 Tomb XX: (a) (photo 1148) from west; (b) (photo 1151) detail of torso and pelvis showing bitumen, from above south

b

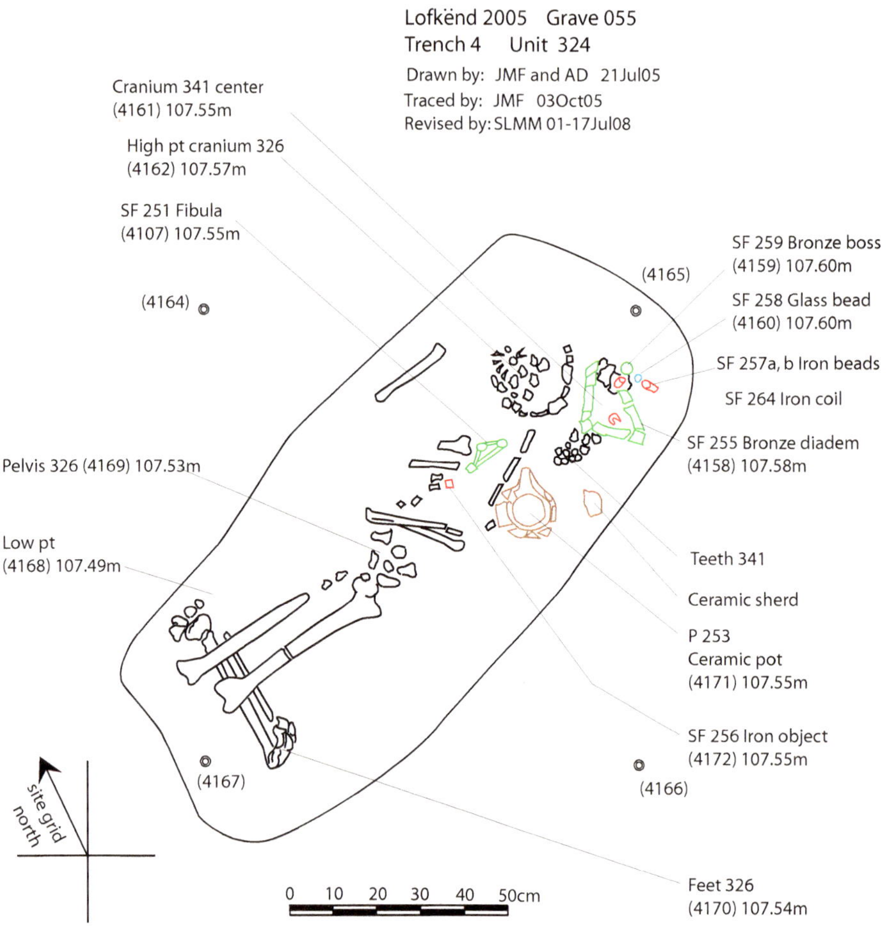

FIGURE 3.63 Plan of Tomb XXI

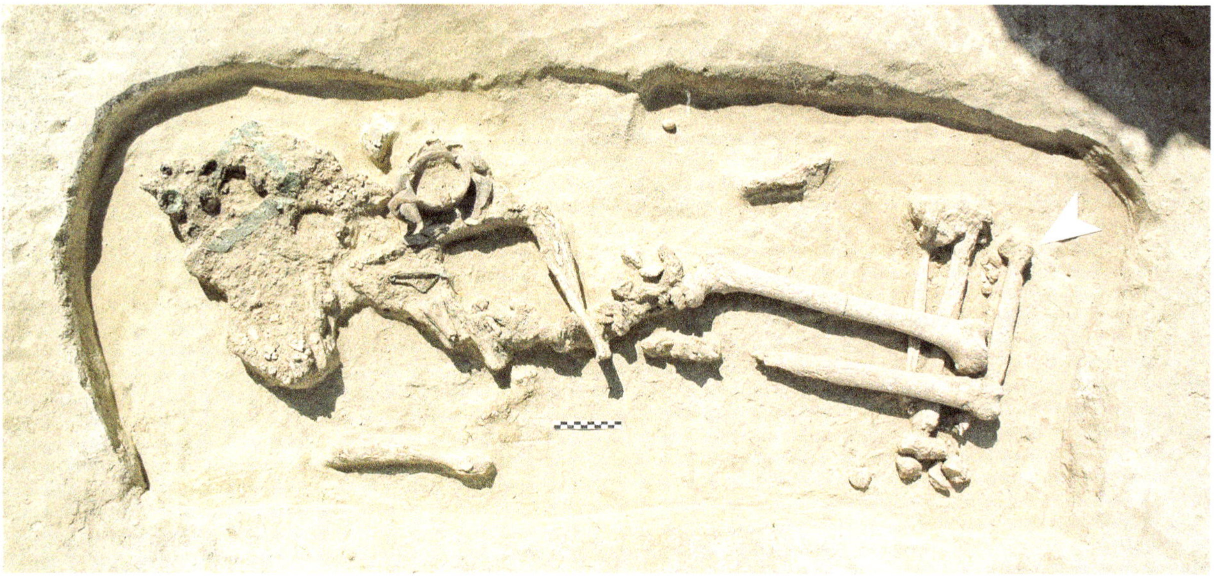

a

FIGURE 3.64 Tomb XXI: (a) (photo 1163), from above northwest; (b, c, d) *on facing page*

b

c

d

Figure 3.64 (*continued from facing page*): Tomb XXI: (b) (photo 1165), in relation to Wall 1 and tumulus edge, from southeast; (c) (photo 1167) detail of **TXXI-1** and **TXXI-2**, from southwest; (d) (photo 1169) detail of headband, **TXXI-3**, and associated finds, from west-northwest

FIGURE 3.65 Tomb XXI: **TXXI-1**

FIGURE 3.66 Tomb XXI: **TXXI-2**

FIGURE 3.67 Tomb XXI: **TXXI-3**

FIGURE 3.68 Tomb XXI: **TXXI-4**

FIGURE 3.69 Tomb XXI: **TXXI-5**

a b

FIGURE 3.70 Tomb XXI: (a–b) **TXXI-6**, **TXXI-7**

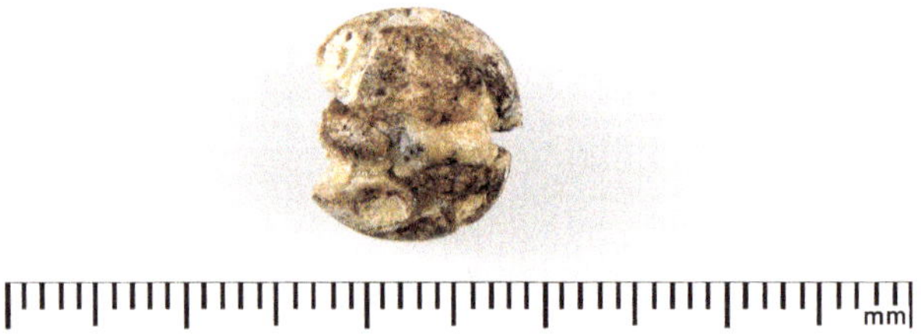

FIGURE 3.71 Tomb XXI: **TXXI-8**

FIGURE 3.72 Tomb XXI: **TXXI-9**

FIGURE 3.73 Tomb XXI: **TXXI-10**

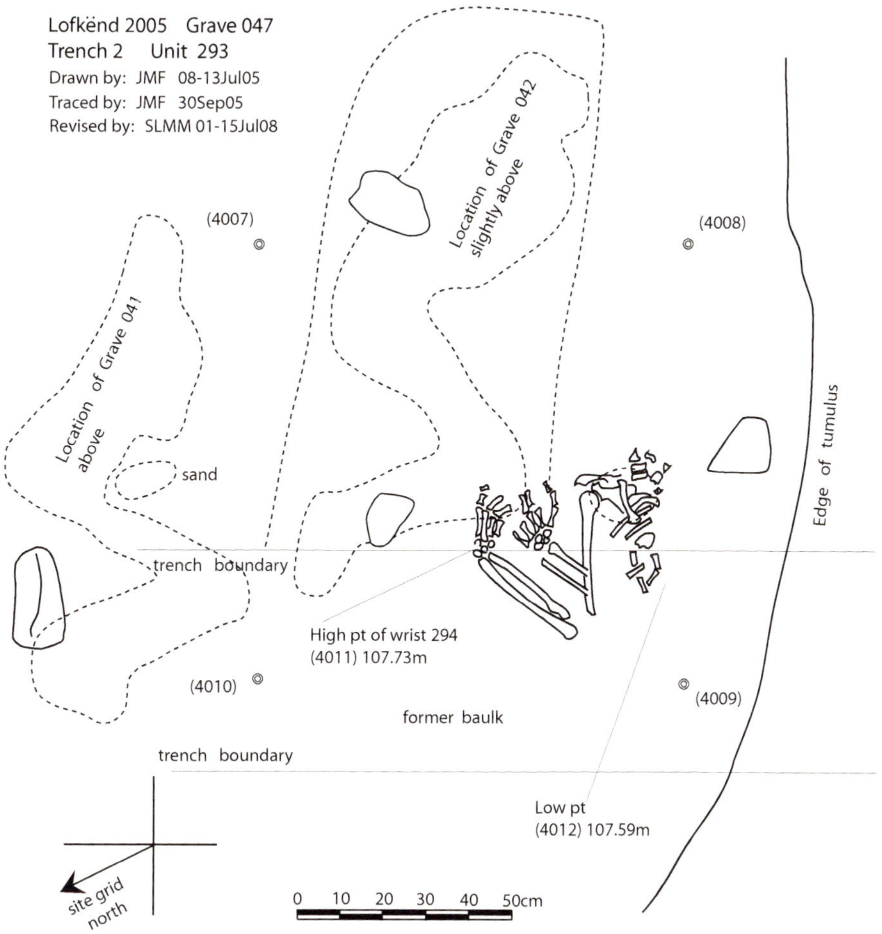

FIGURE 3.74 Plan of Tomb XXII

FIGURE 3.75 Tomb XXII (photo 1110) from above west

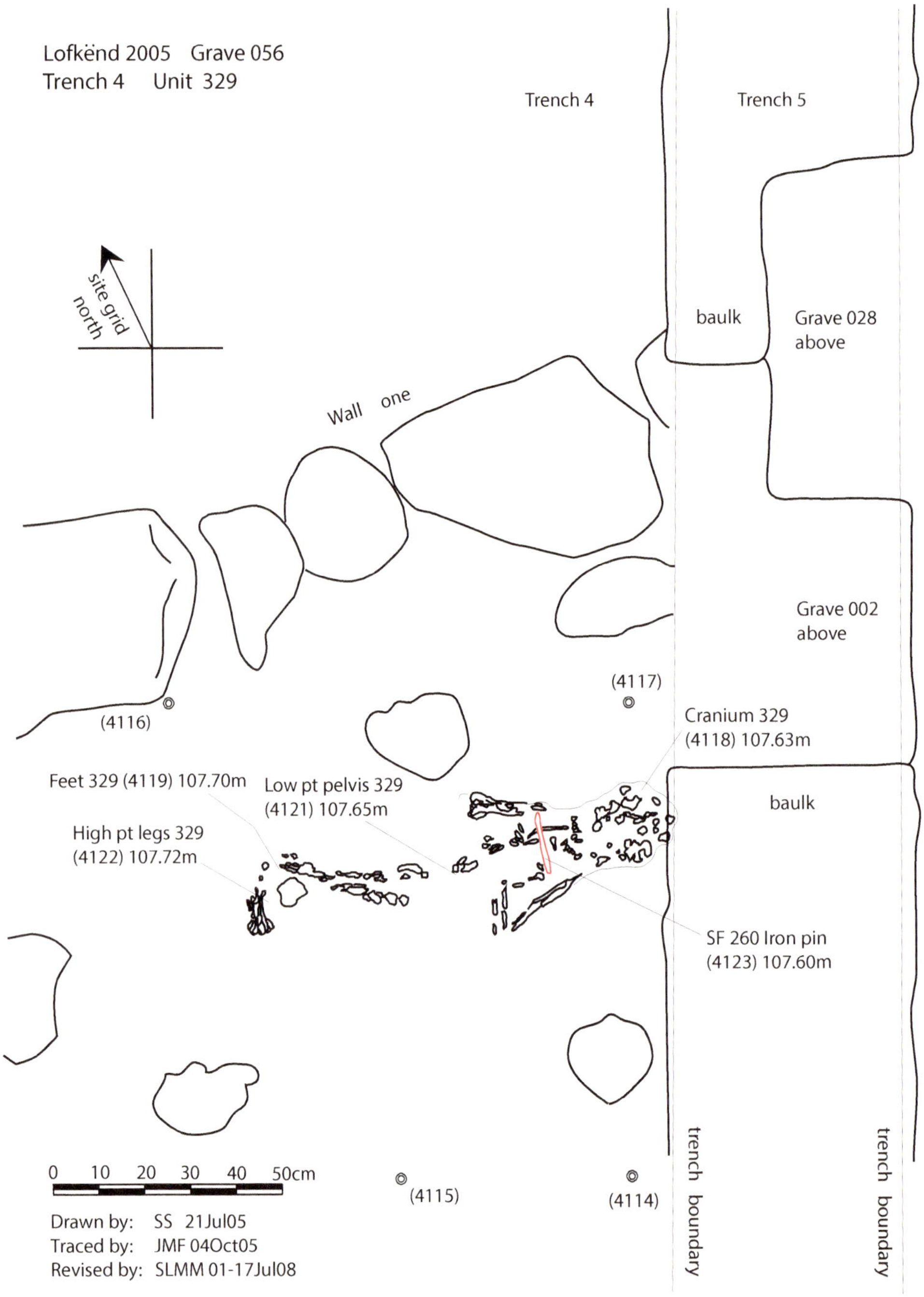

FIGURE 3.76 Plan of Tomb XXIII

FIGURE 3.77 Tomb XXIII (photo 1180) from above west

FIGURE 3.78 Tomb XXIII: **TXXIII-1**

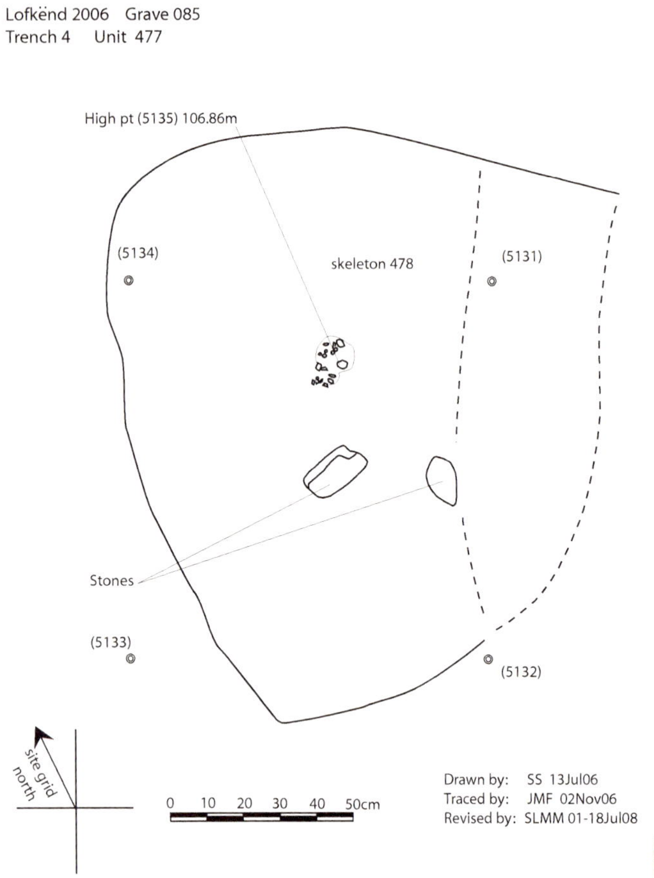

FIGURE 3.79 Plan of Tomb XXIV

FIGURE 3.80 Tomb XXIV (photo 1508) from southwest

a

b

c

FIGURE 3.81 Tomb XXV: (a) (photo 1597) fill and grave cut as first encountered, from west-northwest; (b) Tomb XXV, with legs of Tomb II underneath (photo 1604), as first exposed, from above northwest; (c) Tomb XXV (above) and II (photo 1611) from northwest

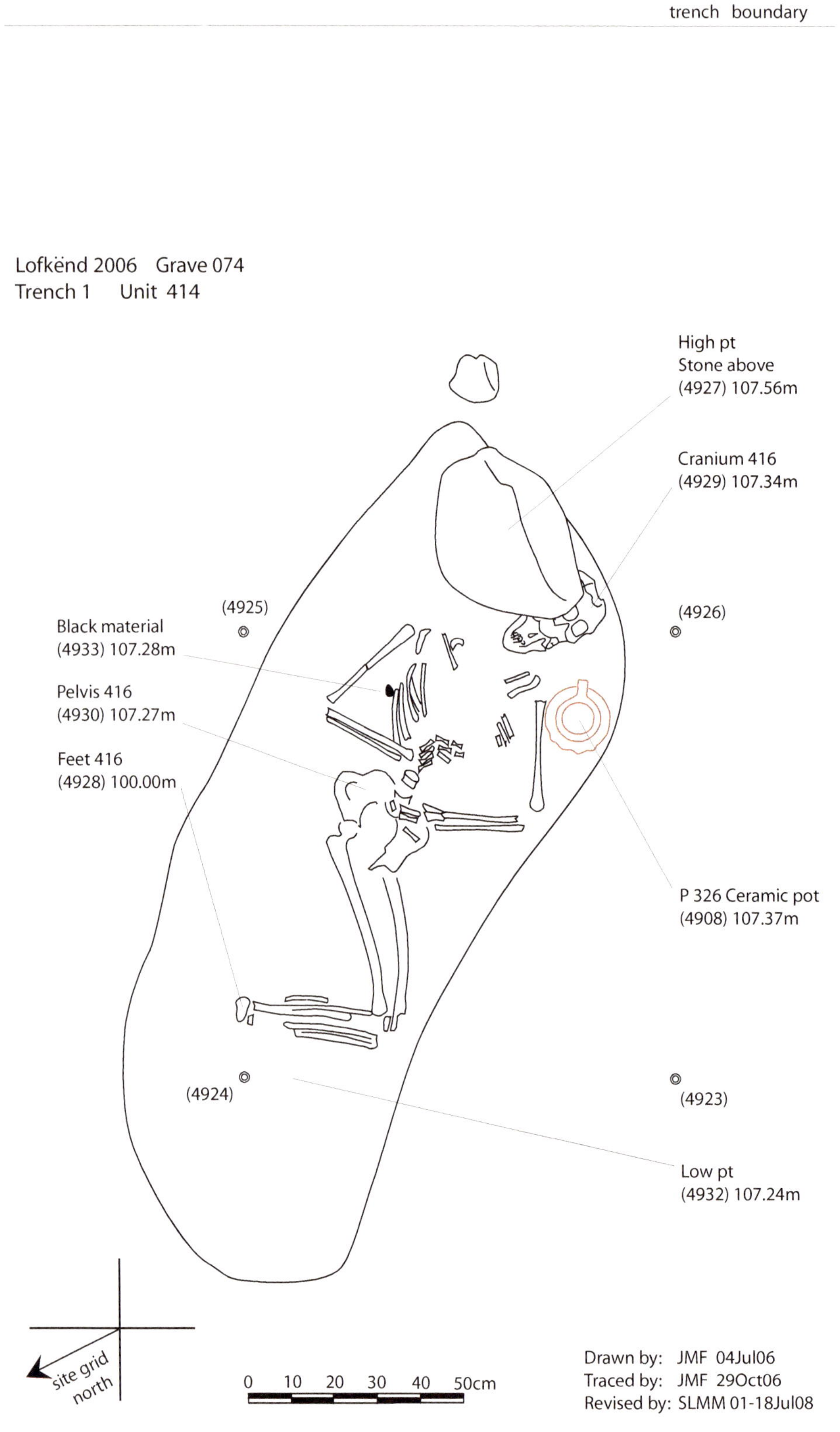

FIGURE 3.82 Plan of Tomb XXVI

a

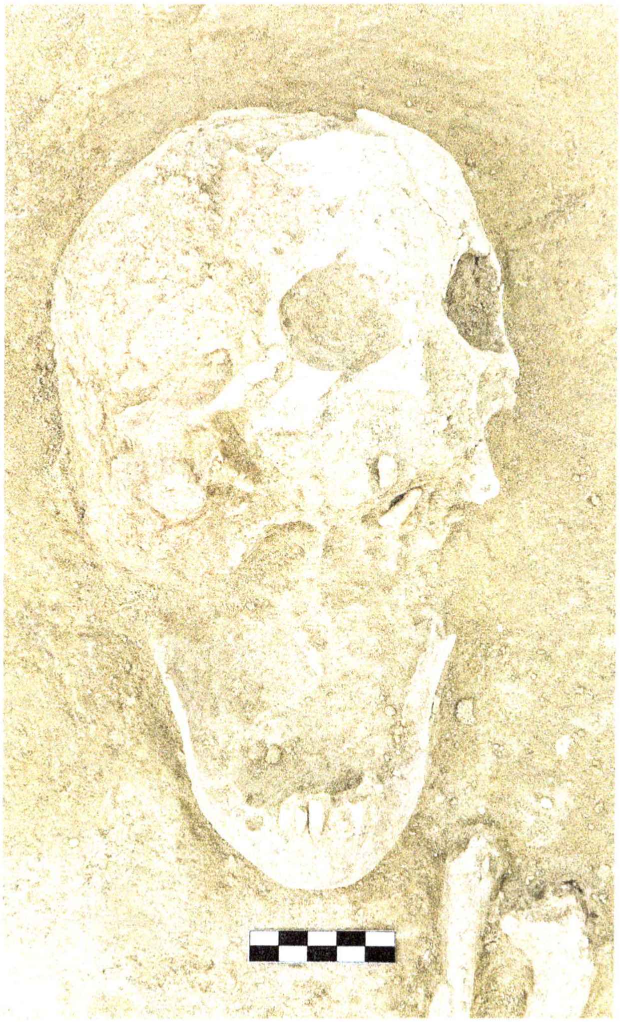

b

FIGURE 3.83 Tomb XXVI: (a) (photo 1379) from northwest;
(b) (photo 1398) detail of cranium after removal of stone, from above northwest

FIGURE 3.84 Tomb XXVI: **TXXVI-1**

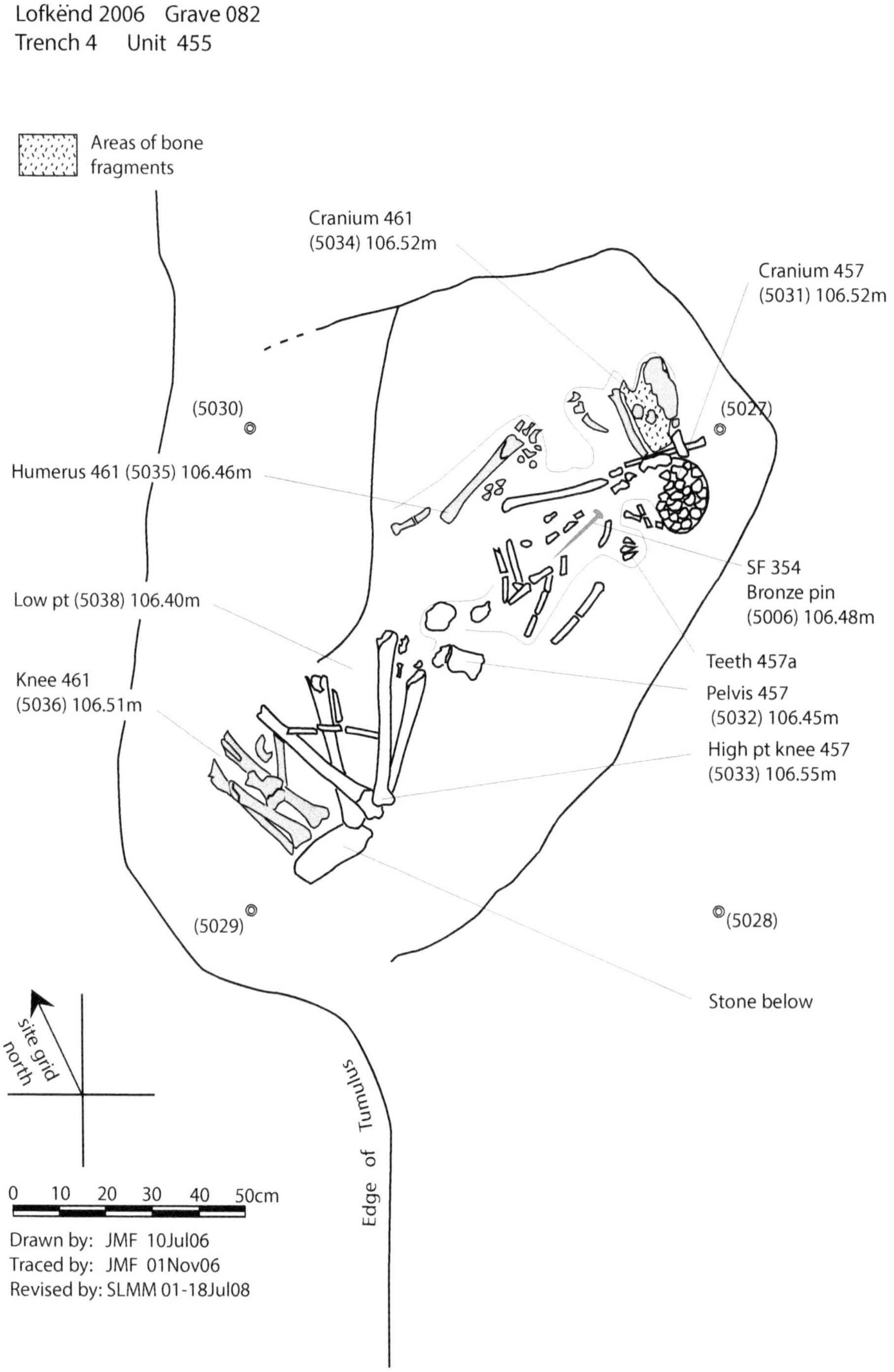

FIGURE 3.85 Plan of Tomb XXVII

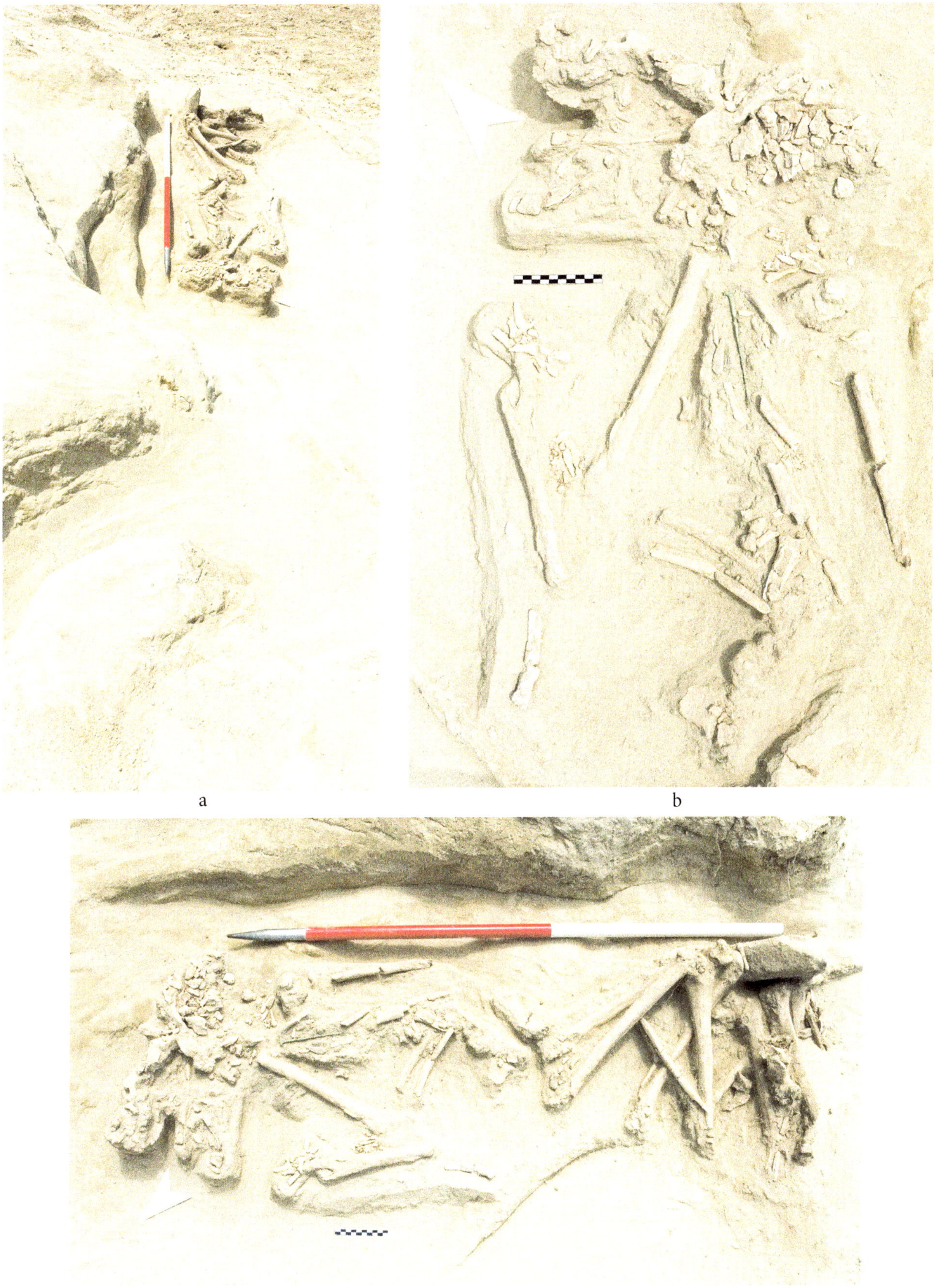

FIGURE 3.86 Tomb XXVII: (a) (photo 1451) from northeast, showing its relationship to the tumulus edge; (b) (photo 1458), detail of cranium and upper body, together with **TXXVII-1**, from above southwest; (c) (photo 1454) from northwest

Figure 3.87 Tomb XXVII: (a) **TXXVII-1**; (b) detail of **TXVII-1**

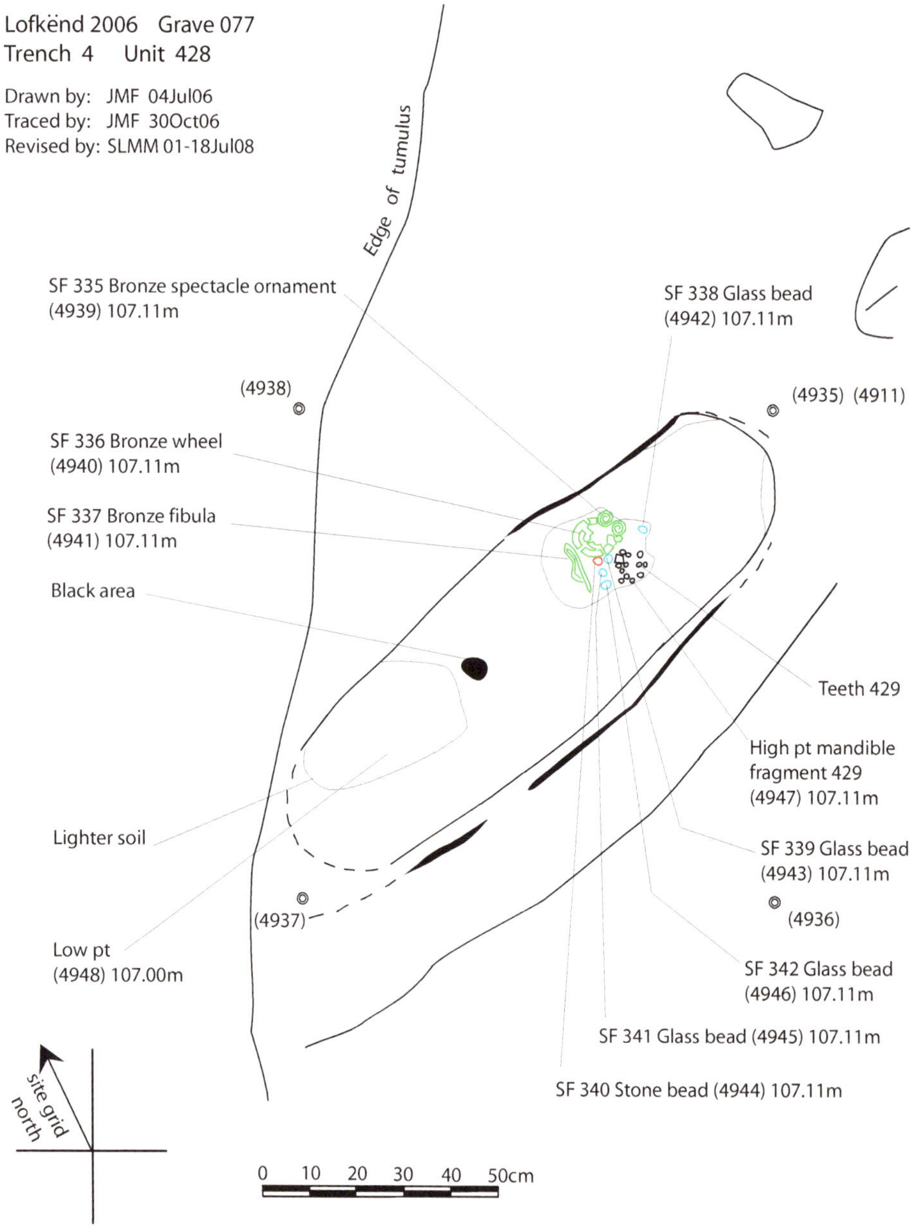

Figure 3.88 Plan of Tomb XXVIII

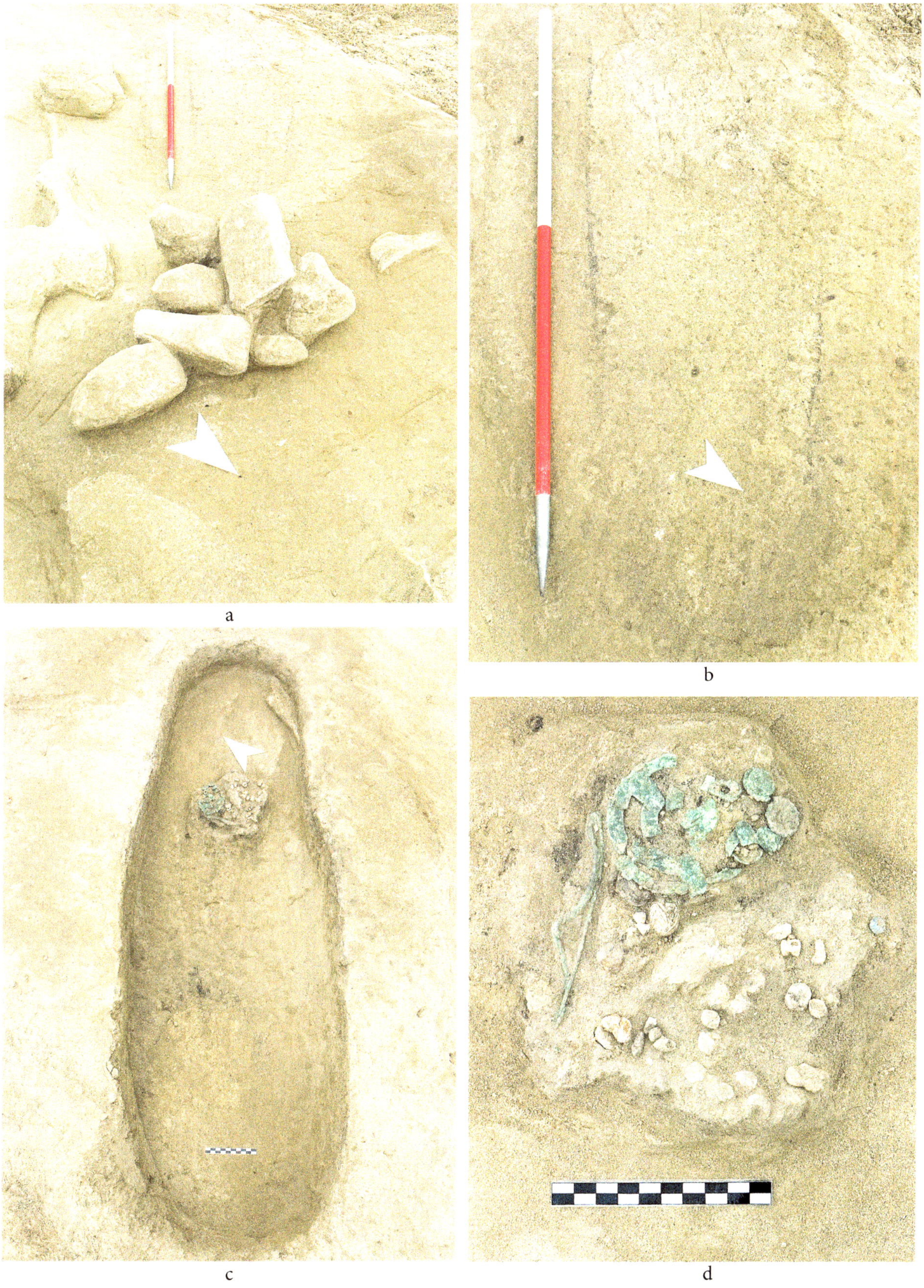

Figure 3.89 Tomb XXVIII: (a) uppermost level of tomb (photo 1372), with stones to the northeast of the tomb, from northeast; (b) pit for Tomb XXVIII (photo 1373) as first encountered, from northeast; (c) tomb fully exposed (photo 1389), from southwest; (d) detail of cranium of Tomb XXVIII with grave goods (photo 1393), from above southeast

FIGURE 3.90 Tomb XXVIII: detail of the cranium of Tomb XXVIII in the process of cleaning in the lab, showing organic pieces thought to be wood or modern roots (photo 2994)

FIGURE 3.91 Tomb XXVIII: detail of organic pieces thought to be wood or modern roots (photo 3086a)

Figure 3.92 Tomb XXVIII: **TXXVIII-1**

Figure 3.93 Tomb XXVIII: **TXXVIII-2**

Figure 3.94 Tomb XXVIII: **TXXVIII-3**

a b c d

Figure 3.95 Tomb XXVIII: (a–b) **TXXVIII-5**; (c) **TXXVIII-4**; (d) **TXXVIII-6**

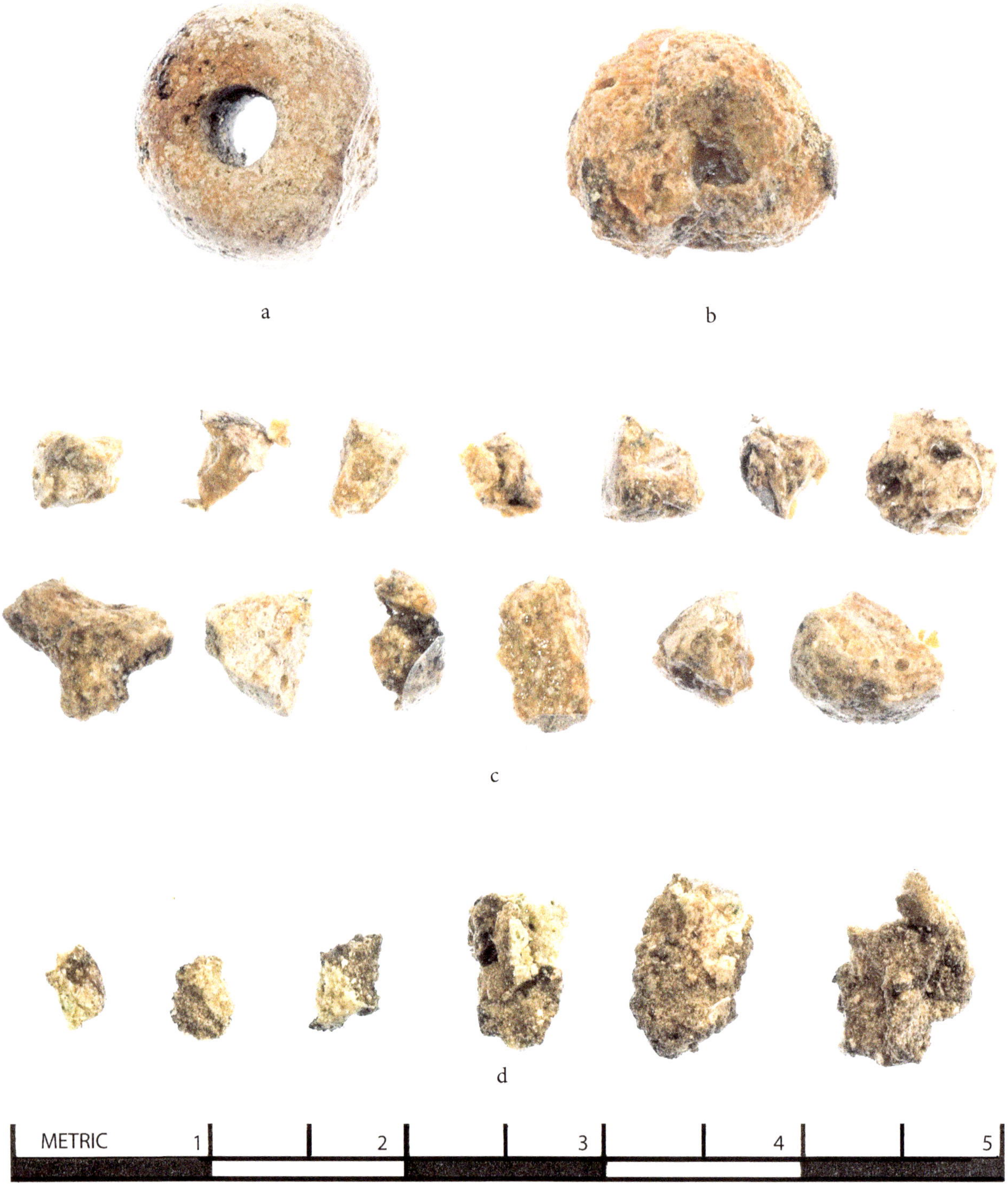

FIGURE 3.96 Tomb XXVIII: (a) **TXXXVIII-7**; (b) **TXXVIII-8**; (c) **TXXVIII-9**; (d) **TXXVIII-10**

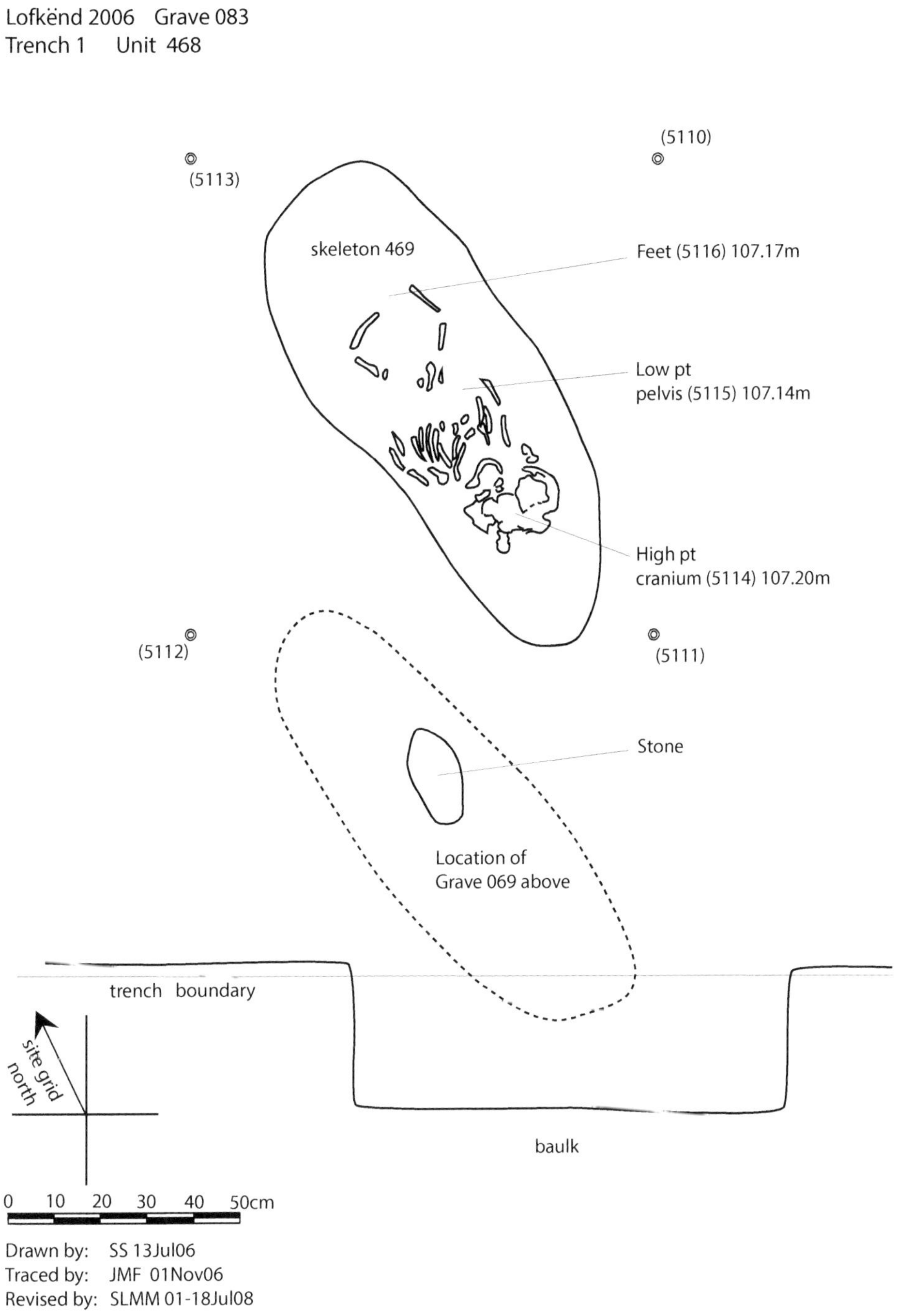

FIGURE 3.97 Plan of Tomb XXIX

a

b

Figure 3.98 Tomb XXIX: (a) (photo 1494) from northwest; (b) (photo 1497) detail of cranium and torso from above northwest

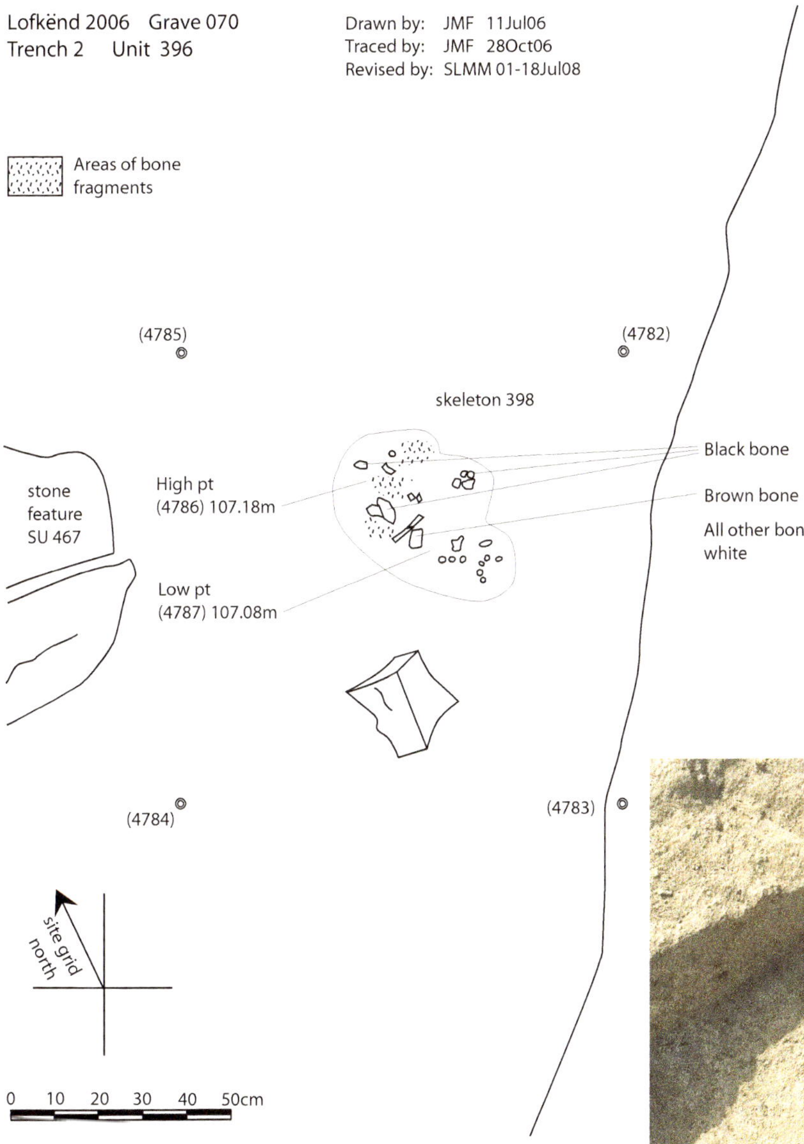

Figure 3.99 Plan of Tomb XXX and surrounding area

Figure 3.100 Tomb XXX: (photo 1341) from above northwest

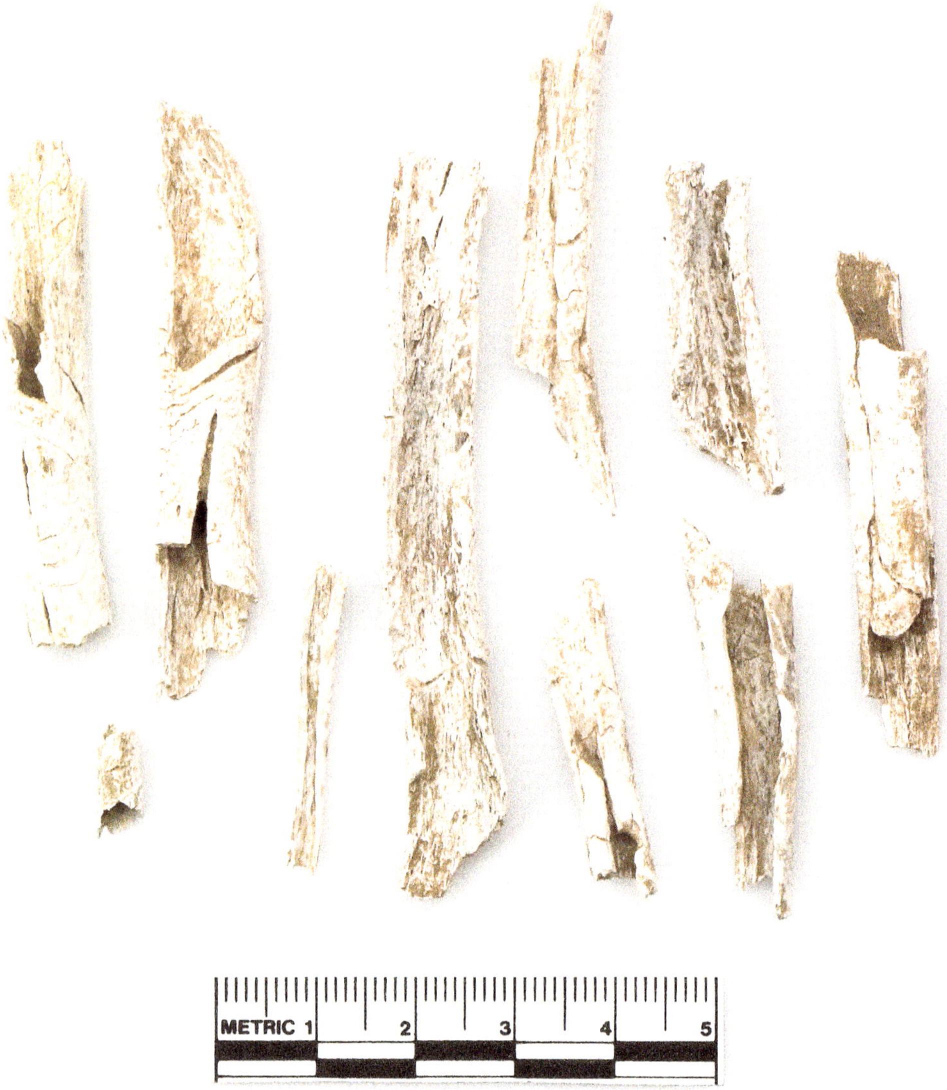

Figure 3.101 Tomb XXX: (photo 2939) cremated long bones of infant, SU 398

Figure 3.102 Tomb XXX: burned olive pip (photo 3134)

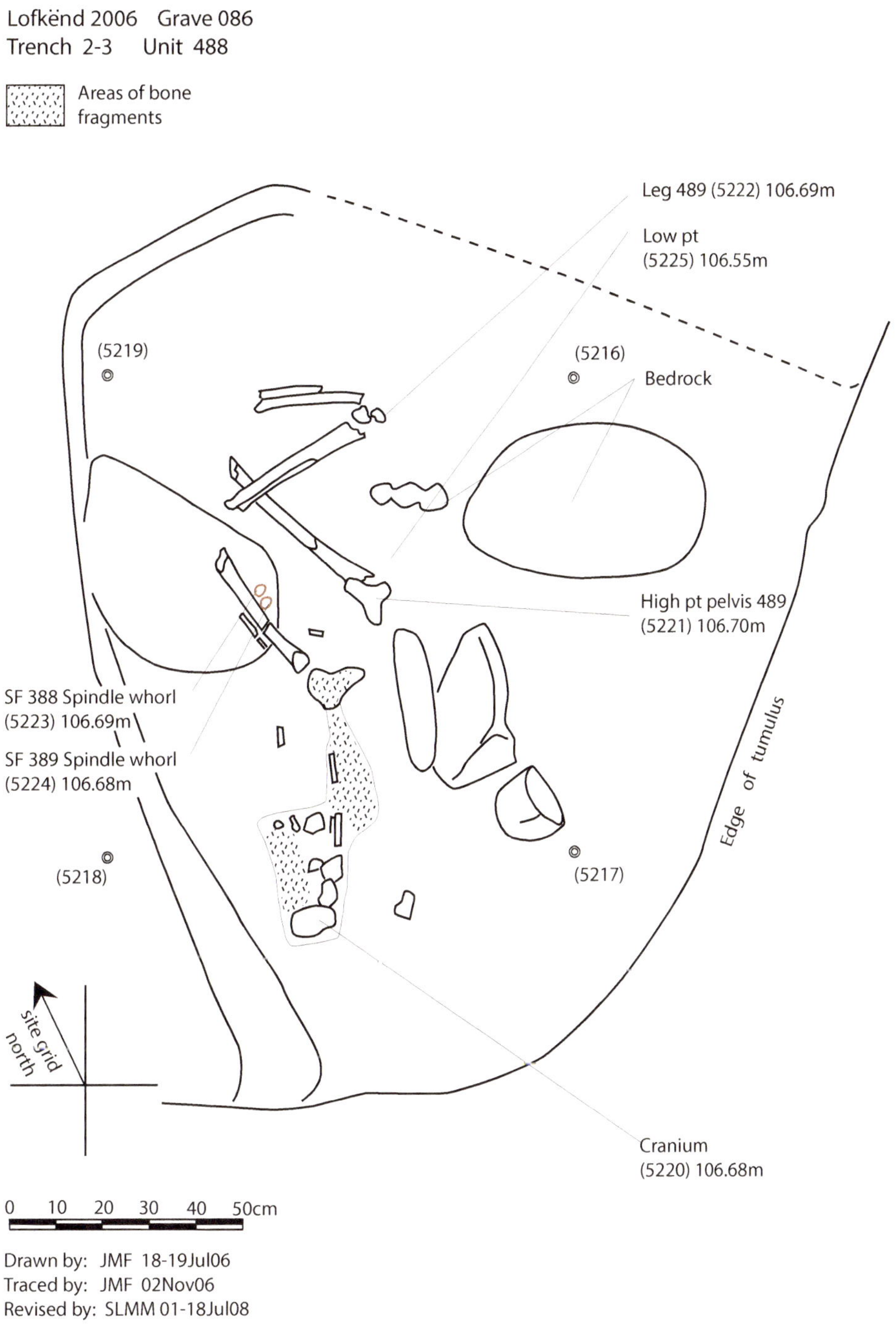

FIGURE 3.103 Plan of Tomb XXXI

a

b

FIGURE 3.104 Tomb XXXI: (a) (photo 1555) from northwest; (b) (photo 1557) from east-northeast, showing **TXXXI-1** and **TXXXI-2** in situ

Figure 3.105 Tomb XXXI: (a–b) **TXXXI-1** and **TXXXI-2**

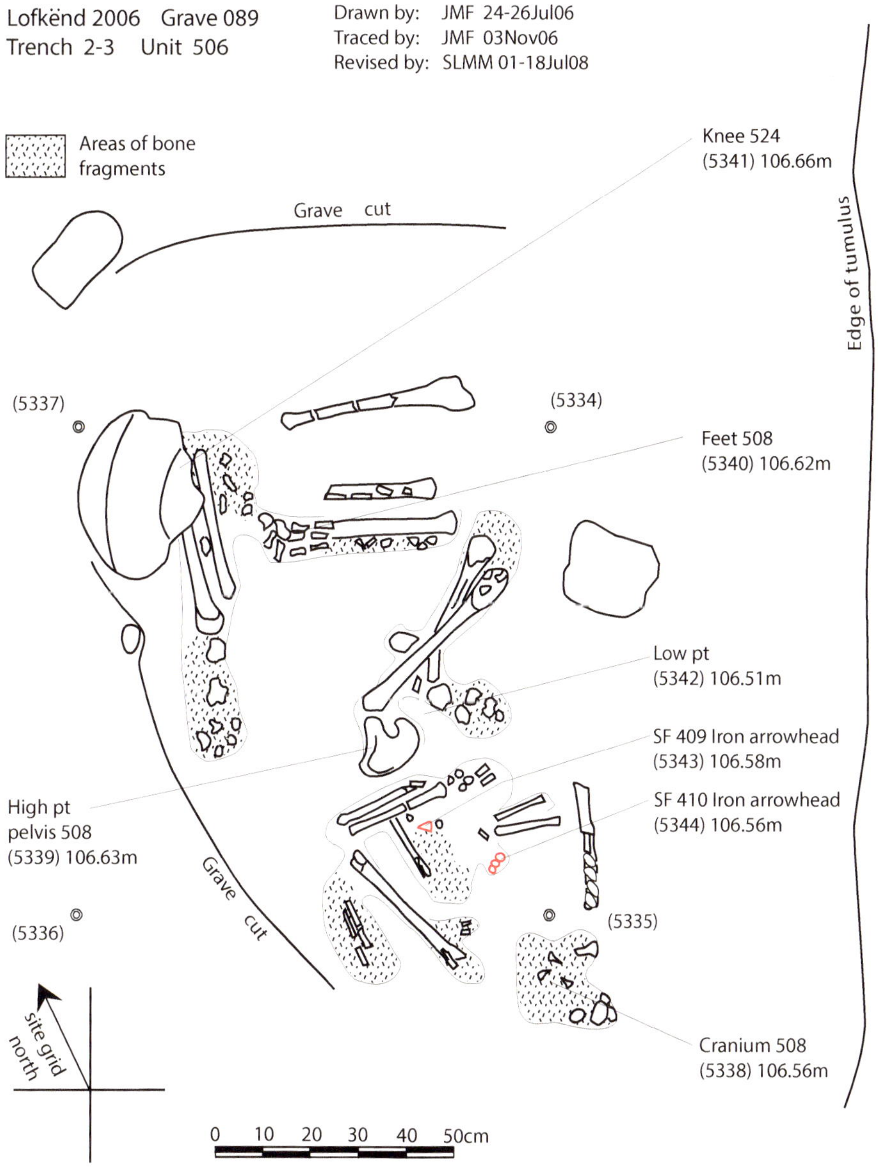

Figure 3.106 Plan of Tomb XXXIII

FIGURE 3.107 Tomb XXXII: (a) (photo 1610) during excavation, from south-southeast; (b) (photo 1619) showing tomb fully exposed, from north-northwest

FIGURE 3.108 Tomb XXXII: (a–b) **TXXXII-1** and **TXXXII-2**

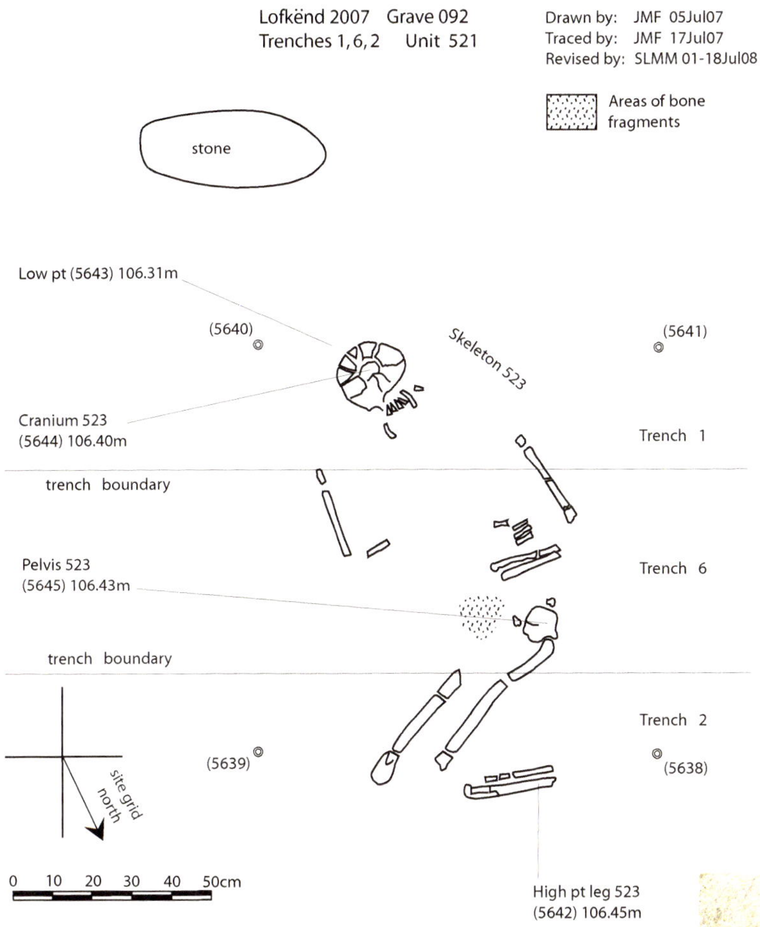

FIGURE 3.109 Plan of Tomb XXXIII

FIGURE 3.110 Tomb XXXIII: (photo 1738) from north-northwest

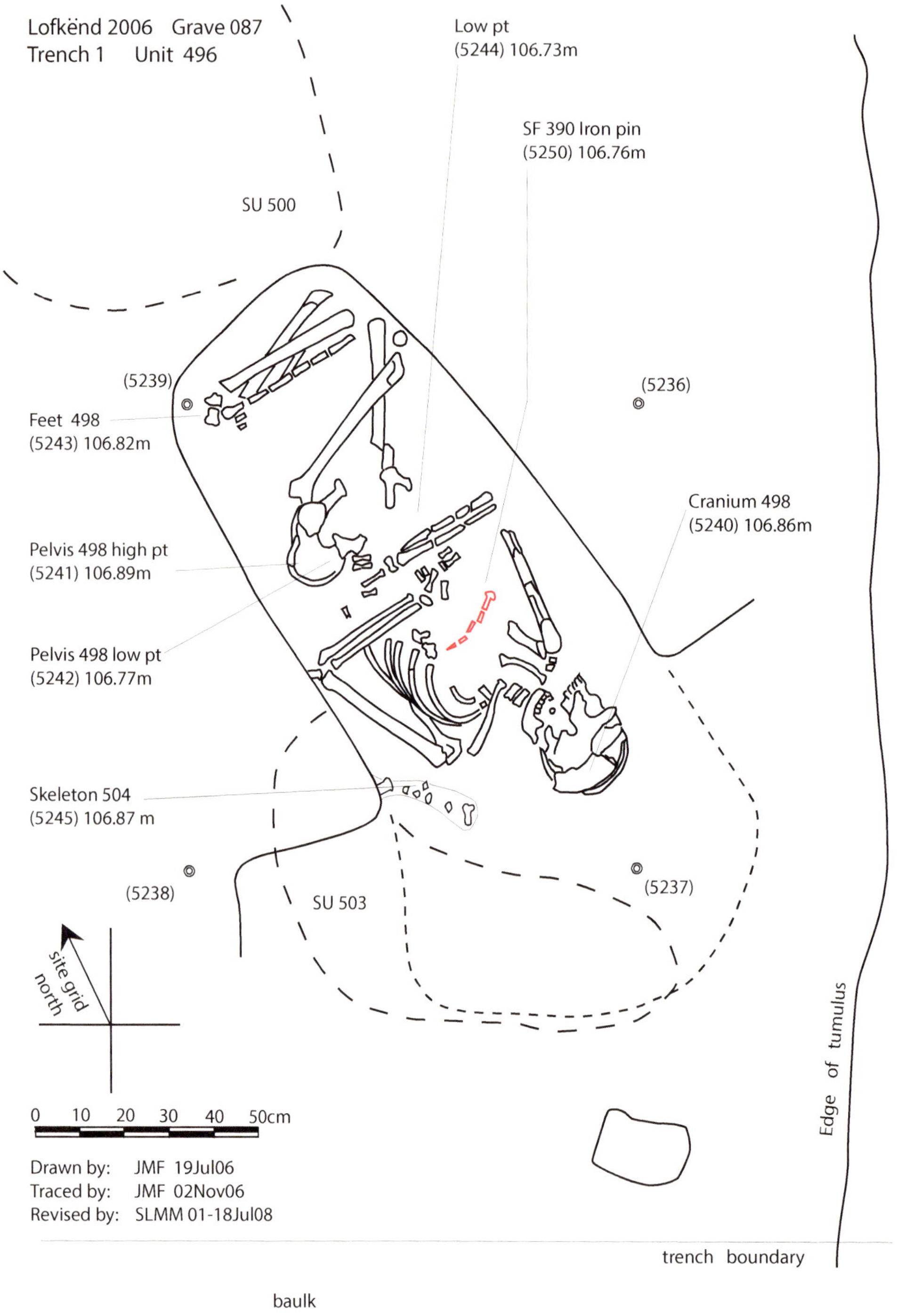

FIGURE 3.111 Plan of Tomb XXXIV

a

FIGURE 3.112 Tomb XXXIV: (a) (photo 1564) from northwest; (b) (photo 1569) detail of torso and cranium of SU 498, together with the poorly preserved remains of another individual, SU 504, from above northwest

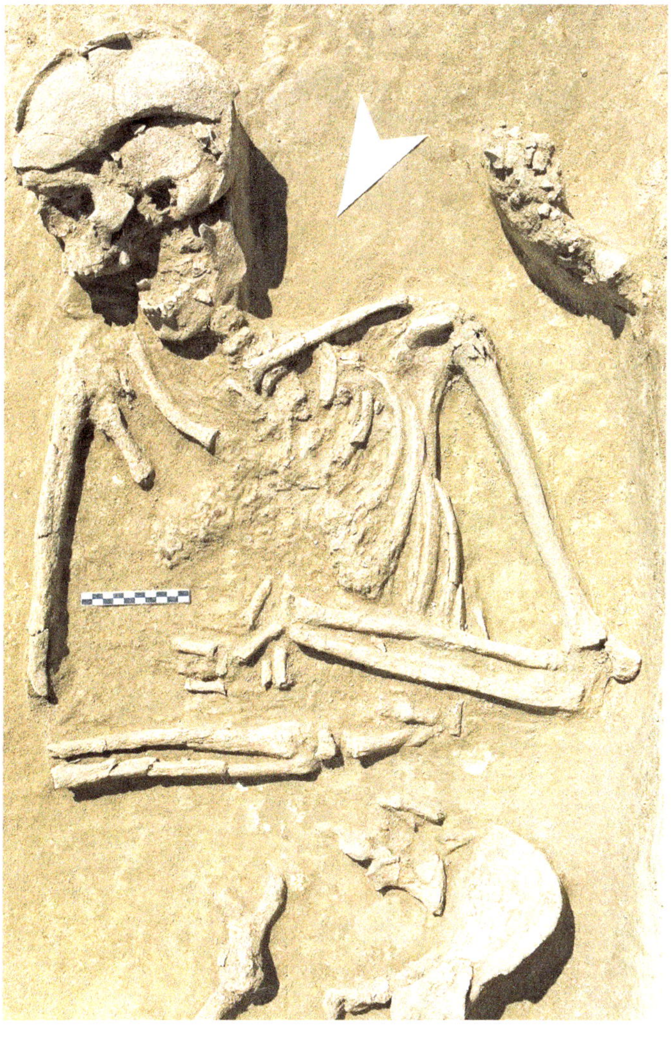

b

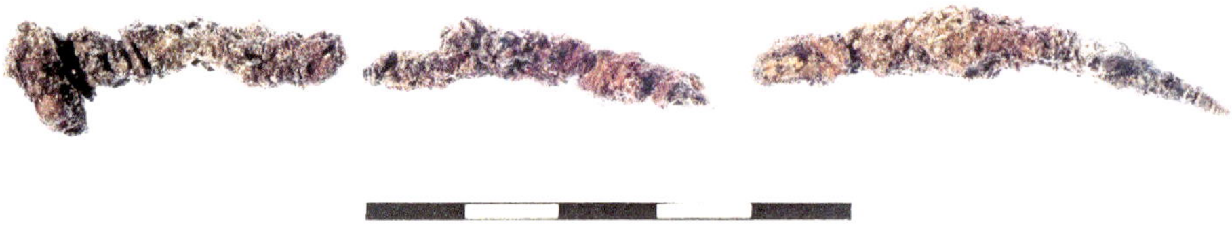

Figure 3.113 Tomb XXXIV: **TXXXIV-1**

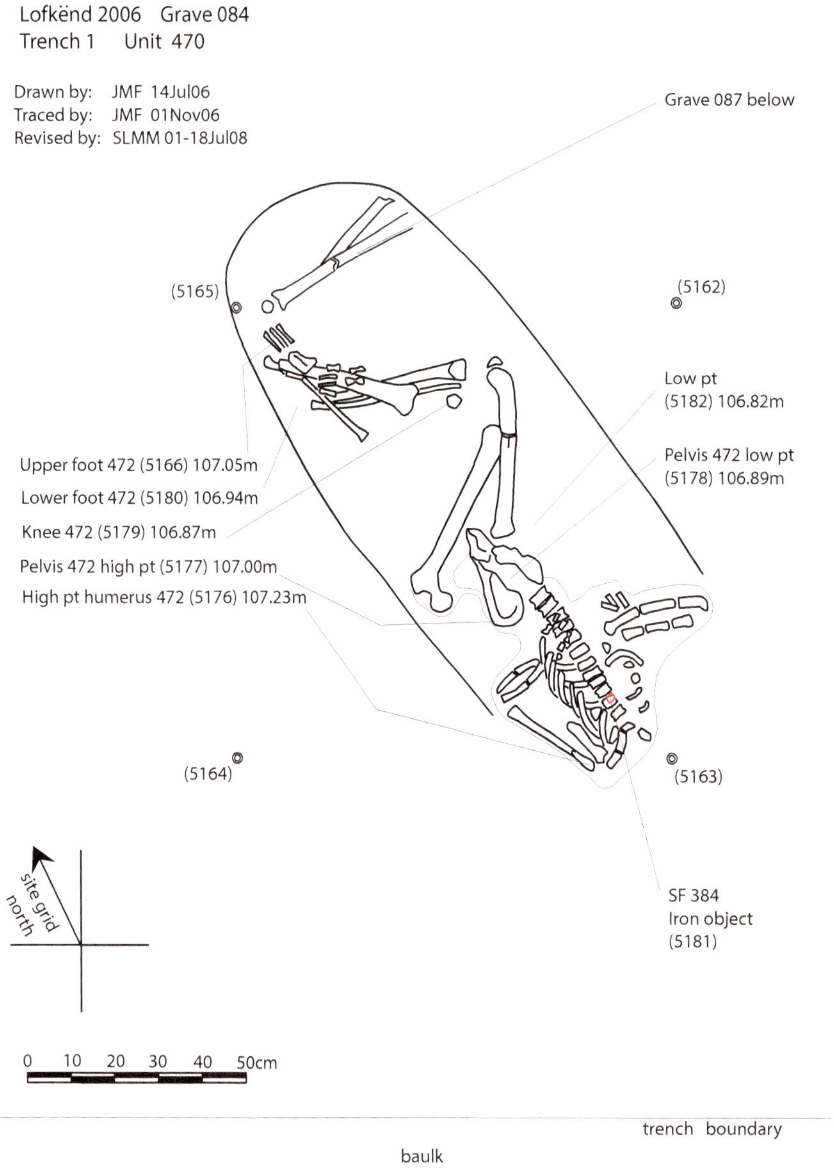

Figure 3.114 Plan of Tomb XXXV

a

b

c

FIGURE 3.115 Tomb XXXV: (a) (photo 1523) in the process of excavation, from northwest; (b) (photo 1536) from northwest after the removal of the feet of SU 472; (c) tomb as fully exposed (photo 1525), from southwest

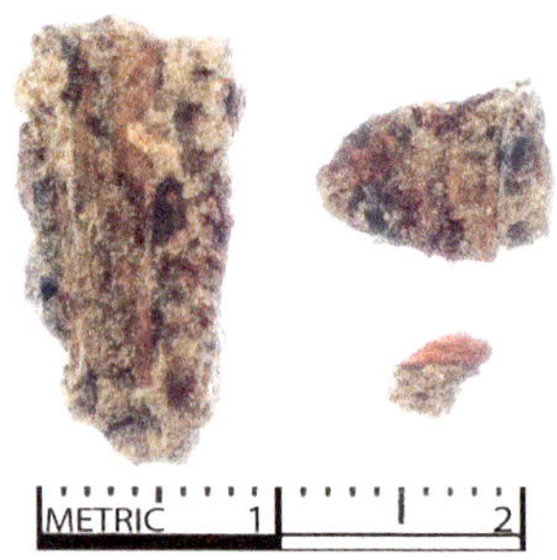

FIGURE 3.116 Tomb XXXV: **TXXXV-1**

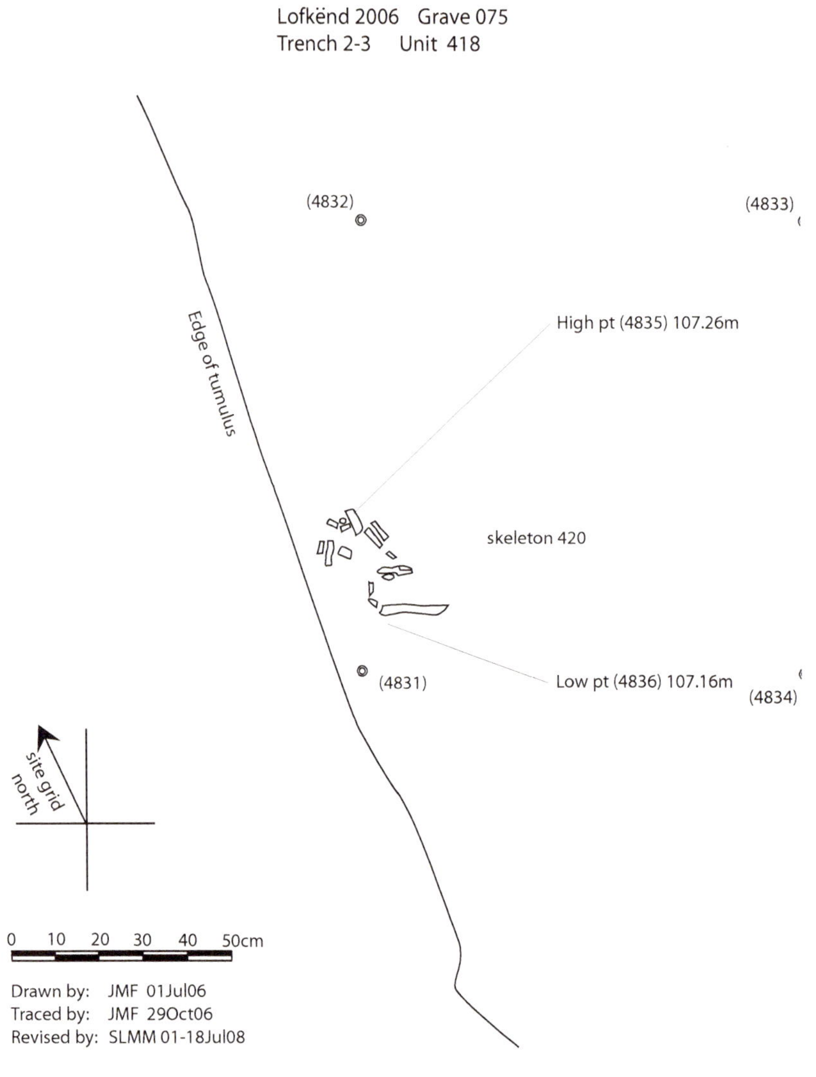

FIGURE 3.117 Plan of Tomb XXXVI

FIGURE 3.118 Tomb XXXVI: (photo 1367) from north

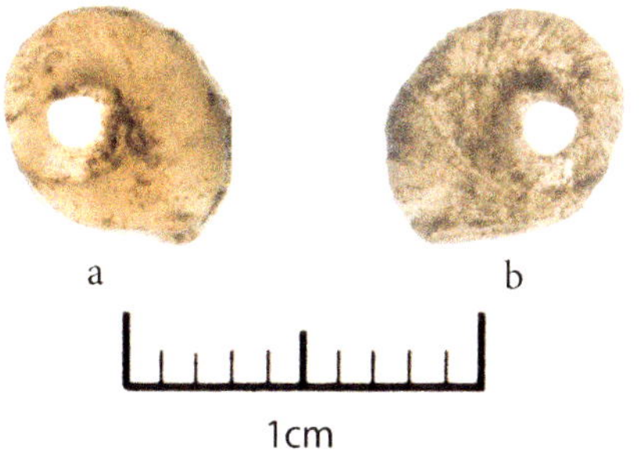

FIGURE 3.119 Tomb XXXVI: (a–b) SF 346

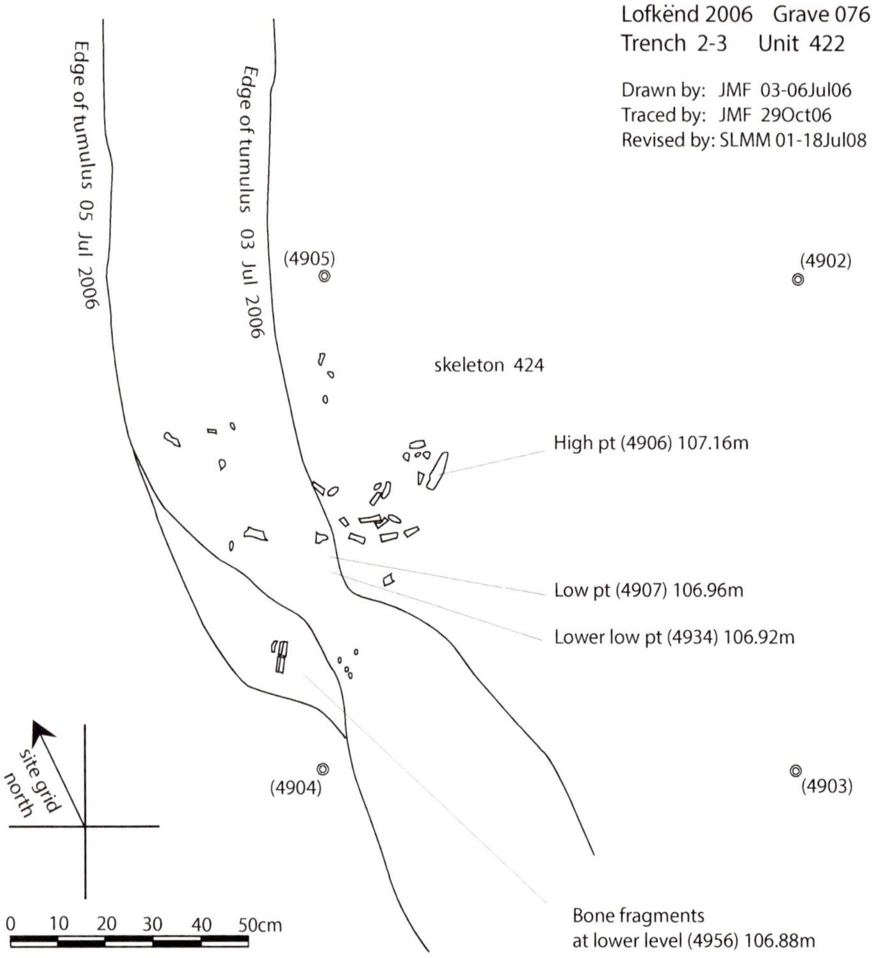

FIGURE 3.120 Plan of Tomb XXXVII

FIGURE 3.121 Tomb XXXVII: (photo 1371) from southeast

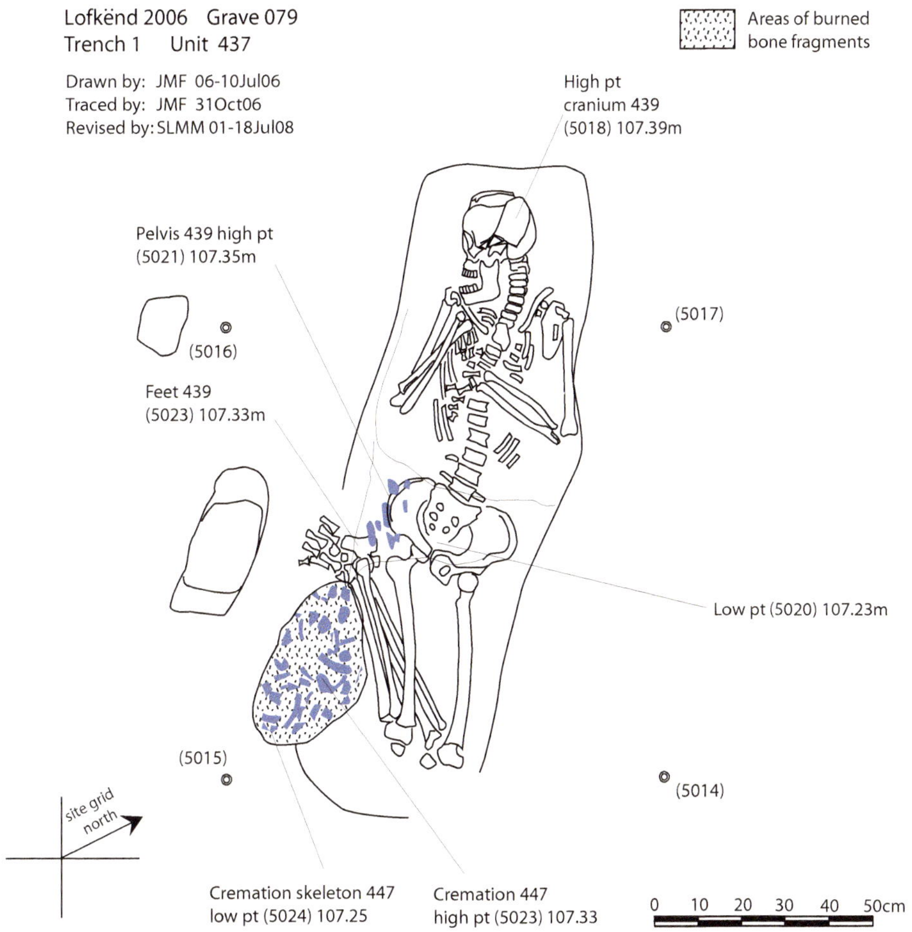

FIGURE 3.122 Plan of Tomb XXXVIII

FIGURE 3.123 Tomb XXXVIII: (photo 1438) in the process of excavation, from east-southeast. The cut edge around the burial is not the grave cut, but a larger area opened around the burial in order to expose the tomb more fully.

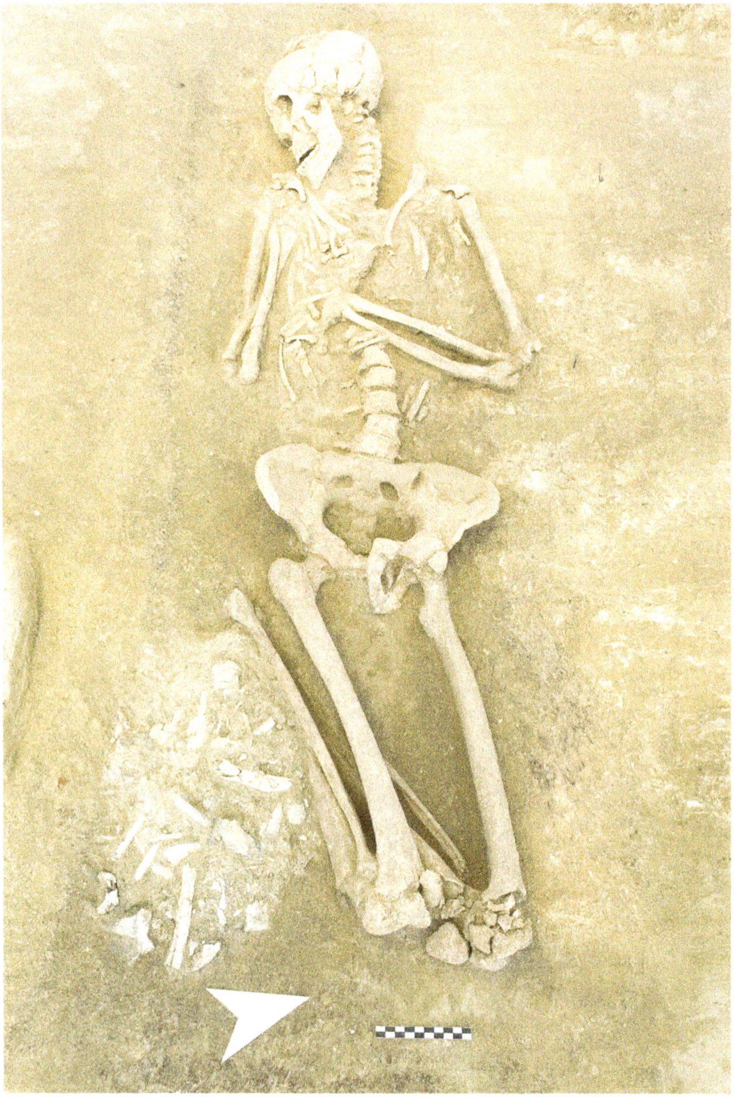

a

FIGURE 3.124 Tomb XXXVIII: (a) (photo 1443) from east; (b) (photo 1445), detail of cremation, from south-southeast

b

Figure 3.125 Tomb XXXVIII: (photo 1460) detail of pelvis of inhumation and position of iron blade point as found, from south

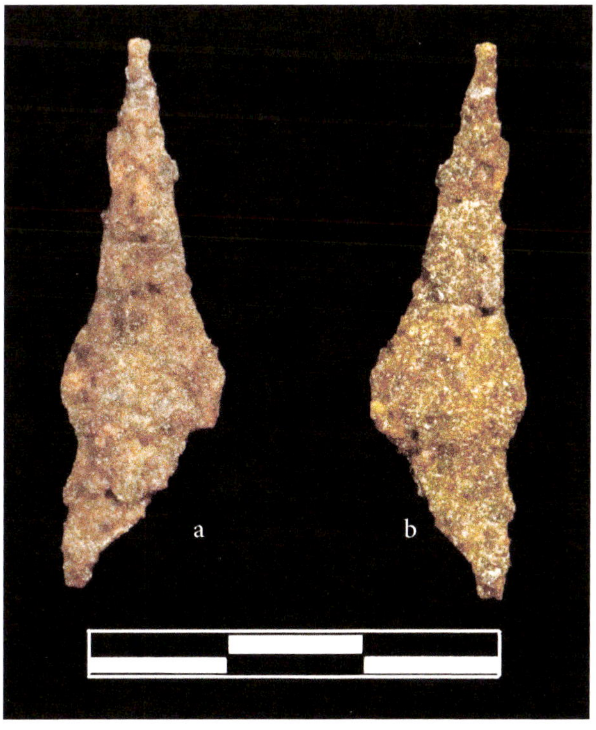

Figure 3.126 Tomb XXXVIII: (a–b) **TXXXVIII-1**

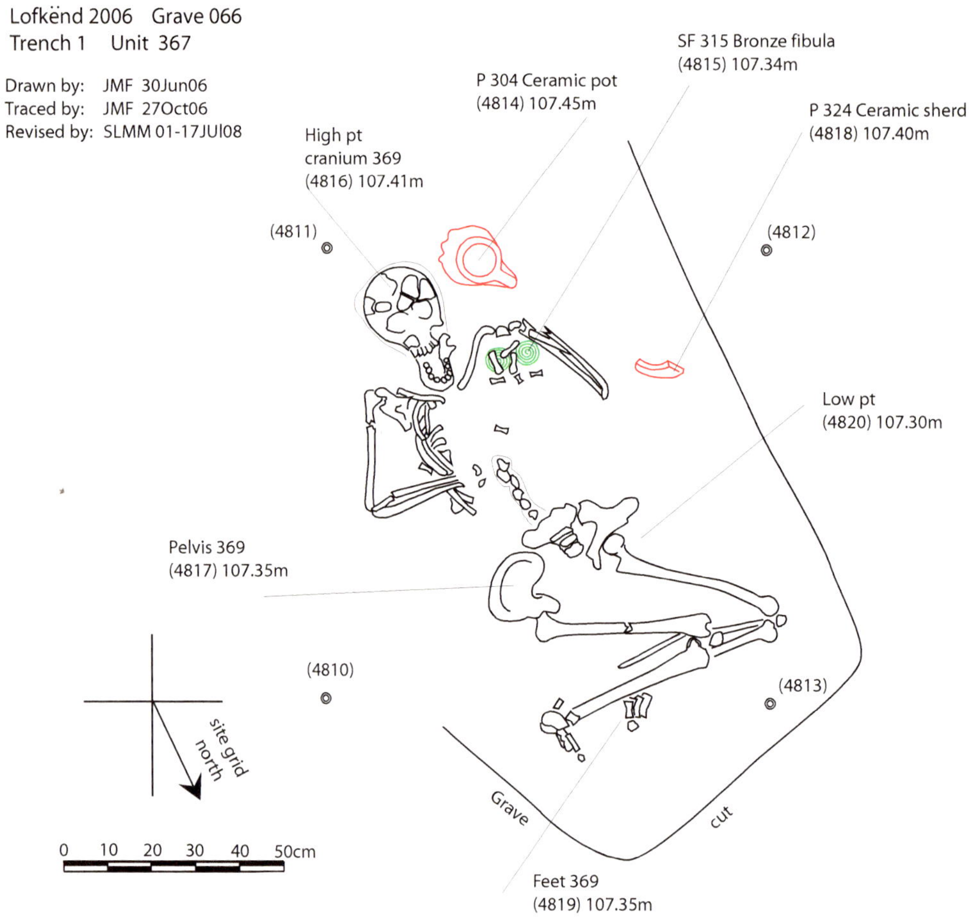

FIGURE 3.127 Plan of Tomb XXXIX

FIGURE 3.128 Tomb XXXIX: (photo 1353) from above northeast

Figure 3.129 Tomb XXXIX: **TXXXIX-1**

Figure 3.130 Tomb XXXIX: **TXXXIX-2**

Figure 3.131 Tomb XXXIX: P 324 (photo 3743), matt-painted body fragment from grave fill

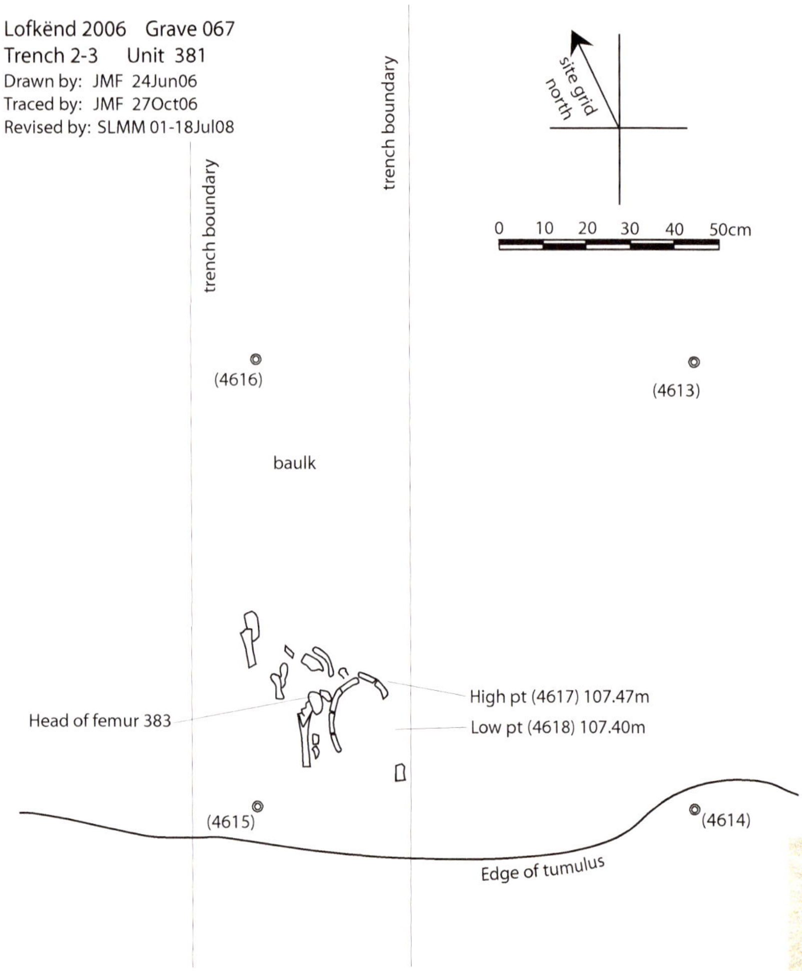

FIGURE 3.132 Plan of Tomb XL

FIGURE 3.133 Tomb XL: (photo 1296) from above west

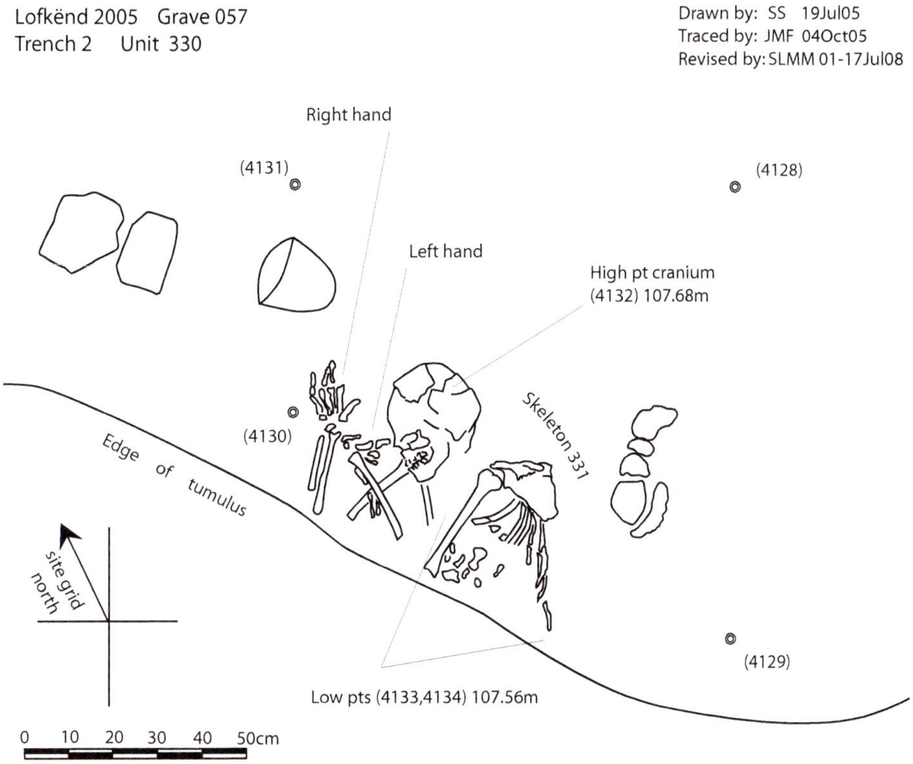

FIGURE 3.134 Plan of Tomb XLI

FIGURE 3.135 Tomb XLI: (photo 1154) from northwest

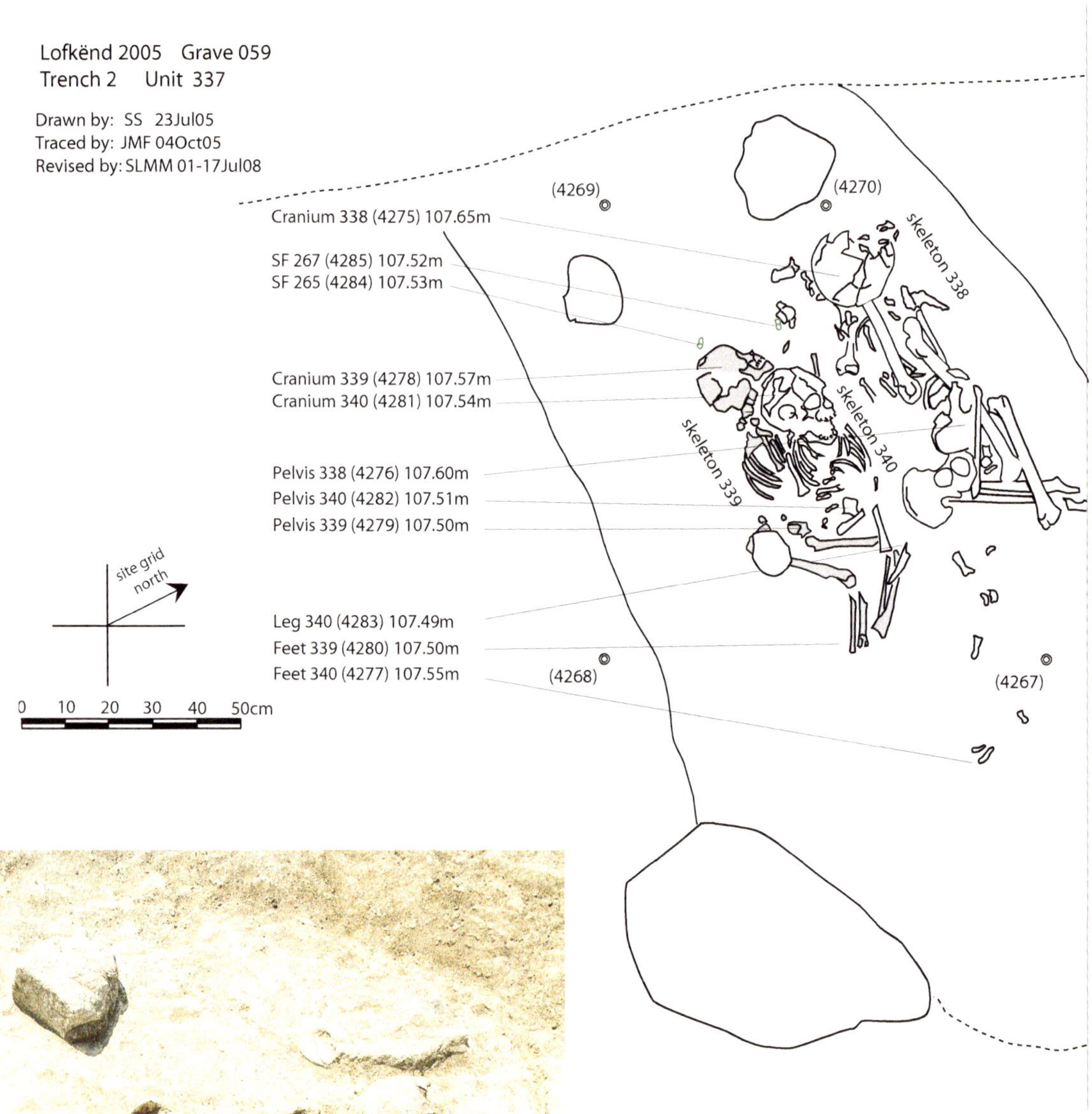

FIGURE 3.136 Plan of Tomb XLII

FIGURE 3.137 Tomb XLII: (photo 1199) from above east-northeast

FIGURE 3.138 Tomb XLII: **TXLII-1**

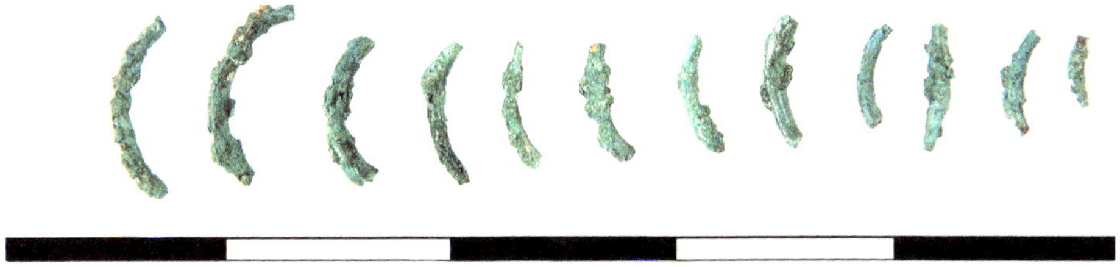

FIGURE 3.139 Tomb XLII: **TXLII-2**

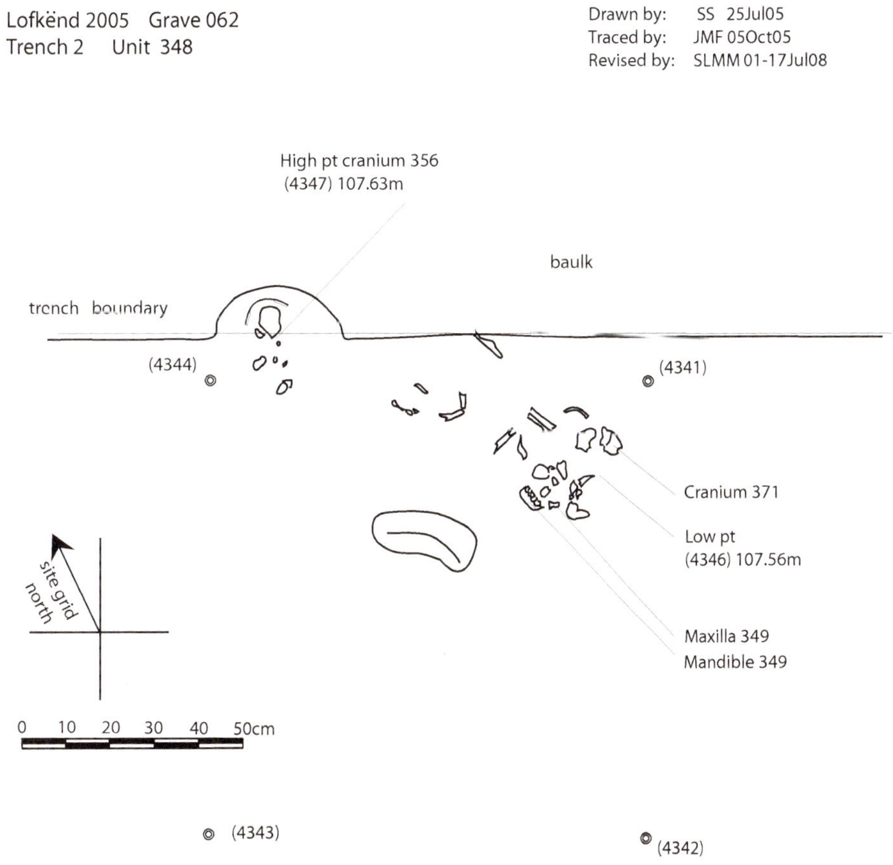

FIGURE 3.140 Plan of Tomb XLIII

a

b

Figure 3.141 Tomb XLIII: (a) (photo 1224) from south; (b) (photo 1225) detail of skeleton, SU 349, from west

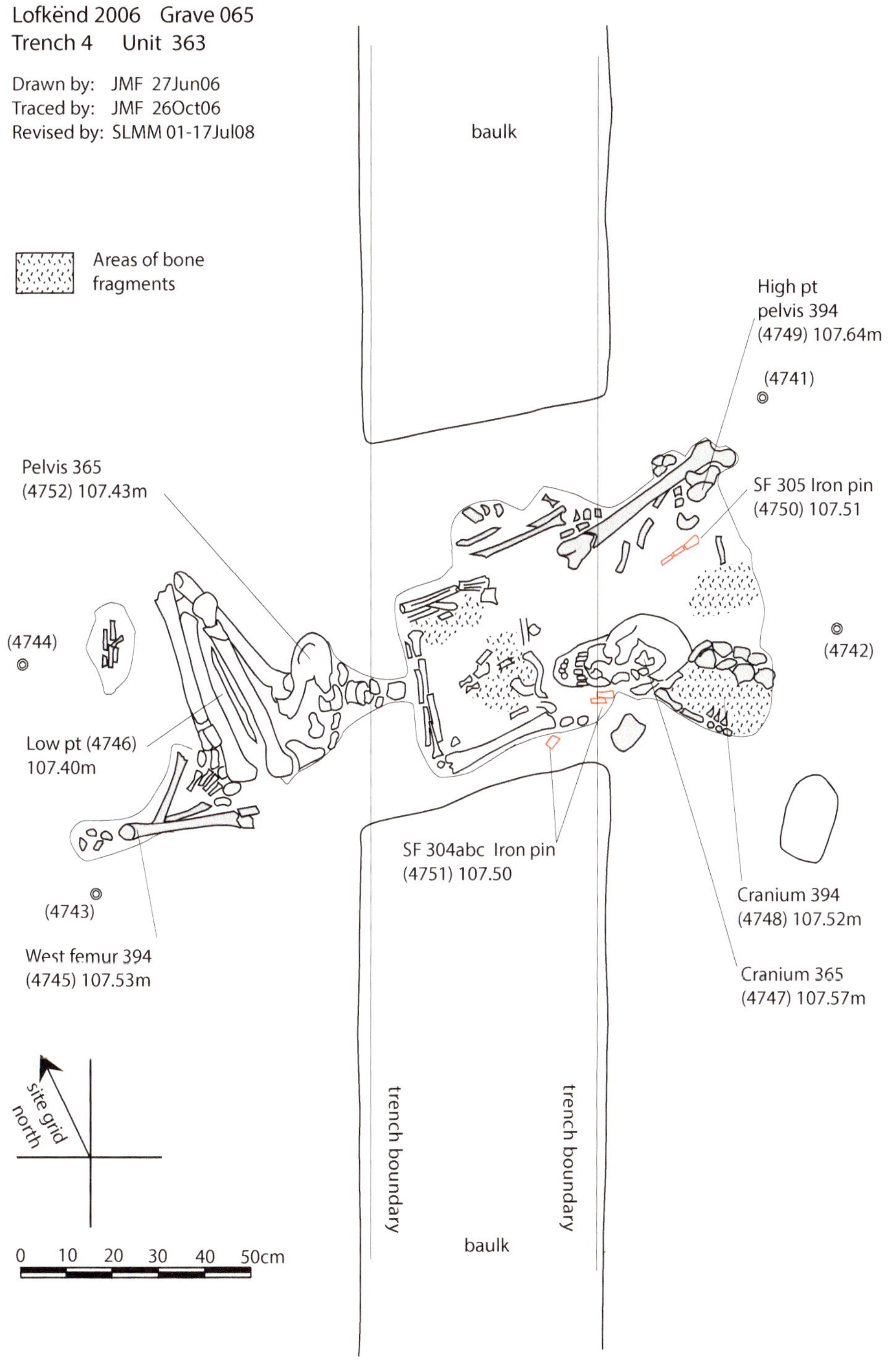

Figure 3.142 Plan of Tomb XLIV

a

b

Figure 3.143 Tomb XLIV: (a) (photo 1324) from west; (b) (photo 1326), detail of upper torso and cranium, from west

FIGURE 3.144 Tomb XLIV: fragments of **TXLIV-1** (photo 2986)

FIGURE 3.145 Tomb XLIV: **TXLIV-2** (photo 3285)

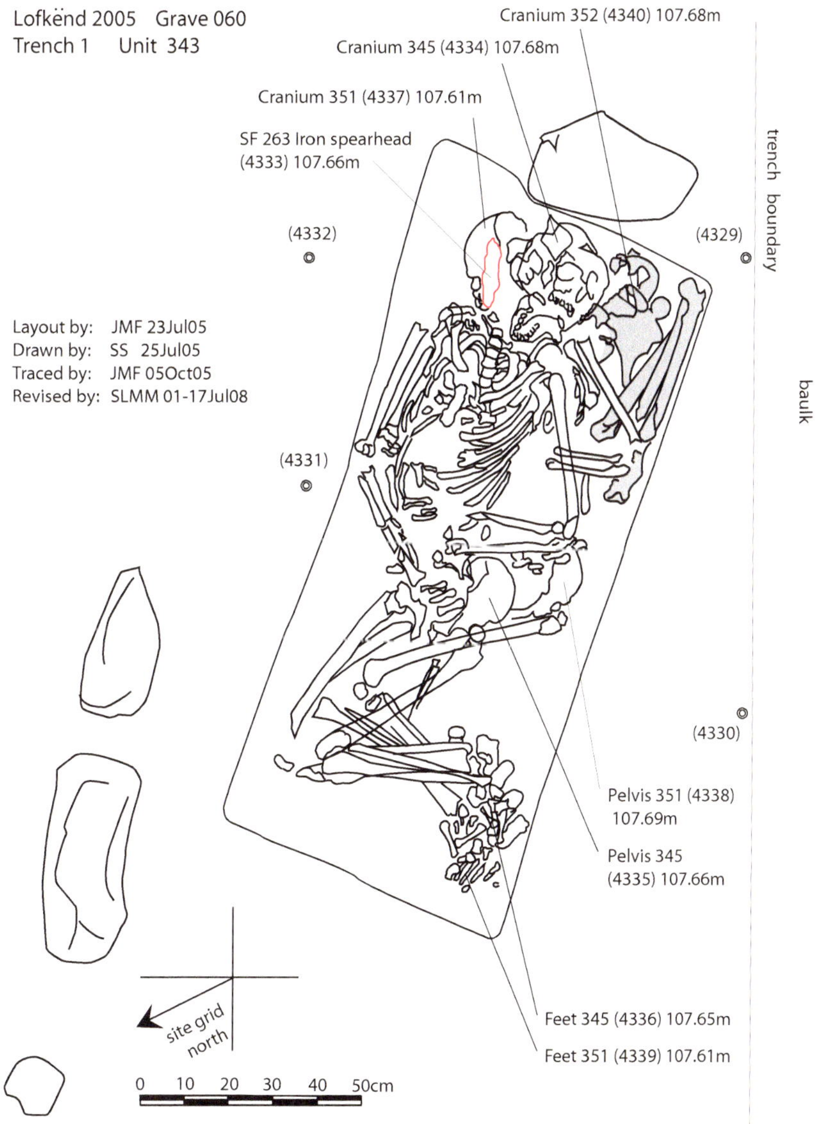

FIGURE 3.146 Plan of Tomb XLV

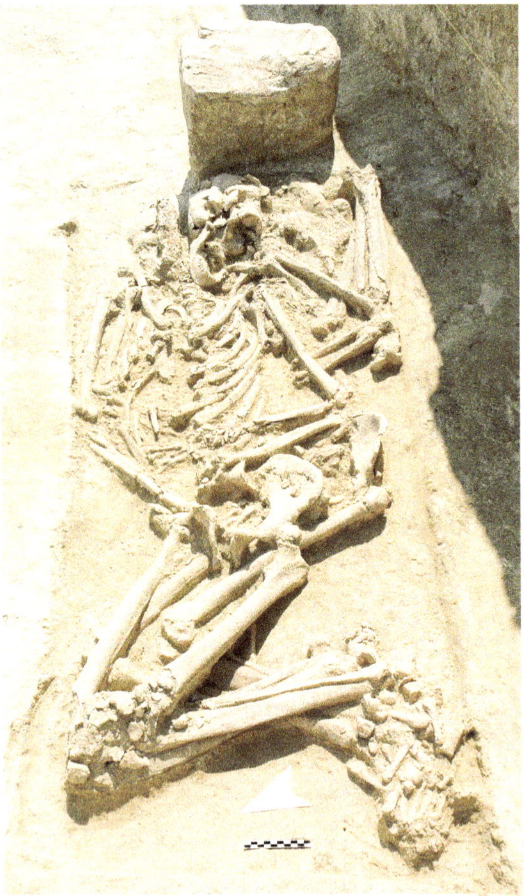

a

FIGURE 3.147 Tomb XLV: (a) (photo 1210) from west-northwest; (b) (photo 1213) detail of **TXLV-1** and crania, from above west-northwest; (c) (photo 1211) from north-northeast

b

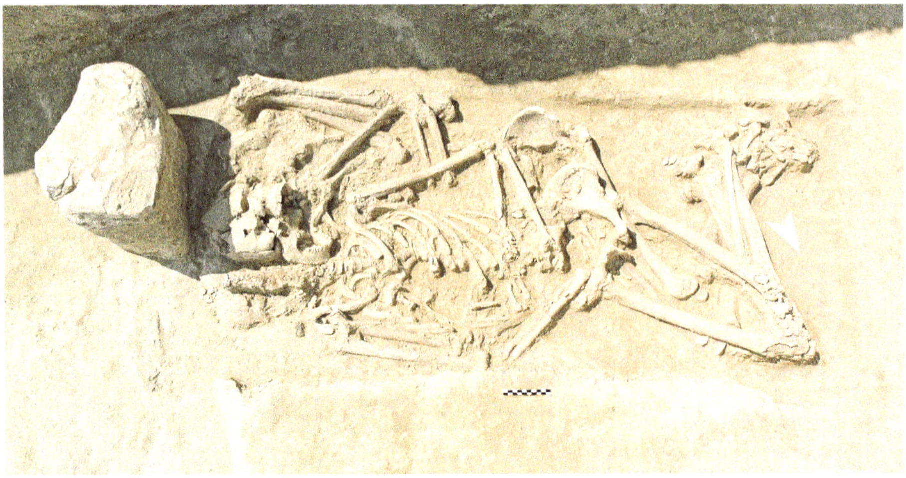

c

Figure 3.148 Tomb XLV: (a–b) **TXLV-1**

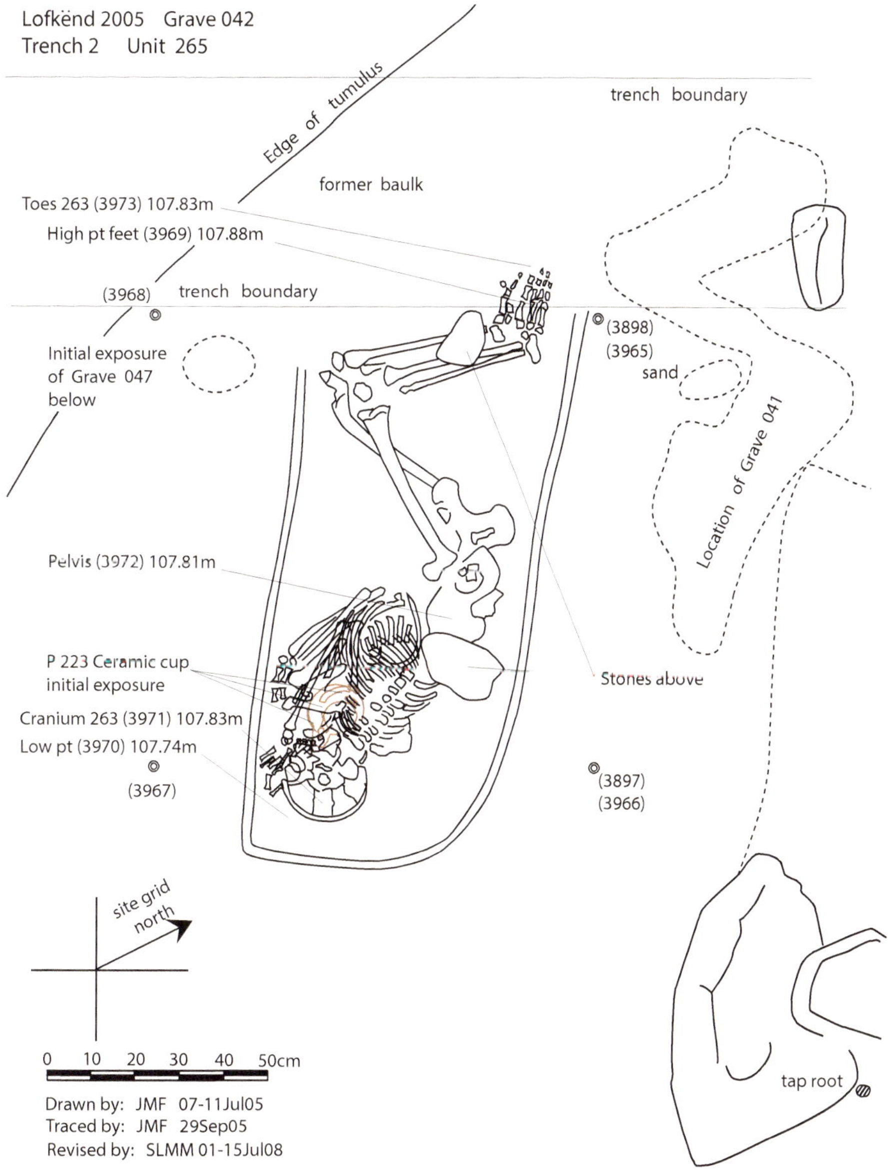

Figure 3.149 Tomb XLVI: (a) plan of Tomb XLVI; (b–d) *on following page*

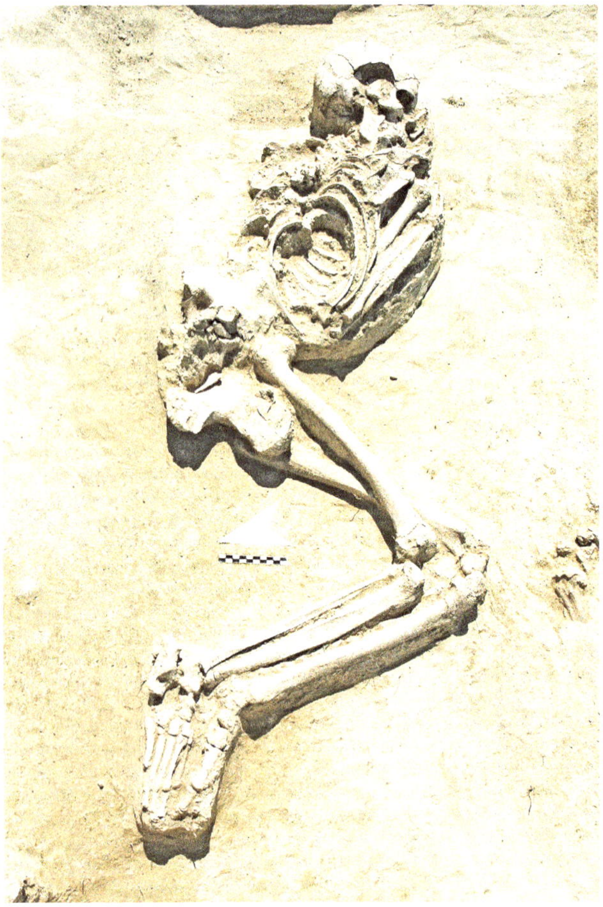

b

c

d

Figure 3.149 (*continued from previous page*). Tomb XLVI: (b) (photo 1072) from above west-northwest; (c) (photo 1076) detail of cranium and torso with **TXLVI-1** from above northwest; (d) (photo 1084) during removal of skeleton, showing **TXLVI-1**, from above west

Figure 3.150 Tomb XLVI: **TXLVI-1**

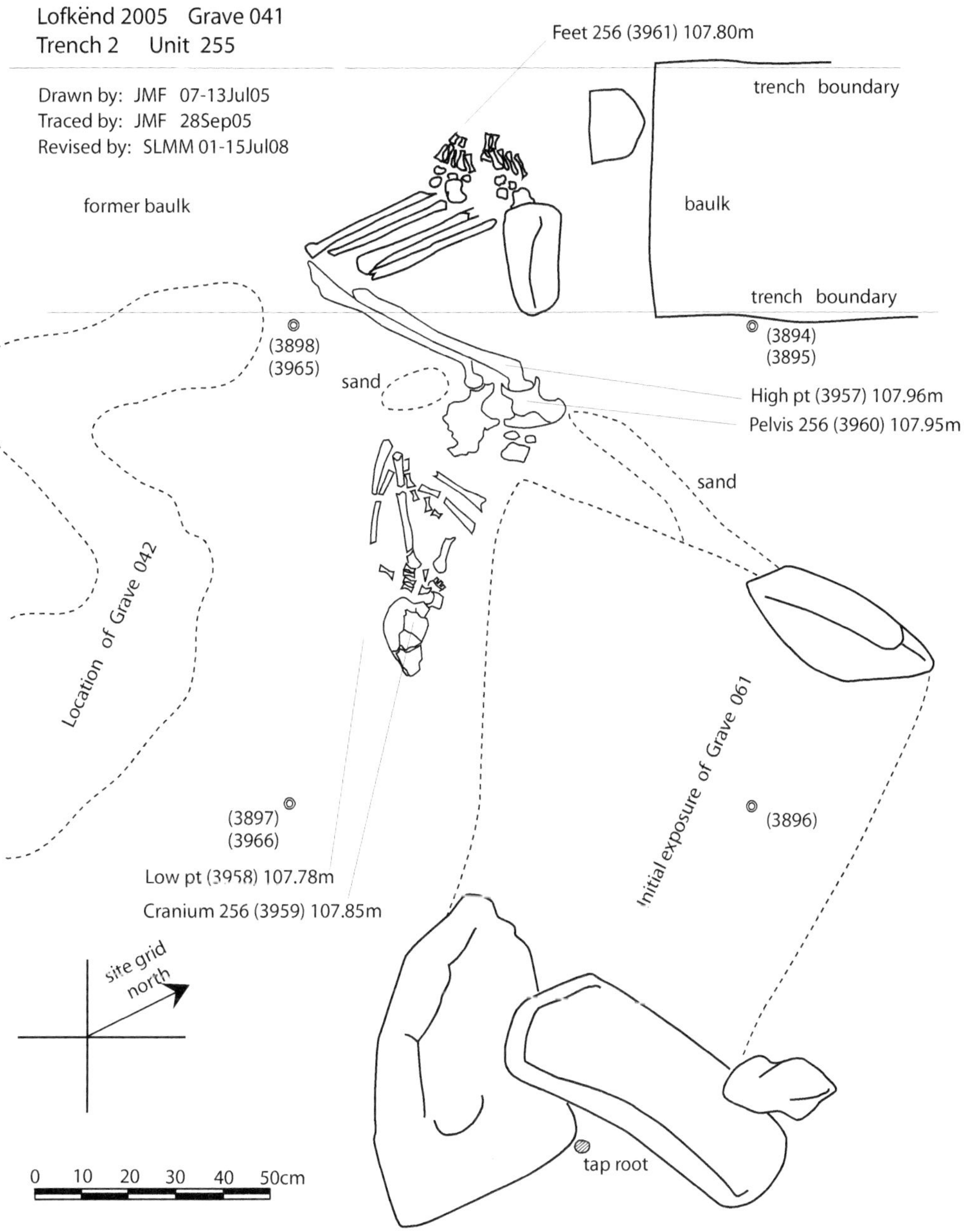

FIGURE 3.151 Plan of Tomb XLVII

Figure 3.152 Tombs XLVII (left) and XLVI (photo 1073) from above west (note cut of Tomb LXIV, which destroyed the back of the cranium and upper torso of Tomb XLVII)

Figure 3.153 Tomb XLVII: (photo 1071) from above west

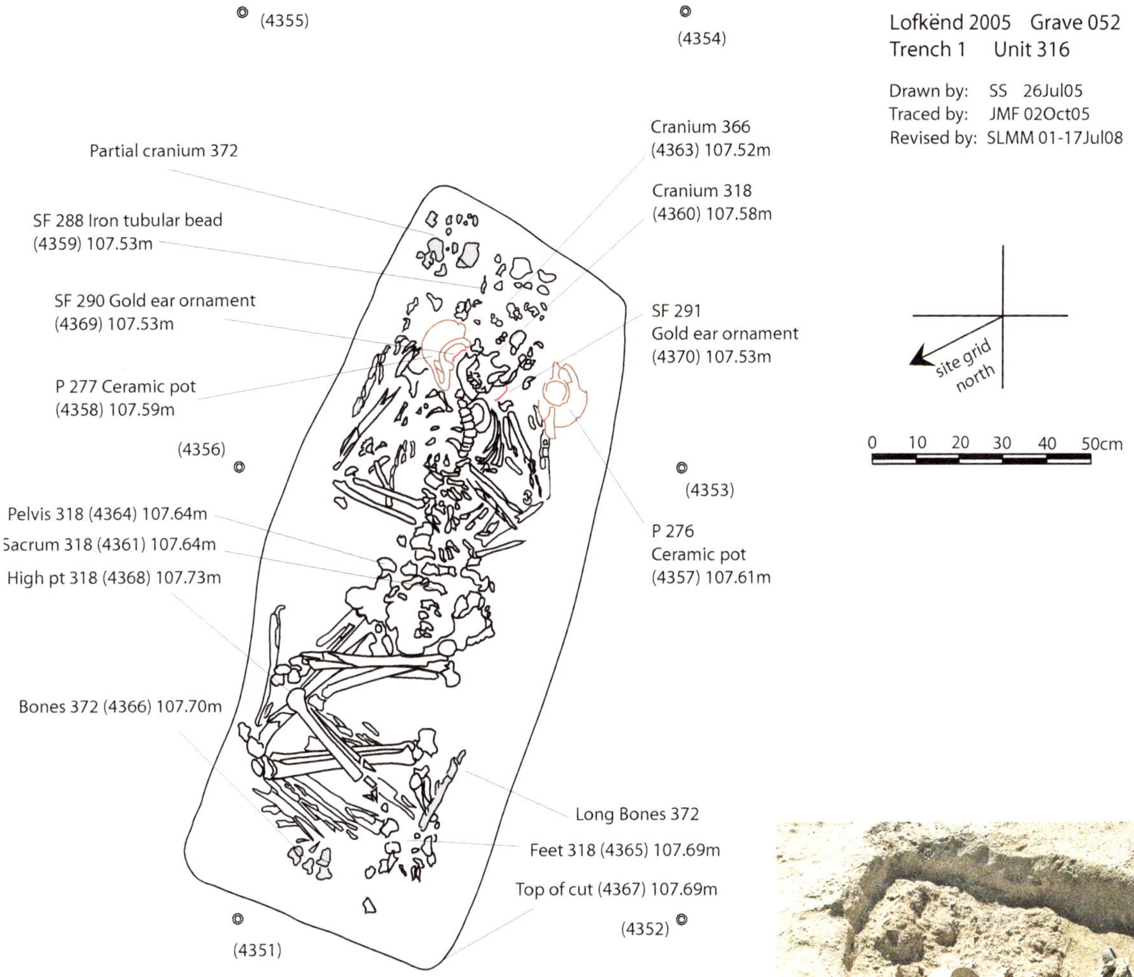

FIGURE 3.154 Plan of Tomb XLVIII

FIGURE 3.155 Tomb XLVIII: (photo 1243) from west-northwest

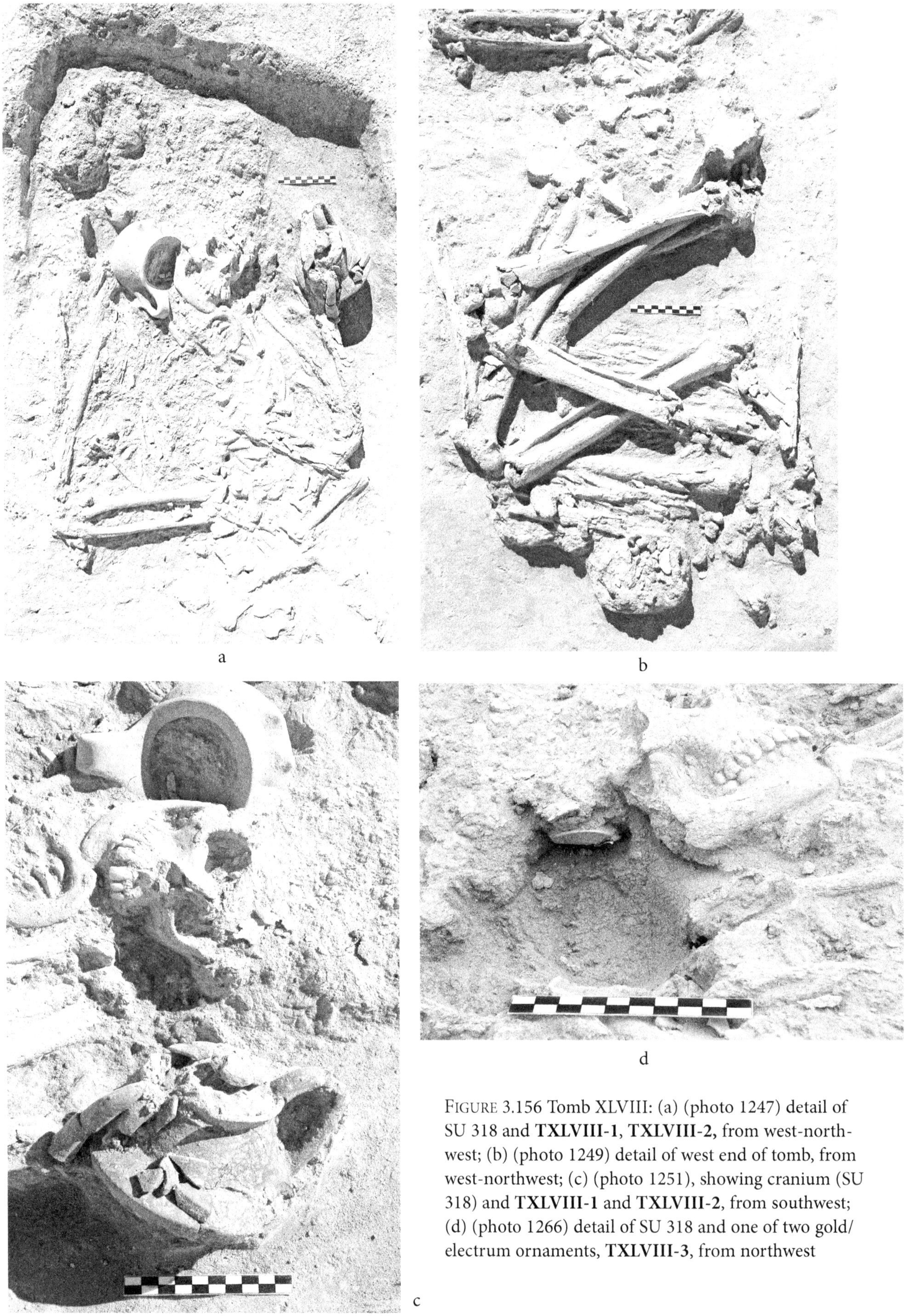

a b c d

FIGURE 3.156 Tomb XLVIII: (a) (photo 1247) detail of SU 318 and **TXLVIII-1**, **TXLVIII-2,** from west-northwest; (b) (photo 1249) detail of west end of tomb, from west-northwest; (c) (photo 1251), showing cranium (SU 318) and **TXLVIII-1** and **TXLVIII-2**, from southwest; (d) (photo 1266) detail of SU 318 and one of two gold/electrum ornaments, **TXLVIII-3**, from northwest

FIGURE 3.157 Tomb XLVIII: (a) **TXLVIII-1**; (b) **TXLVIII-2**

FIGURE 3.158 Tomb XLVIII: **TXLVIII-3** and **TXLVIII-4**

FIGURE 3.159 Tomb XLVIII: **TXLVIII-5**

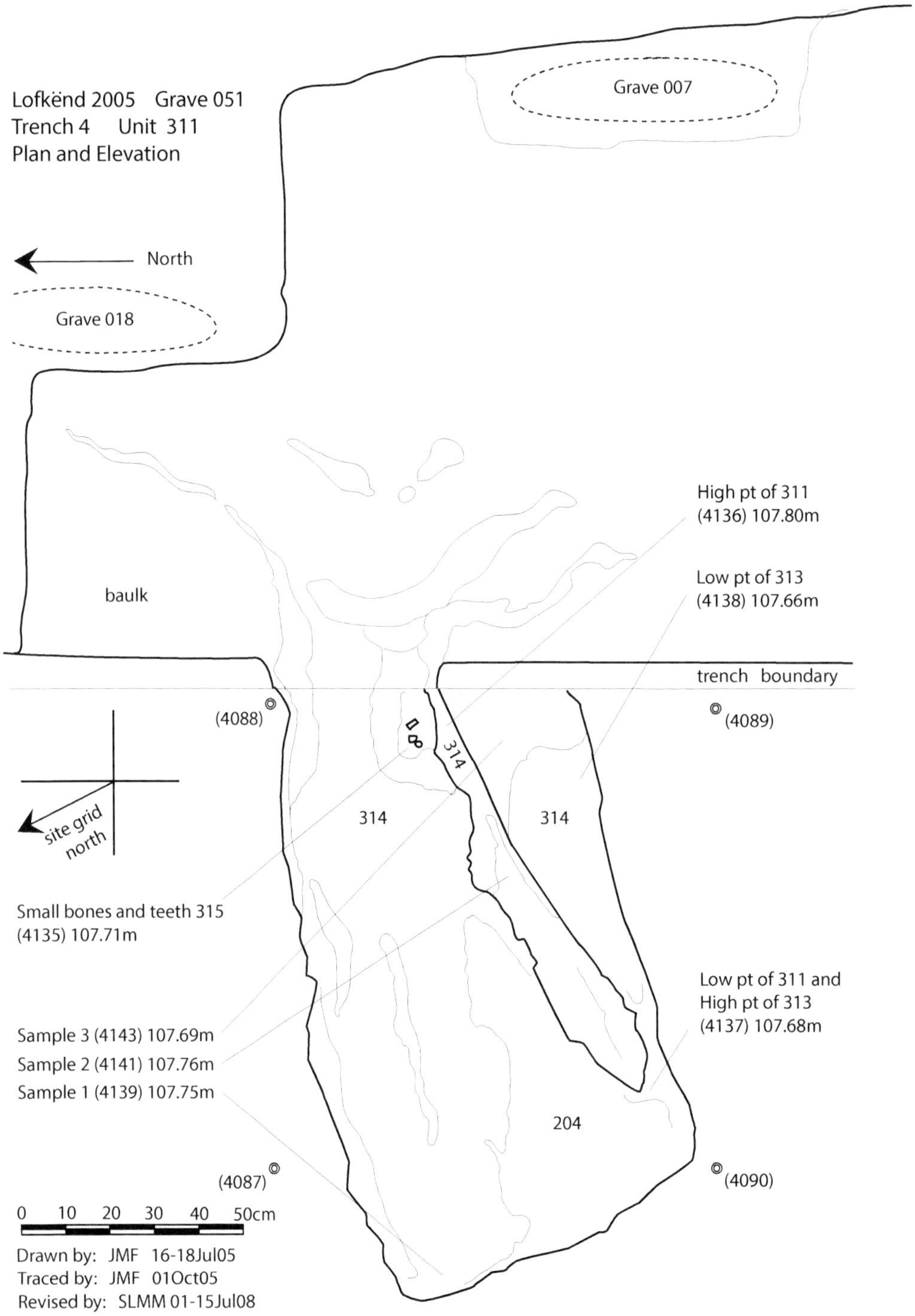

FIGURE 3.160 Tomb XLIX: plan (bottom), with projected elevation at top

a

b

FIGURE 3.161 Tomb XLIX: (a) (photo 1132), upper level, from west;
(b) (photo 1137) lower level with poorly preserved human remains, from west

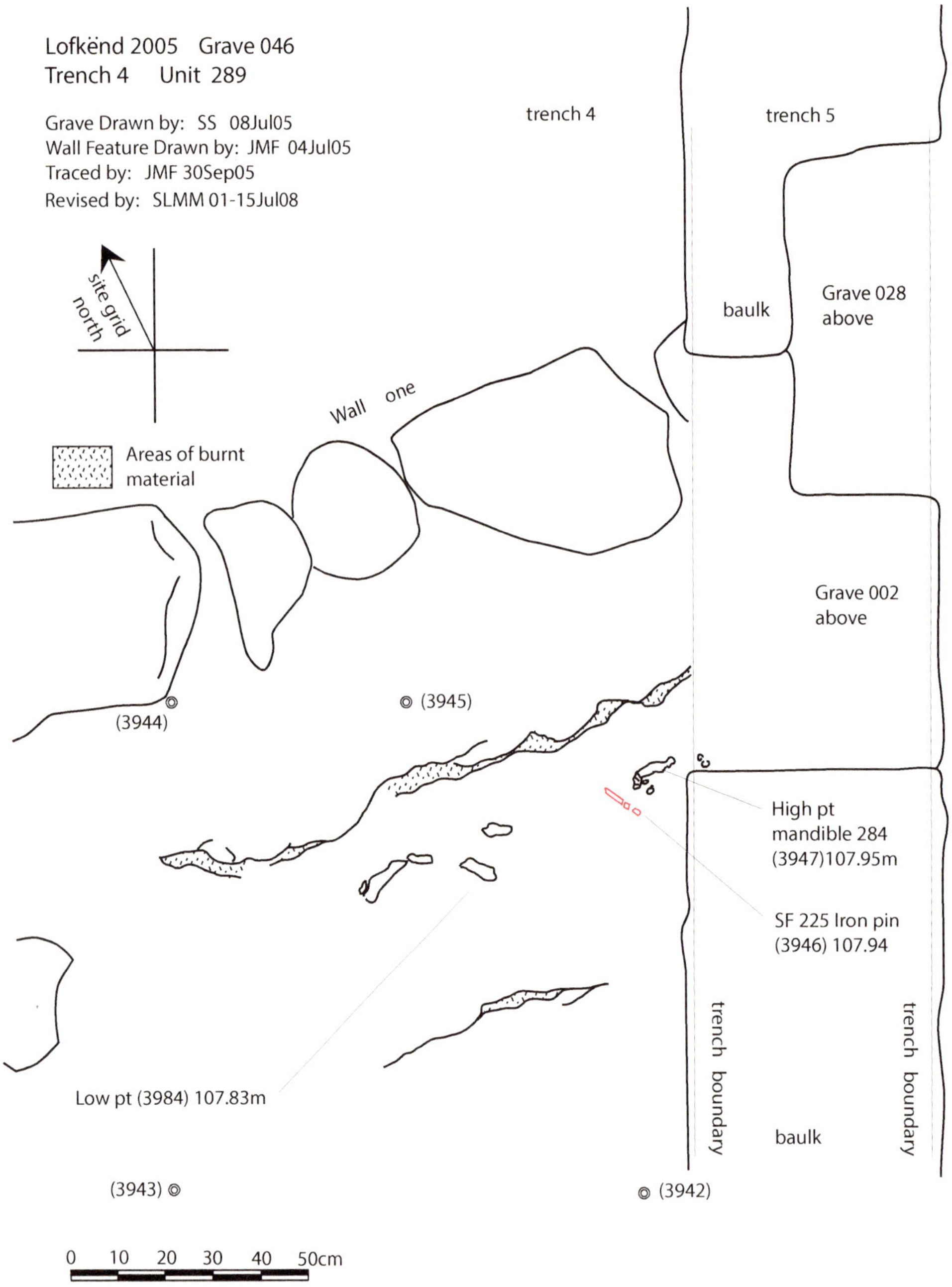

FIGURE 3.162 Plan of Tomb L

a b

Figure 3.163 Tomb L: (a) Stage 1 (photo 1062) from above west-southwest; (b) Stage 2 (photo 1081) from above west-southwest

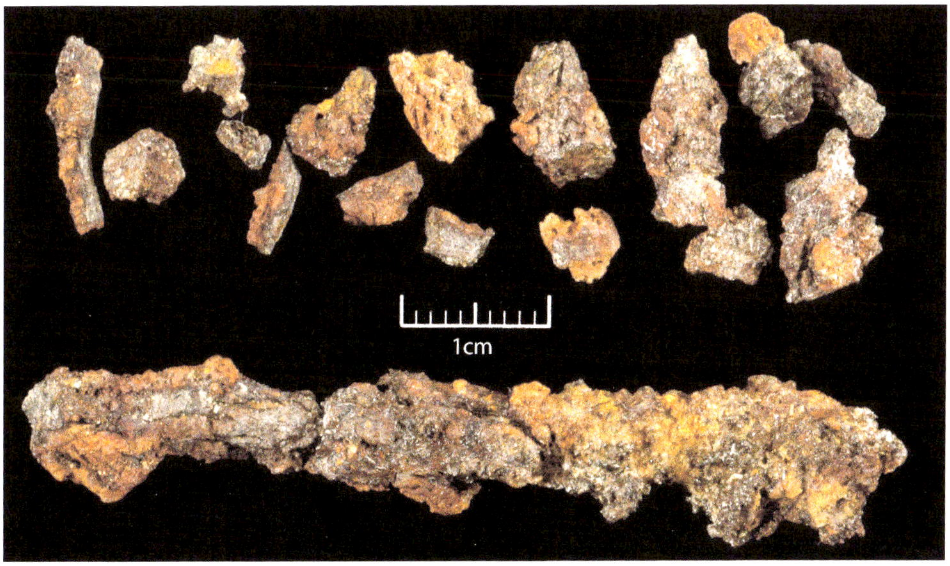

Figure 3.164 Tomb L: **TL-1**

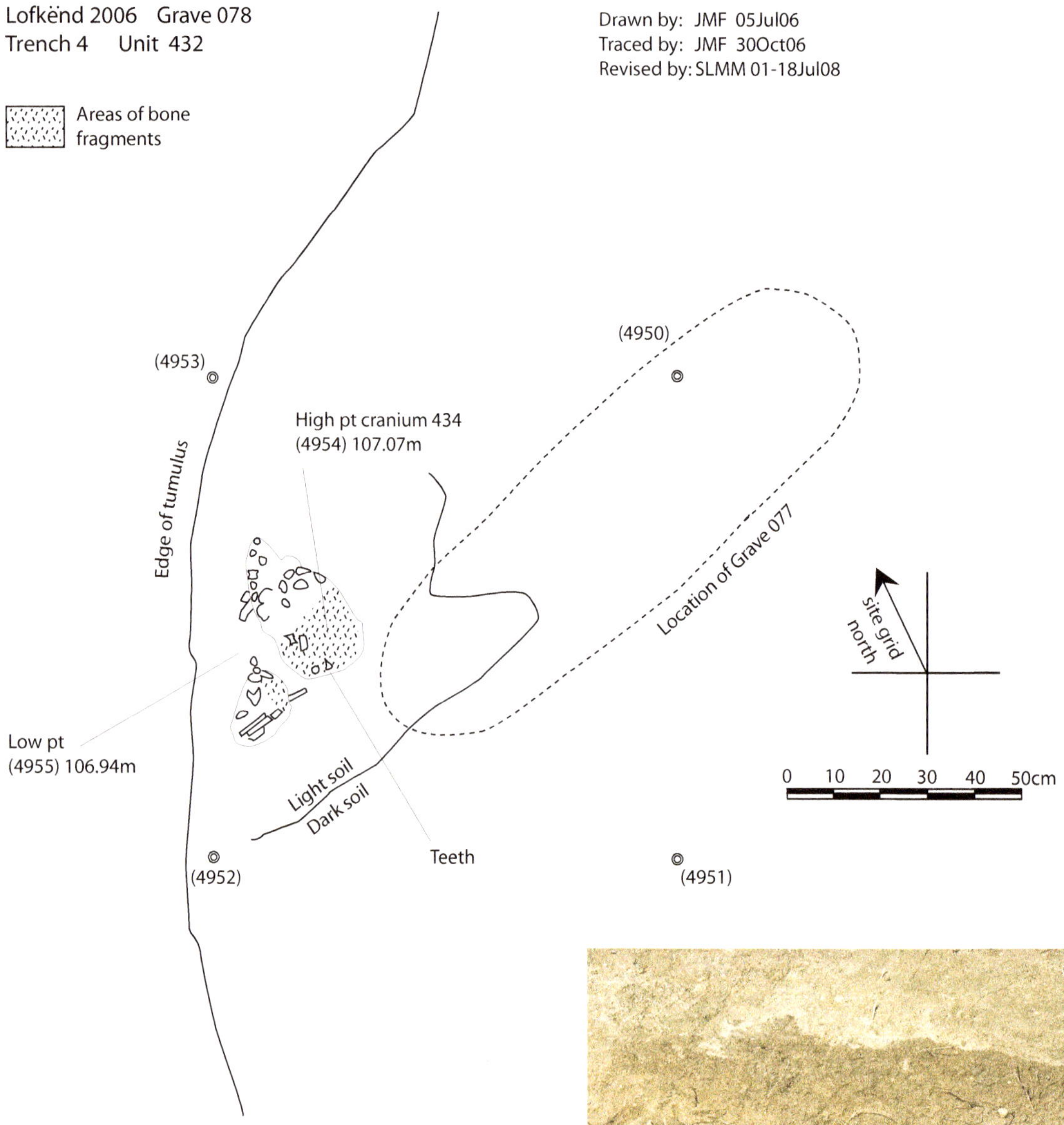

FIGURE 3.165 Plan of Tomb LI

FIGURE 3.166 Tomb LI (photo 1400) from east

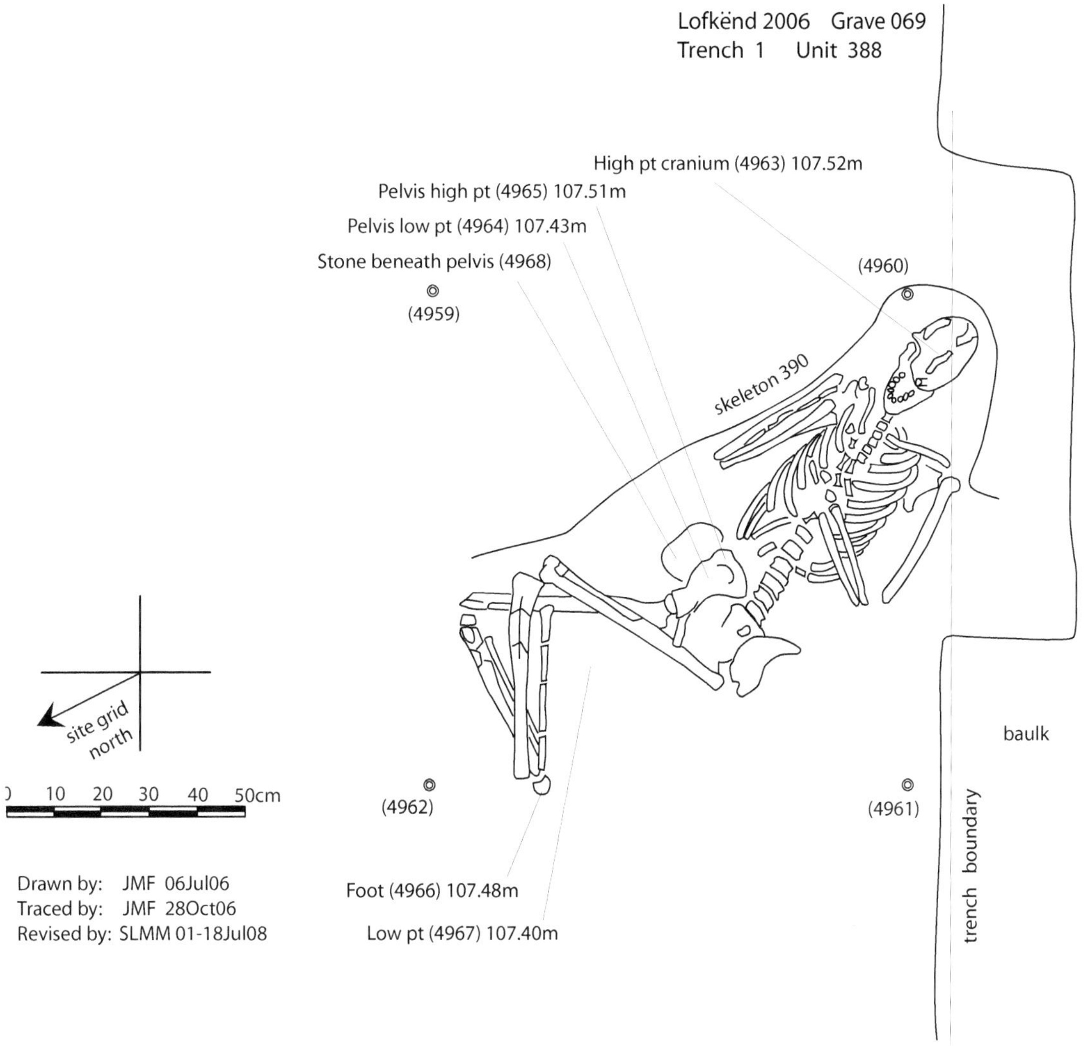

Figure 3.167 Plan of Tomb LII

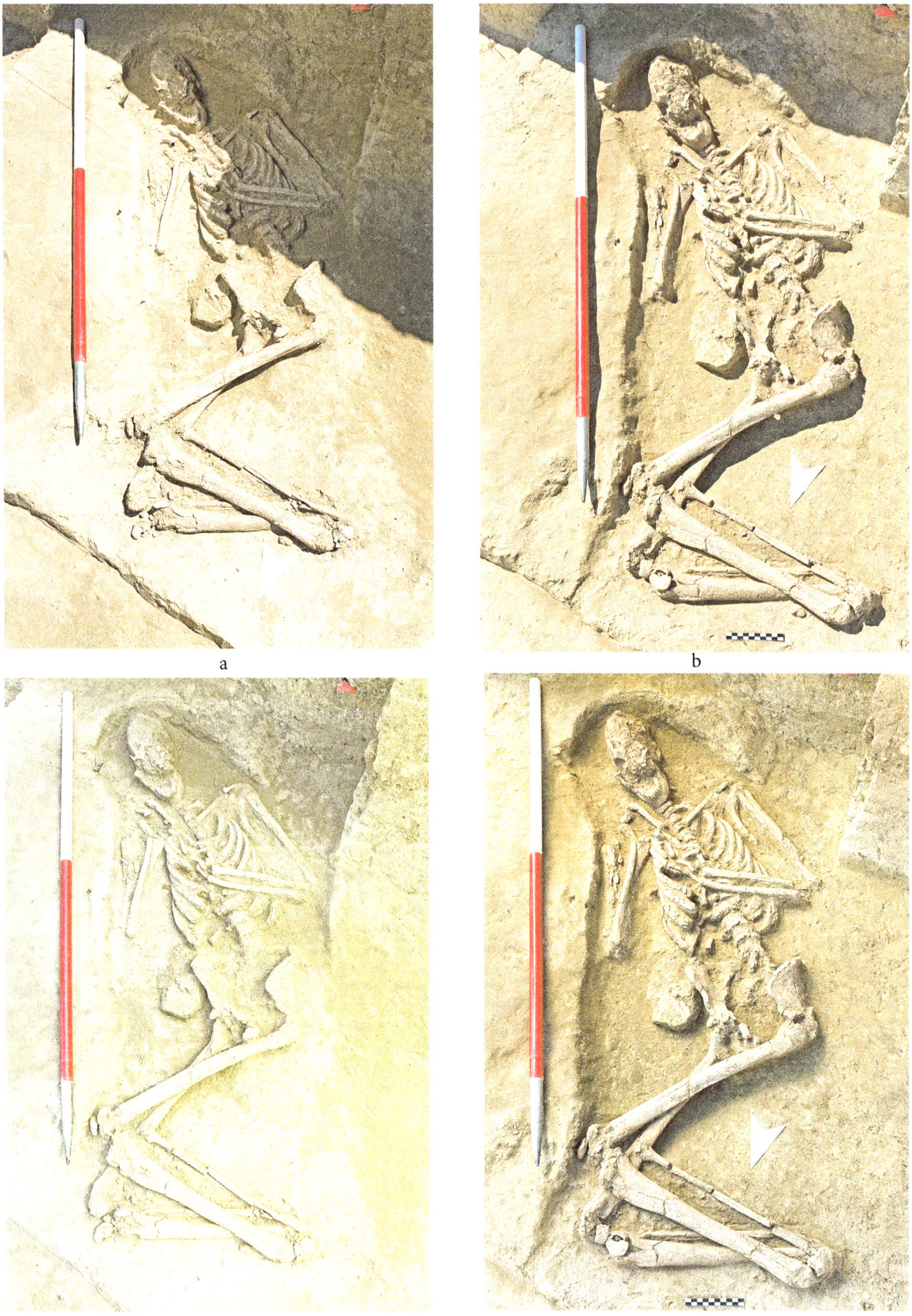

a b c d

Figure 3.168 Tomb LII: (a–d) four shots all from northwest, experimenting with different light conditions

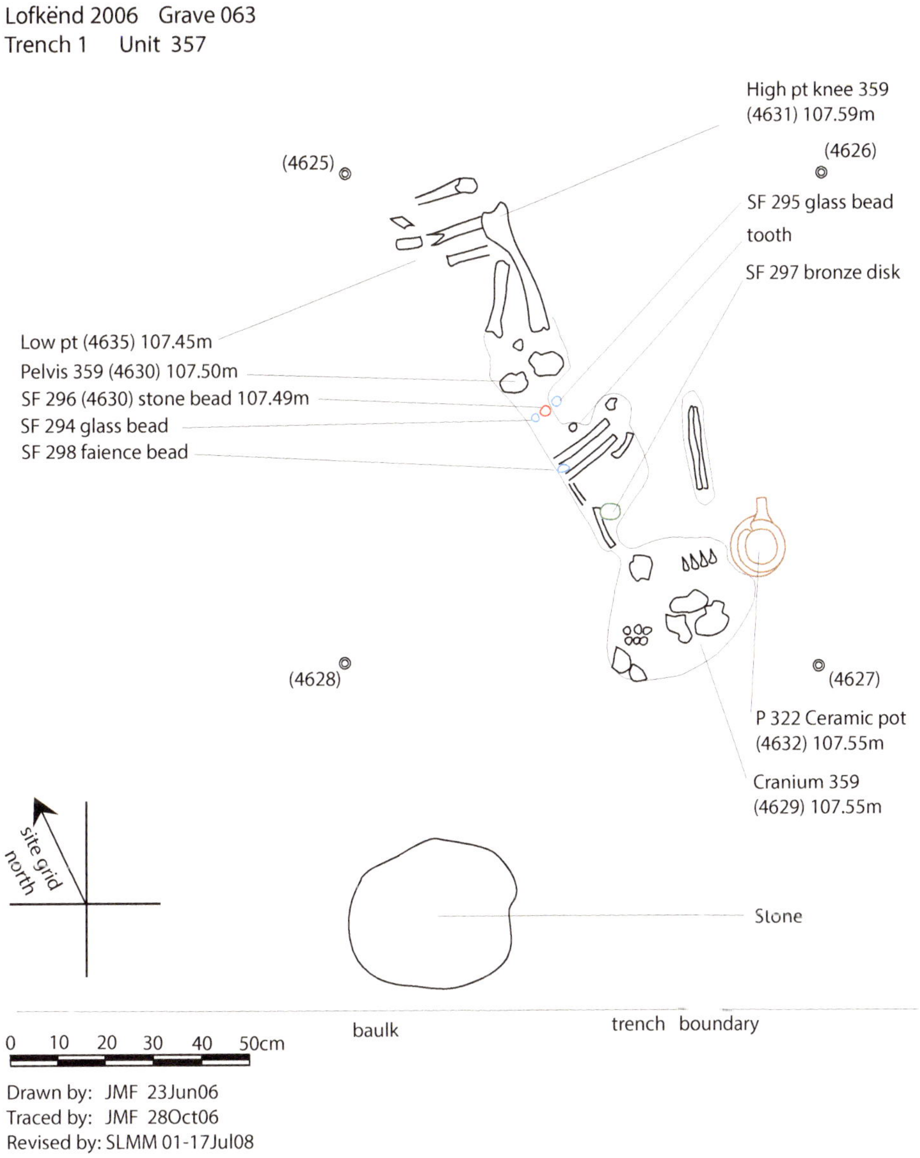

Figure 3.169 Plan of Tomb LIII

a b

FIGURE 3.170 Tomb LIII: (a) (photo 1297) from above north-northwest; (b) (photo 1298), detail of beads in situ, from above east-northeast

FIGURE 3.171 Tomb LIII: **TLIII-1**

FIGURE 3.172 Tomb LIII: **TLIII-2**

FIGURE 3.173 Tomb LIII: **TLIII-3**

FIGURE 3.174 Tomb LIII: **TLIII-4**

FIGURE 3.175 Tomb LIII: **TLIII-5**

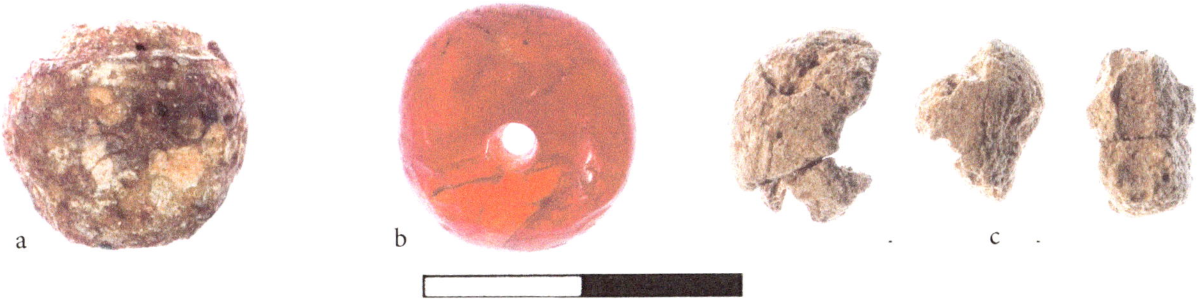

FIGURE 3.176 Tomb LIII: (a) **TLIII-6**; (b) **TLIII-7**; (c) **TLIII-8**

FIGURE 3.177 Tomb LIII: **TLIII-9**

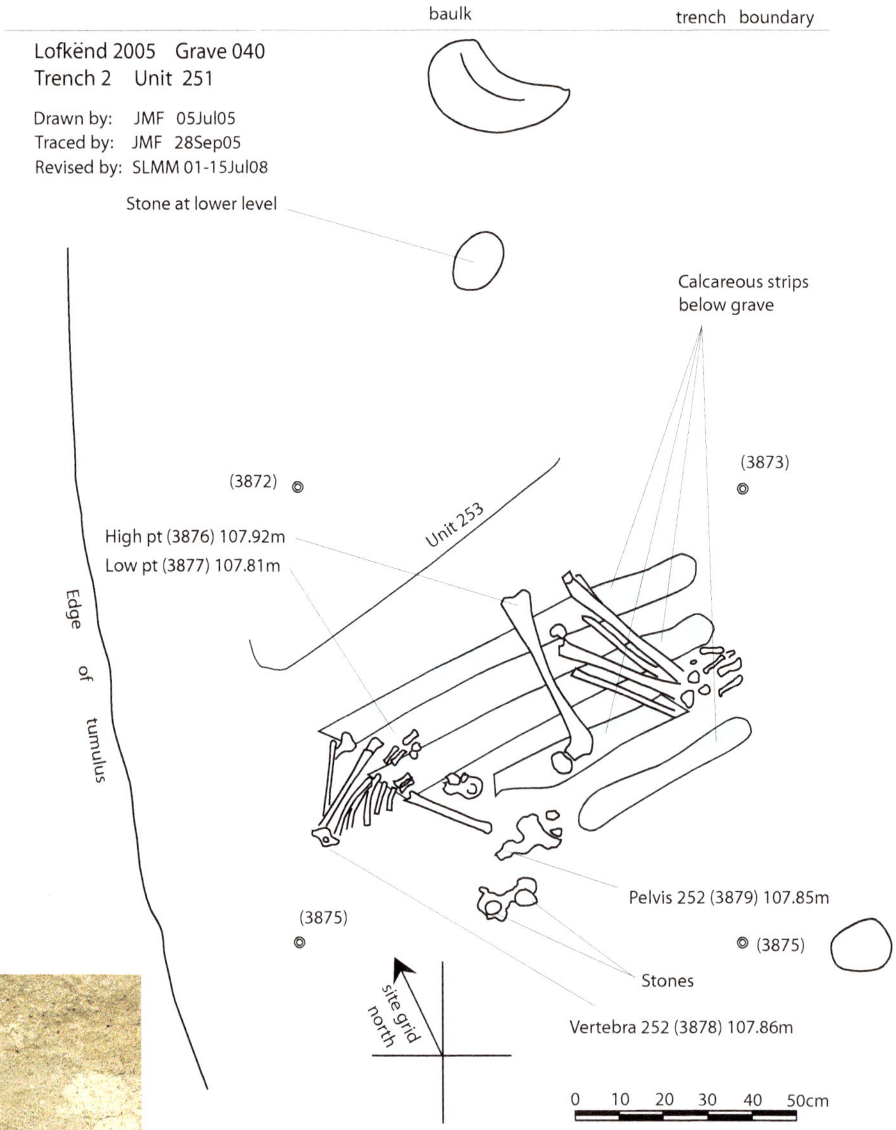

FIGURE 3.178 Plan of Tomb LIV

FIGURE 3.179 Tomb LIV: (photo 1016) from northeast, with tumulus edge above

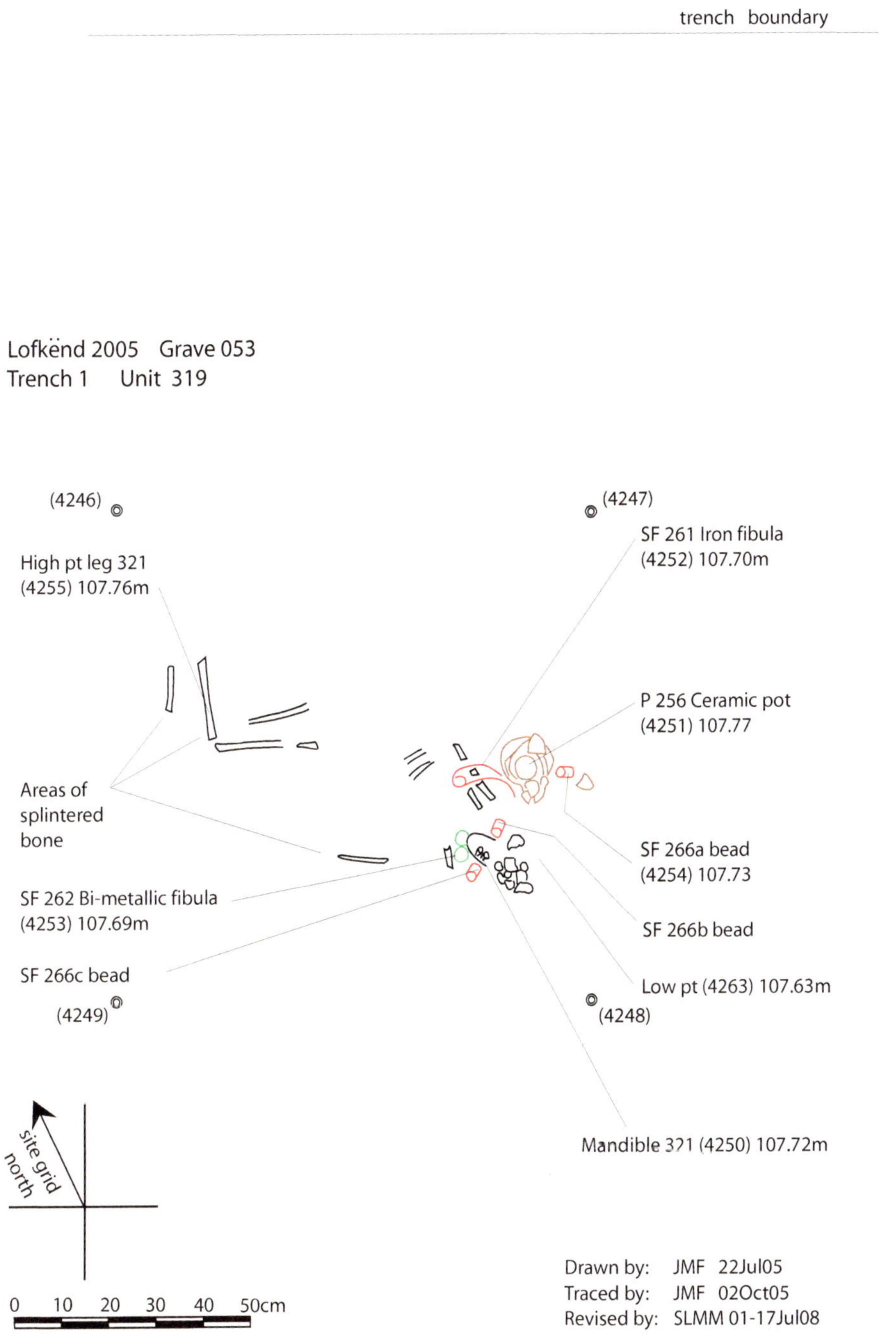

FIGURE 3.180 Plan of Tomb LV

a

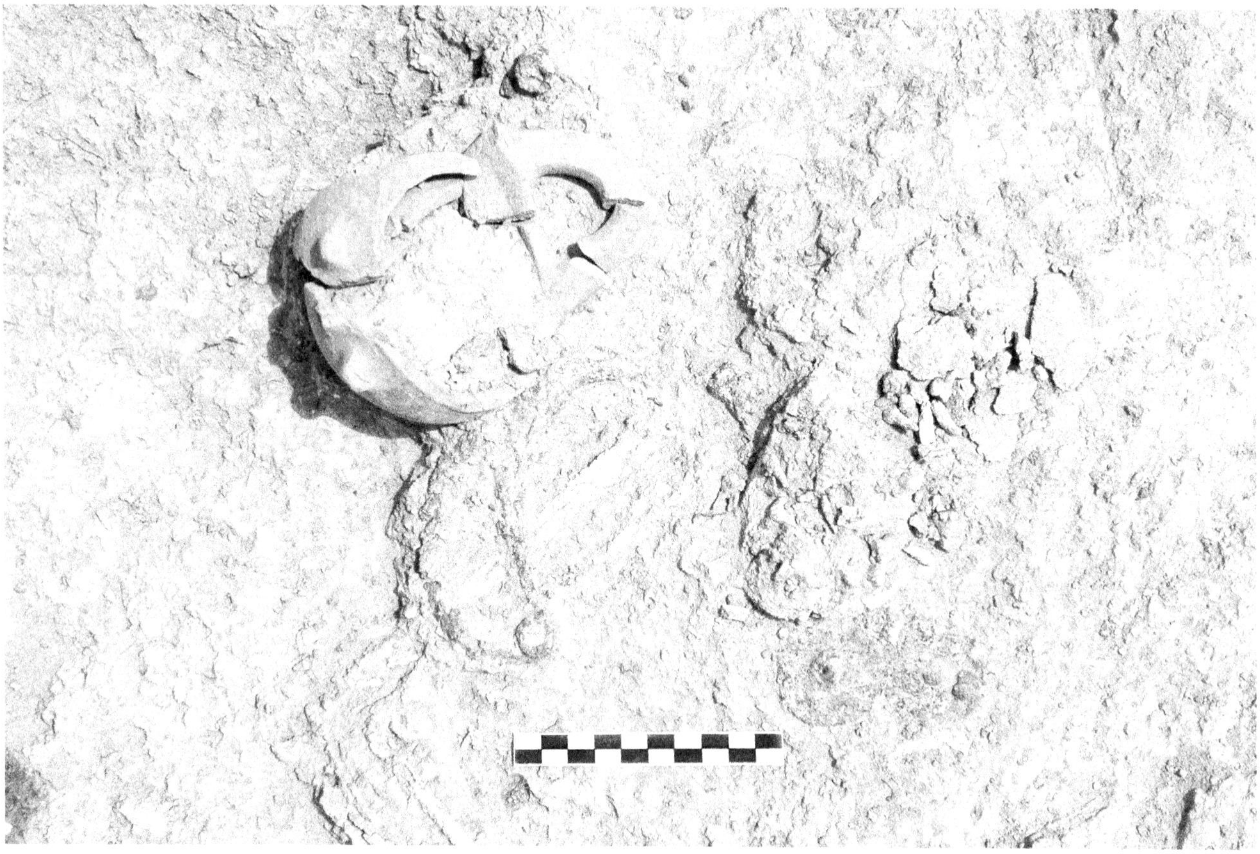

b

FIGURE 3.181 Tomb LV: (a) (photo 1195) from north;
(b) (photo 1196) detail of cranium and grave goods, from above west

Figure 3.182 TLV: **TLV-1**

Figure 3.183 Tomb LV: **TLV-2**

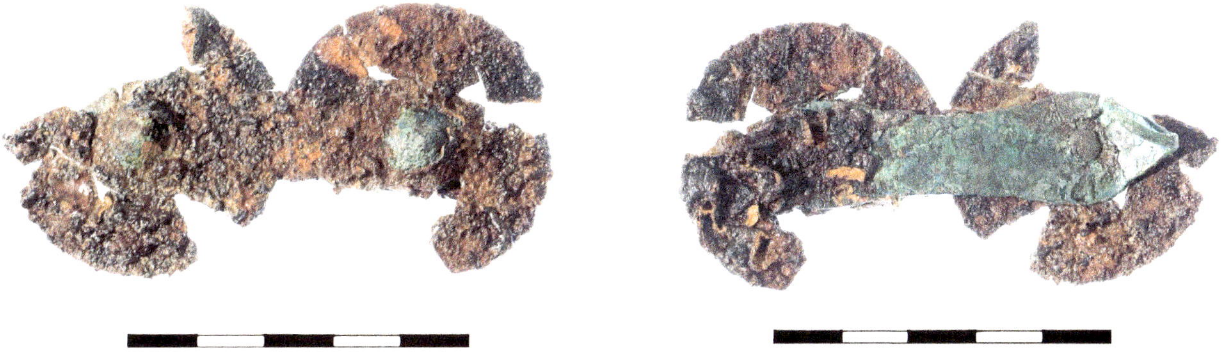

Figure 3.184 Tomb LV: (a–b) **TLV-3**

Figure 3.185 Tomb LV: (a) **TLV-4**; (b) **TLV-5**; (c) **TLV-6**; (d) **TLV-7**

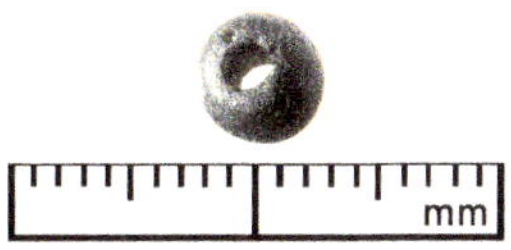

Figure 3.186 Tomb LV: **TLV-8**

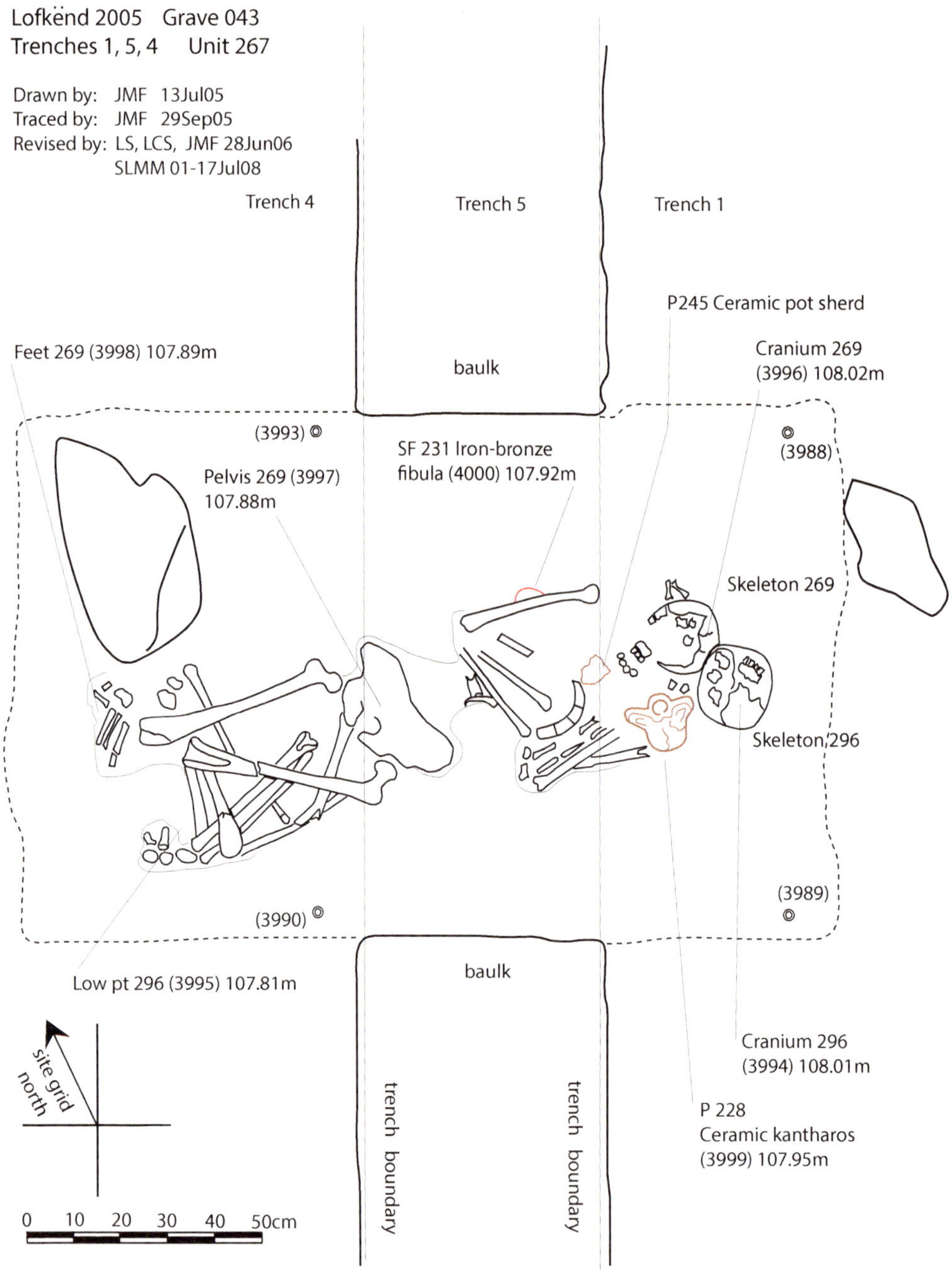

FIGURE 3.187 Plan of Tomb LVI

a

b

c

FIGURE 3.188 Tomb LVI: (a) (photo 1093) from above south; (b) (photo 1103) detail of **TLVI-1**, from south-southeast; (c) (photo 1107) detail of **TLVI-2**, from above south

Figure 3.189 Tomb LVI: **TLVI-1**

Figure 3.190 Tomb LVI: **TLVI-2**

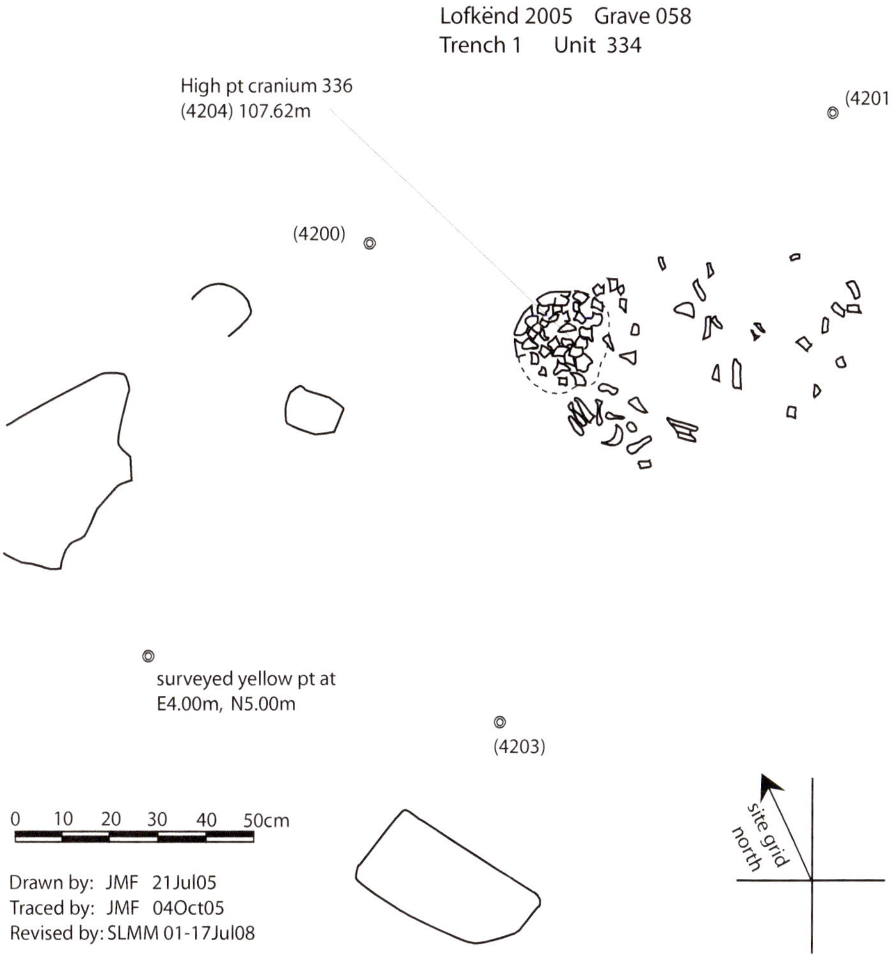

FIGURE 3.191 Plan of Tomb LVII

FIGURE 3.192 Tomb LVII: (photo 1174) from west

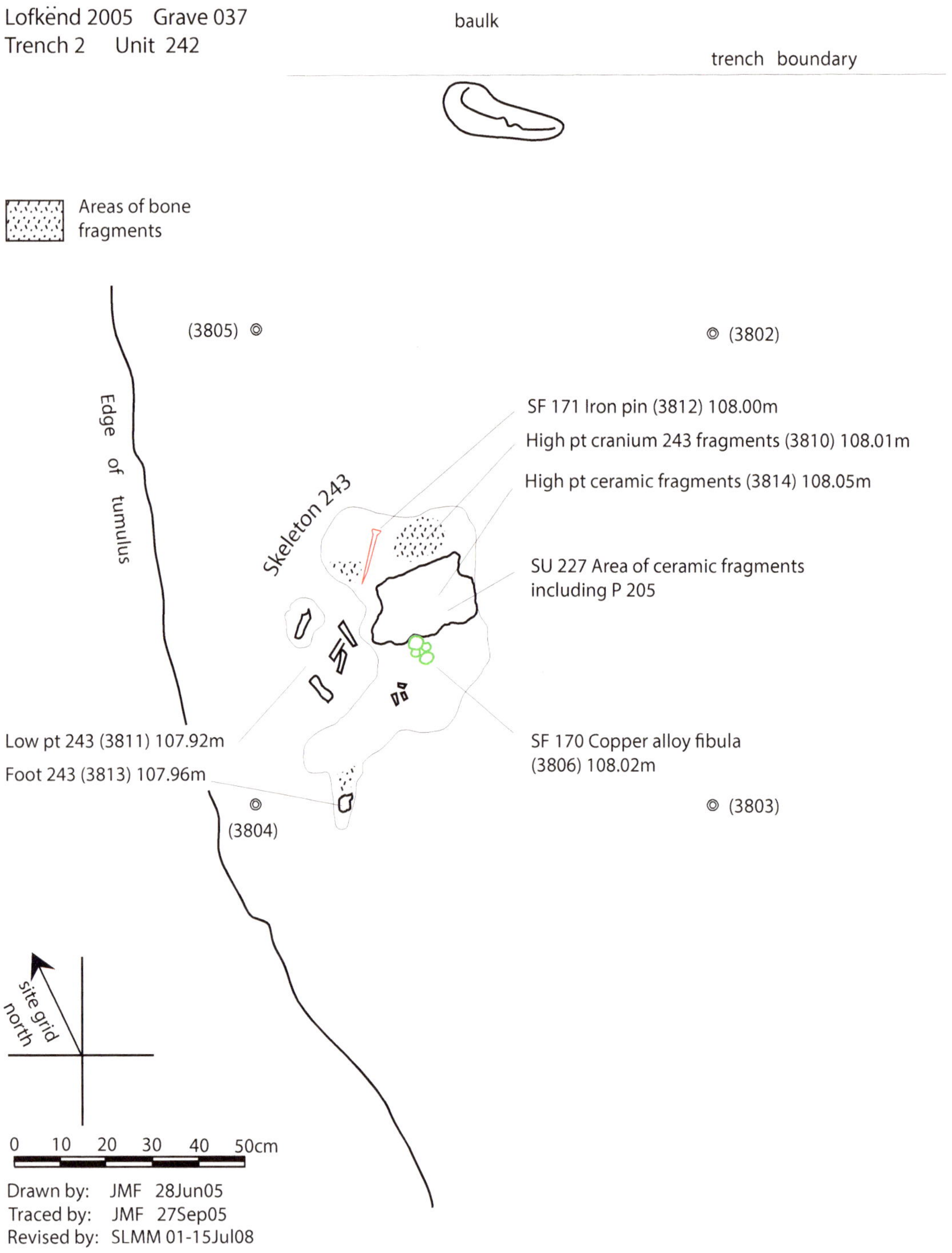

FIGURE 3.193 Plan of Tomb LVIII

a

FIGURE 3.194 Tomb LVIII: (a) (photo 892) detail from above east; (b) (photo 893) from east showing tumulus edge

b

FIGURE 3.195 Tomb LVIII: **TLVIII-1**

FIGURE 3.196 Tomb LVIII: **TLVIII-2**

FIGURE 3.197 Tomb LVIII: **TLVIII-3**

FIGURE 3.198 Tomb LVIII: **TLVIII-4**

FIGURE 3.199 Tomb LVIII: P205
(part of SU 227)

Figure 3.200 Tomb LVIII: fragments of pottery from SU 227 (photo 3437)

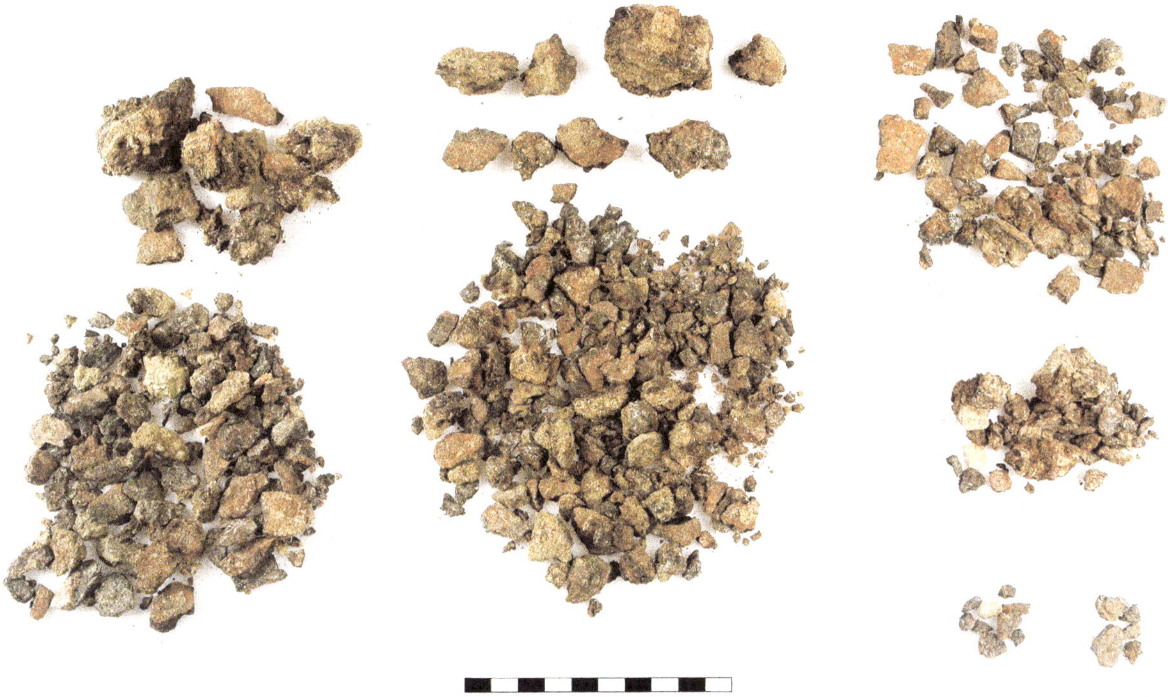

Figure 3.201 Tomb LVIII: fragments of fire-affected clay, as preserved (photo 4006)

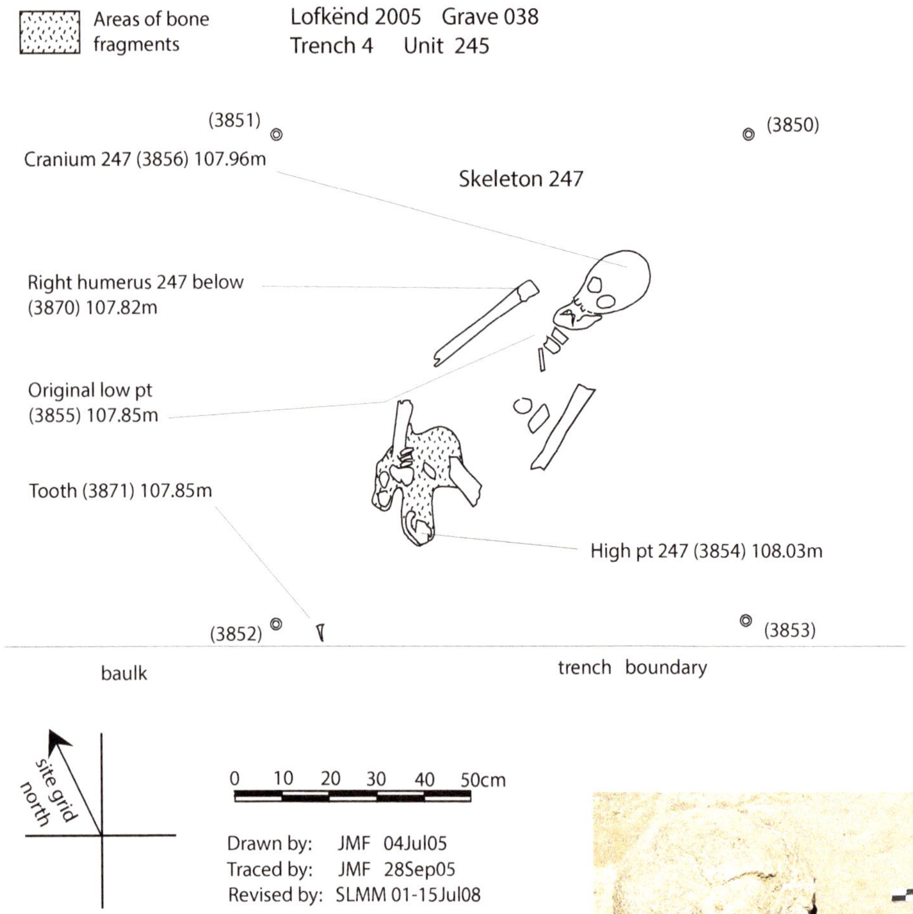

FIGURE 3.202 Plan of Tomb LIX

FIGURE 3.203 Tomb LIX: (photo 1003) from above south

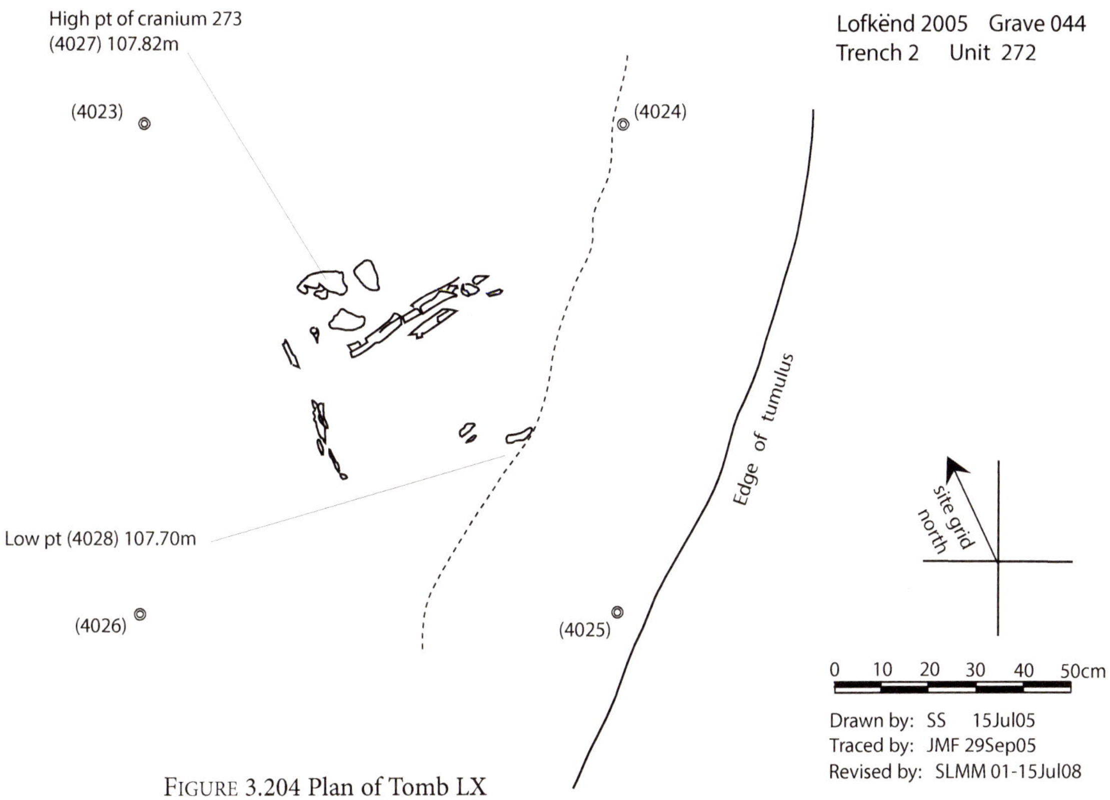

FIGURE 3.204 Plan of Tomb LX

FIGURE 3.205 Tomb LX: (photo 1121) from above east

FIGURE 3.206 Iron blade, SF 230 (photo 3952) found in topsoil in the immediate vicinity of Tomb LX

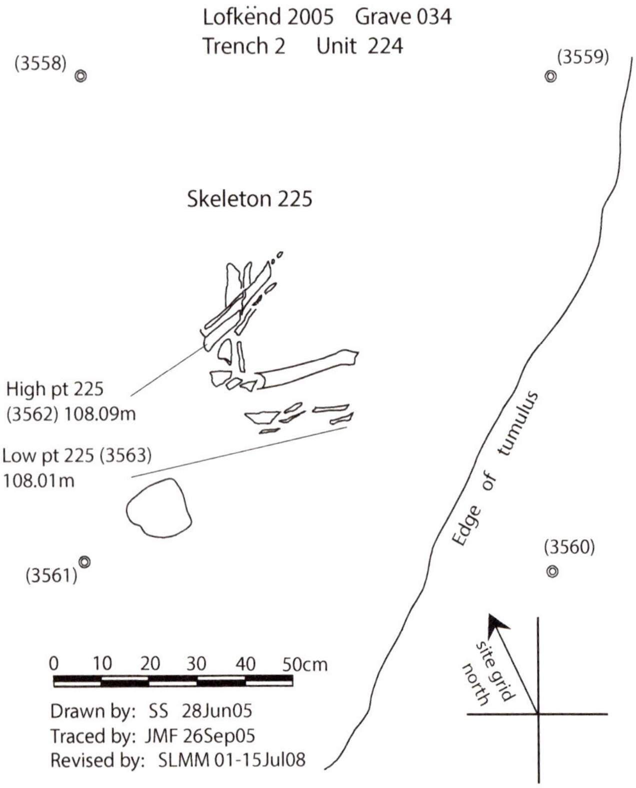

FIGURE 3.207 Plan of Tomb LXI

FIGURE 3.208 Tomb LXI: (photo 839) from northwest, showing tumulus edge

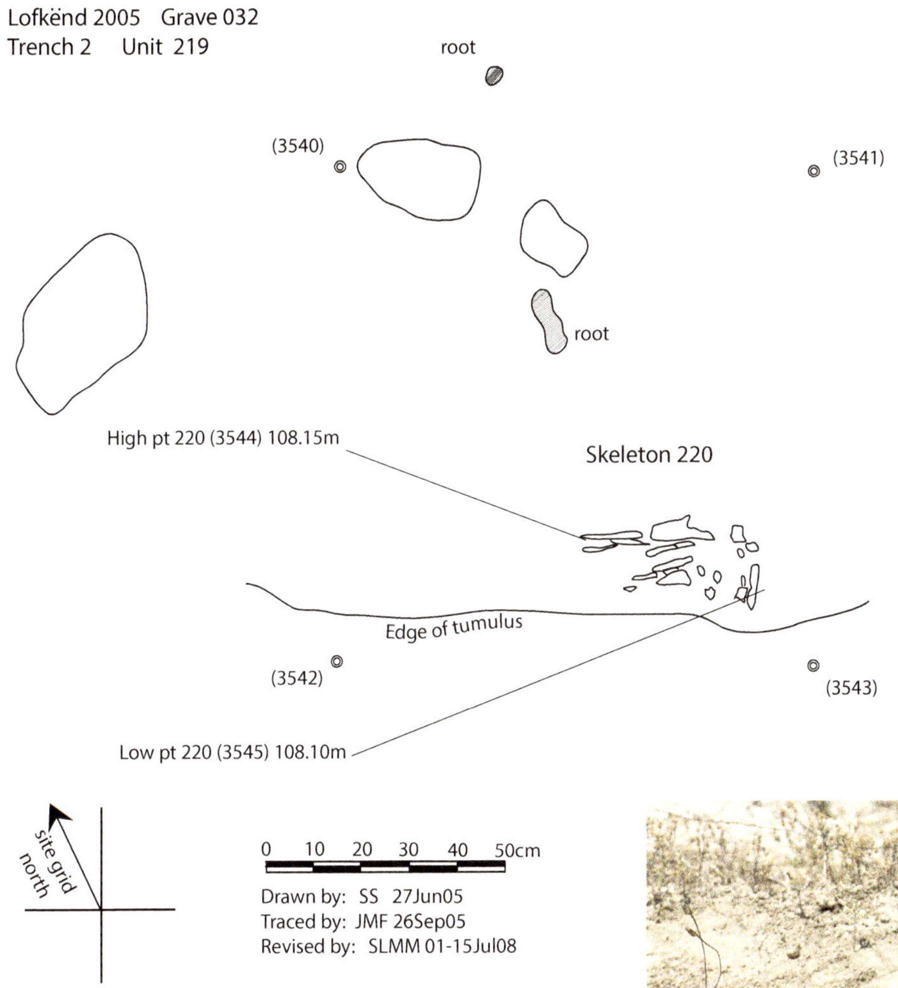

FIGURE 3.209 Plan of Tomb LXII

FIGURE 3.210 Tomb LXII: (photo 825) from north, showing tumulus edge

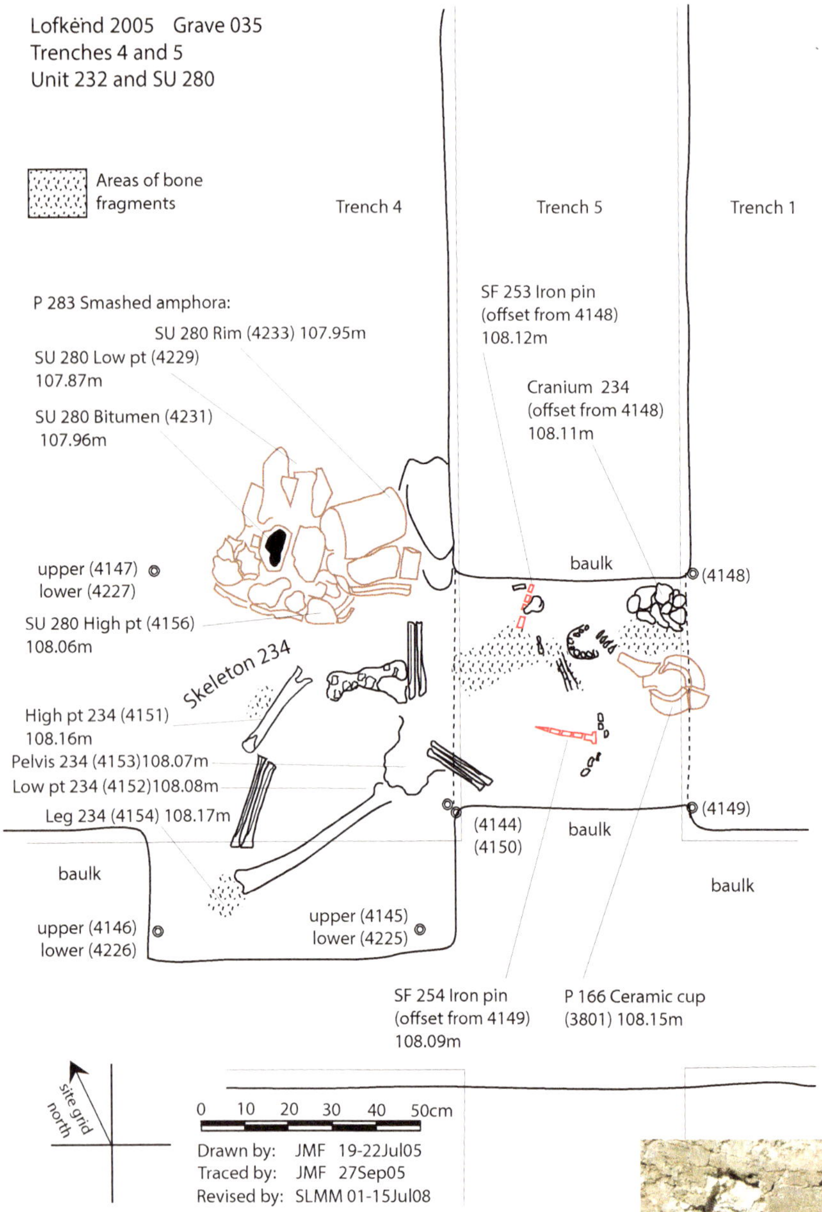

FIGURE 3.211 Plan of Tomb LXIII

FIGURE 3.212 Tomb LXIII: (photo 888) showing **TLXIII-1** in situ in Sector 1, from east

a

b

FIGURE 3.213 Tomb LXIII: (a) (photo 1161) as exposed by tunneling through baulk, from west; (b) (photo 1160) as exposed by tunneling, from east

FIGURE 3.214 Coarse pot containing bitumen (photo 1183) near Tomb LXIII, from west

FIGURE 3.215 Tomb LXIII: **TLXIII-1**

Figure 3.216 Tomb LXIII: **TLXIII-2**

Figure 3.217 Tomb LXIII: **TLXIII-3**

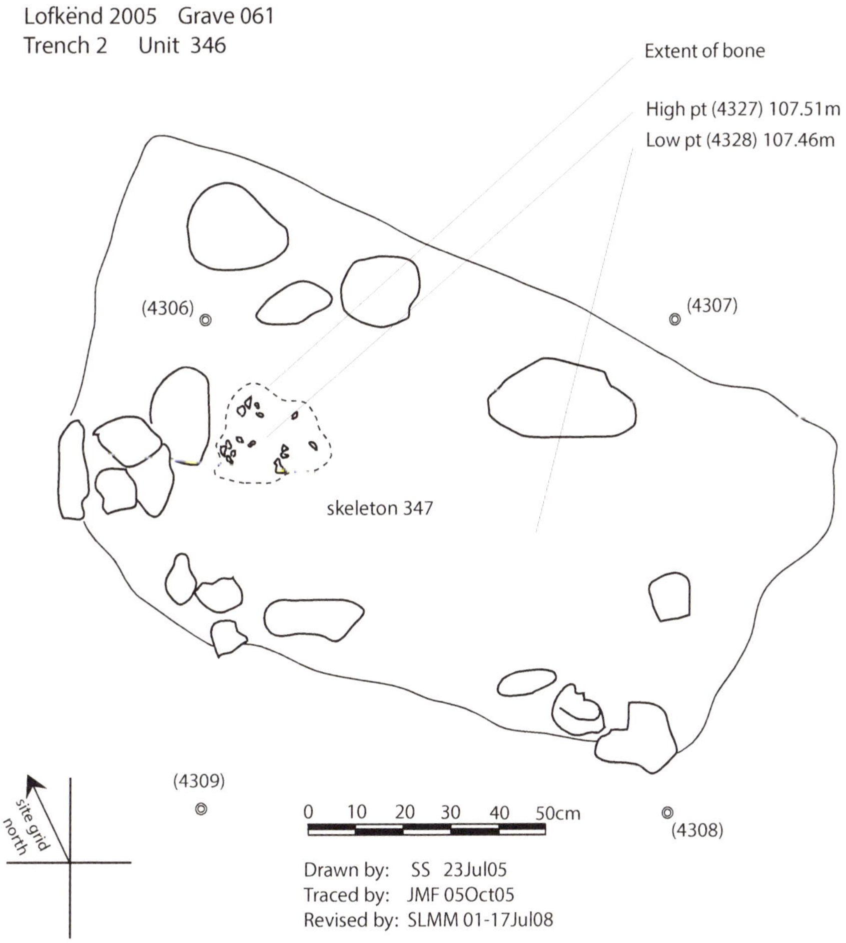

Figure 3.218 Plan of Tomb LIV

a

b

FIGURE 3.219 Tomb LXIV: (a) pit for tomb as first exposed (photo 1023), with Tomb XLVIII in the process of excavation, from northeast; (b) (photo 1204) tomb as fully exposed, from east

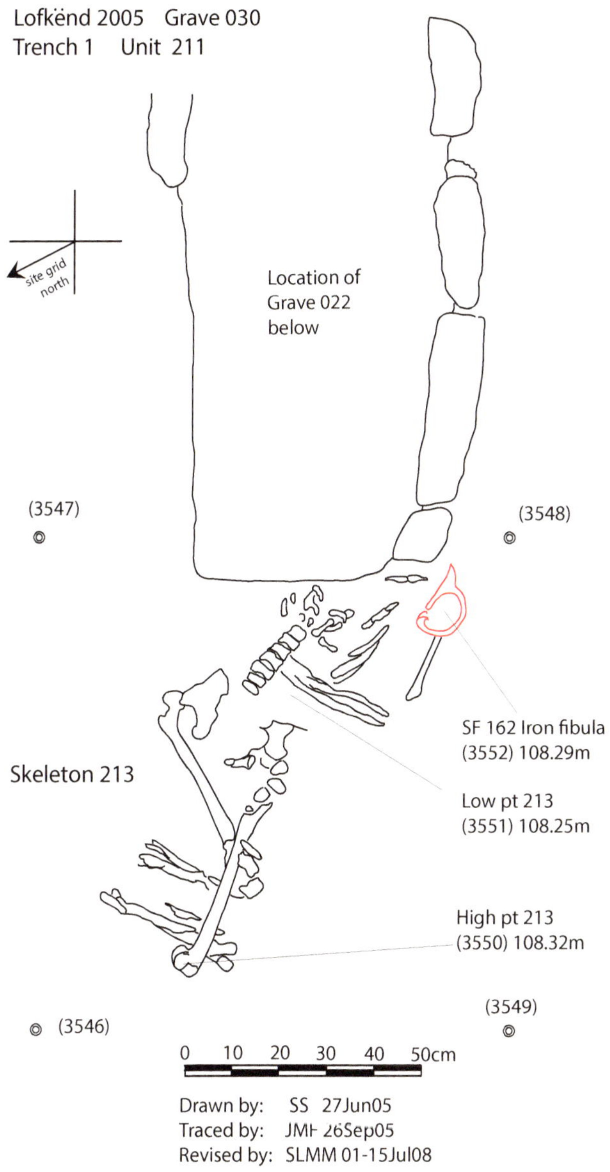

FIGURE 3.220 Plan of Tomb LXV

a

(*Right: top and bottom*). FIGURE 3.221 Tomb LXV: (a) (photo 832) having been cut by the modern Tomb LXXXVI, from west; (b) (photo 817) detail of iron fibula **TLXV-1** in situ, from southeast

b

FIGURE 3.222 Tomb LXV: **TLXV-1**

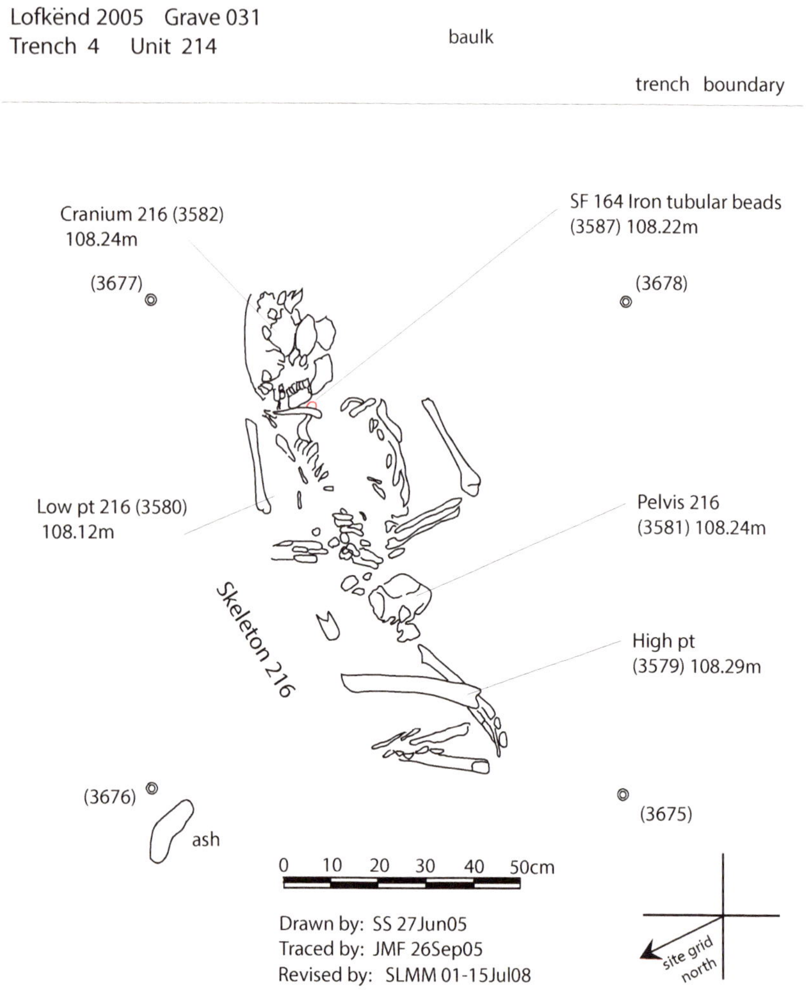

FIGURE 3.223 Plan of Tomb LXVI

FIGURE 3.224 Tomb LXVI: (photo 846) from above west

FIGURE 3.225 Tomb LXVI: (a–b) **TLXVI-1** and **TLXVI-2**

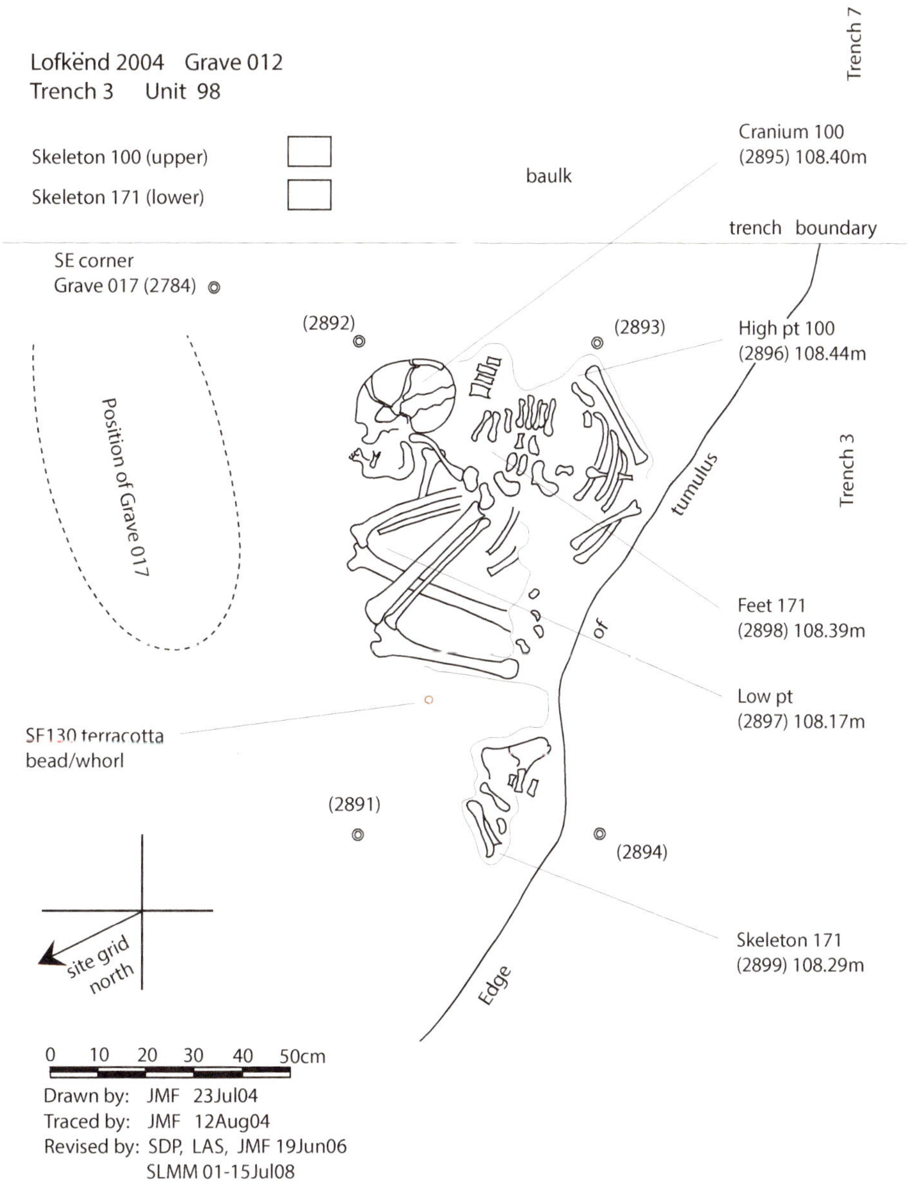

FIGURE 3.226 Plan of Tomb LXVII

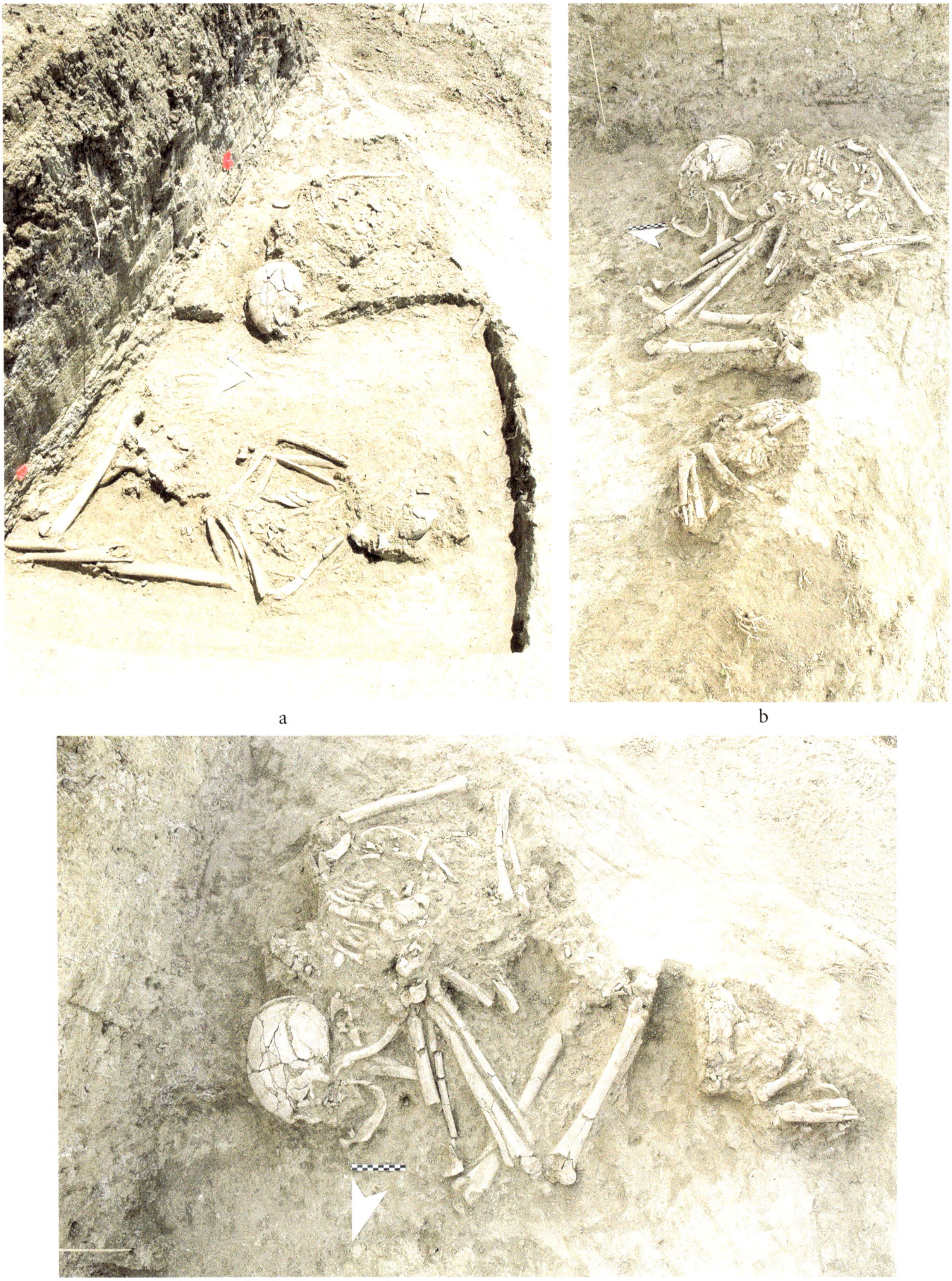

FIGURE 3.227 Tomb LXVII: (a) Tombs LXX (foreground) and LXVII (photo 323) from northwest; (b) Tomb LXVII (photo 440) from north; (c) (photo 441) from west

Figure 3.228 Tomb LXVII: **TLXVII-1**

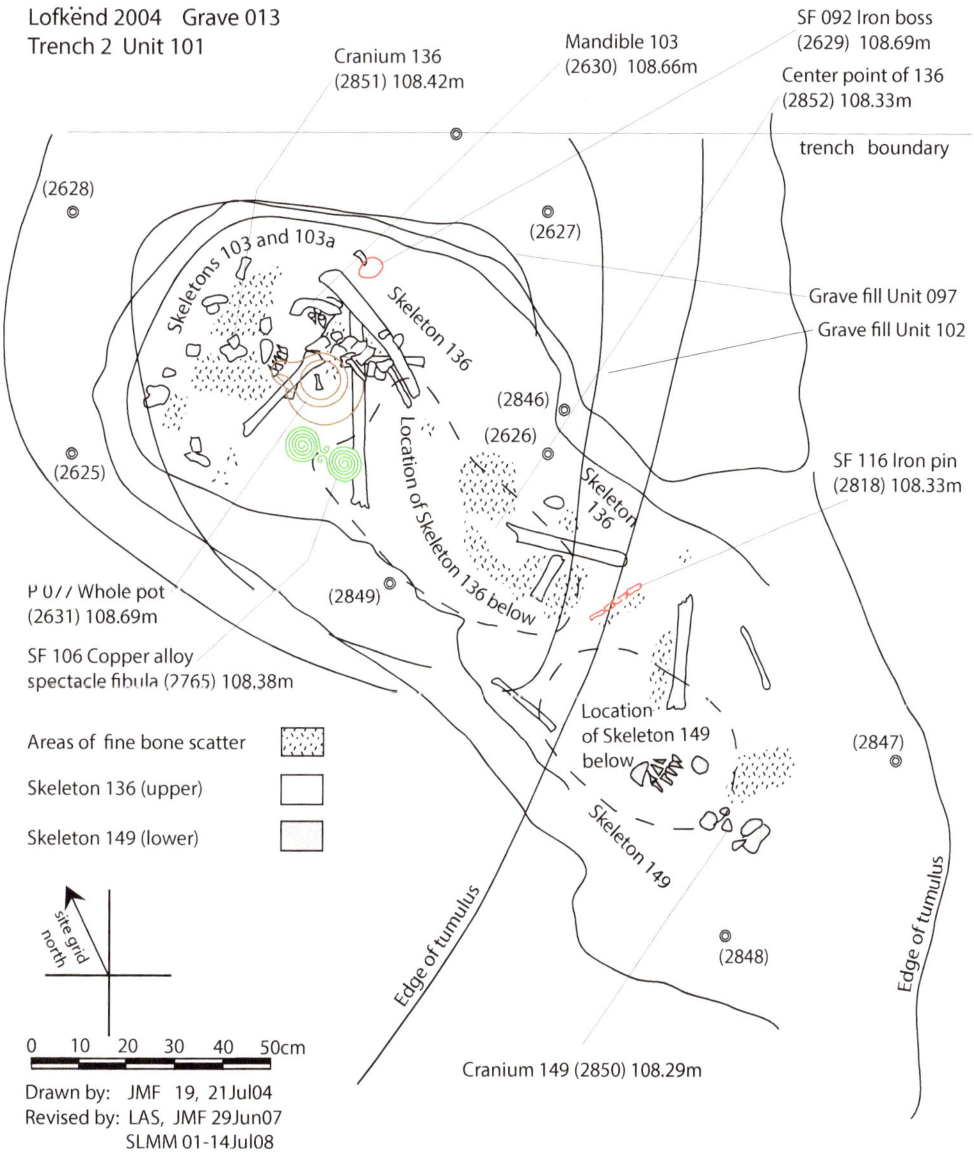

Figure 3.229 Plan of Tomb LXVIII

a

b

c

d

FIGURE 3.230 Tomb LXVIII: (a) (photo 234) uppermost cranium as encountered, with pot (**TLXVIII-1**) and iron boss (**TLXVIII-2**) from south; (b) (photo 272) after further clearing of the northwest portion of tomb, showing bronze fibula (**TLXVIII-3**), from north; (c) (photo 373) after further cleaning, showing iron pin (**TLXVIII-4**), from southwest; (d) (photo 405) lowest level of tomb as exposed prior to lifting skeleton, from west

Figure 3.231 Tomb LXVIII: **TLXVIII-1**

Figure 3.232 Tomb LXVIII: **TLXVIII-2**

Figure 3.233 Tomb LXVIII: **TLXVIII-3**

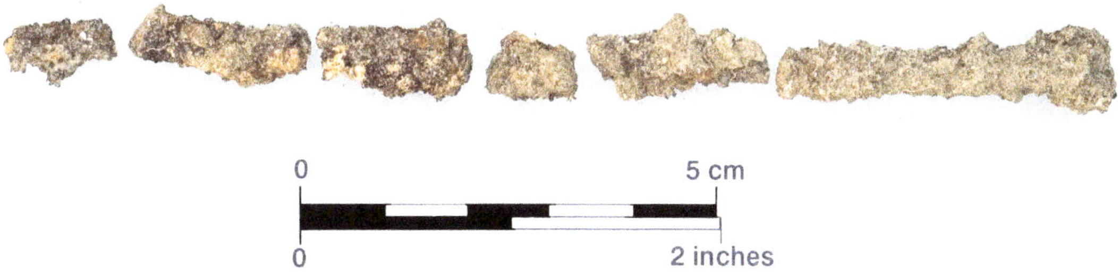

Figure 3.234 Tomb LXVIII: **TLXVIII-4**

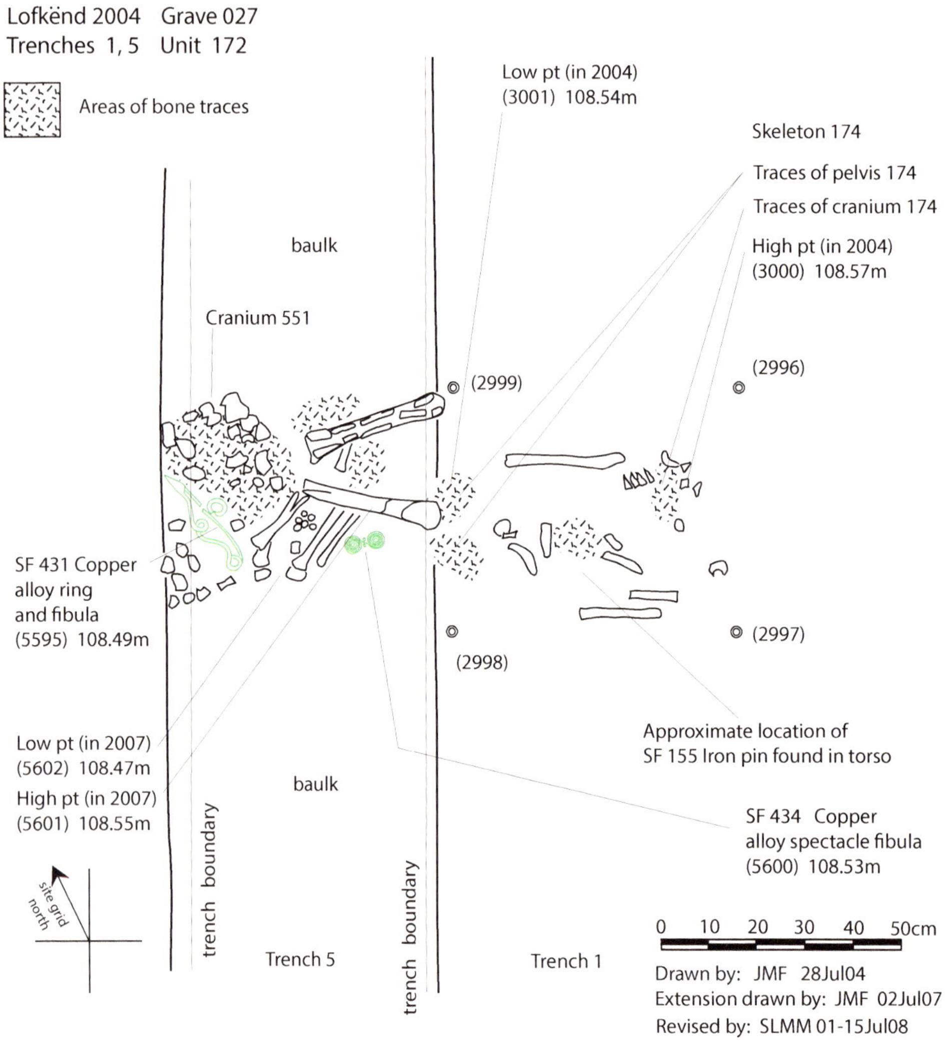

FIGURE 3.235 Plan of Tomb LXIX

a

b

Figure 3.236 Tomb LXIX: (a) (photo 465) as exposed in 2004, from above west; (b) (photo 1707), portion of tomb as uncovered in baulk in 2007, from above south

FIGURE 3.237 Tomb LXIX: **TLXIX-1**

FIGURE 3.238 Tomb LXIX: **TLXIX-2**

FIGURE 3.239 Tomb LXIX: **TLXIX-3**

FIGURE 3.240 Tomb LXIX: **TLXIX-4**

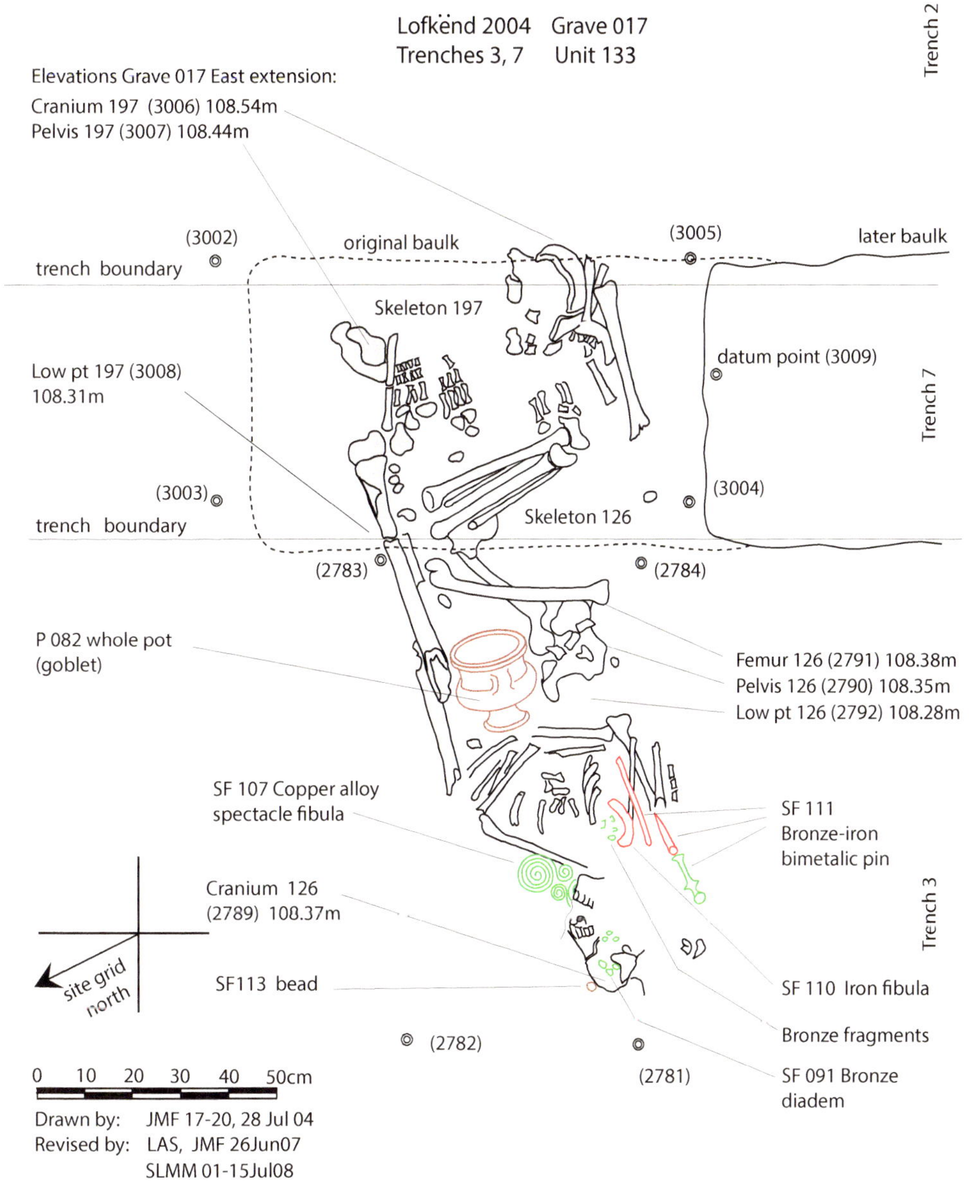

FIGURE 3.241 Plan of Tomb LXX

FIGURE 3.242 Tomb LXX: (a) (photo 282) during early stage of excavation showing grave goods associated with adolescent female, from east-northeast; (b) (photo 321) after removal of grave goods, from east-northeast; (c) (photo 403) showing discoloration below skeleton, perhaps from textile, from east-northeast; (d) (photo 653) showing continuation of tomb in Sector 7 and second (male) individual, from north

FIGURE 3.243 Tomb LXX: **TLXX-1**

FIGURE 3.244 Tomb LXX: **TLXX-2**

FIGURE 3.245 Tomb LXX: (a) **TLXX-2a**; (b) **TLXX-2b**

FIGURE 3.246 Tomb LXX: **TLXX-3**

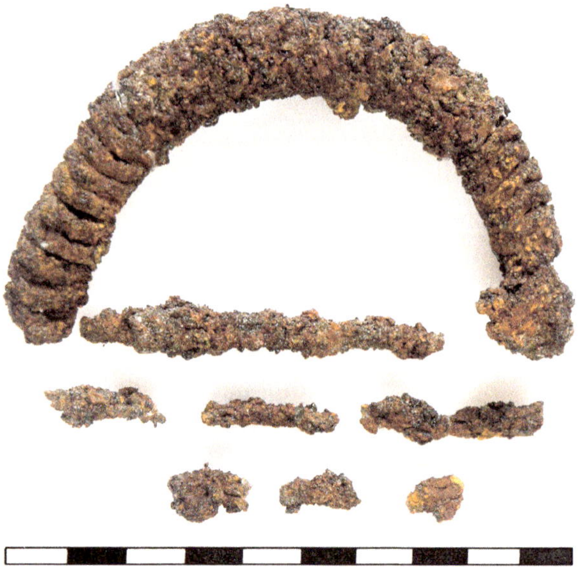

Figure 3.247 Tomb LXX: **TLXX**-4

Figure 3.248 Tomb LXX: (a–b) **TLXX-5**

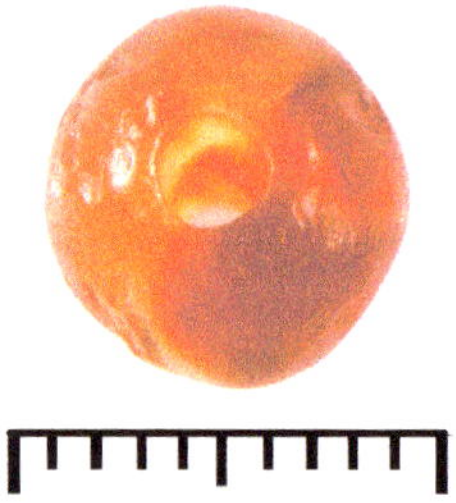

Figure 3.249 Tomb LXX: **TLXX-6**

Figure 3.250 Tomb LXX: **TLXX-7**

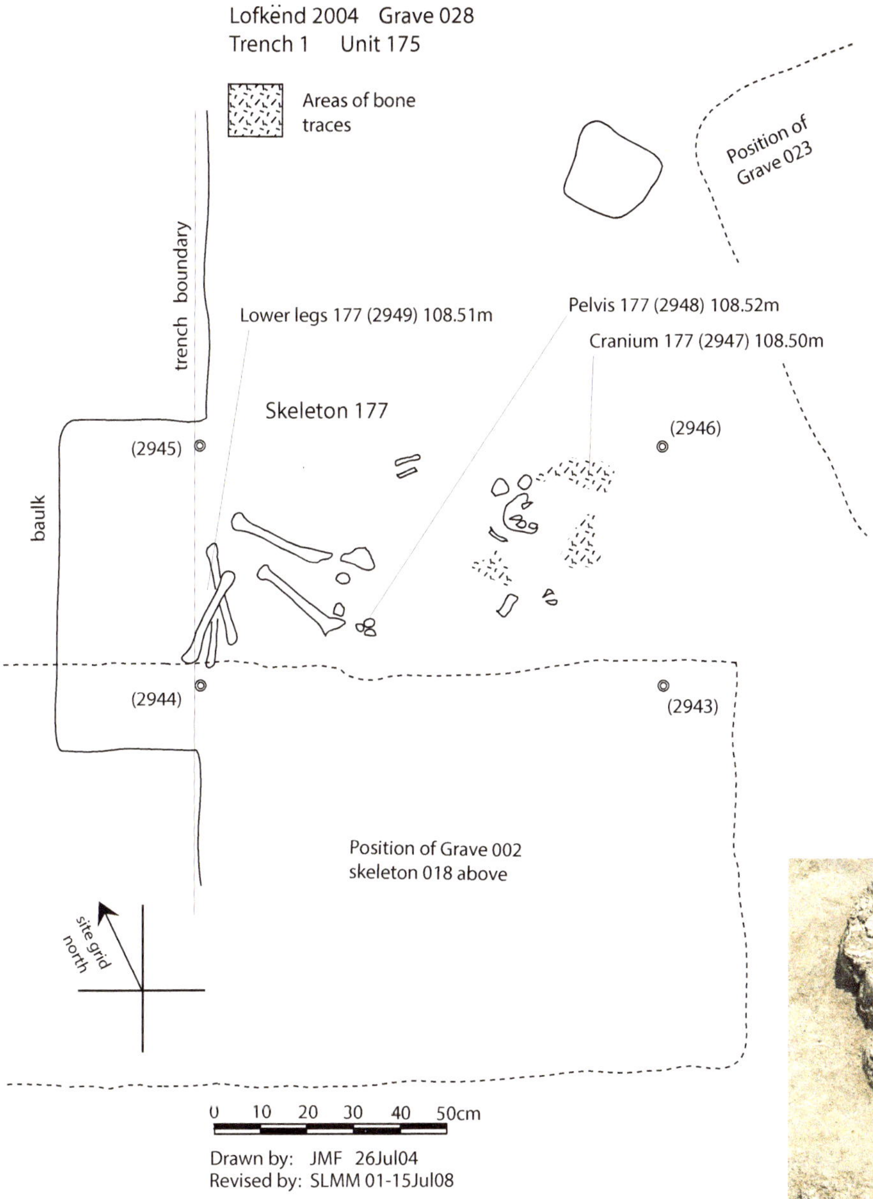

FIGURE 3.251 Plan of Tomb LXXI

FIGURE 3.252 Tomb LXXI (photo 467) from above west

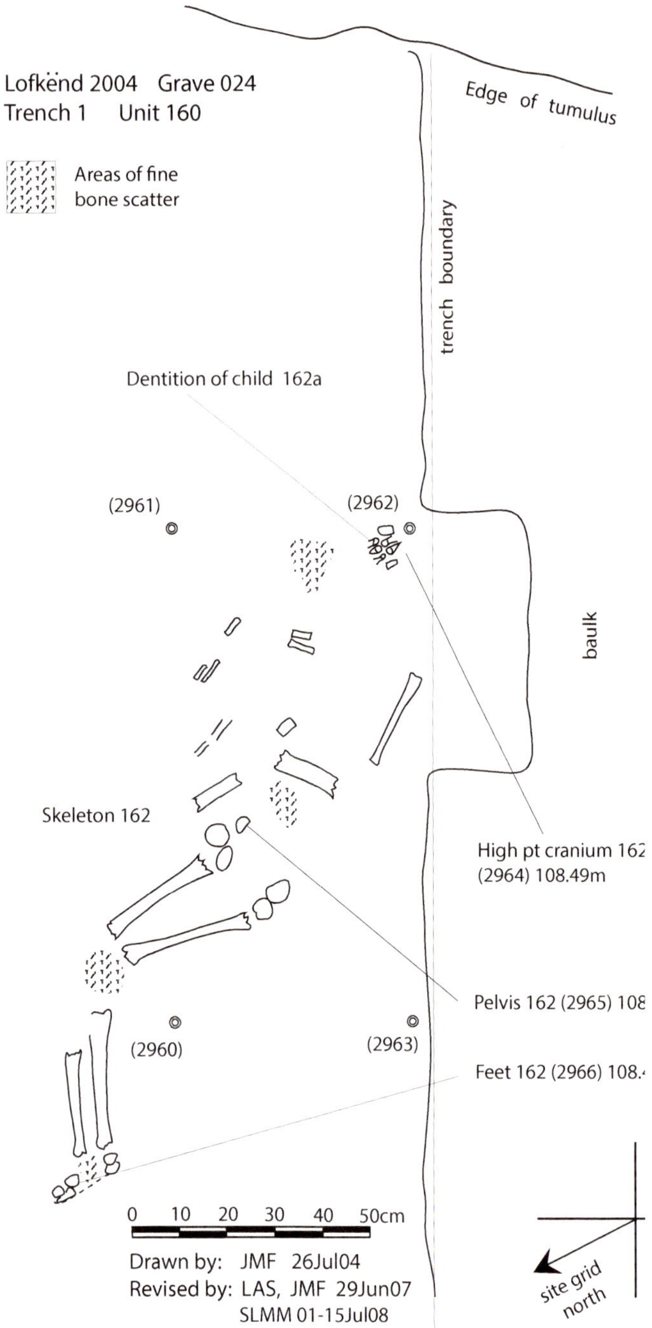

FIGURE 3.253 Plan of Tomb LXXII

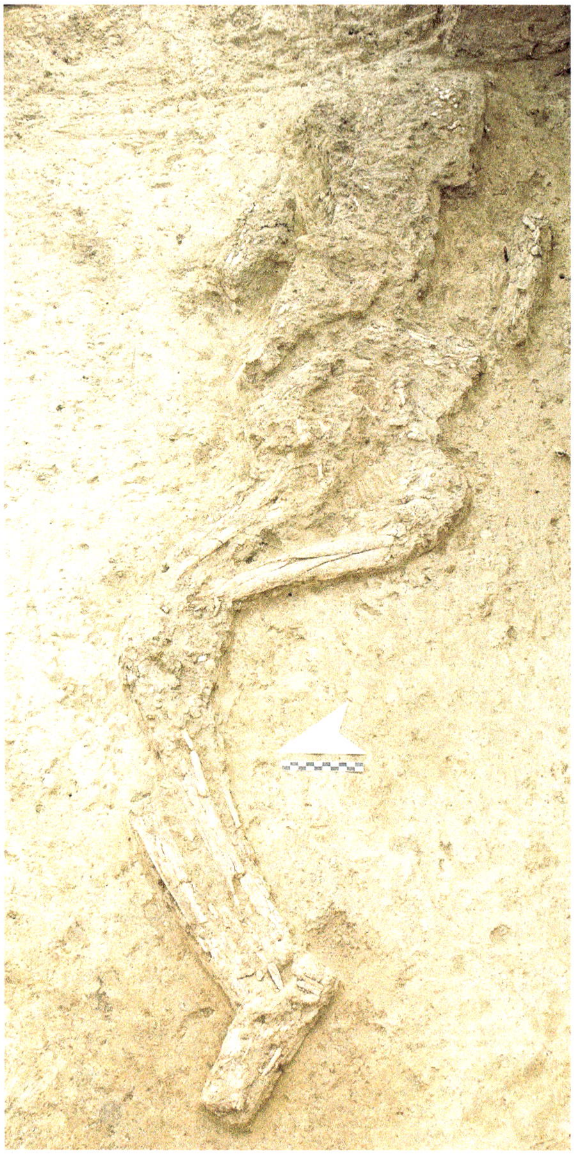

a

b

(*Right, top and bottom)* FIGURE 3.254 Tomb LXXII: (a) (photo 481) from west-northwest; (b) (photo 435) detail of juvenile teeth encountered at uppermost level of tomb, from above north

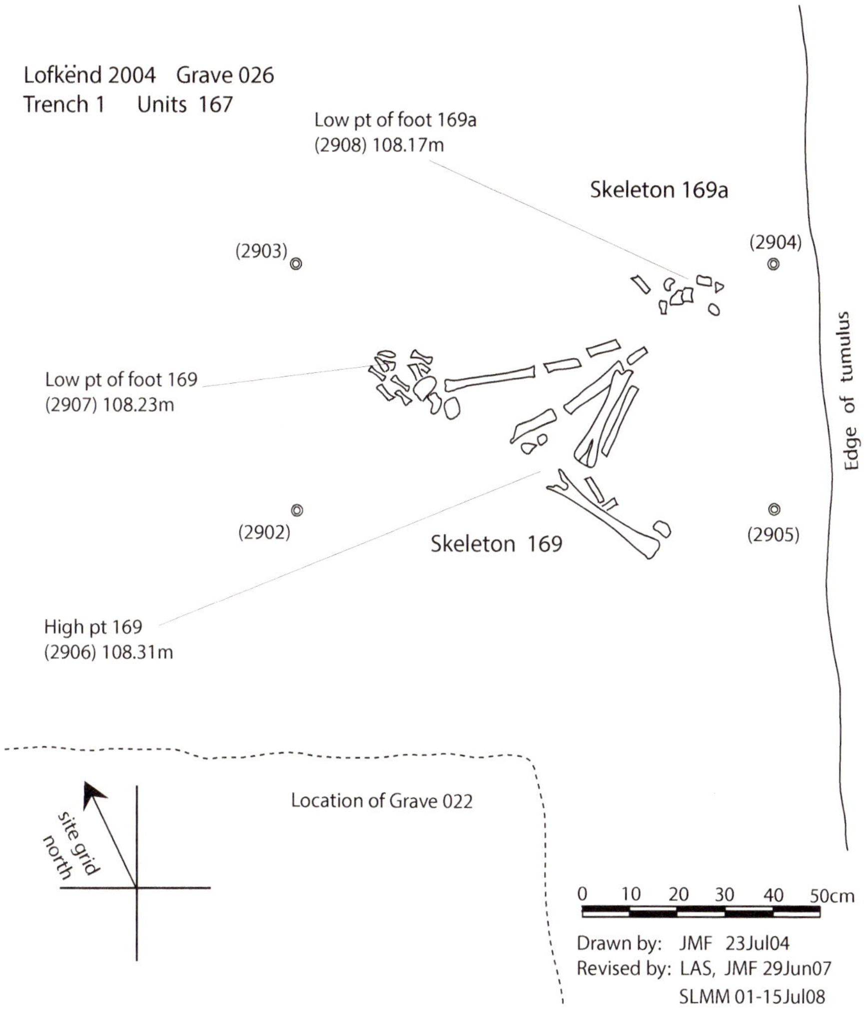

Figure 3.255 Plan of Tomb LXXIII

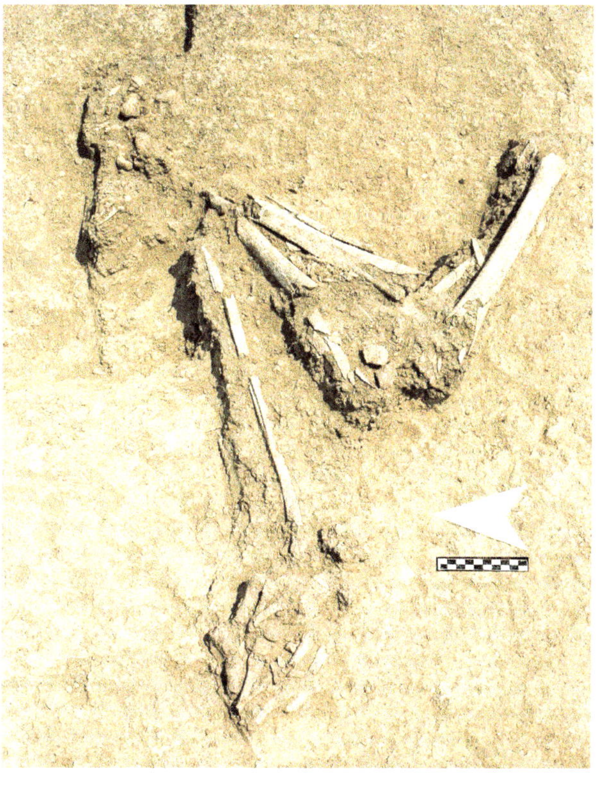

Figure 3.256 Tomb LXXIII (photo 449) from west

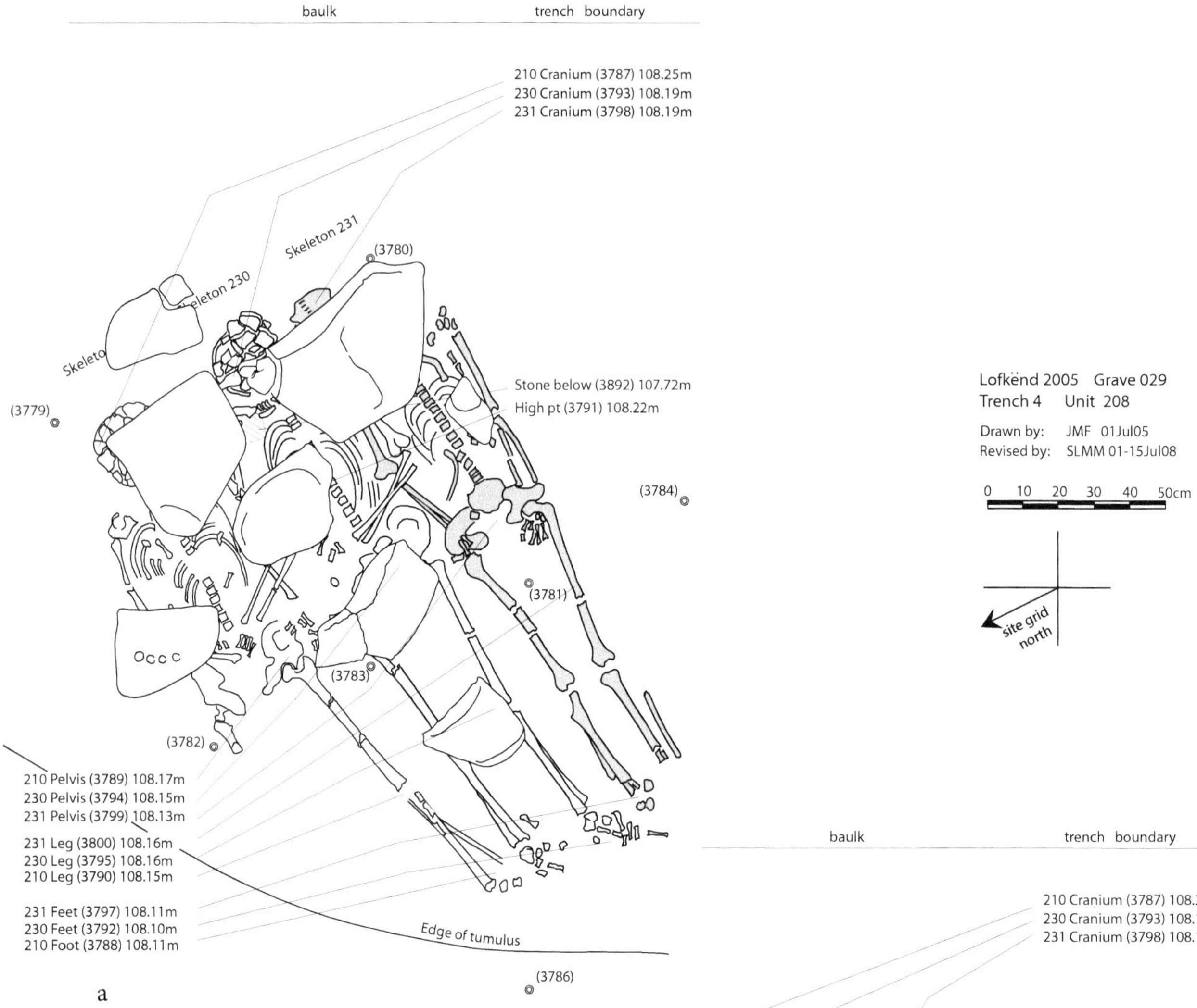

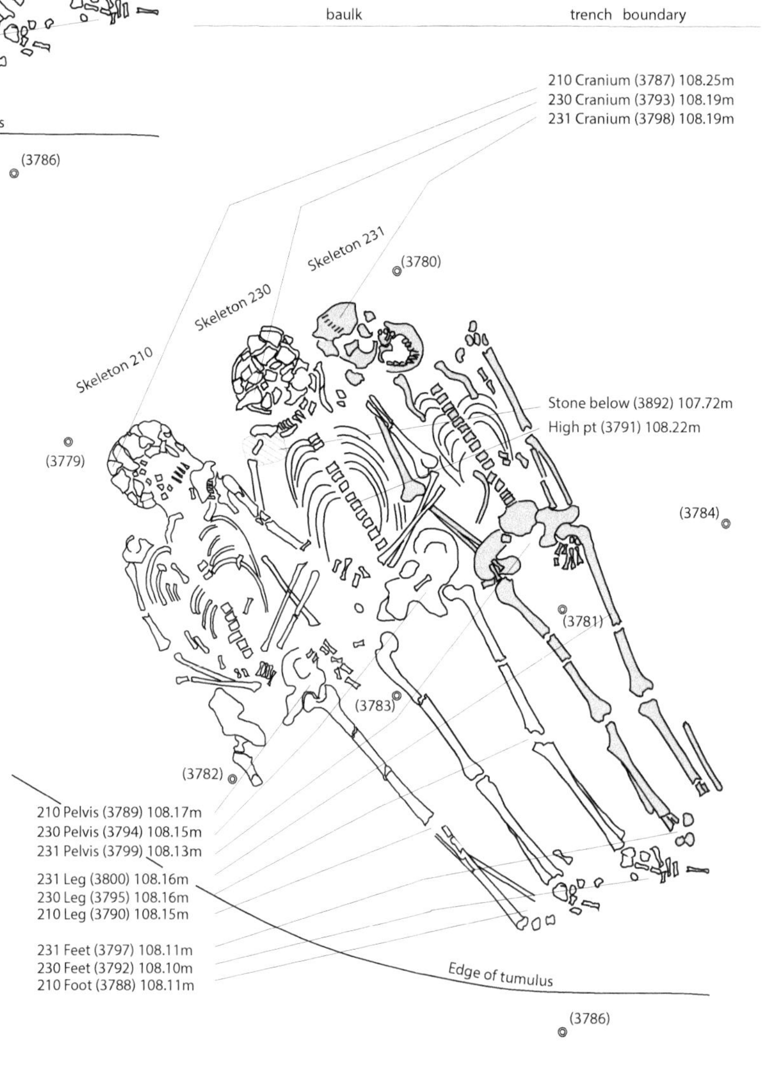

FIGURE 3.257 Plans of Tomb LXXIV: (a) Stage 1; (b) Stage 2

a

b

c

FIGURE 3.258 Tomb LXXIV: (a) (photo 841) during excavation showing cover stones, from southwest; (b) tomb as fully exposed (photo 871) from northeast; (c) as previous, (photo 873) from southwest

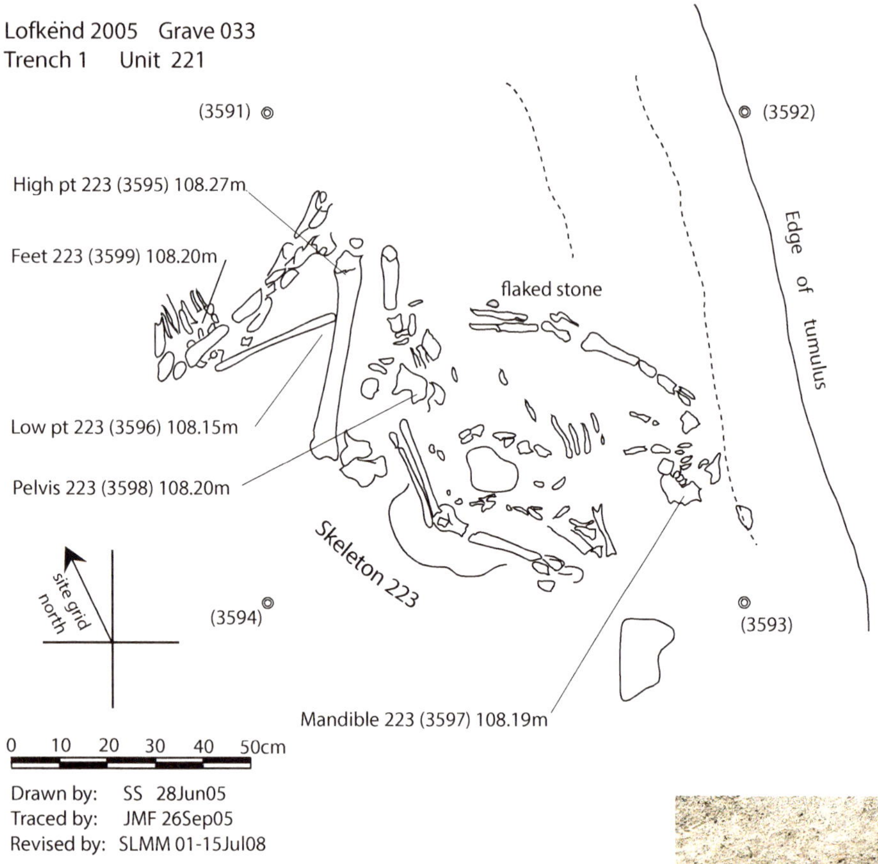

FIGURE 3.259 Plan of Tomb LXXV

FIGURE 3.260 Tomb LXXV (photo 850) from west-northwest

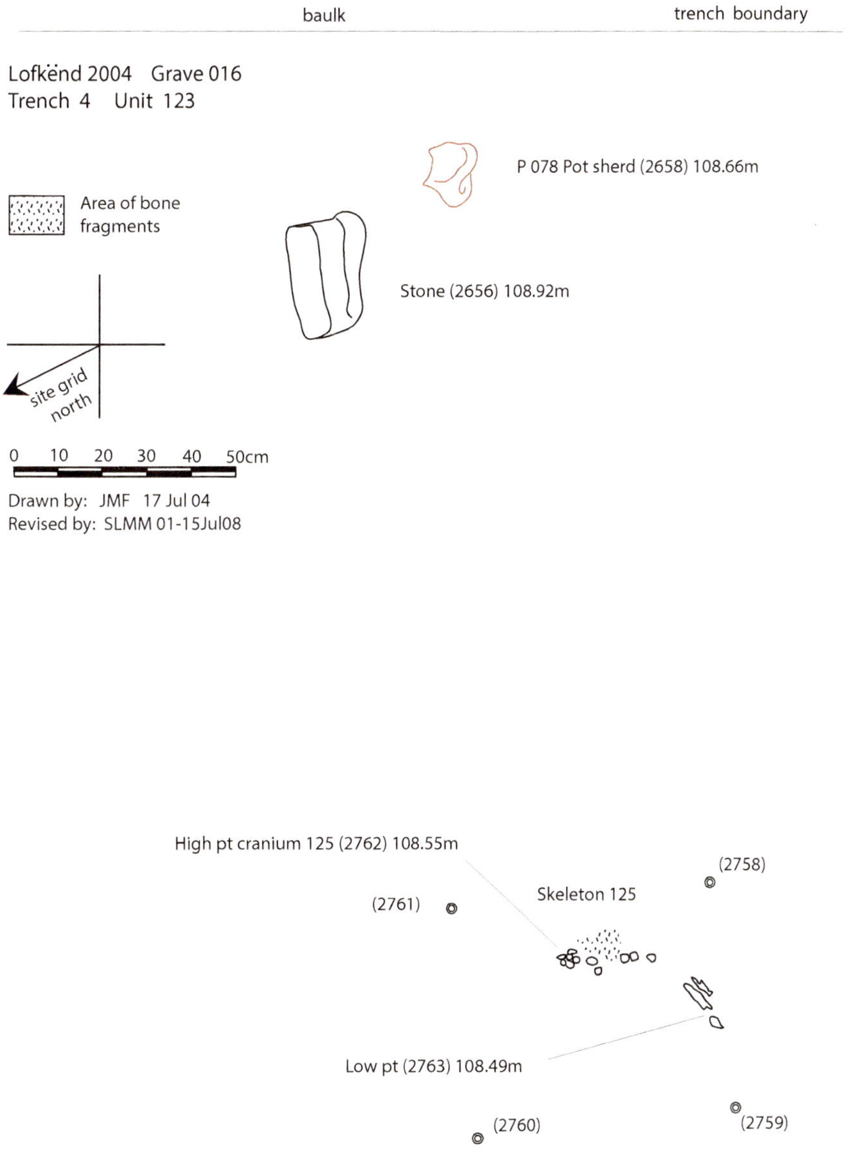

FIGURE 3.261 Plan of Tomb LXXVI as preserved

FIGURE 3.262 Tomb LXXVI (photo 261) from above west

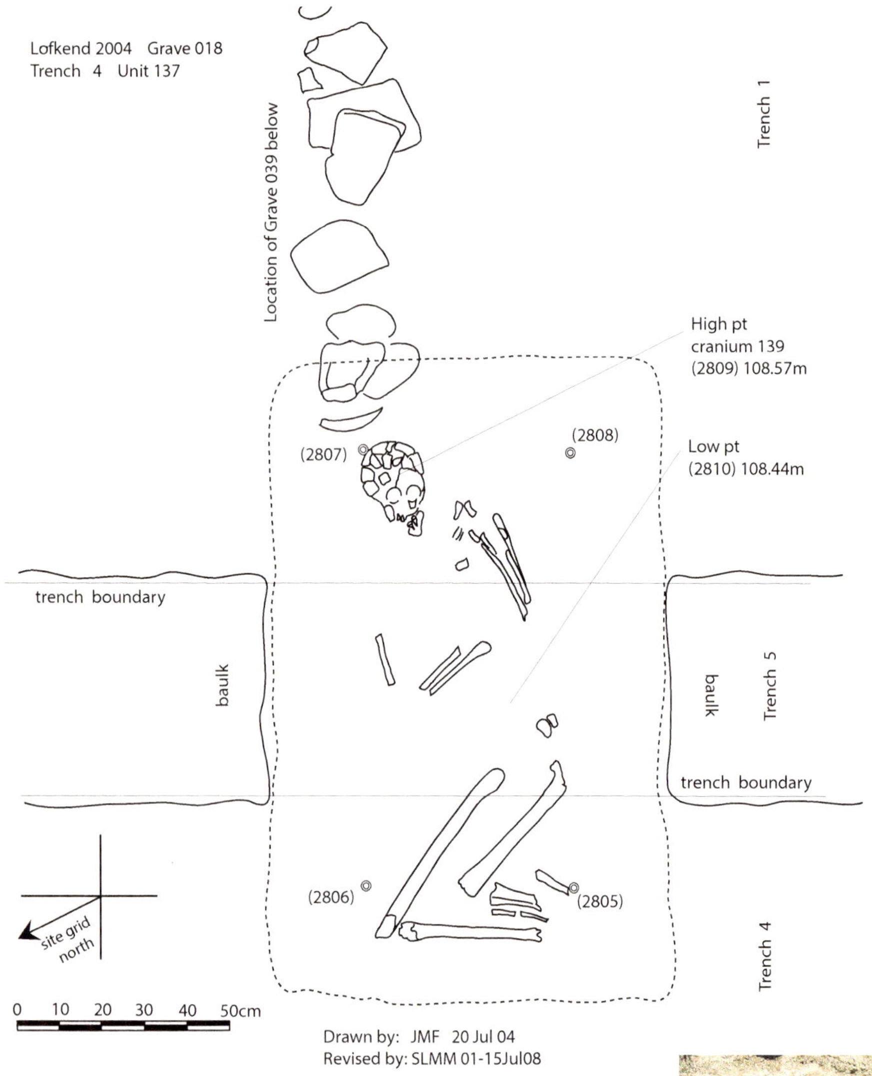

FIGURE 3.263 Plan of Tomb LXXVII

FIGURE 3.264 Tomb LXXVII (photo 362) from west (with excavation cut)

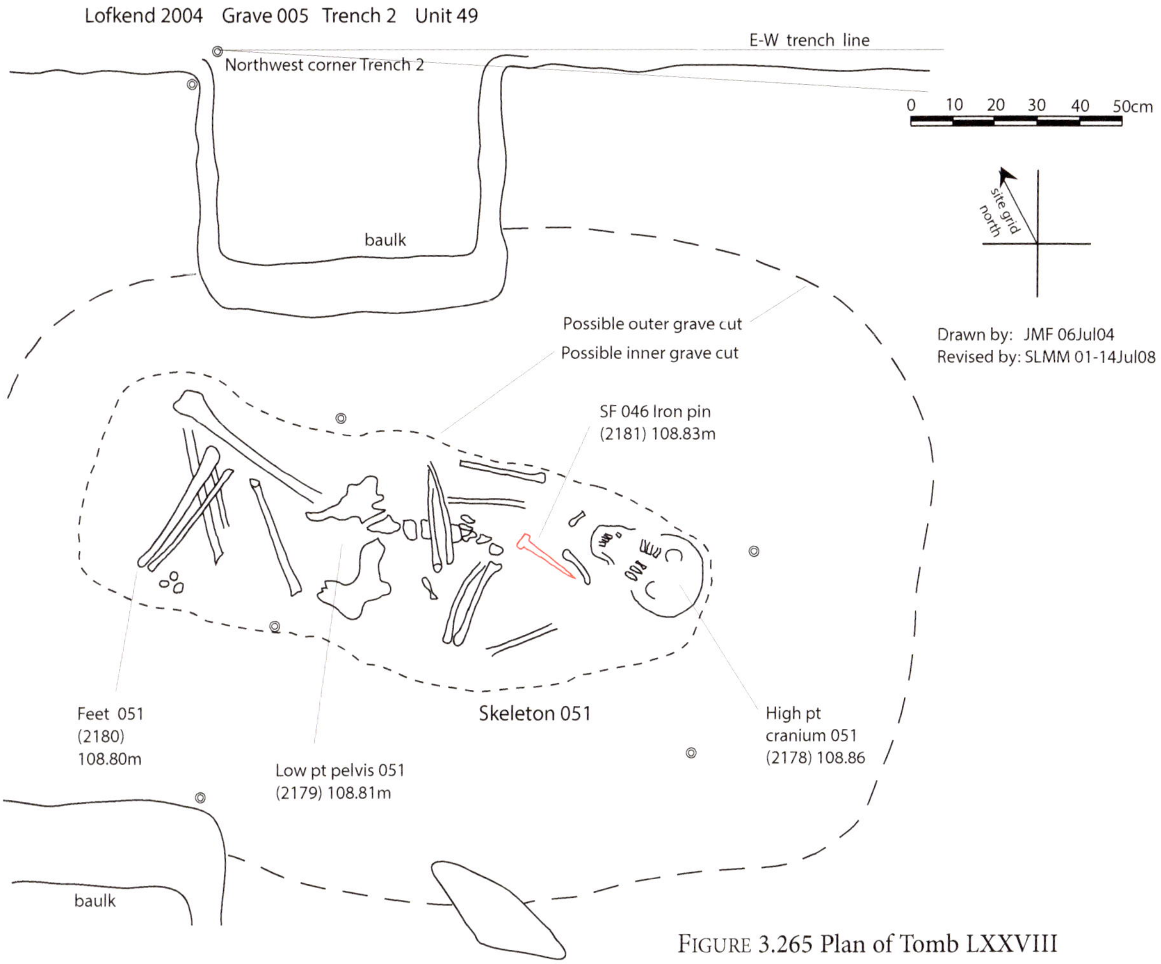

FIGURE 3.265 Plan of Tomb LXXVIII

a   b

FIGURE 3.266 Tomb LXXVIII: (a) (photo 183) from west; (b) (photo 180), detail from west

FIGURE 3.267 Tomb LXXVIII: **TLXXVIII-1**

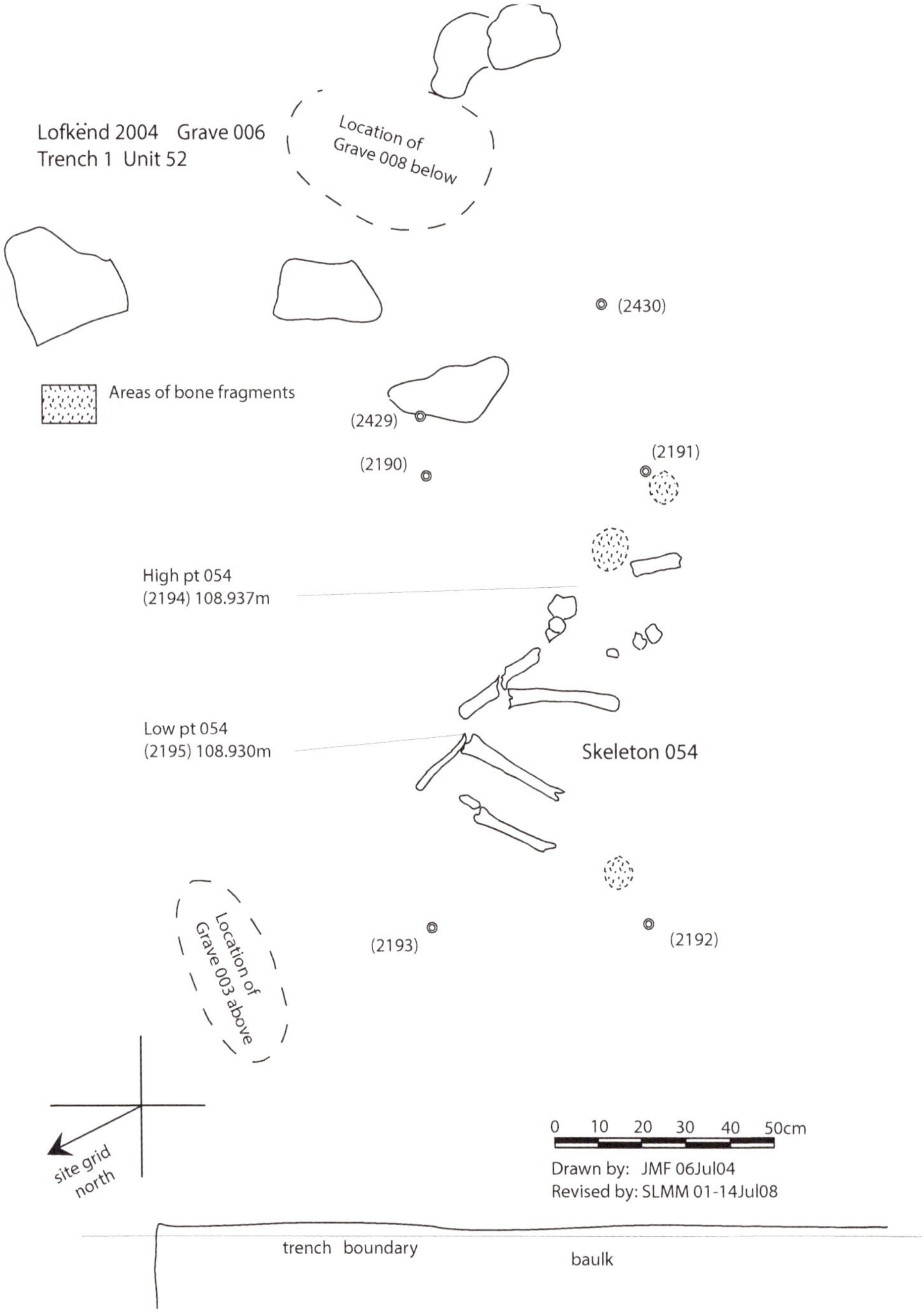

FIGURE 3.268 Plan of Tomb LXXIX

Figure 3.269 Tomb LXXIX (photo 186) from southwest

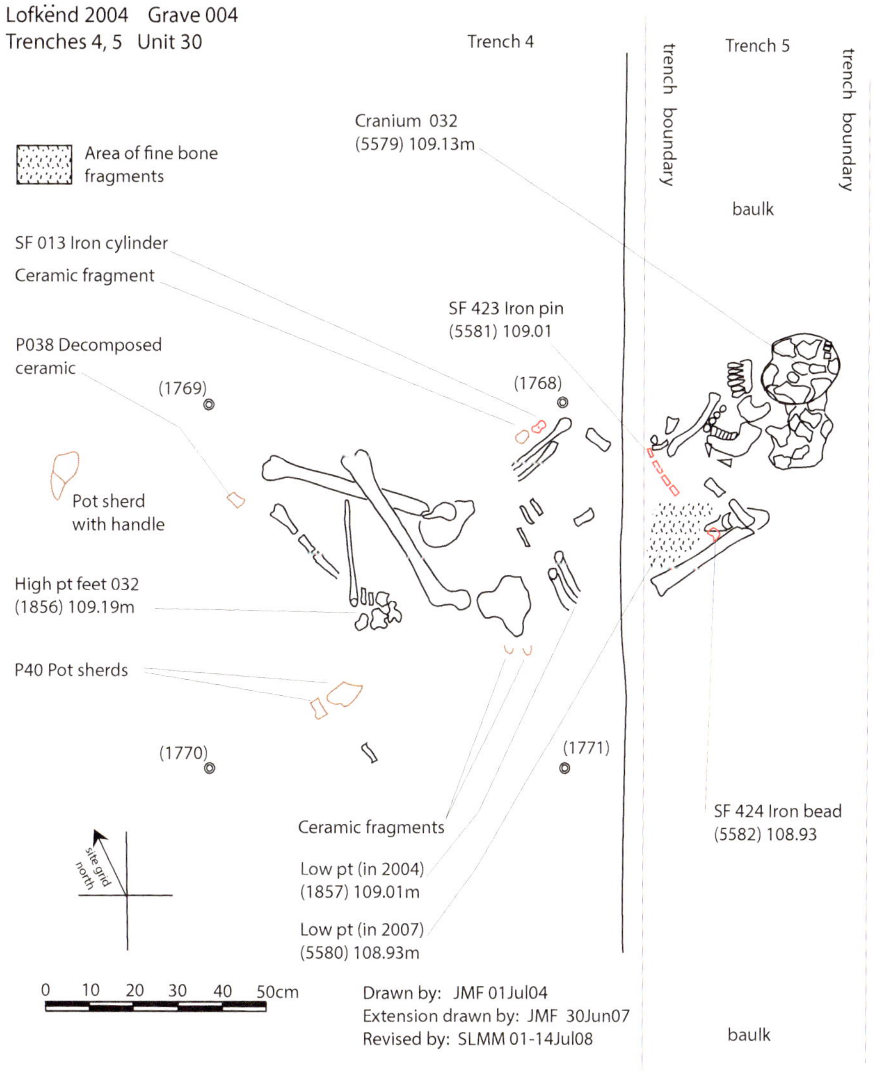

Figure 3.270 Plan of Tomb LXXX

a

b

FIGURE 3.271 Tomb LXXX:
(a) (photo 167) postcranial bones as encountered in 2004, from north;
(b) (photo 1694) cranium and upper torso as exposed in 2007, from south

FIGURE 3.272 Tomb LXXX: **TLXXX-1**

FIGURE 3.273 Tomb LXXX: **TLXXX-2**

FIGURE 3.274 Tomb LXXX: **TLXXX-3**

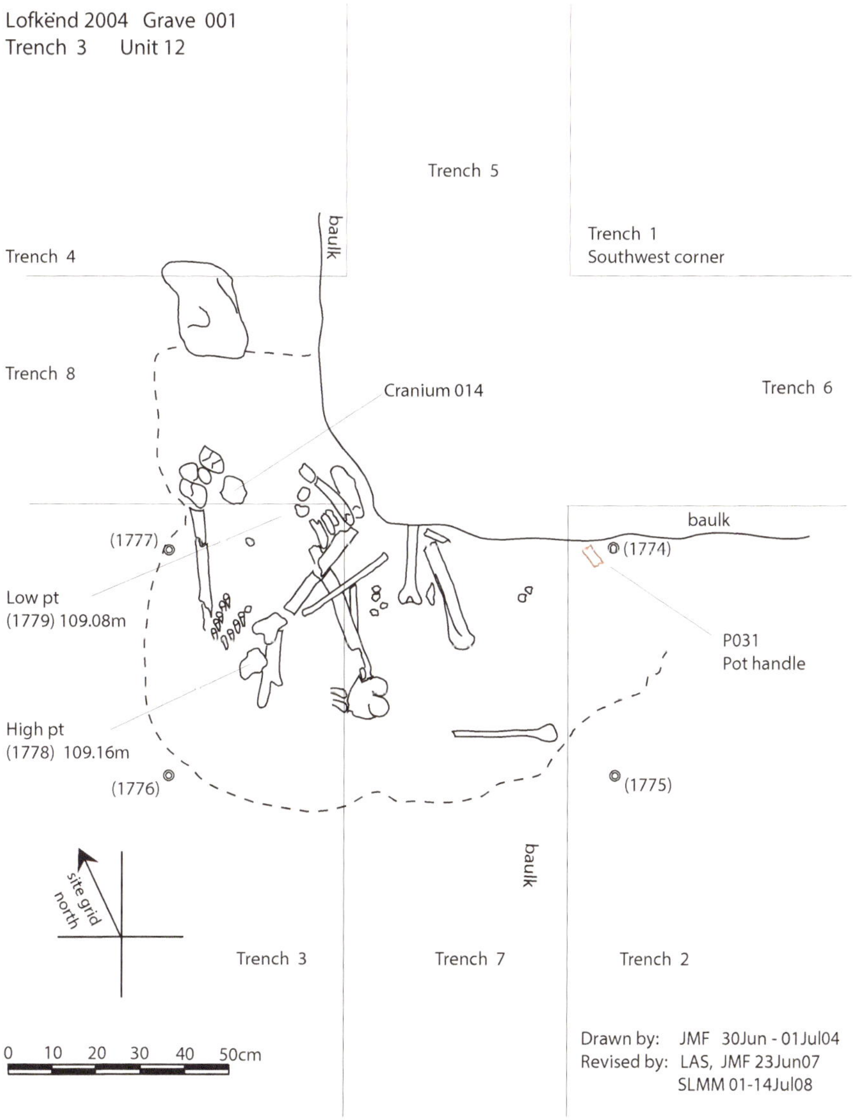

FIGURE 3.275 Plan of Tomb LXXXI

FIGURE 3.276 Tomb LXXXI (photo 143) from south

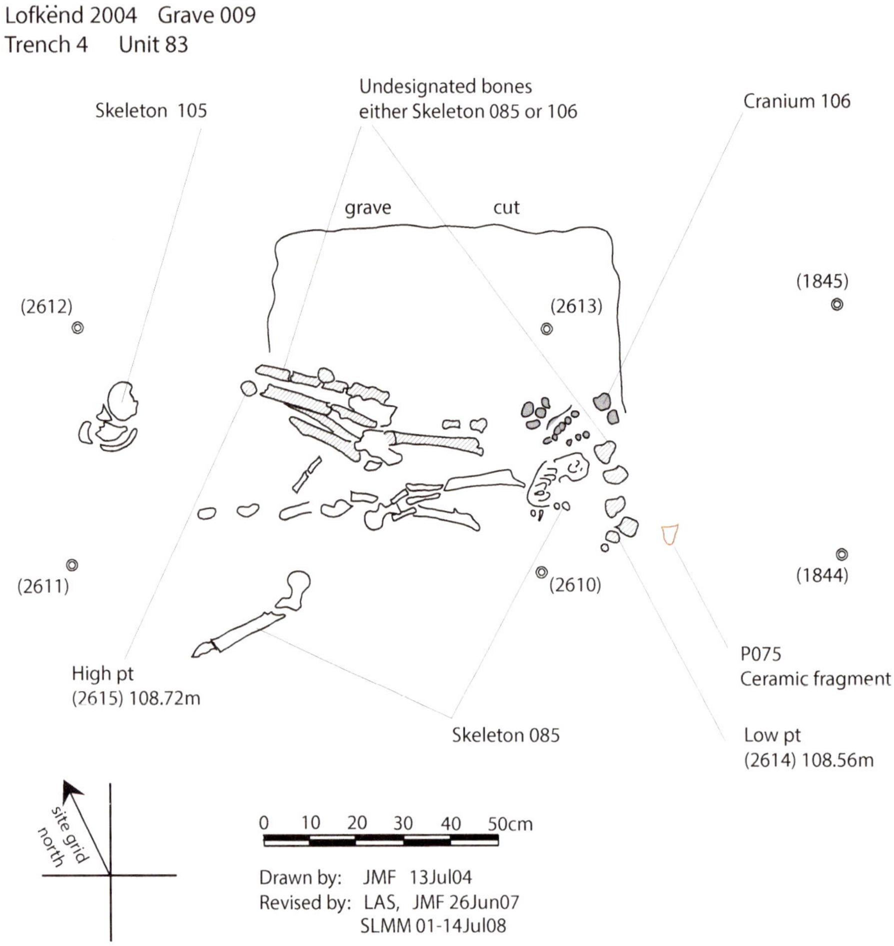

FIGURE 3.277 Plan of Tomb LXXXII

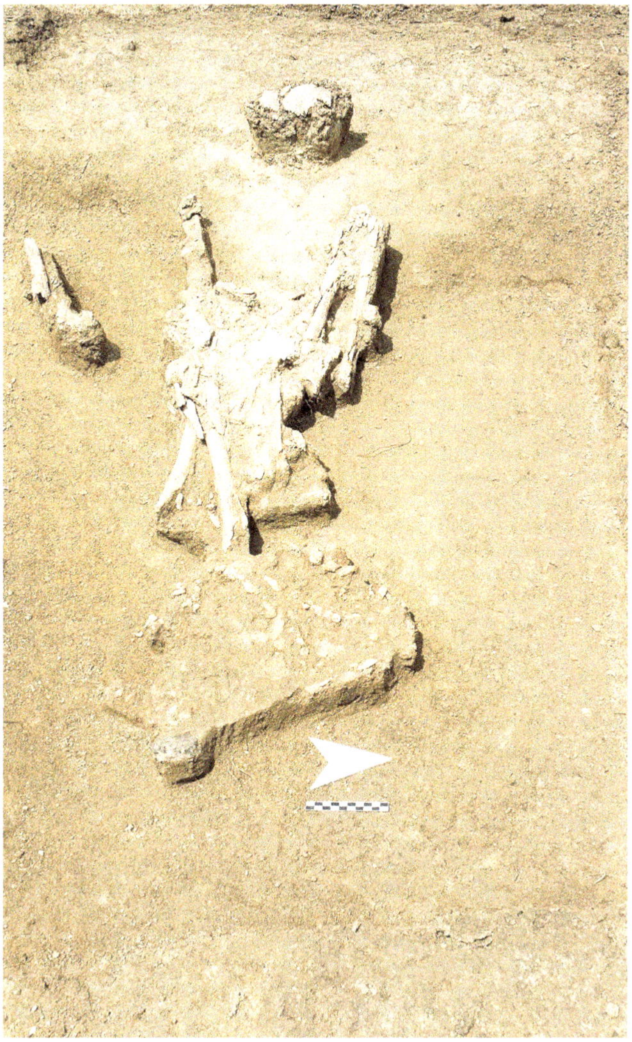

Figure 3.278 Tomb LXXXII (a) from east (photo 222); (b) from west (photo 223)

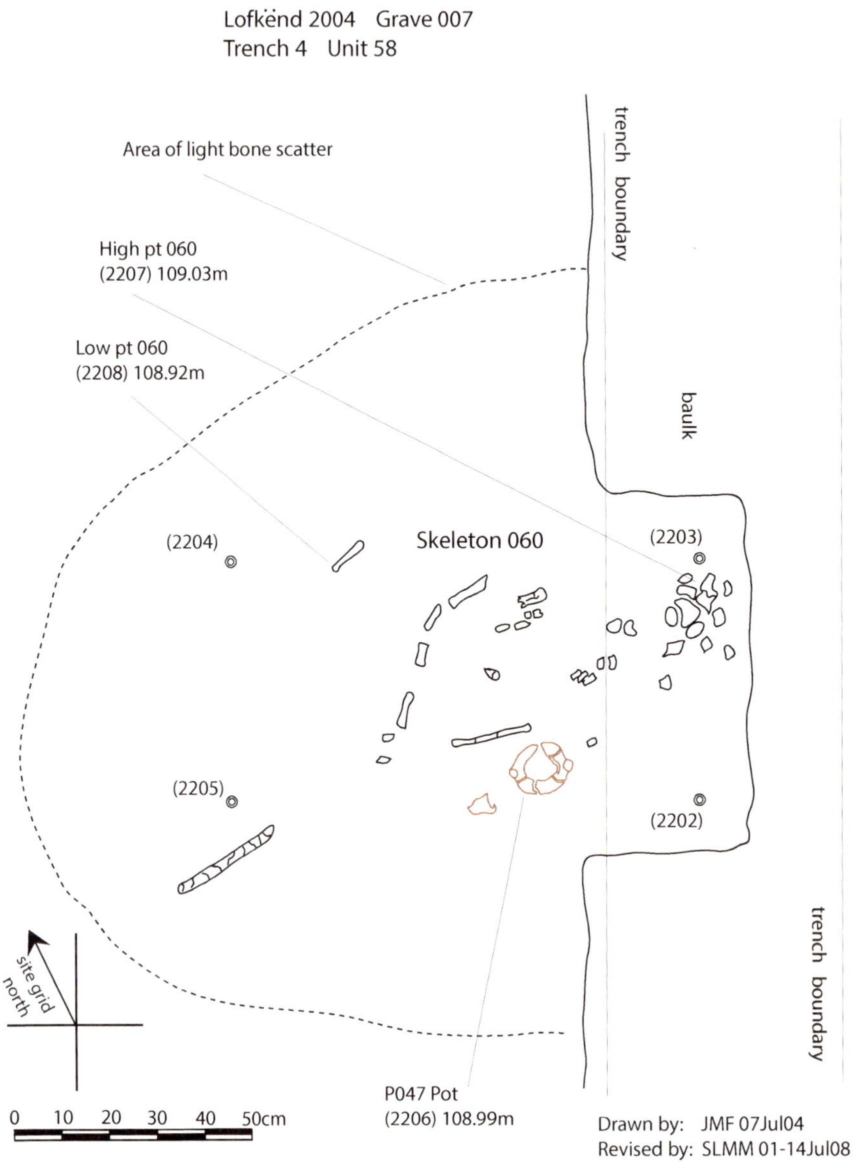

FIGURE 3.279 Plan of Tomb LXXXIII

a

b

Figure 3.280 Tomb LXXXIII: (a) (photo 191) from north; (b) (photo 194), detail from above east

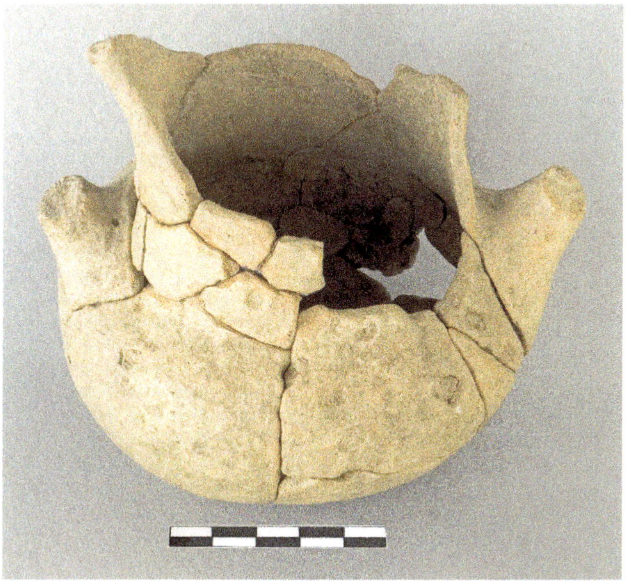

Figure 3.281 Tomb LXXXIII: **TLXXXIII-1**

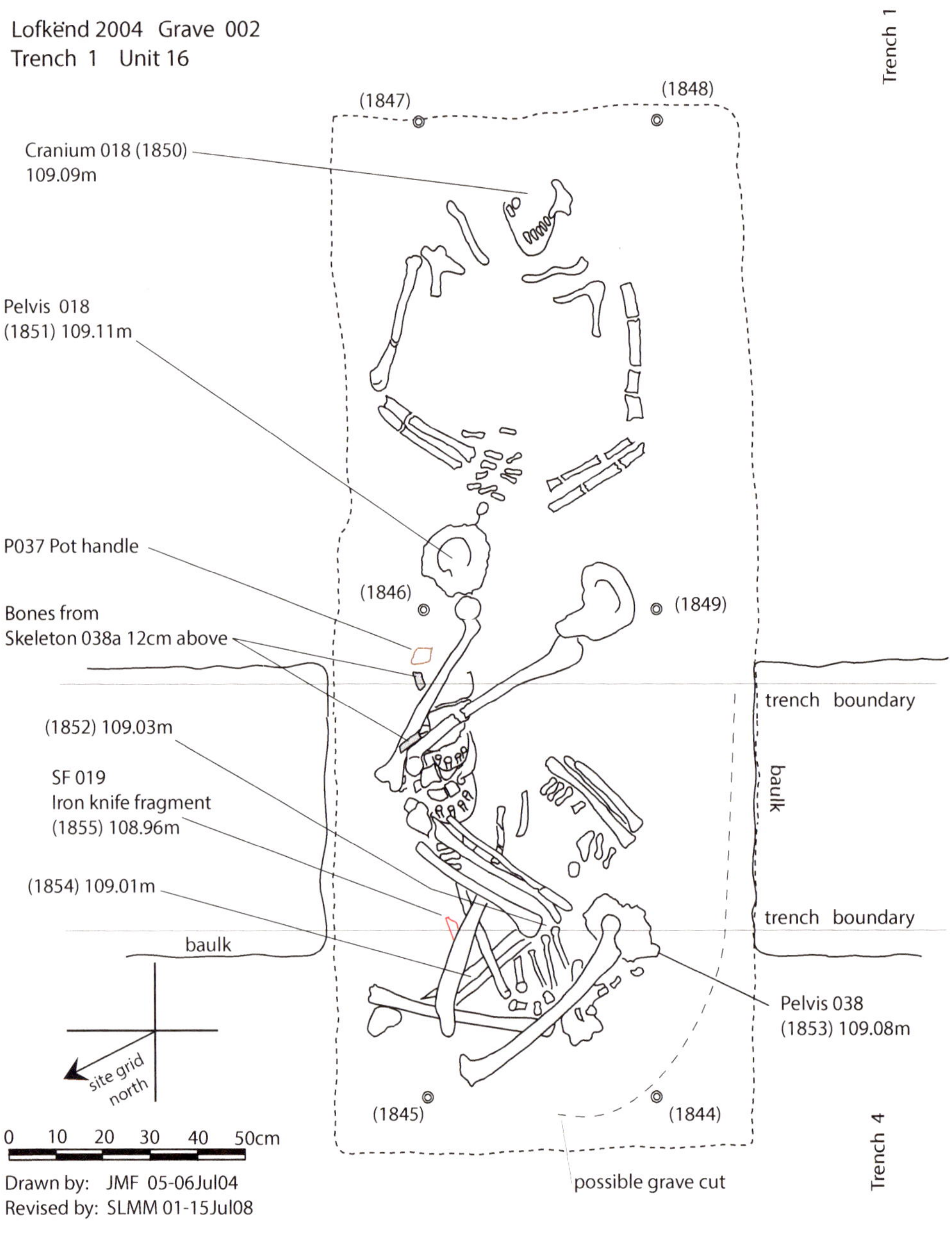

FIGURE 3.282 Plan of Tomb LXXXIV

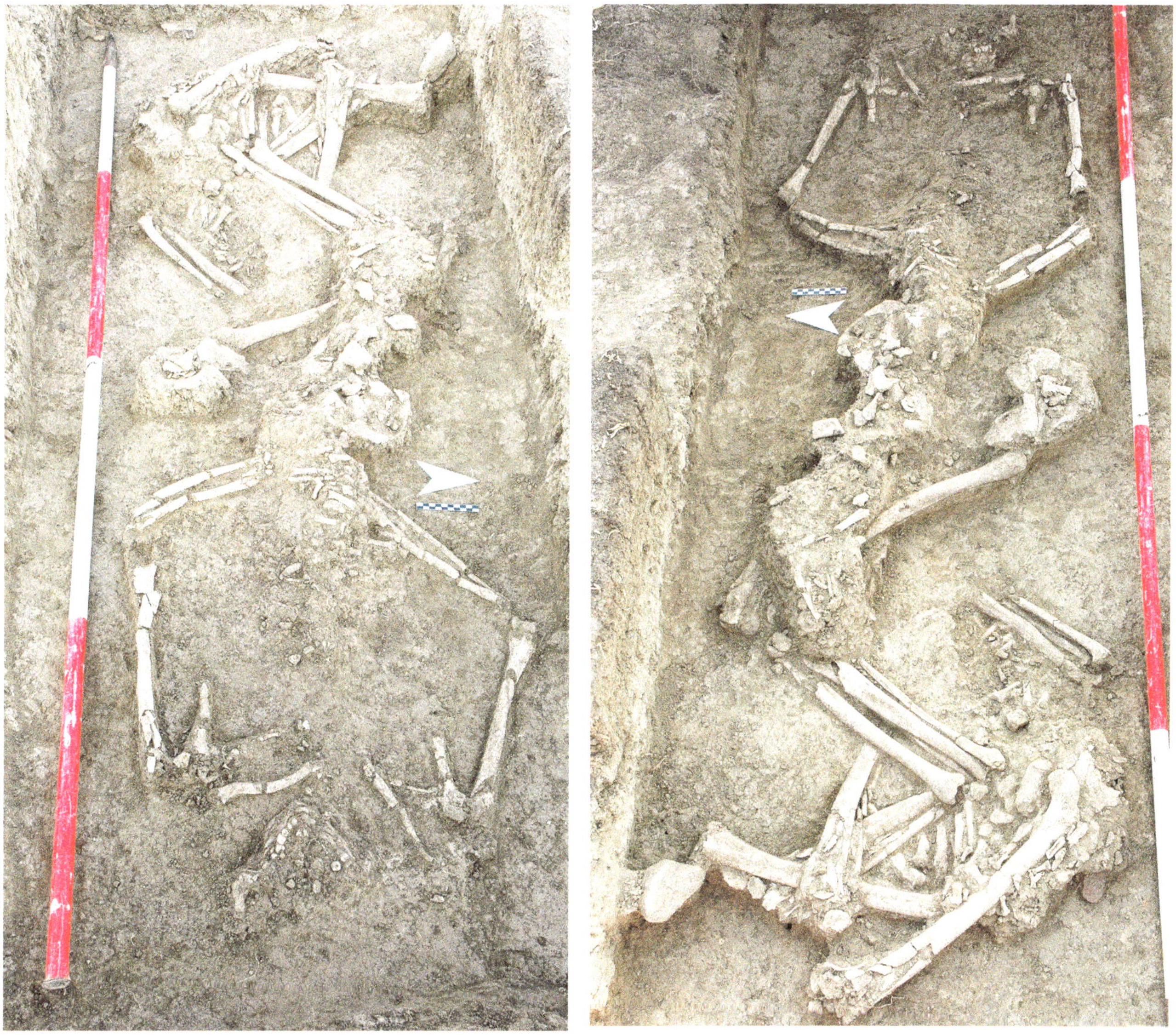

a b

Figure 3.283 Tomb LXXXIV: (a) (photo 153) from east; (b) (photo 154) from west

Figure 3.284 Tomb LXXXIV: **TLXXXIV-1**

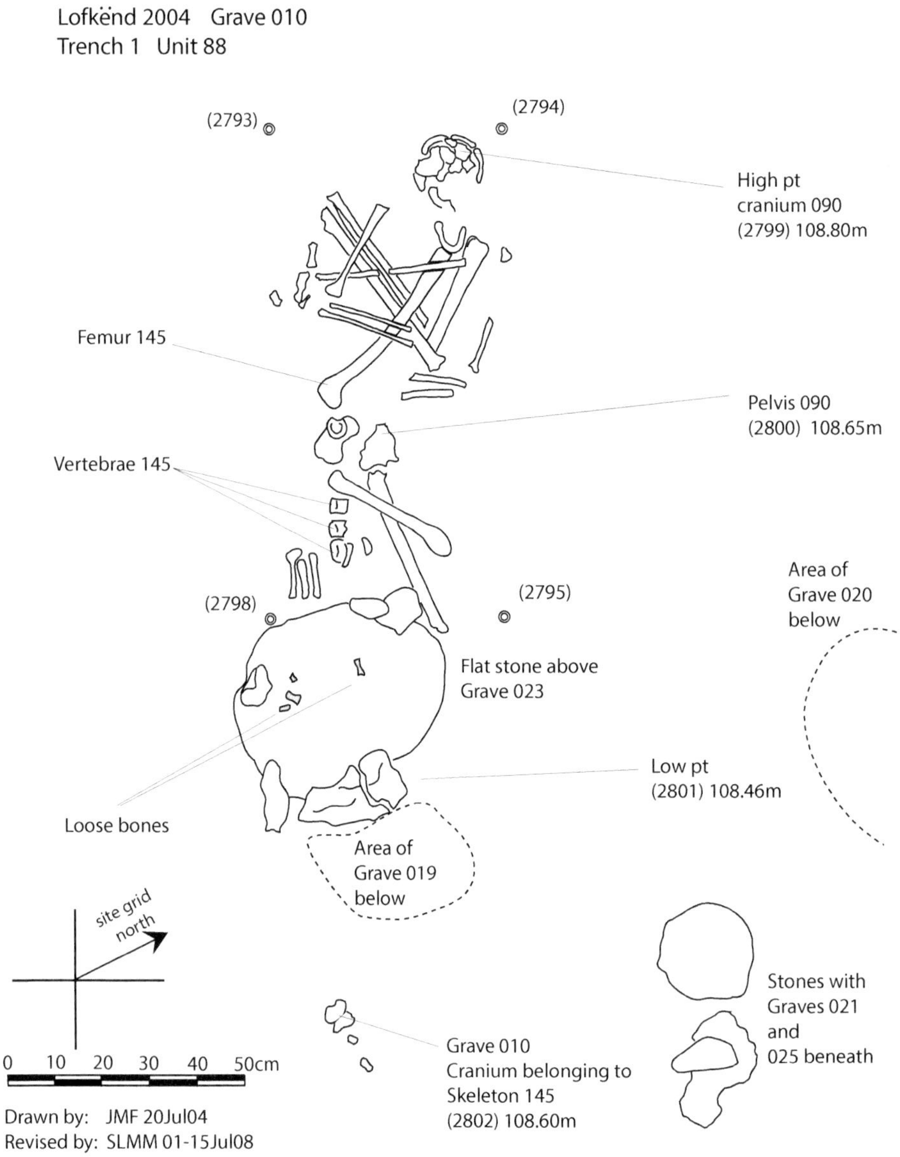

FIGURE 3.285 Plan of Tomb LXXXV

a

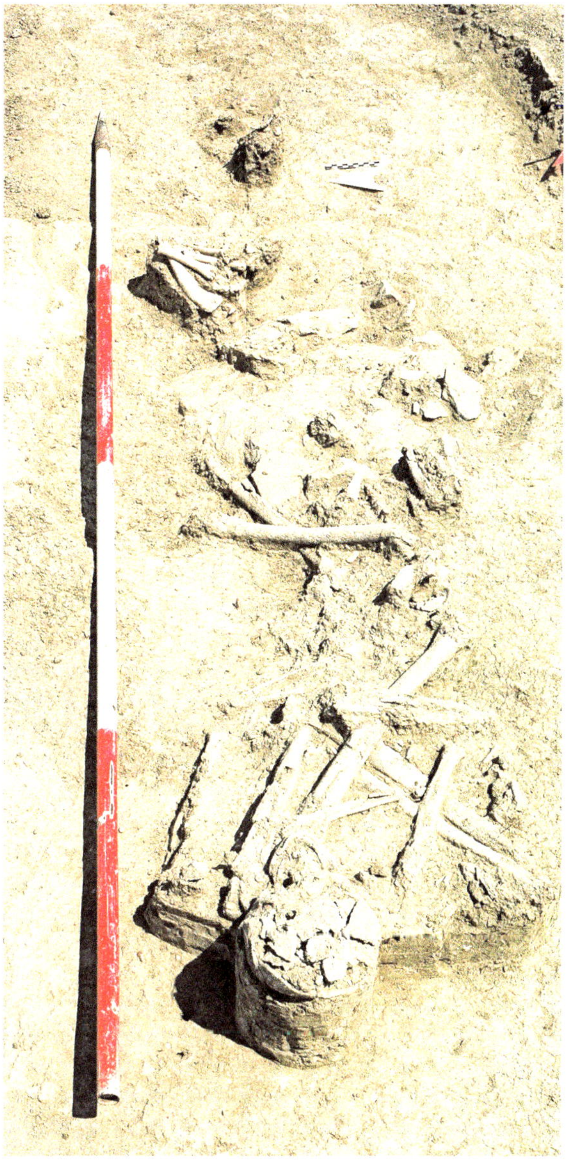
b

Figure 3.286 Tomb LXXXV: (a) (photo 325) from east; (b) (photo 327) from west

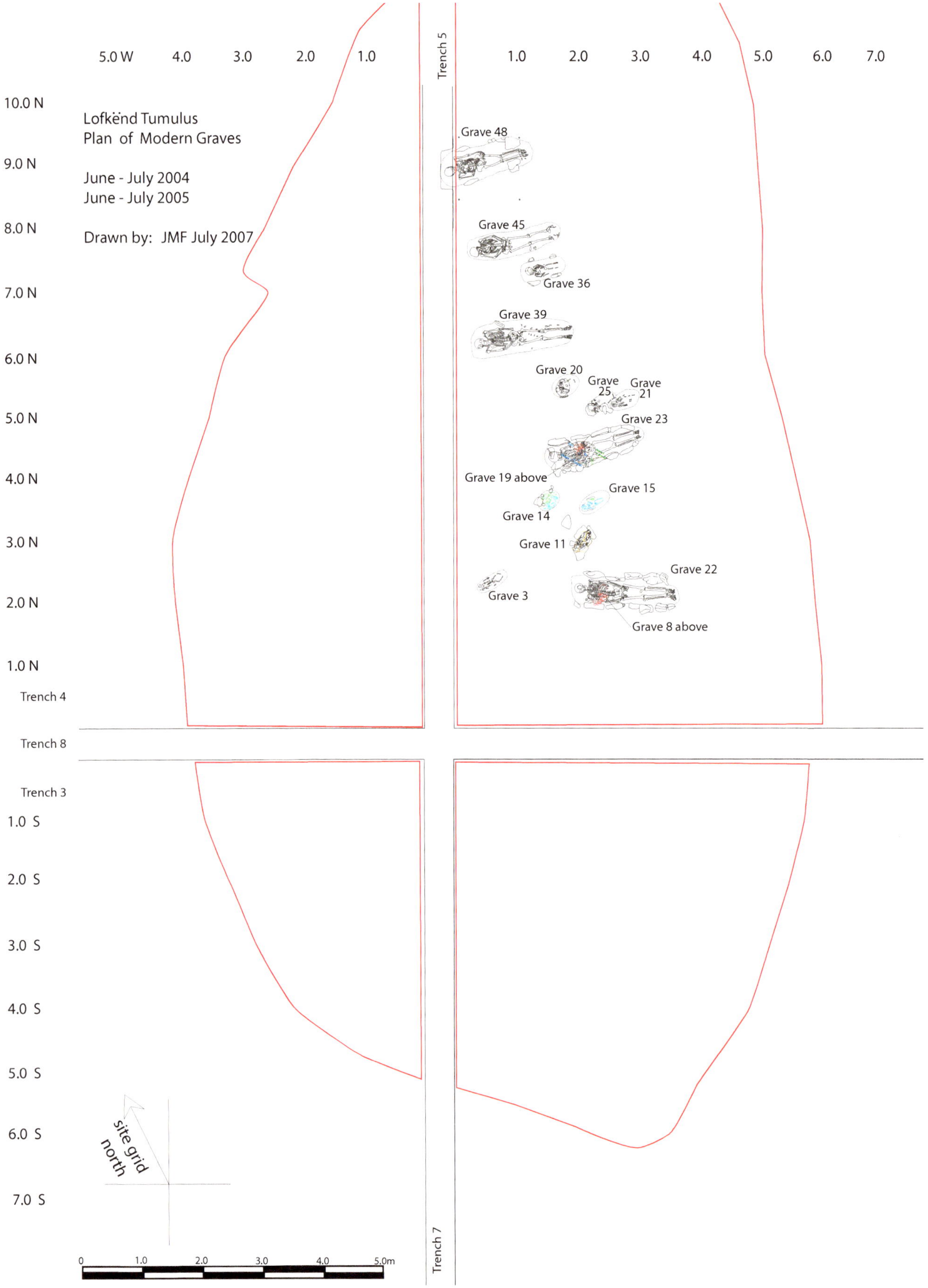

FIGURE 3.287 Plan of modern graves

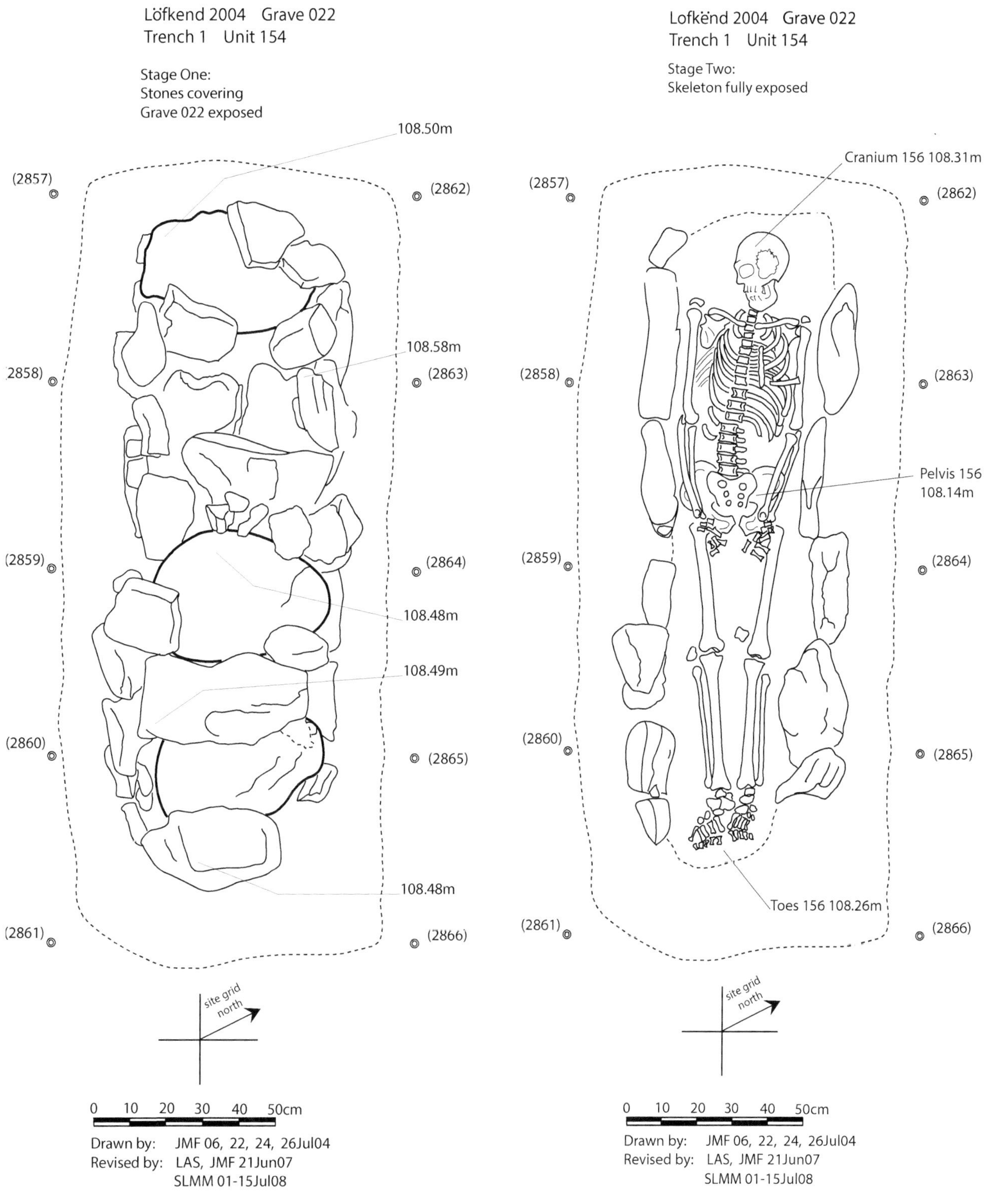

a b

FIGURE 3.288 Plans of Tomb LXXXVI: (a) Stage 1; (b) Stage 2

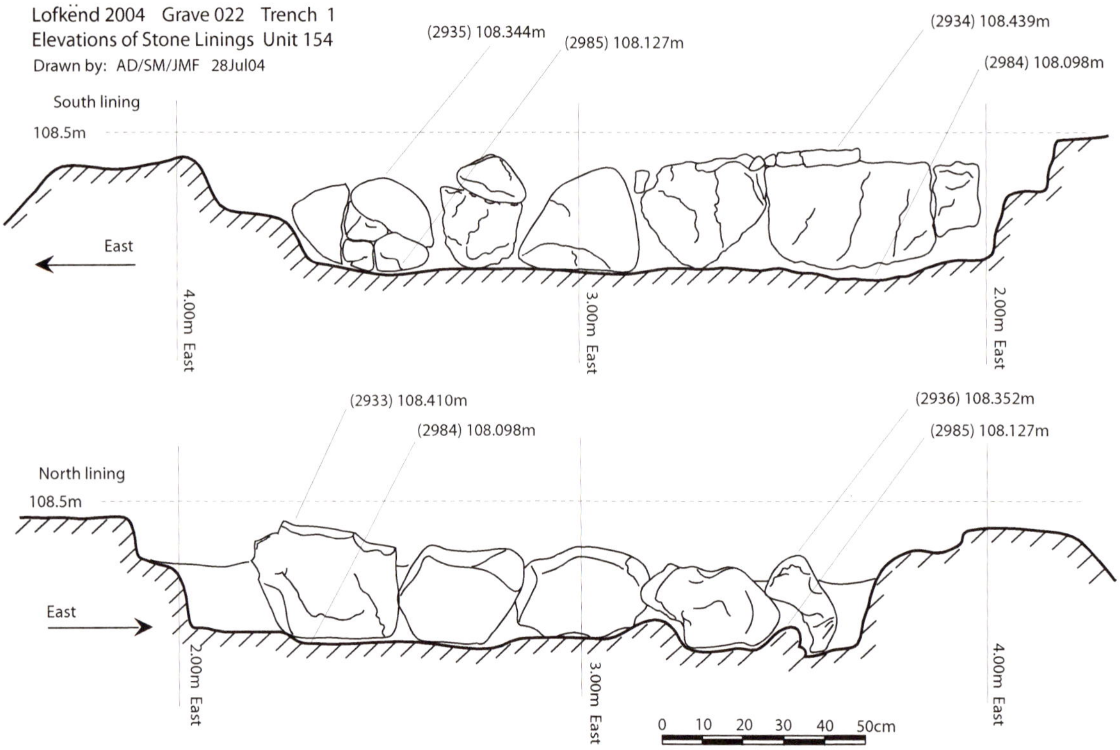

FIGURE 3.289 Tomb LXXXVI: elevations of south and north walls of tomb

a

b

FIGURE 3.290 Tomb LXXXVI: (a) (photo 415) with cover stones, from west; (b) (photo 450) partially uncovered, from east; (c, d) *on facing page*

c

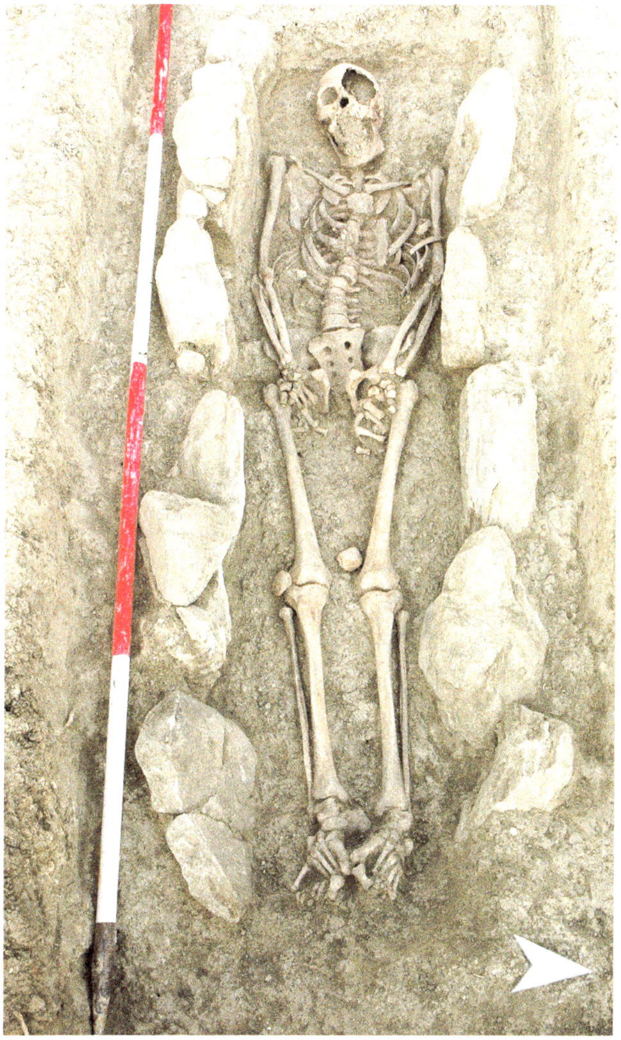

d

Figure 3.290 (*continued*). Tomb LXXXVI: (c) (photo 471) partially uncovered, from east; (d) (photo 476) showing skeleton uncovered, from east

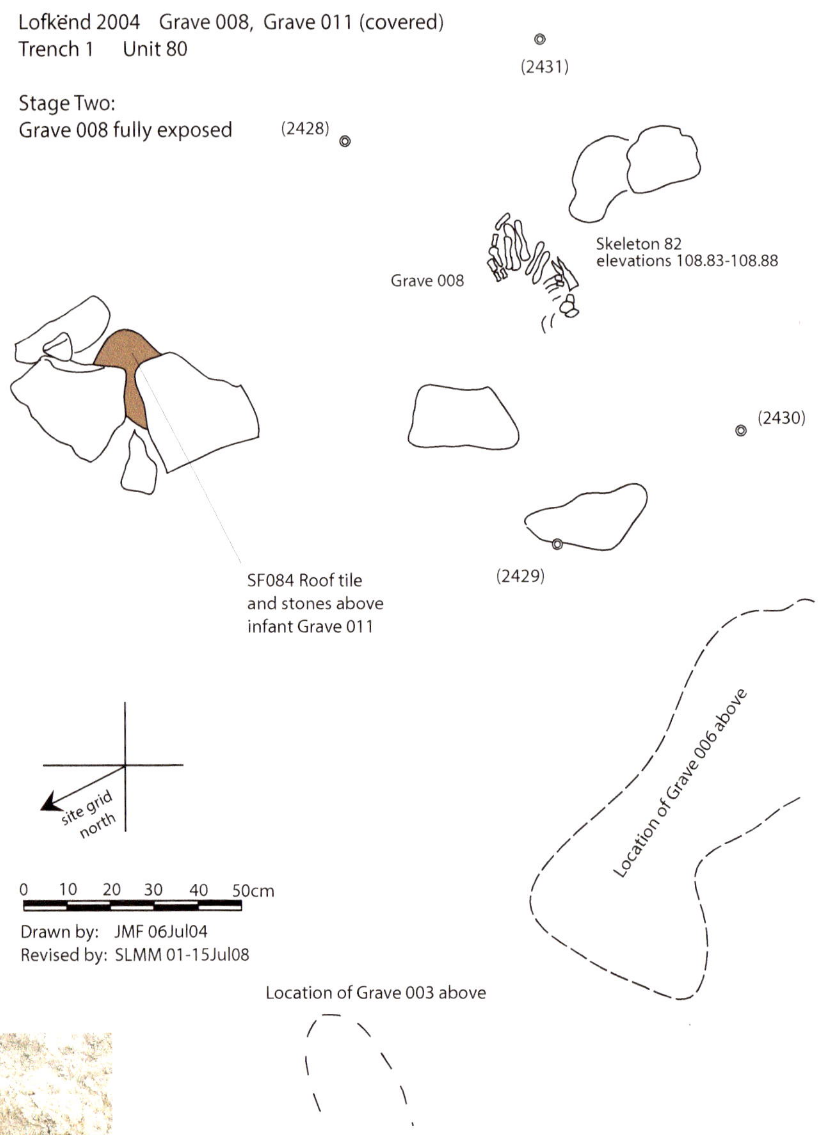

Figure 3.291 Plan of Tomb LXXXVII and nearby features

Figure 3.292 Tomb LXXXVII (photo 206) from above west

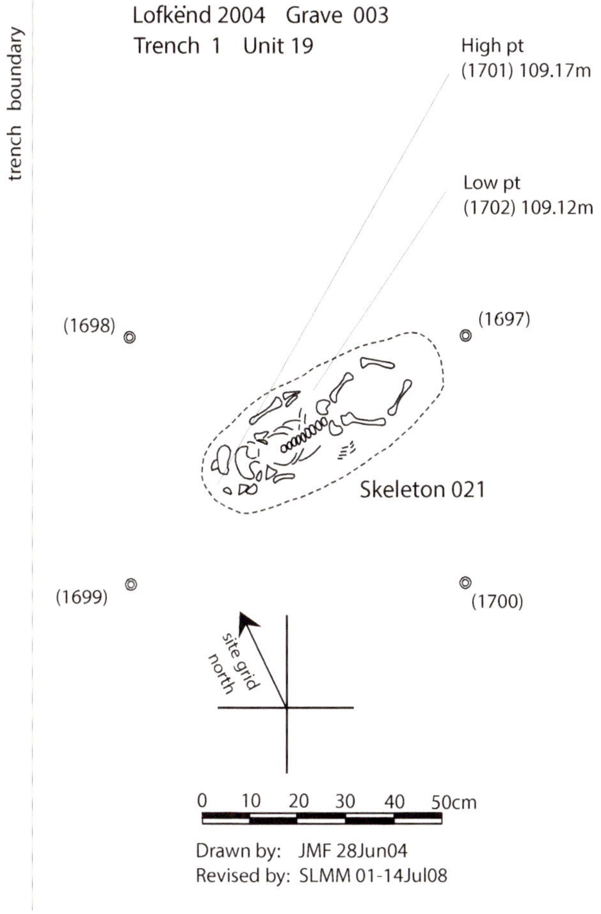

Figure 3.293 Plan of Tomb LXXXVIII

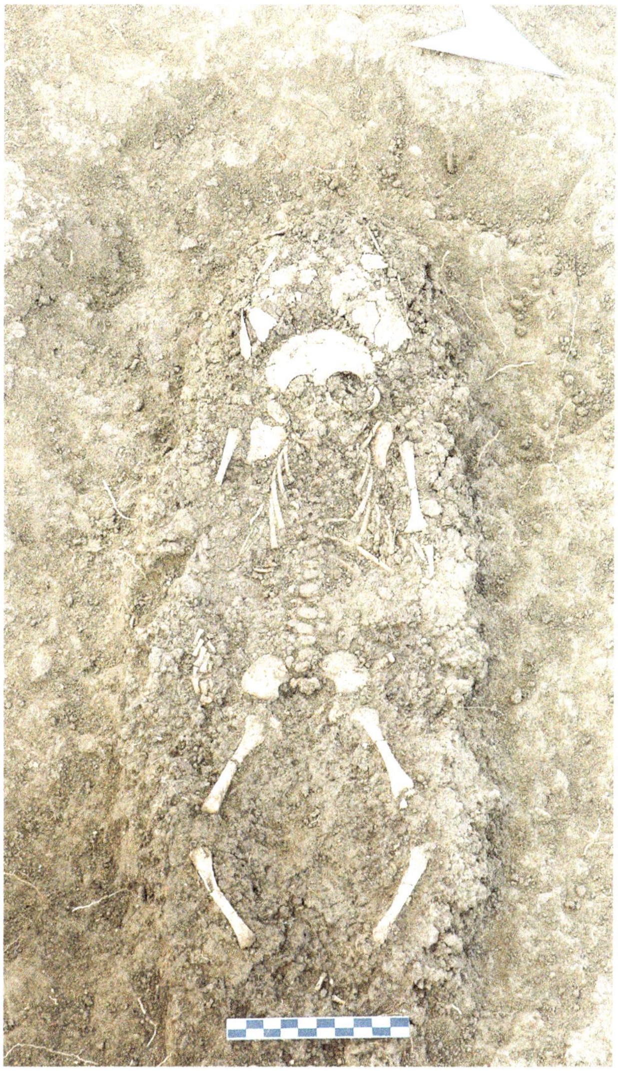

a

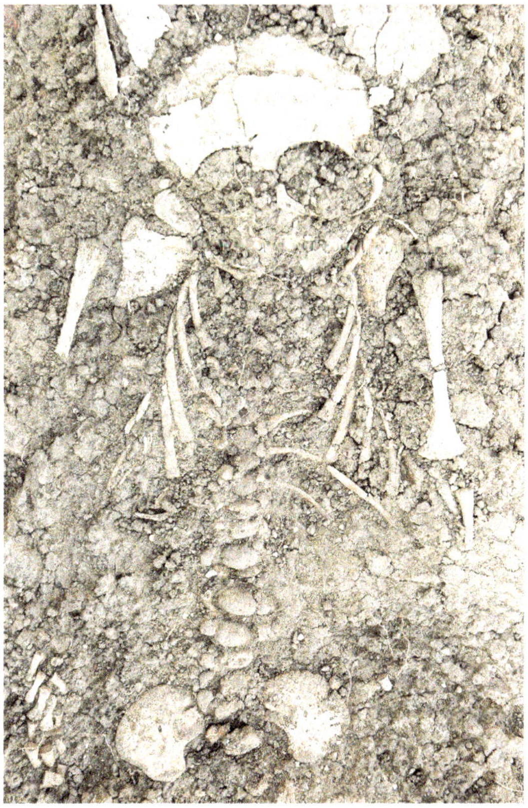

b

Figure 3.294 Tomb LXXXVIII: (a) (photo 133) from east-northeast; (b) (photo 136), detail from above east-northeast

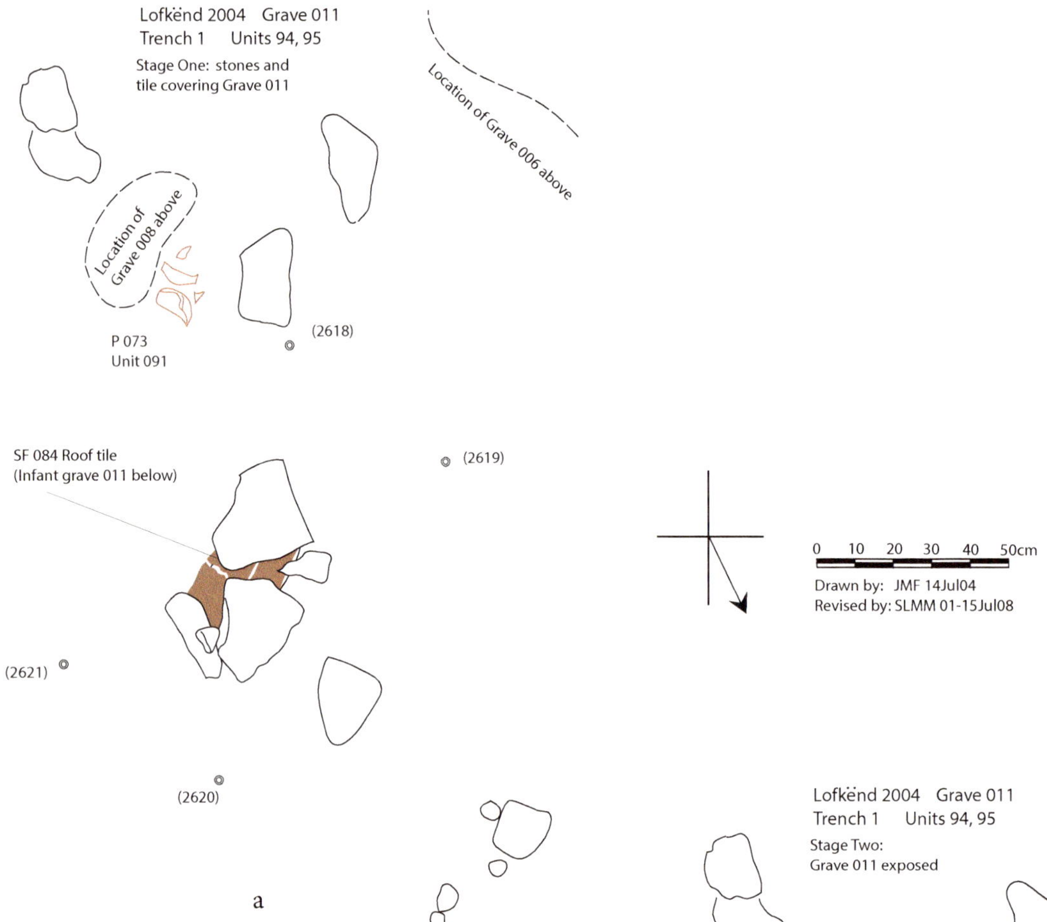

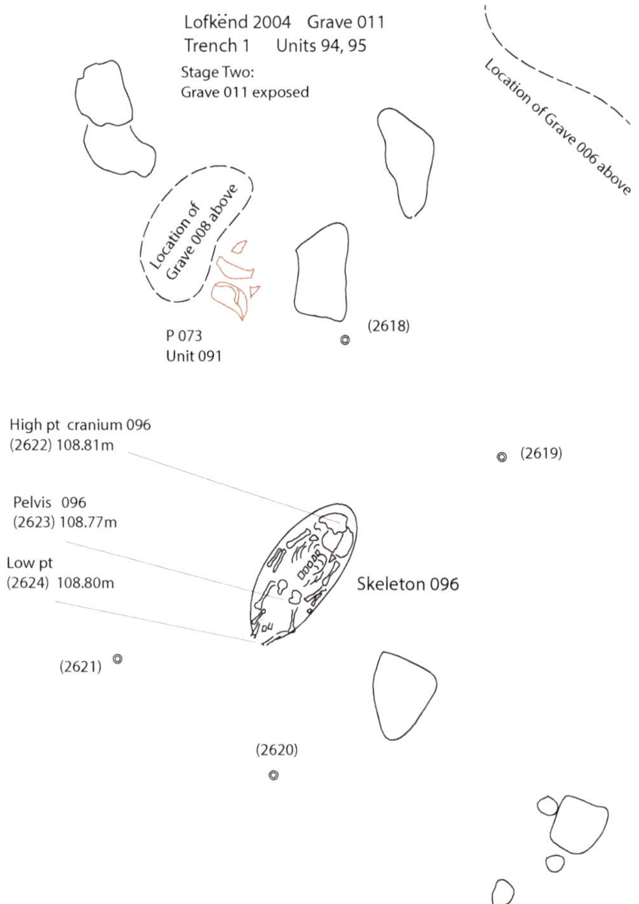

FIGURE 3.295 Plans of Tomb LXXXIX: (a) Stage 1 (covered); (b) Stage 2 (skeleton exposed)

a

b

FIGURE 3.296 Tomb LXXXIX: (a) (photo 203) showing cover stones overlying roof tile; (b) (photo 209) showing roof tile after removal of cover stones

FIGURE 3.297 Tomb LXXXIX: (photo 229) showing exposed infant skeleton after removal of cover stones and roof tile

FIGURE 3.298 Fragmentary roof tile, **TLXXXIX-1** (photo 519)

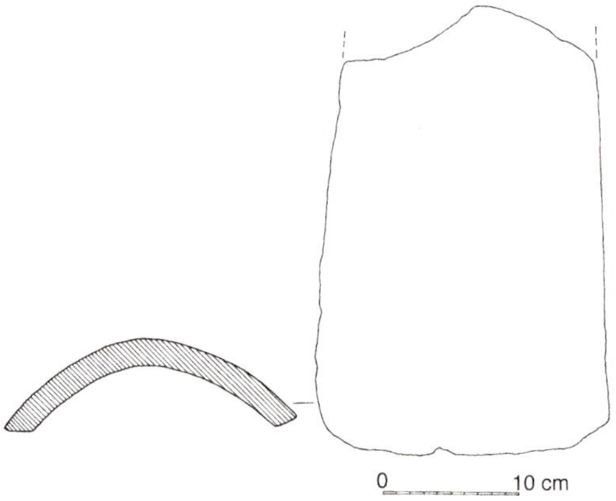

FIGURE 3.299 Fragmentary roof tile, **TLXXXIX-1** (drawing, sheet 22)

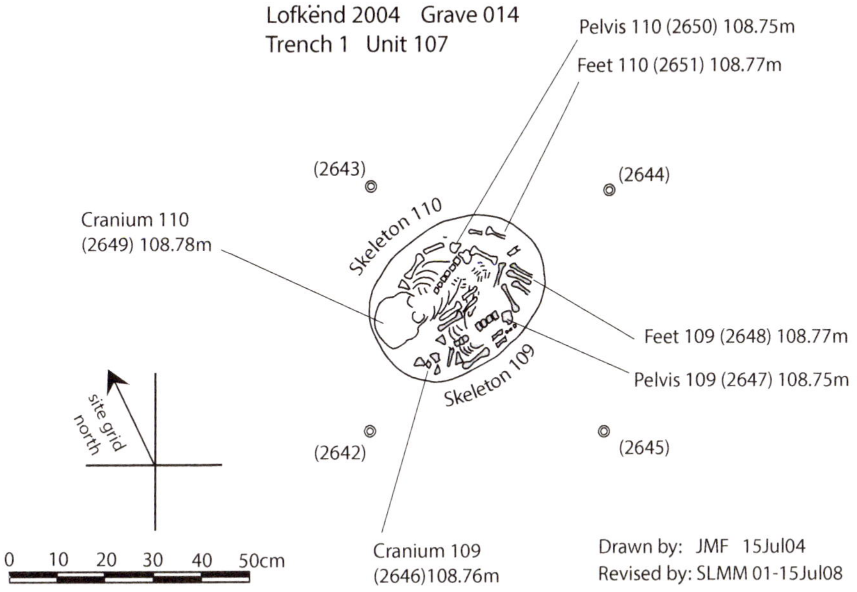

FIGURE 3.300 Plan of Tomb XC

a b

FIGURE 3.301 Tomb XC: (a) (photo 248) from northeast; (b) (photo 250) from southwest

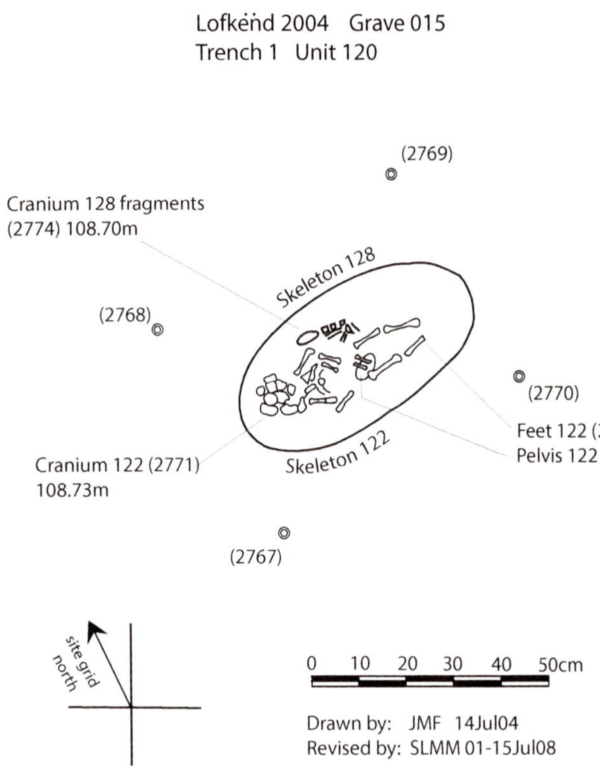

FIGURE 3.302 Plan of Tomb XCI

FIGURE 3.303 Tomb XCI (photo 263) from east-northeast

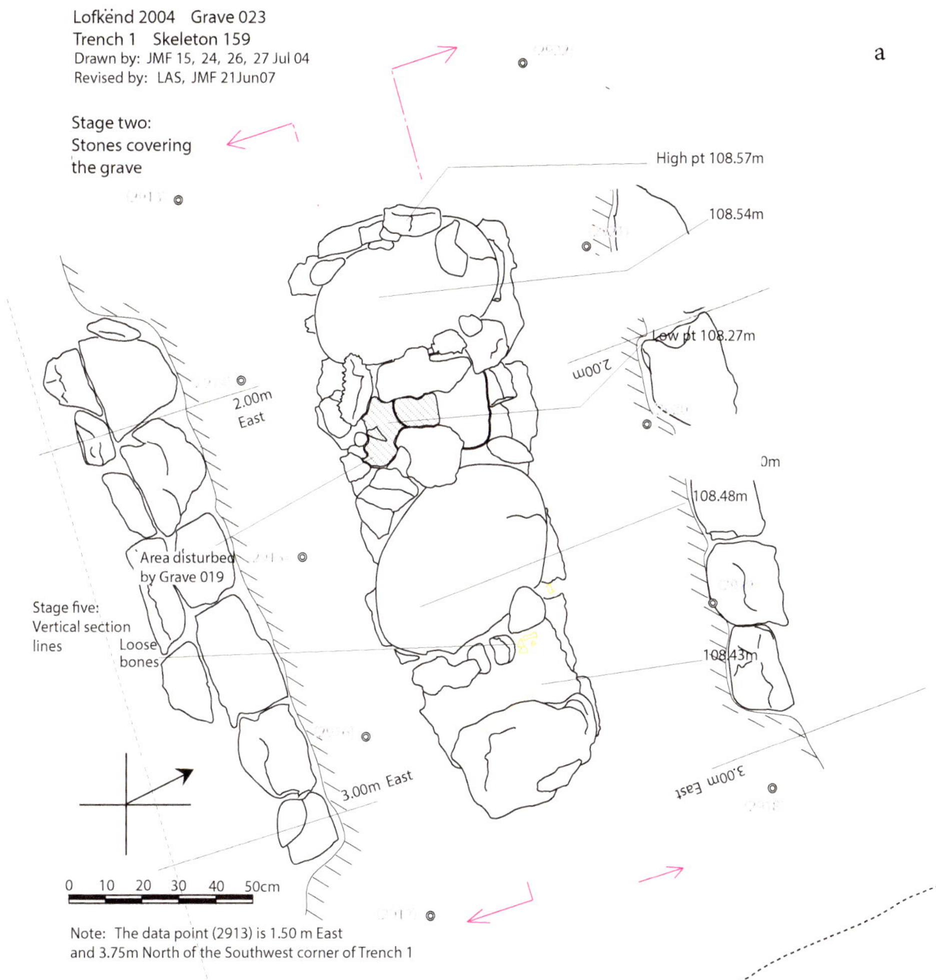

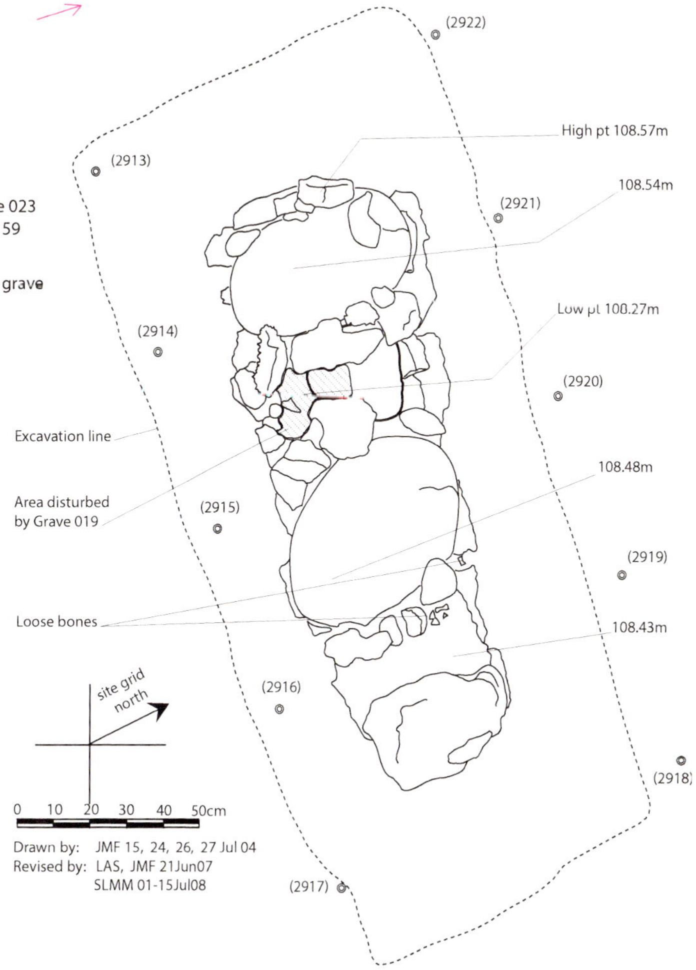

b

Figure 3.304 Tomb XCII: (a) Initial plan of tomb, with cover stones and elevations; (b) Stage 1, with cover stones in place; (c) *on next page*

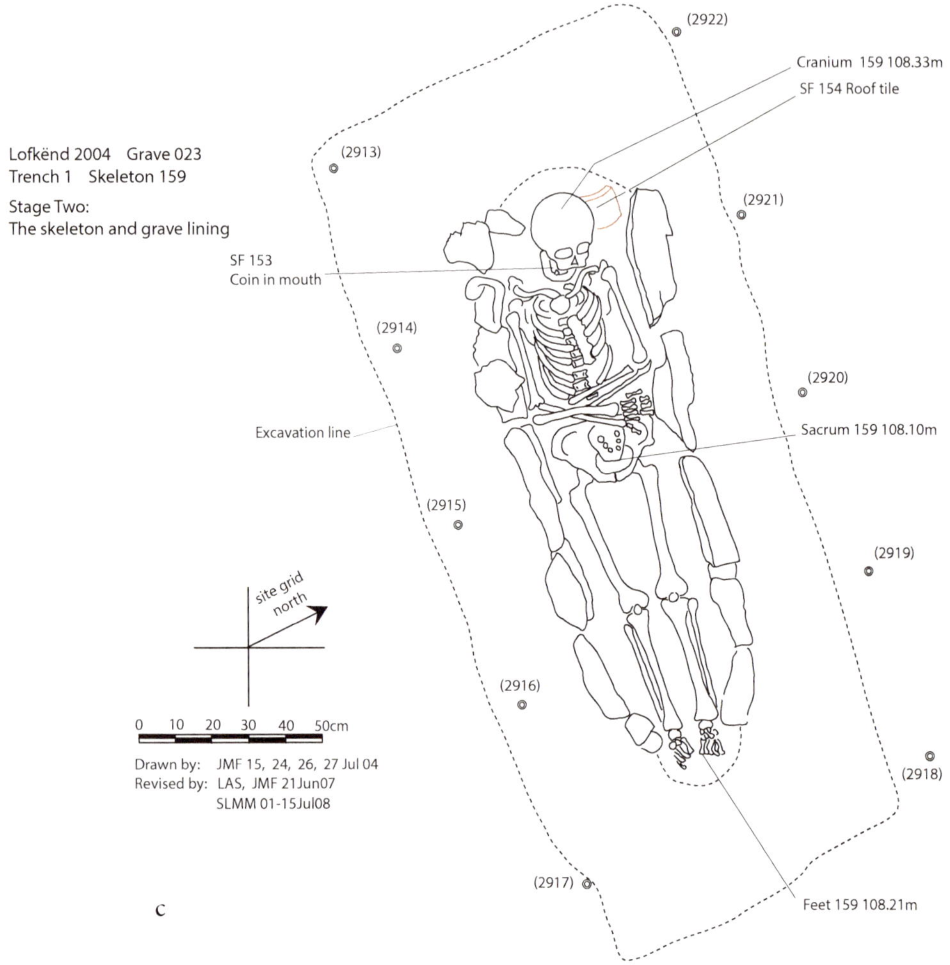

FIGURE 3.304 *(Continued from previous page)*. Tomb XCII: (c) Stage 2, with cover stones removed

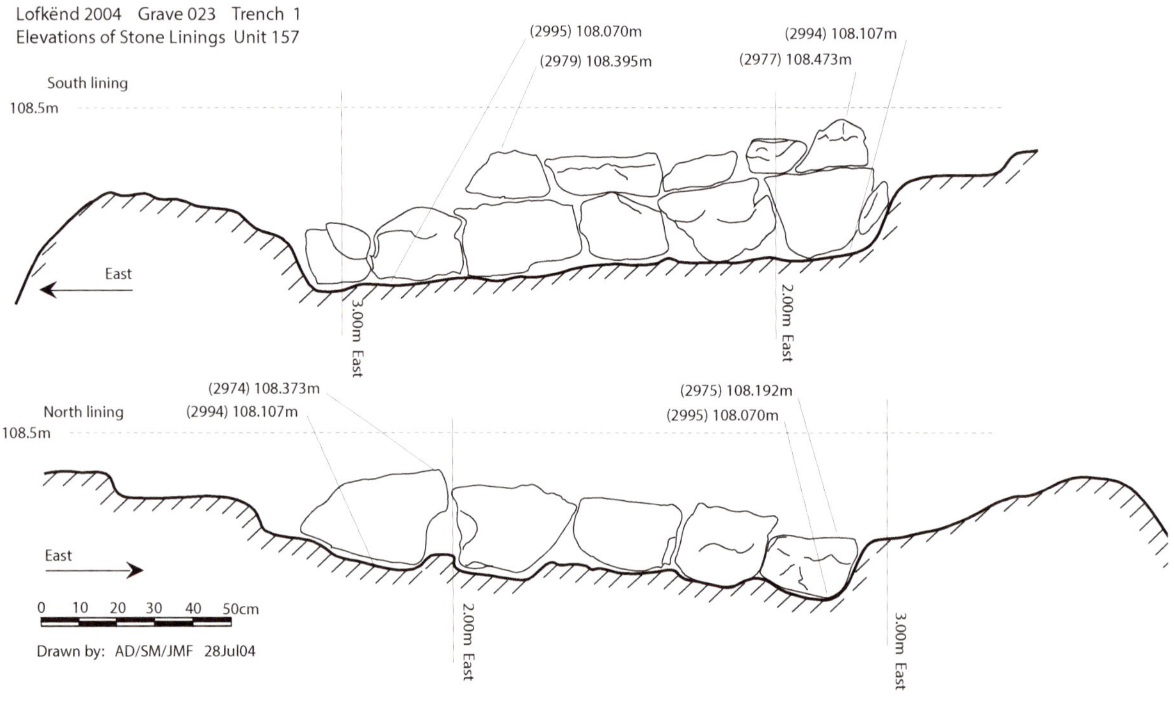

FIGURE 3.305 Tomb XCII: elevations of south and north walls of tomb

a

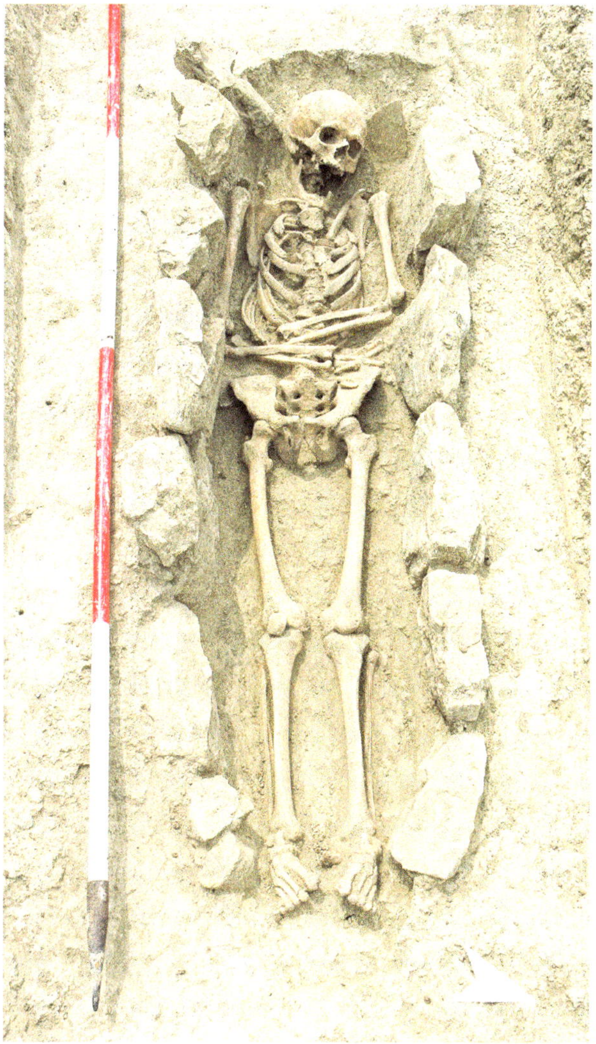

b

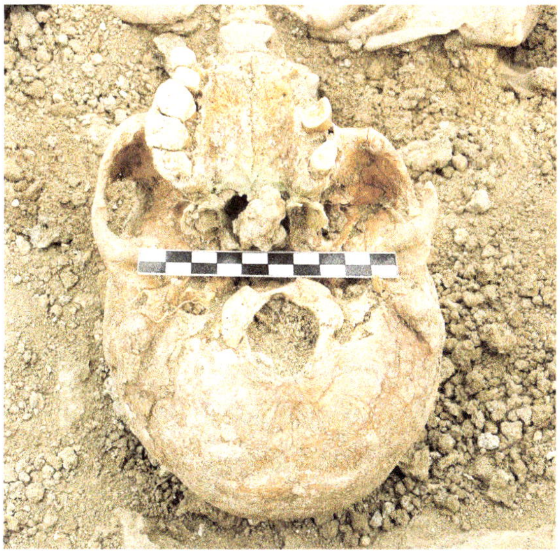

c

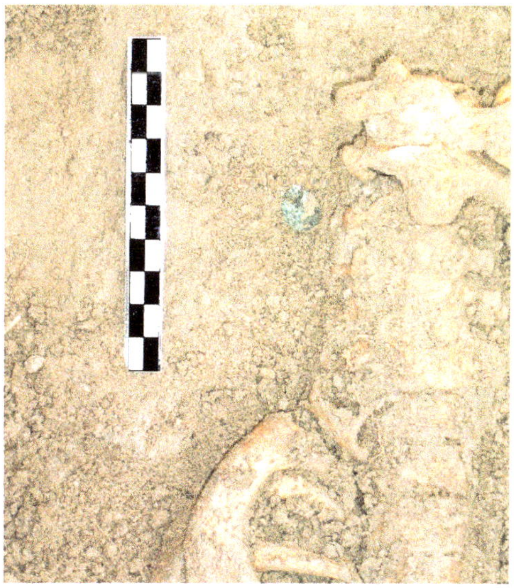

d

FIGURE 3.306 Tomb XCII: (a) (photo 433) covered, from east-northeast; (b) (photo 491) uncovered, from east-northeast; (c) cranium (photo 495), showing discoloration on maxilla from copper-silver alloy coin placed originally in mouth; (d) (photo JKP 003), showing **TXCII-1** as found in situ

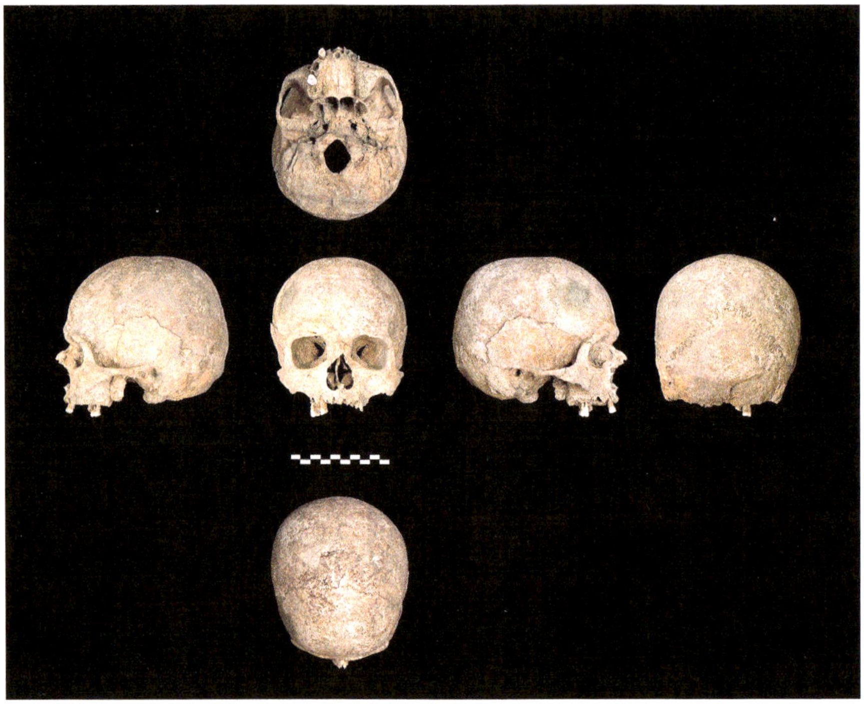

FIGURE 3.307 Tomb XCII: composite views (photo 2066) of the cranium of SU 159

a b

FIGURE 3.308 Tomb XCII: (a) **TXCII-1** (one side) (photo 700); (b) **TXCII-1** (opposite side) (photo 701b)

a b

FIGURE 3.309 Tomb XCII: (a–b) **TXCII-2**

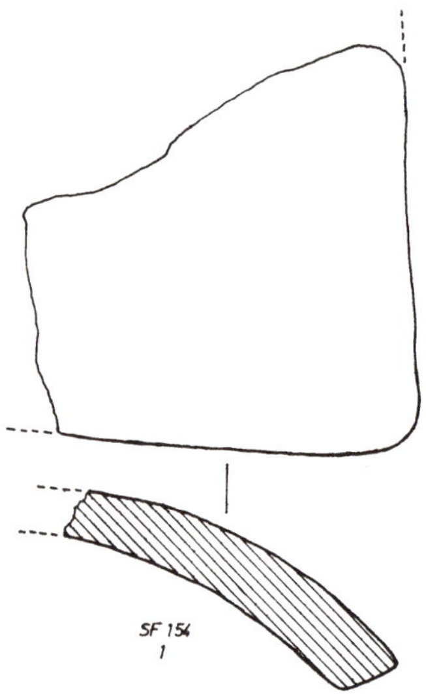

FIGURE 3.310 Tomb XCII: **TXCII-2**

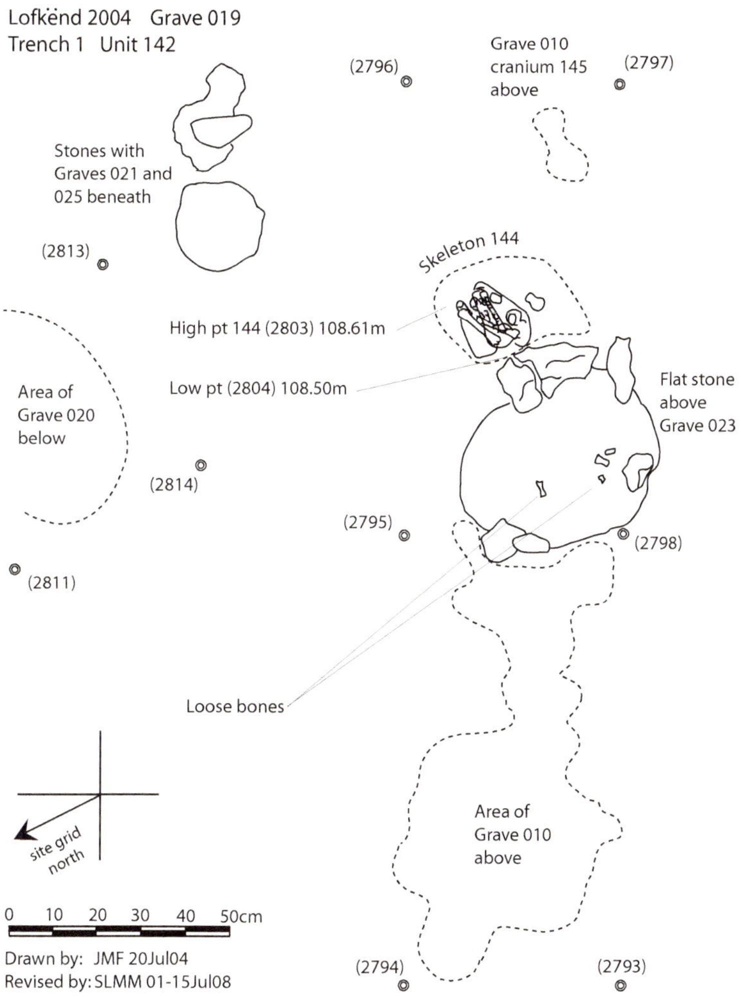

FIGURE 3.311 Plan of Tomb XCIII

a

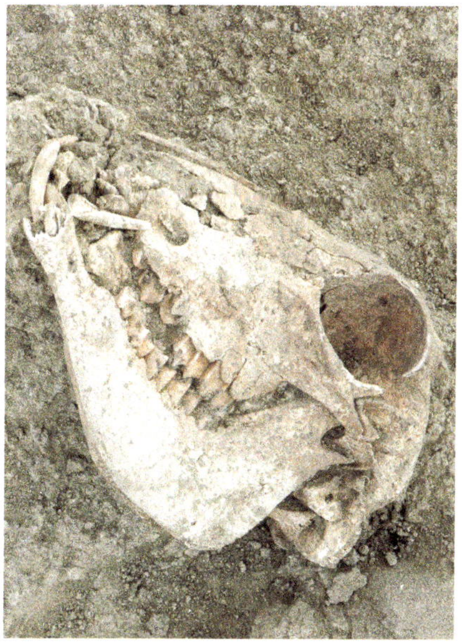

b

FIGURE 3.312 Tomb XCIII: (a) (photo 331) as first encountered from west; (b) (photo 370) showing sheep skull after removal of postcranial bones, from above west

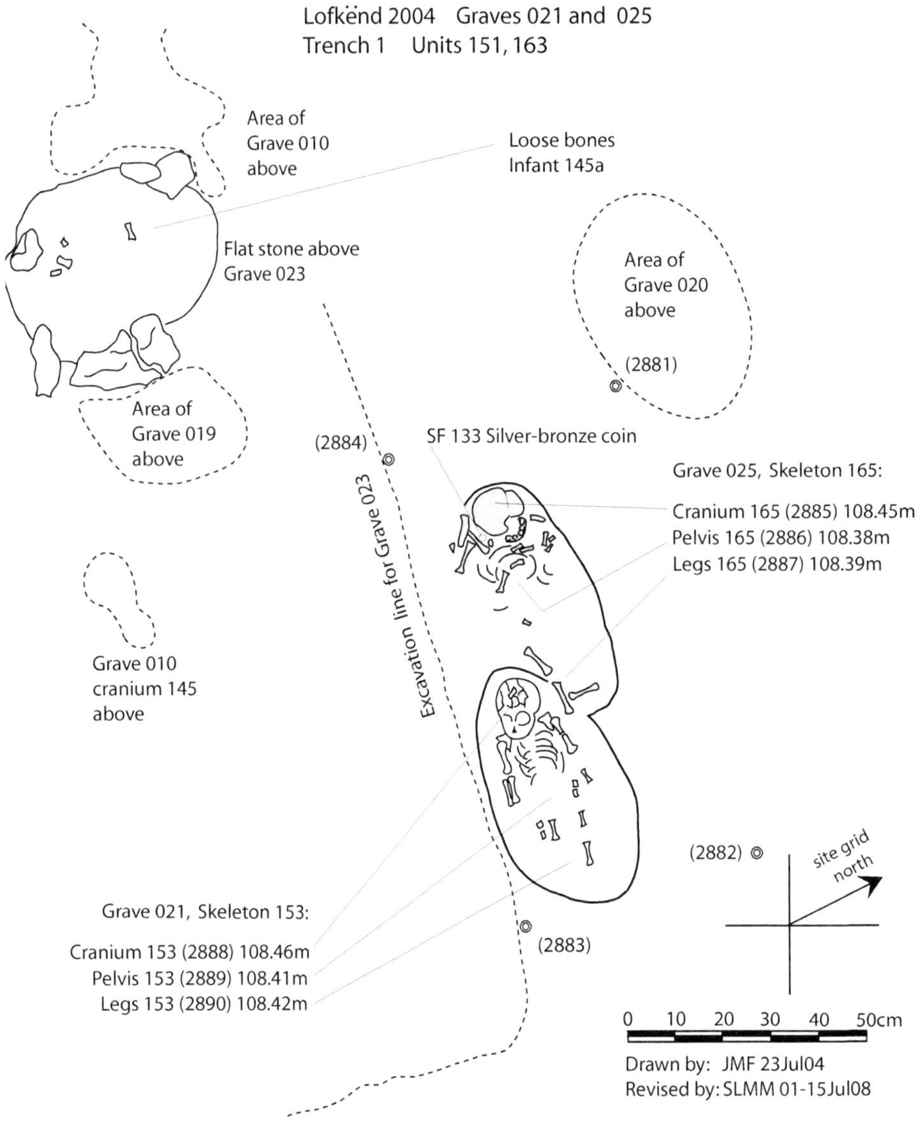

FIGURE 3.313 Plan of Tombs XCIV and XCV

a

b

(*Above and right*): FIGURE 3.314 Tomb XCIV: (a) covered (photo 428) from east; (b) uncovered (photo 430) from east; (c) detail of coin, **TXCIV-1** (photo 455), as found with the removal of bones of the upper torso, from above east-southeast

c

FIGURE 3.315 Tomb XCIV: (a) TXCIV-1 (one side) (photo 696); (b) **XCIV-1** (opposite side) (photo 698)

FIGURE 3.316 Tombs XCV and XCIV (photo 425), from east, with Tomb XCIV as covered with stones

FIGURE 3.317 Tombs XCV and XCIV (photo 432) fully exposed from east-northeast in relation to Tomb XCII

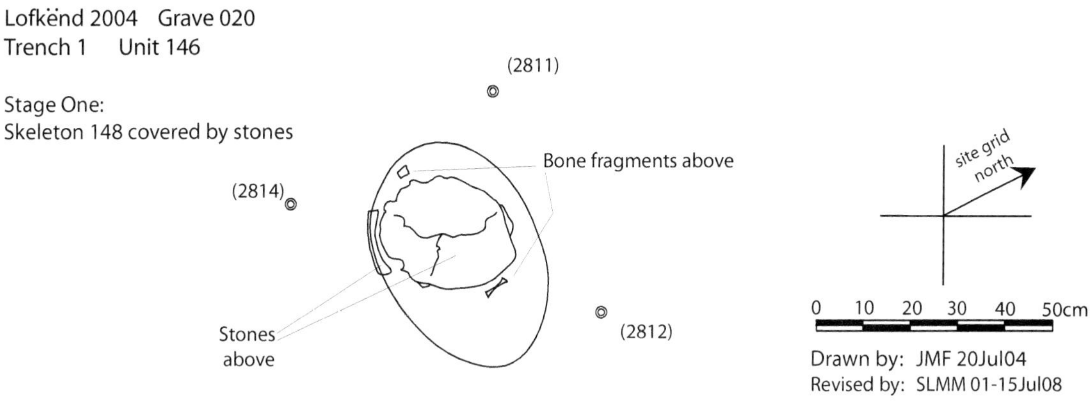

a

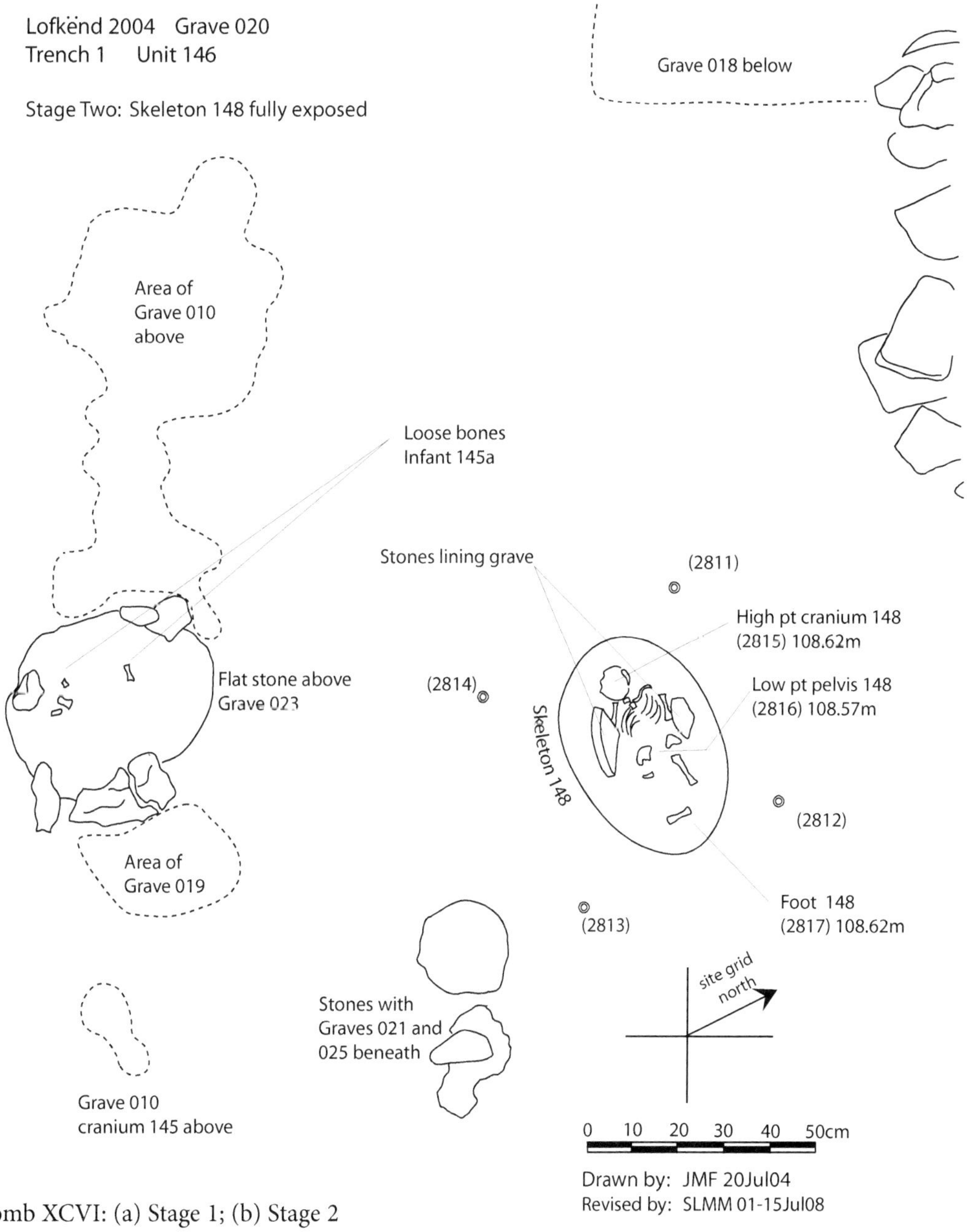

b

FIGURE 3.318 Plans of Tomb XCVI: (a) Stage 1; (b) Stage 2

a

FIGURE 3.319 Tomb XCVI: (a) (photo 333) cover stone from east; (b) tomb as exposed (photo 366) from east-northeast

b

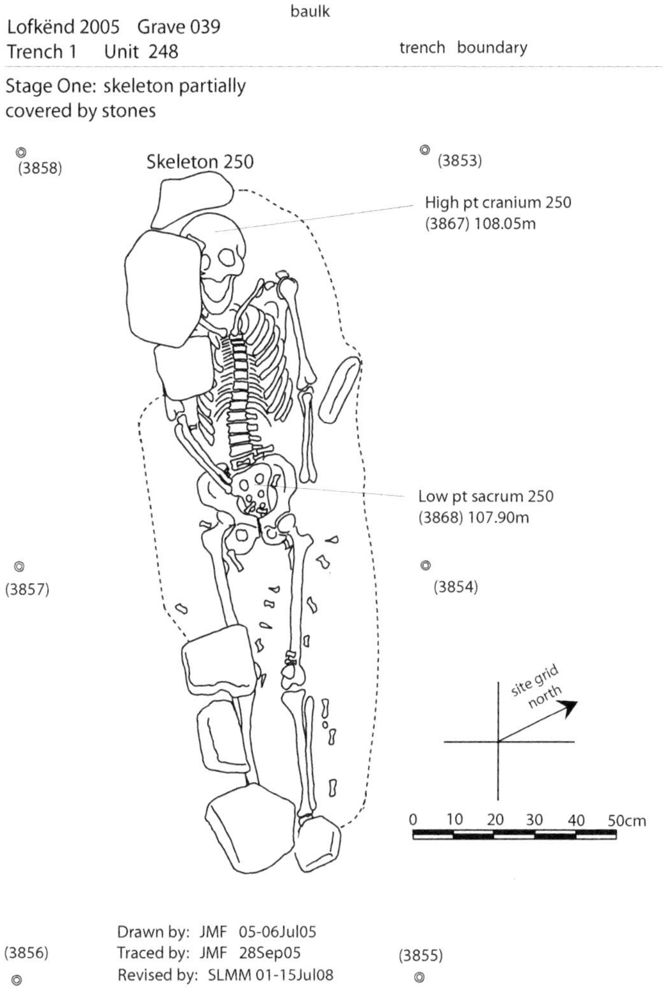

a

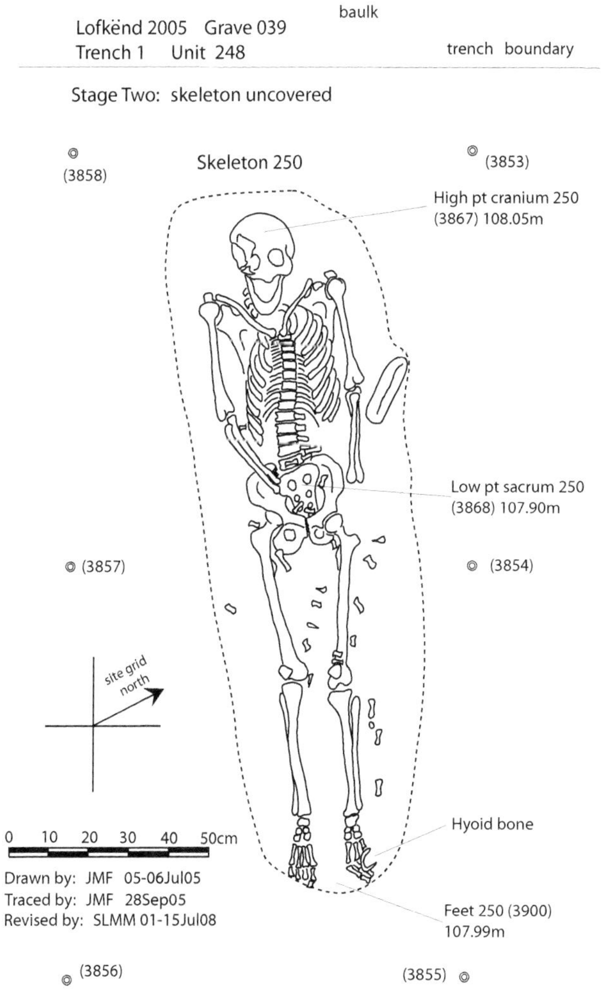

b

FIGURE 3.320 Plans of Tomb XCVII: (a) Stage 1; (b) Stage 2

a

Figure 3.321 Tomb XCVII: (a) cover stones for Tomb XCVII as encountered in 2004 (foreground), with Tomb LXXVII in background (photo 334) from east; (b) detail of cover stones of Tomb XCVII as encountered in 2004 (photo 335) from east

b

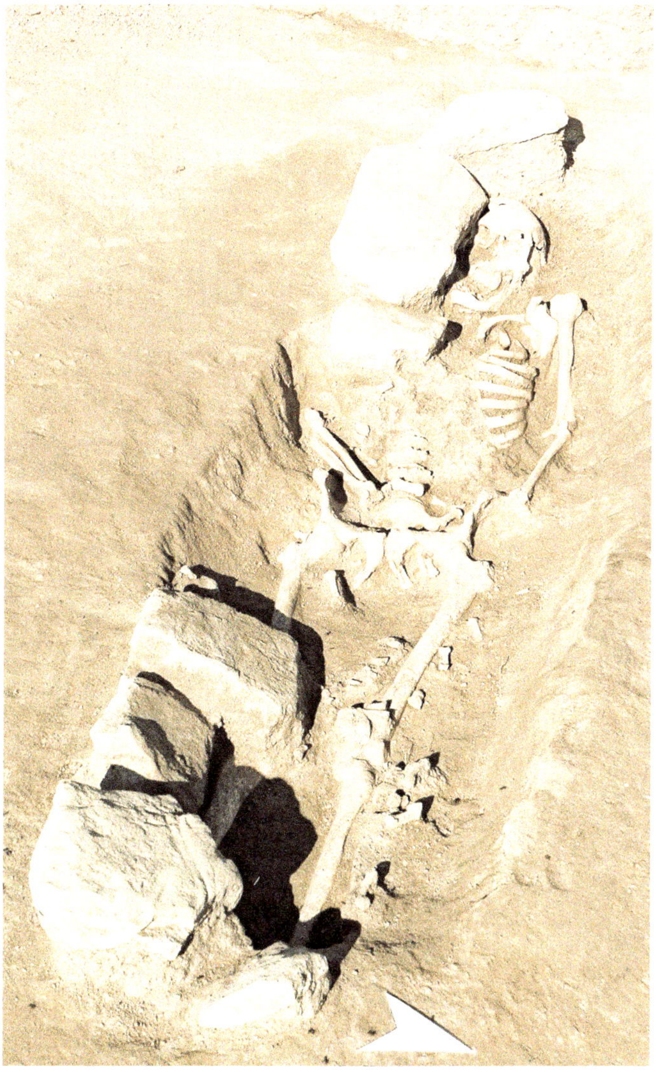

a

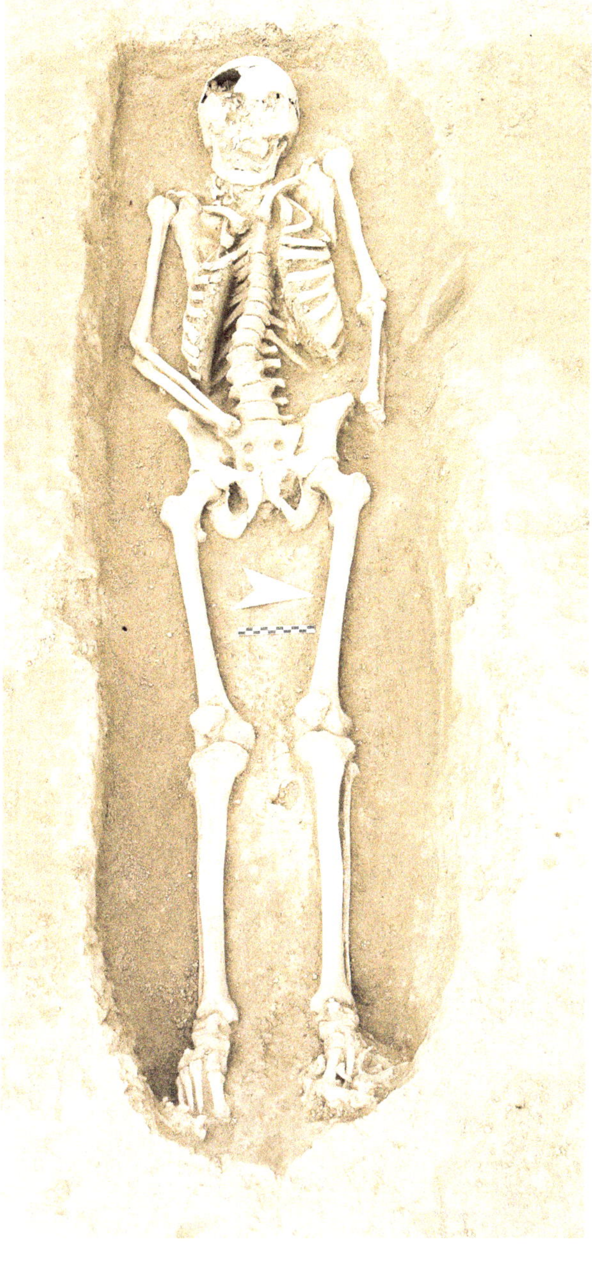

b

Figure 3.322 Tomb XCVII: (a) (photo 1008) as covered, from east-northeast; (b) (photo 1028) with skeleton fully exposed, from east

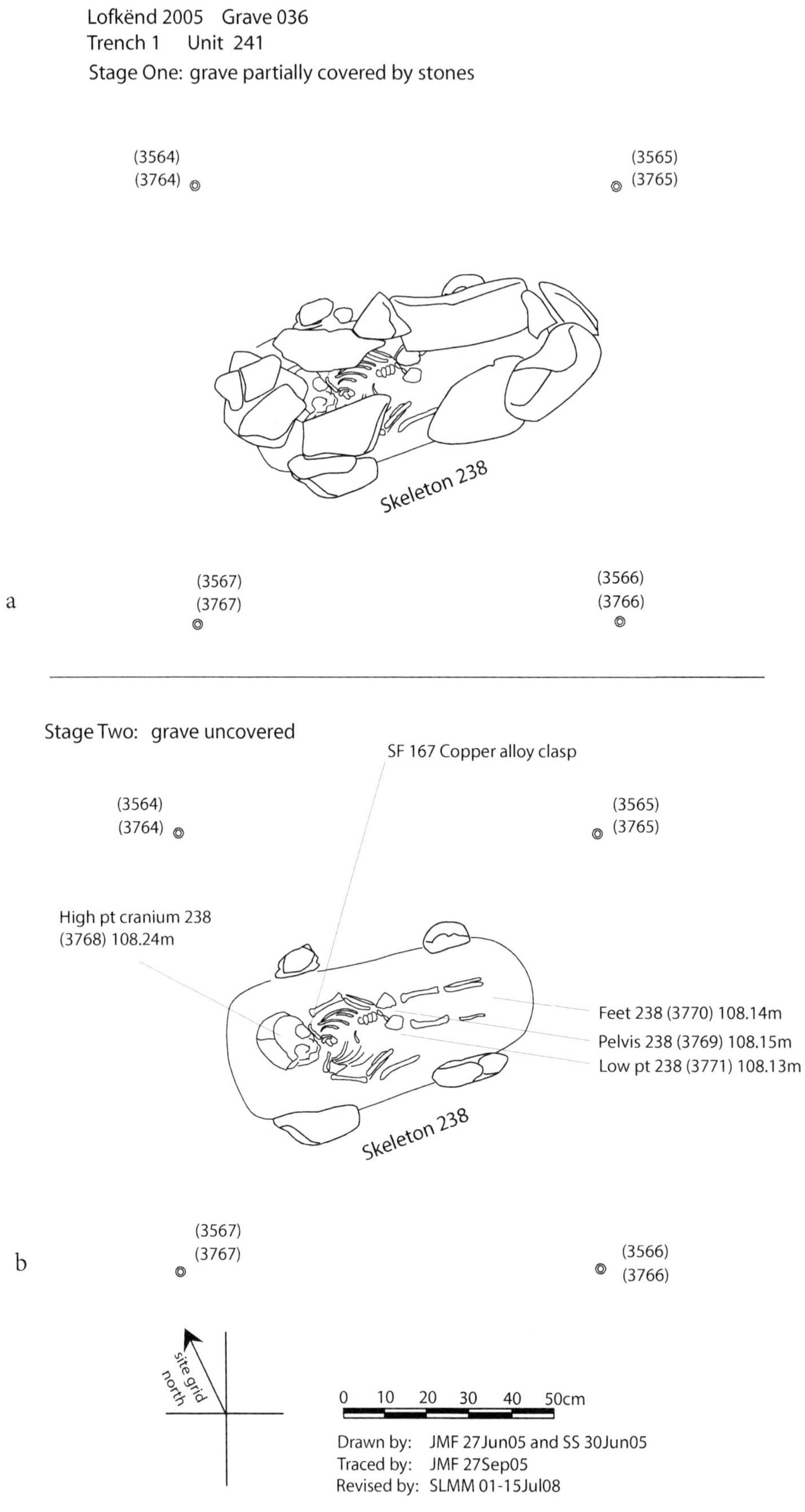

FIGURE 3.323 Plans of Tomb XCVIII: (a) Stage 1; (b) Stage 2

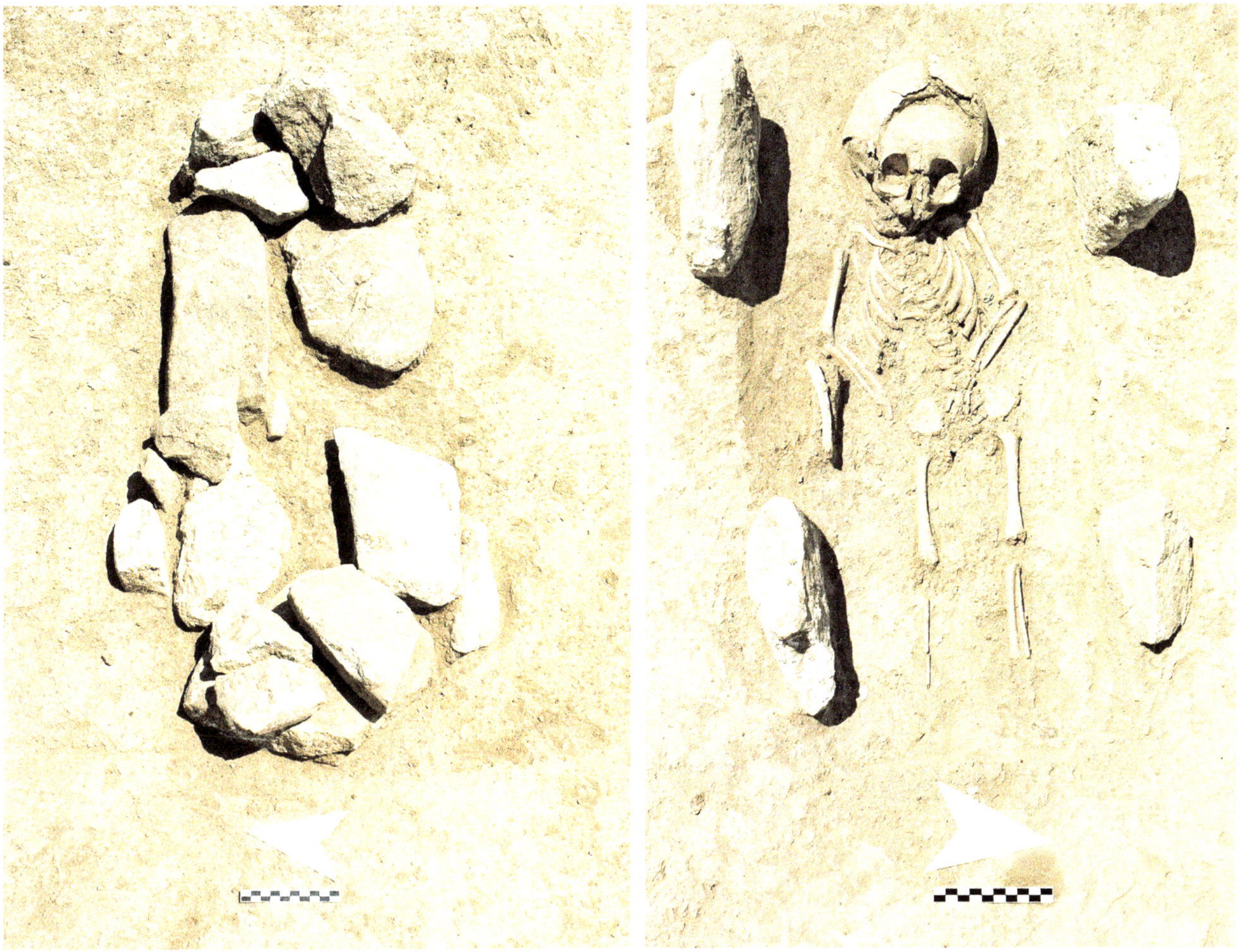

a b

Figure 3.324 Tomb XCVIII: (a) (photo 834) with cover stones, from west; (b) (photo 861) from east

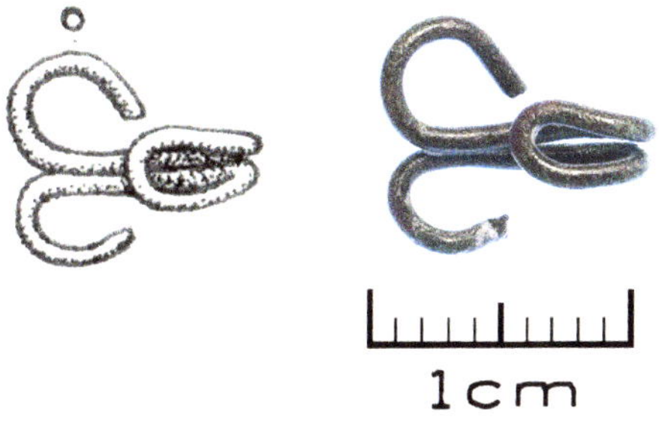

Figure 3.325 Tomb XCVIII: (a–b) **TXCVIII-1**

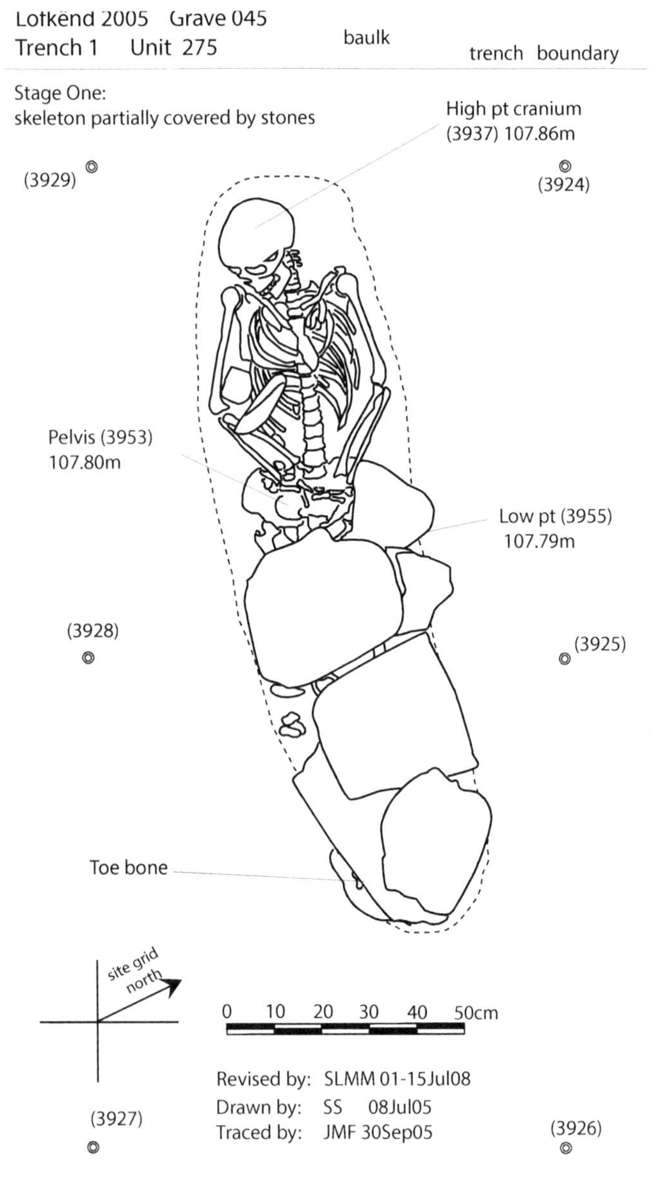

a

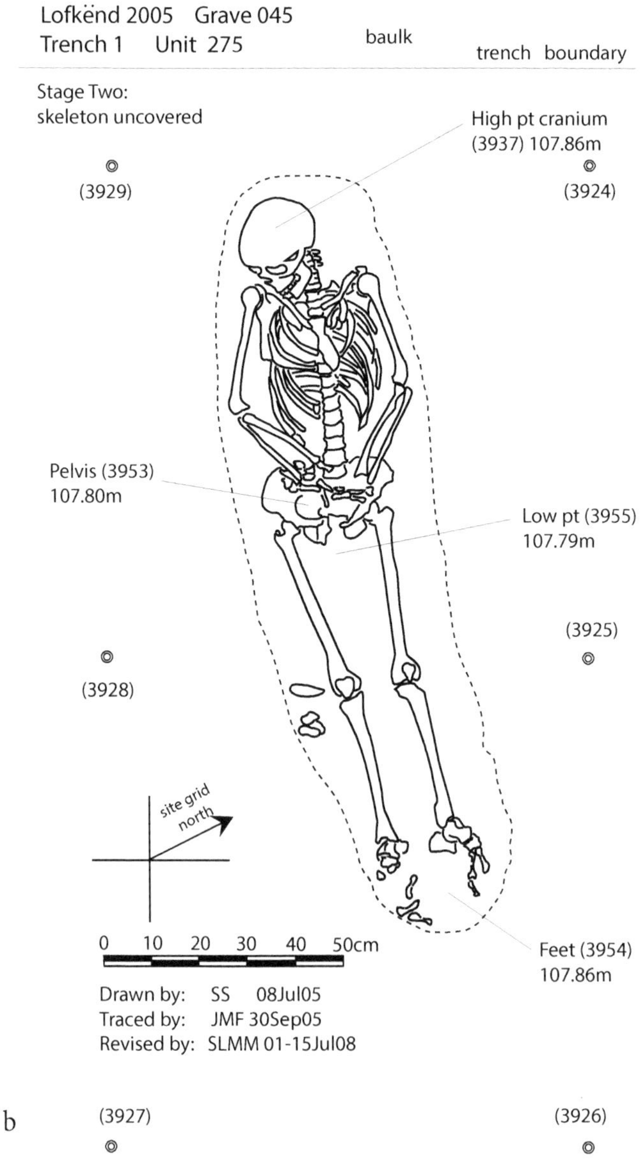

b

FIGURE 3.326 Plans of Tomb XCIX:
(a) Stage 1; (b) Stage 2

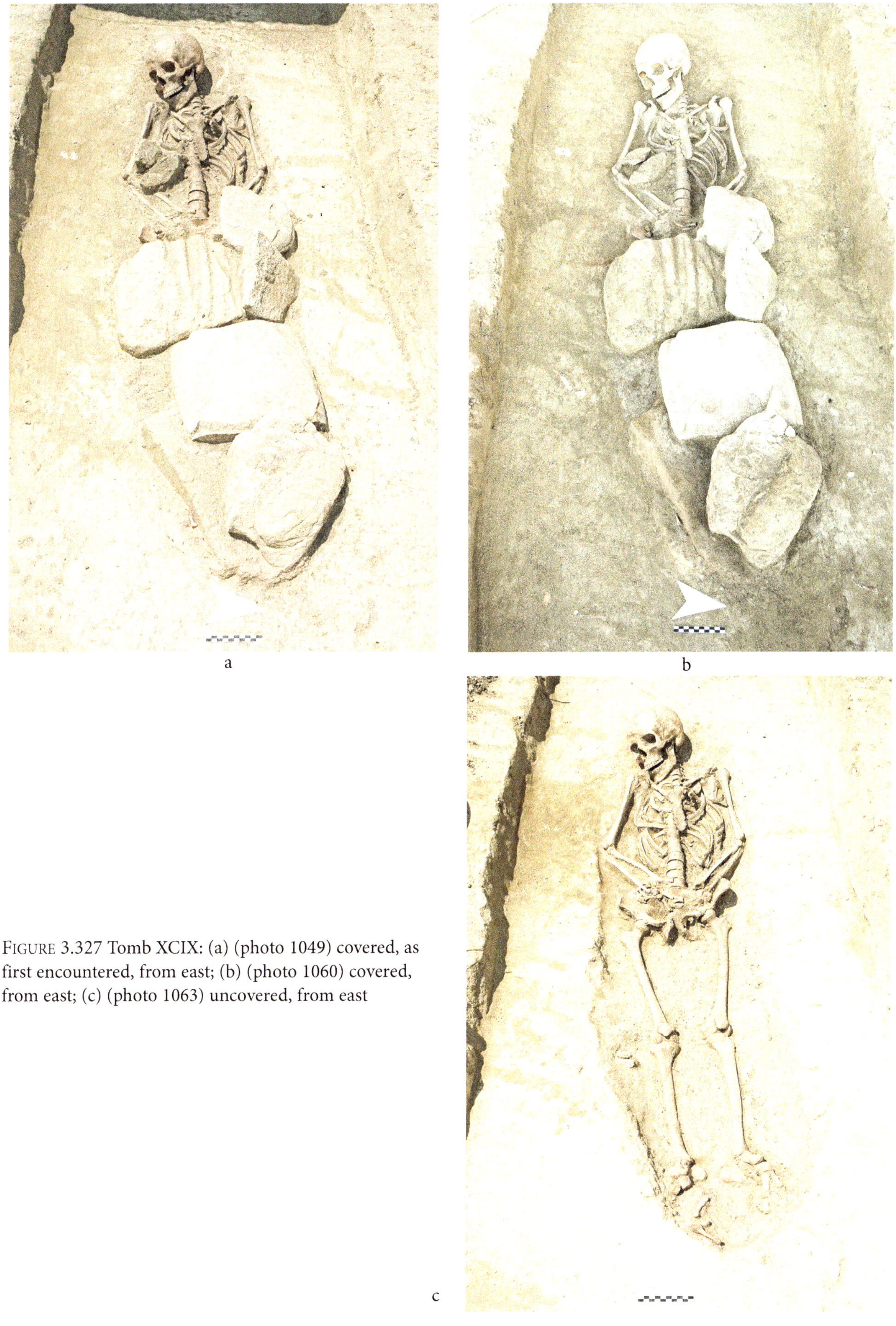

a

b

c

FIGURE 3.327 Tomb XCIX: (a) (photo 1049) covered, as first encountered, from east; (b) (photo 1060) covered, from east; (c) (photo 1063) uncovered, from east

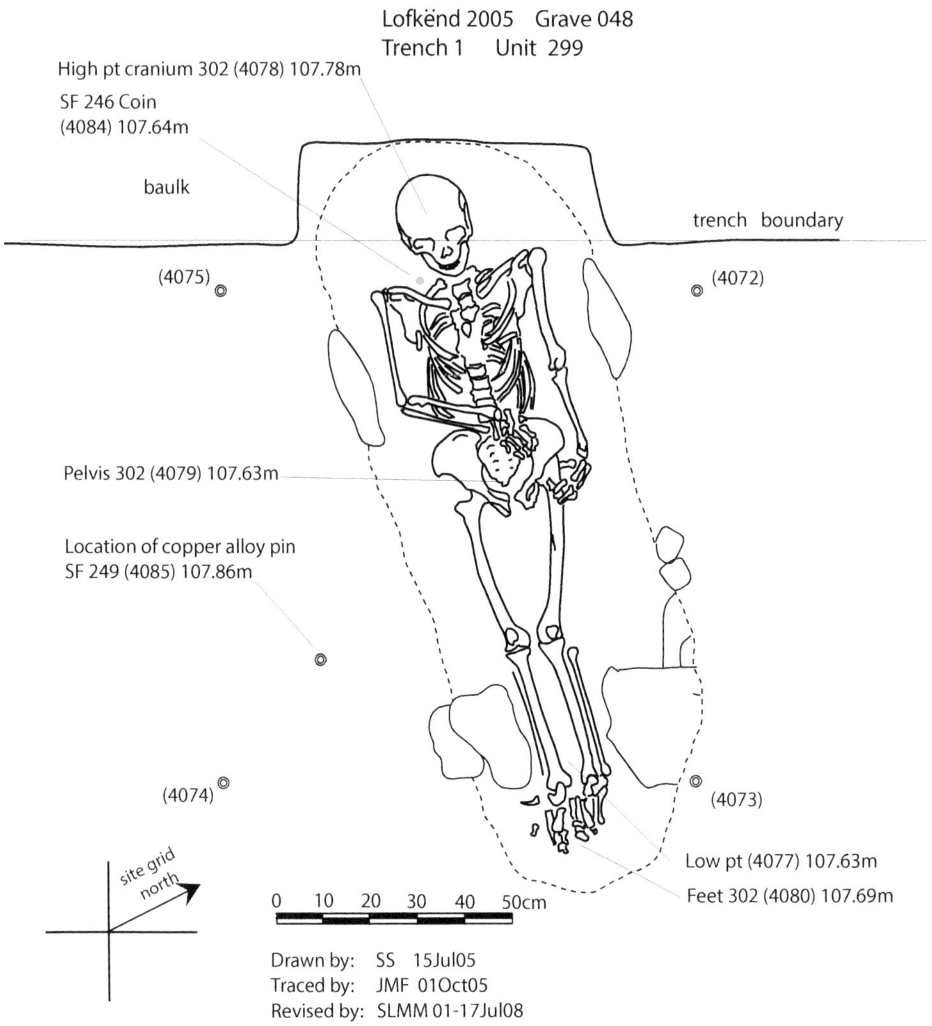

FIGURE 3.328 Plan of Tomb C

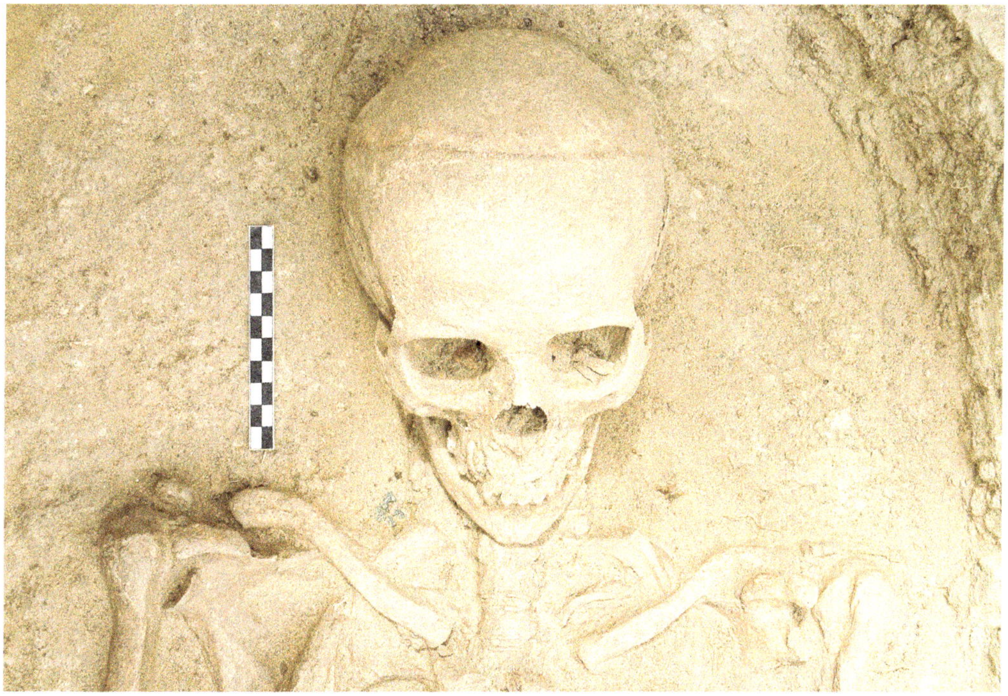

a

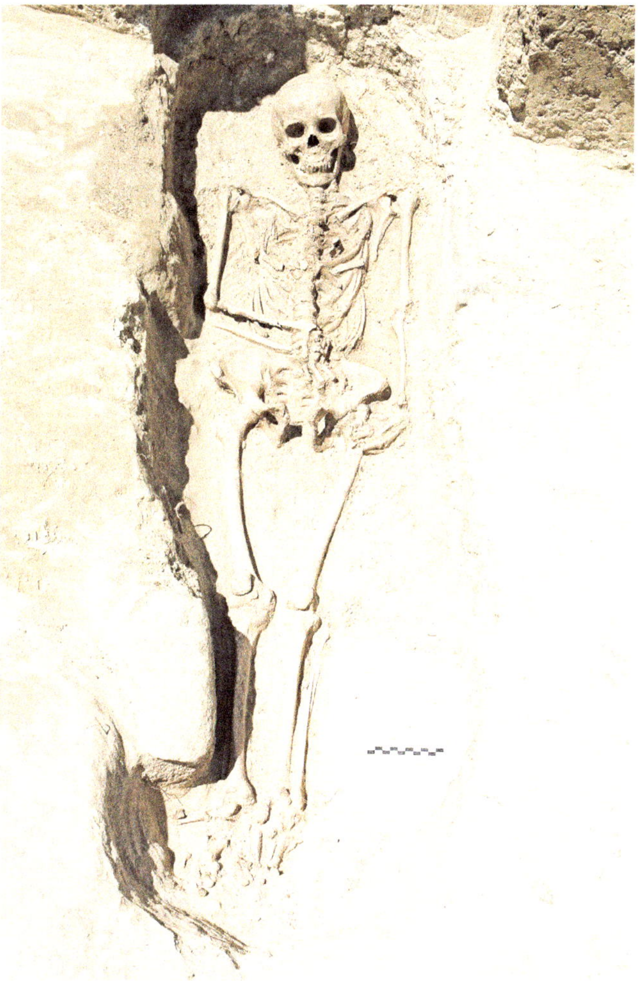

b

FIGURE 3.329 Tomb C: (a) (photo 1116) detail of cranium showing **TC-1,** from east-southeast; (b) (photo 1123) from east-northeast

FIGURE 3.330 Tomb C: **TC-1**

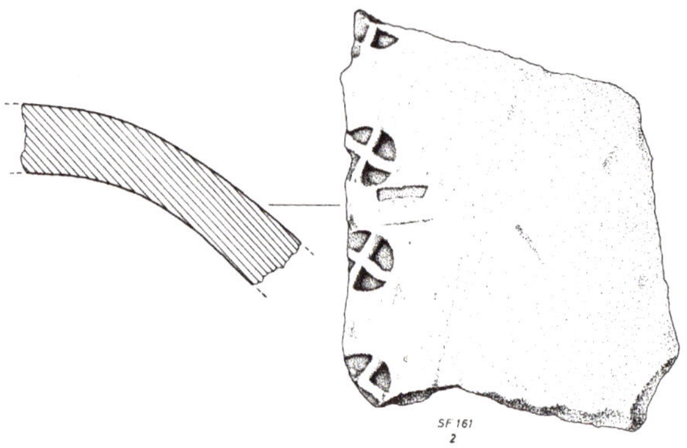

FIGURE 3.331 Fragment of modern stamped roof tile **3/1** (SF 161)

FIGURE 3.332 Fragment of roof tile **3/1** (SF 161): (a) (photo 2209) upperside with stamped decoration; (b) (photo 2210) underside

FIGURE 3.333 Modern roof tiles: (a) uppersides of modern roof tiles, including **3/2** (SF 382) (right) from the village of Ngrançija, ca. AD 1900 (photo 3357); (b) as previous, undersides

Chapter 4 Illustrations

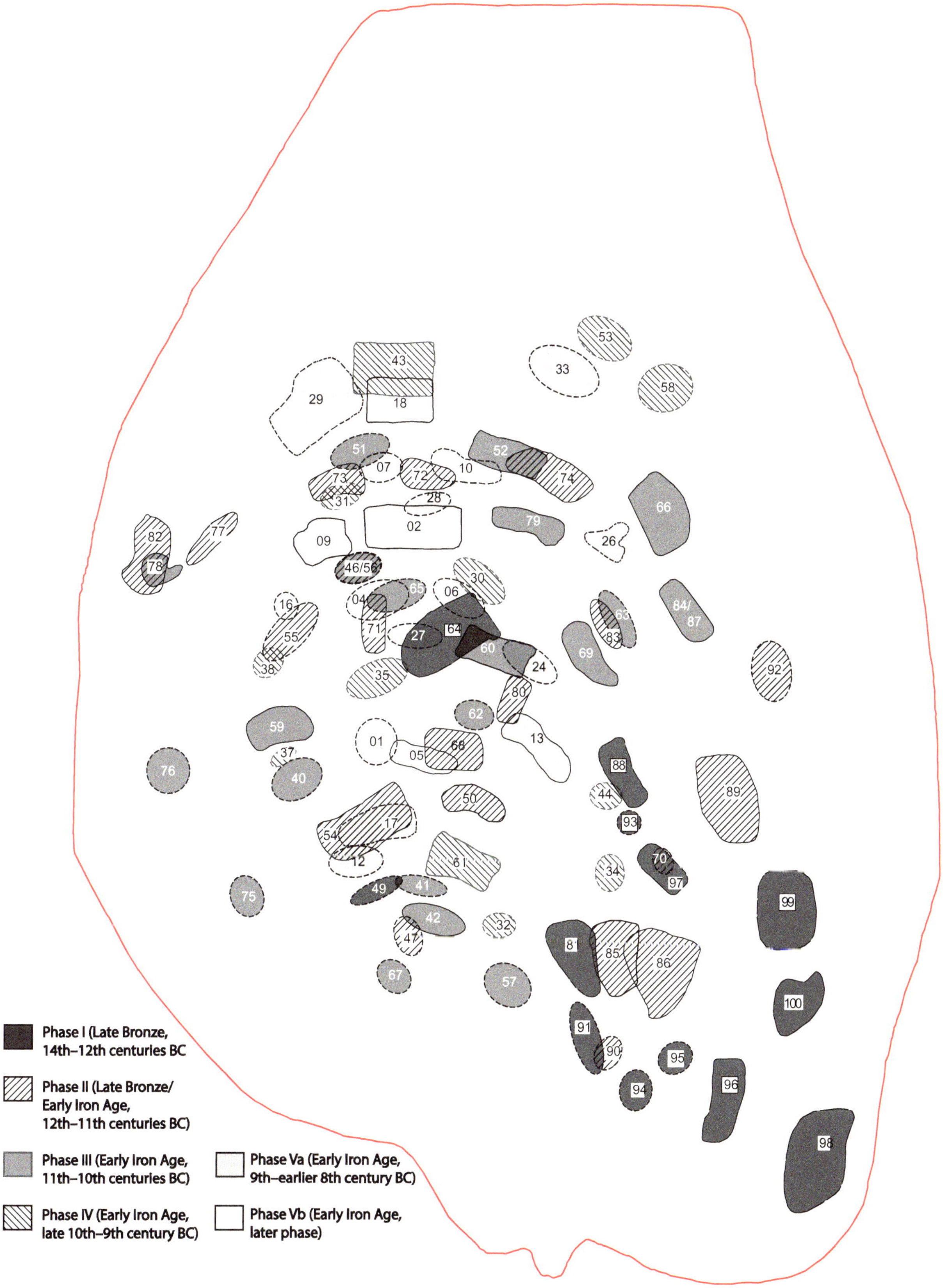

FIGURE 4.1 The Lofkënd tombs, rendered schematically (six prehistoric phases)

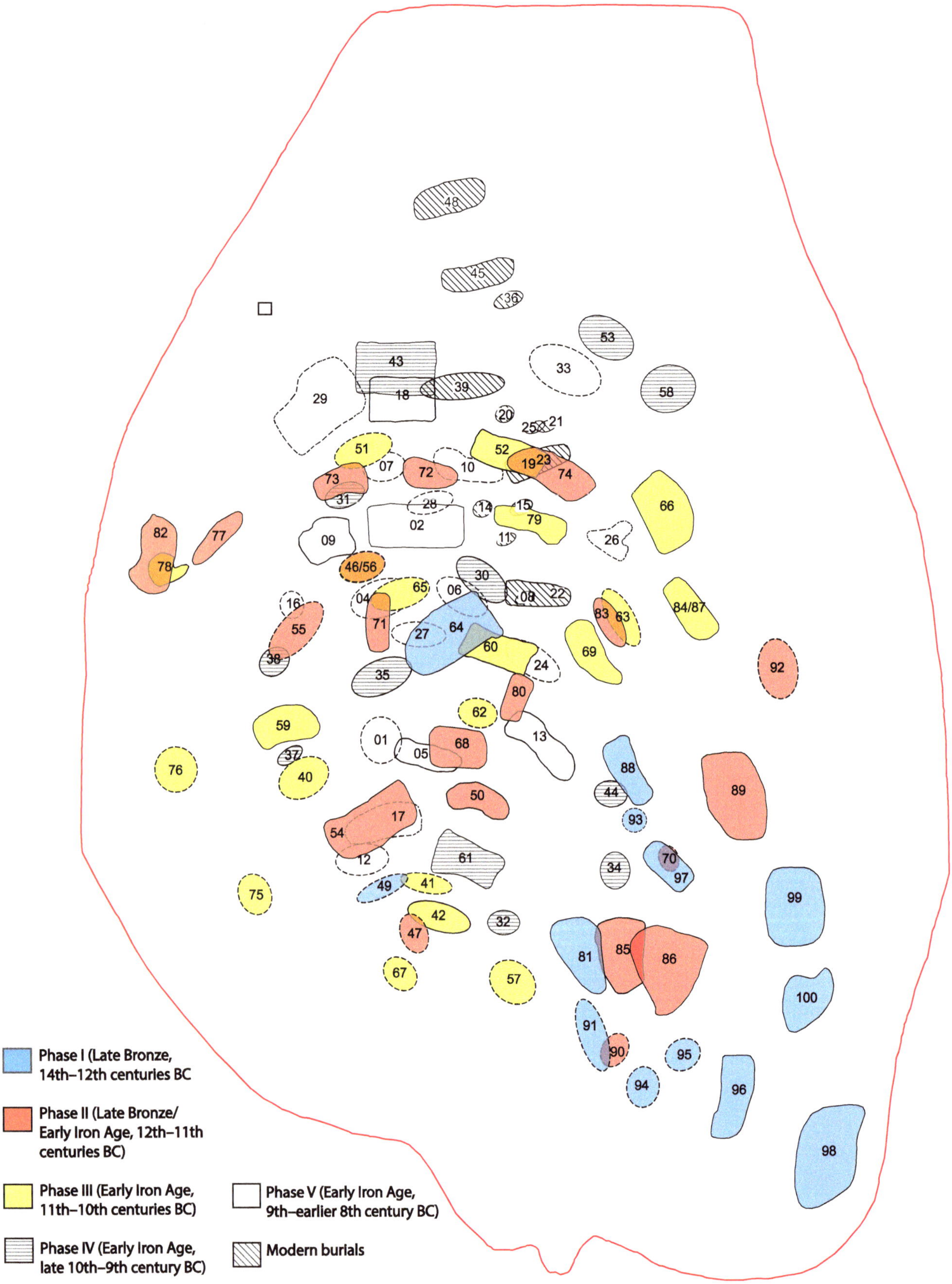

Figure 4.2 The Lofkënd tombs arranged according to five prehistoric phases, collapsing Phases Va and Vb into a single one (plus the modern phase)

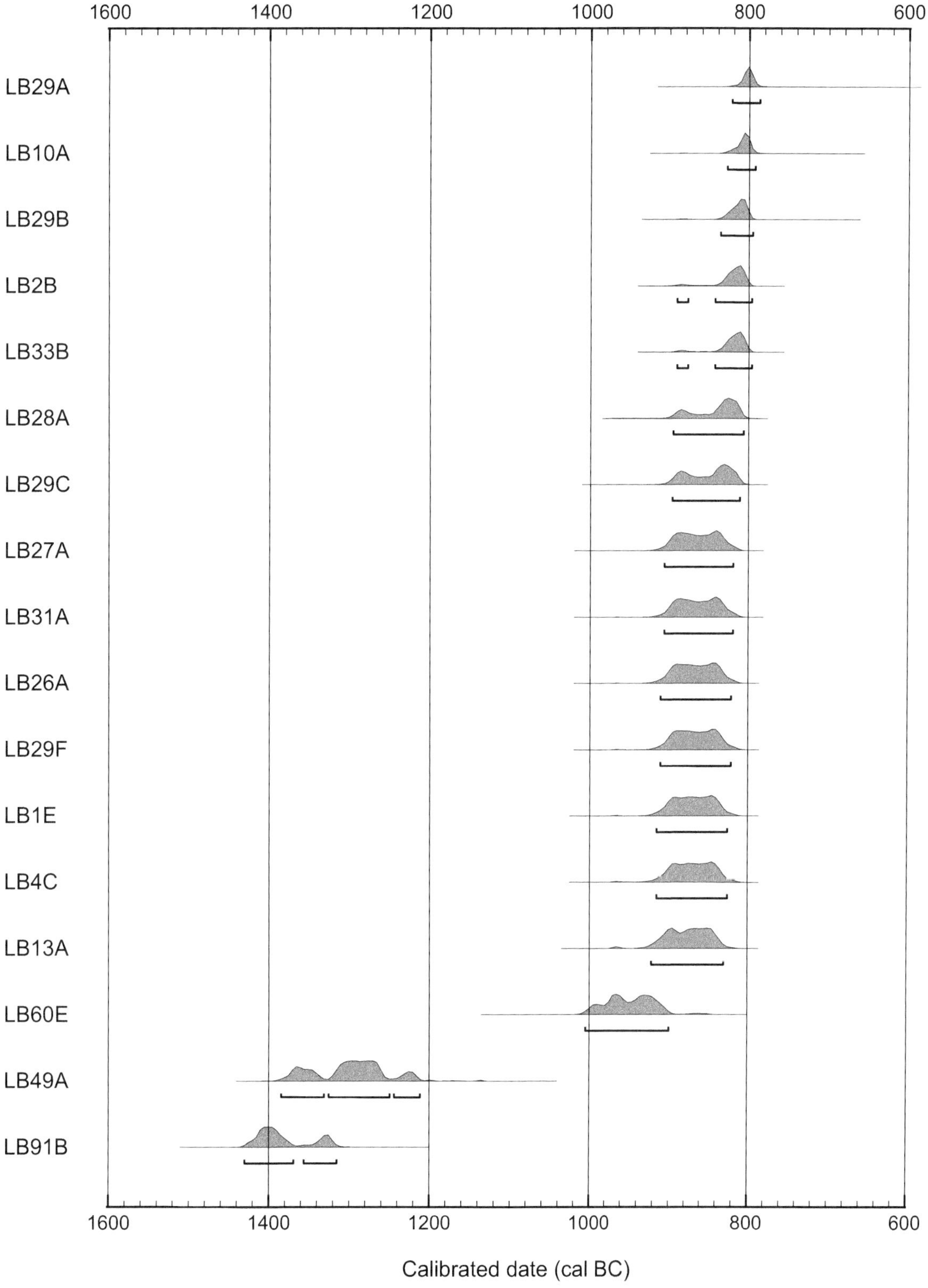

Figure 4.3 Chronological distribution of calibrated dates for human bone samples as generated by OxCal 4.1.3 (Bronk Ramsey 2009). Brackets denote 2-σ spread. The probability distributions are given in **Figures 4.5–4.11**.

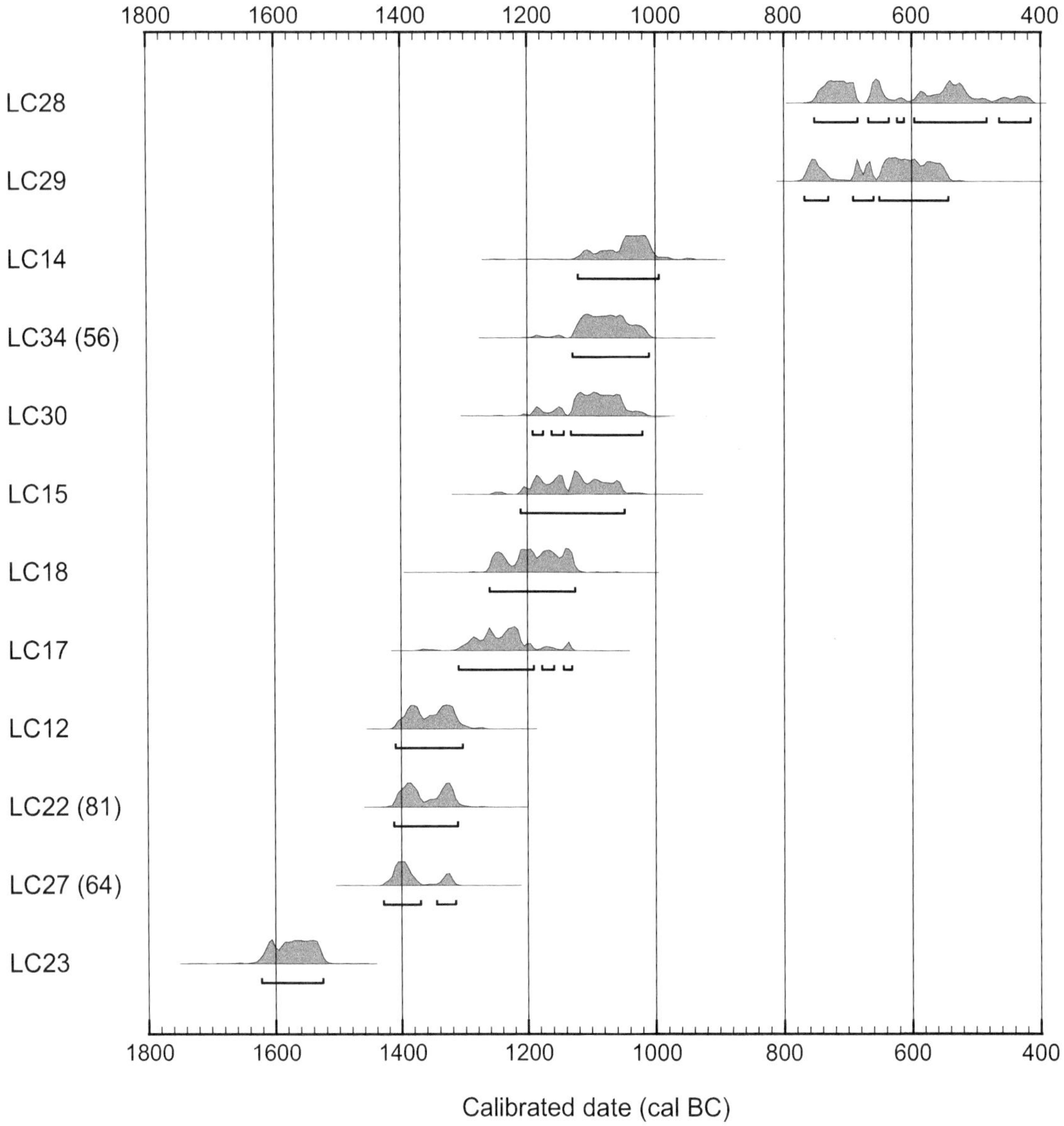

FIGURE 4.4 Chronological distribution of calibrated dates for charcoal samples as generated by OxCal 4.1.3 (Bronk Ramsey 2009). Number in parentheses indicates associated grave number. Brackets denote 2-σ spread. The probability distributions are given in **Figures 4.5–4.11**.

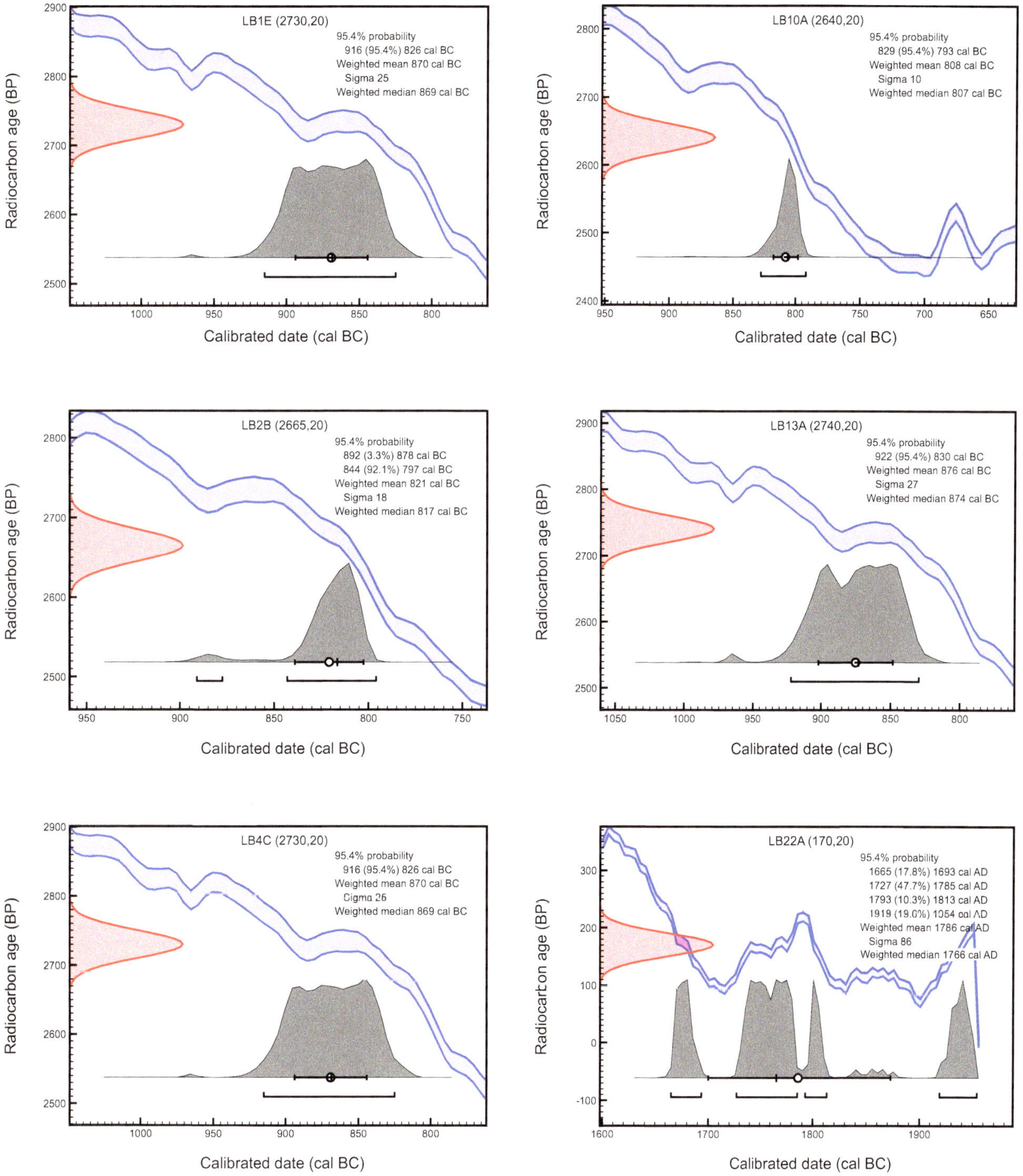

Figure 4.5 Calibrated radiocarbon results for samples LB1E, LB2B, LB4C, LB10A, LB13A, and LB22A using OxCal 4.1.3 (Bronk Ramsey 2009). Double lines denote the IntCal04 terrestrial calibration curve with 1-σ envelope (Reimer et al. 2004). Insets denote the results of Bayesian statistical analysis. Circle denotes weighted mean; cross denotes weighted median.

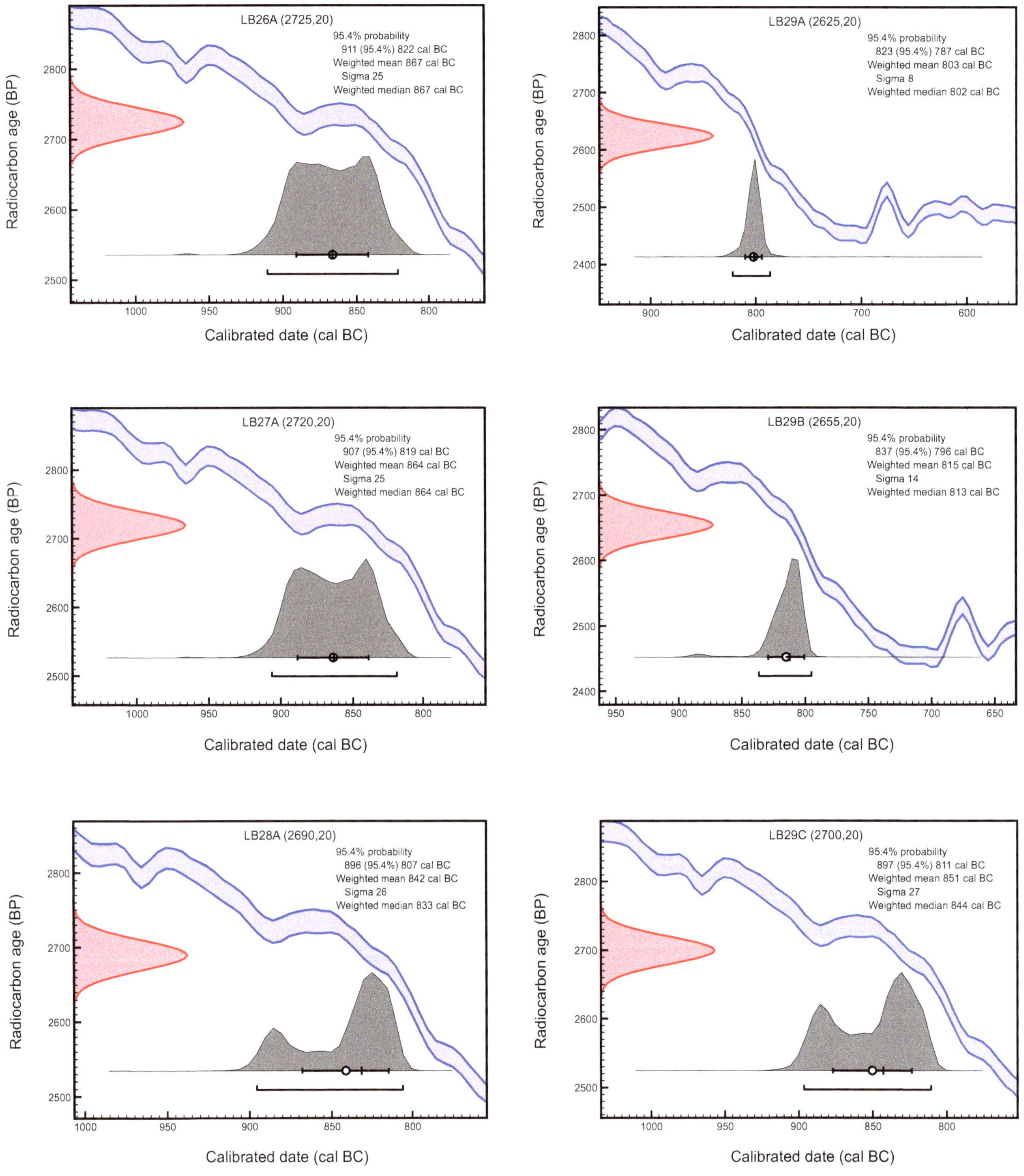

FIGURE 4.6 Calibrated radiocarbon results for samples LB26A, LB27A, LB28A, LB29A, LB29B, and LB29C using OxCal 4.1.3 (Bronk Ramsey 2009). See caption for **Figure 4.5**.

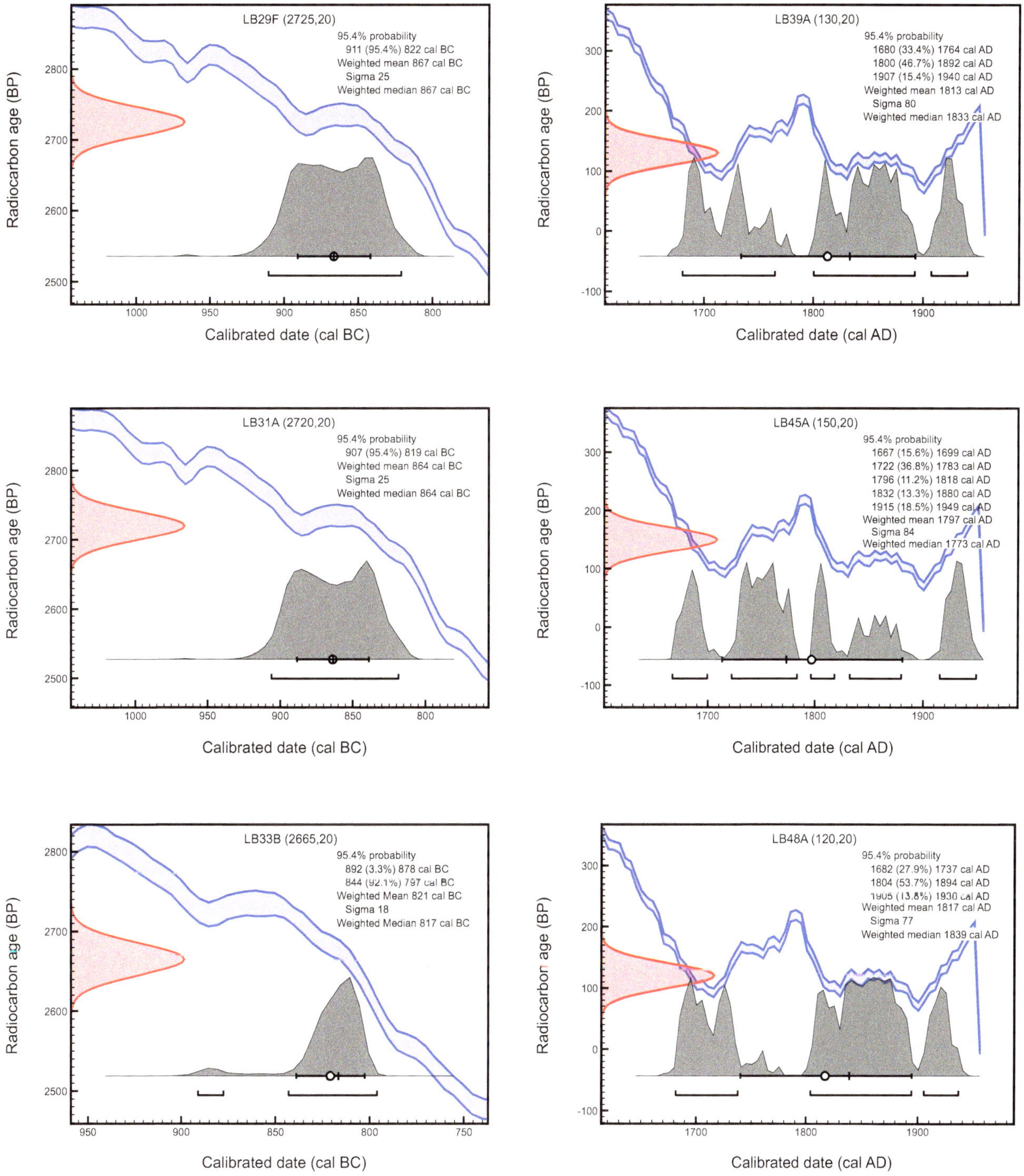

FIGURE 4.7 Calibrated radiocarbon results for samples LB29F, LB31A, LB33B, LB39A, LB45A, and LB48A using OxCal 4.1.3 (Bronk Ramsey 2009). See caption for **Figure 4.5**.

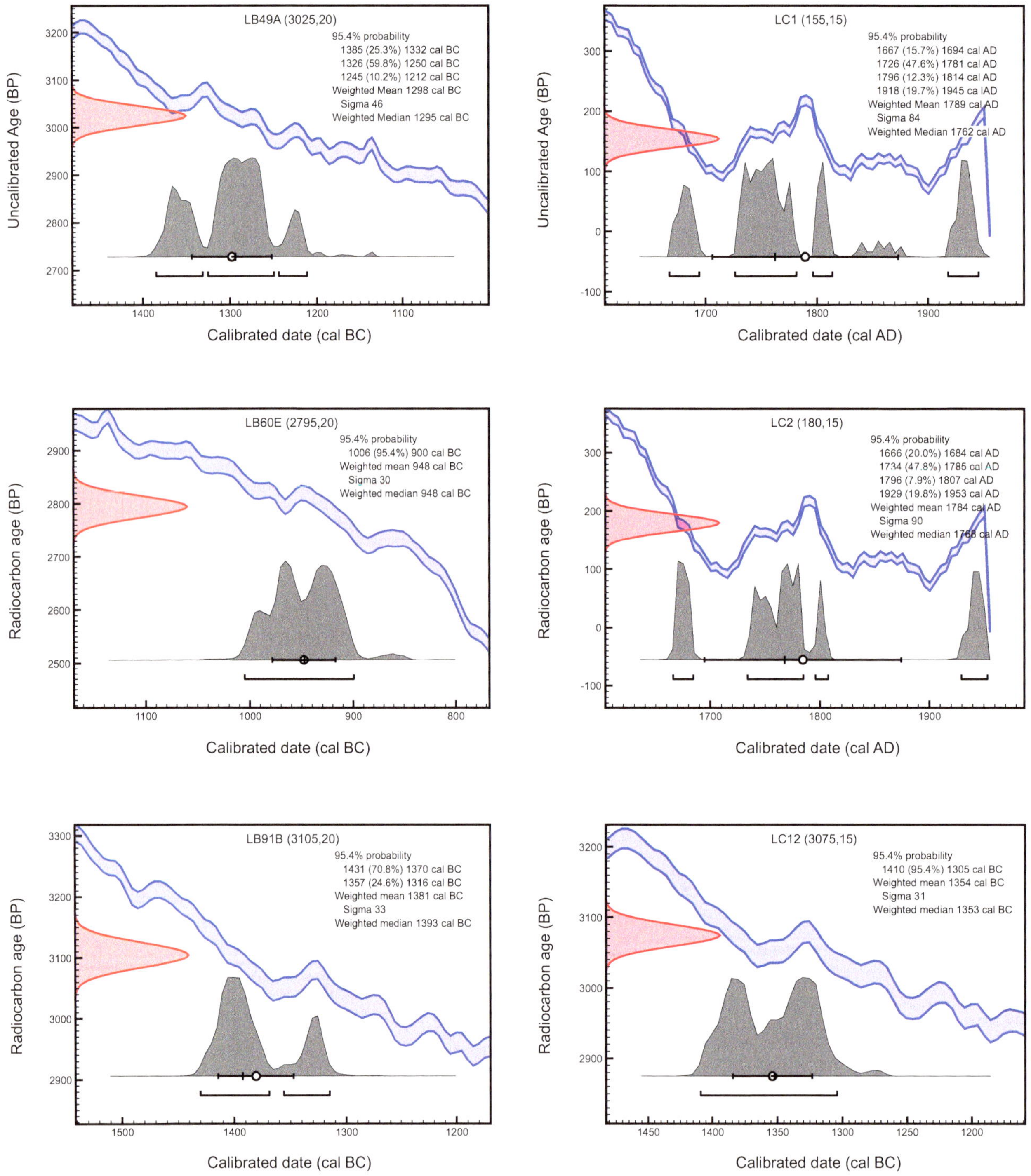

FIGURE 4.8 Calibrated radiocarbon results for samples LB49A, LB60E, LB91B, LC1, LC2, and LC12, using OxCal 4.1.3. See caption for **Figure 4.5**.

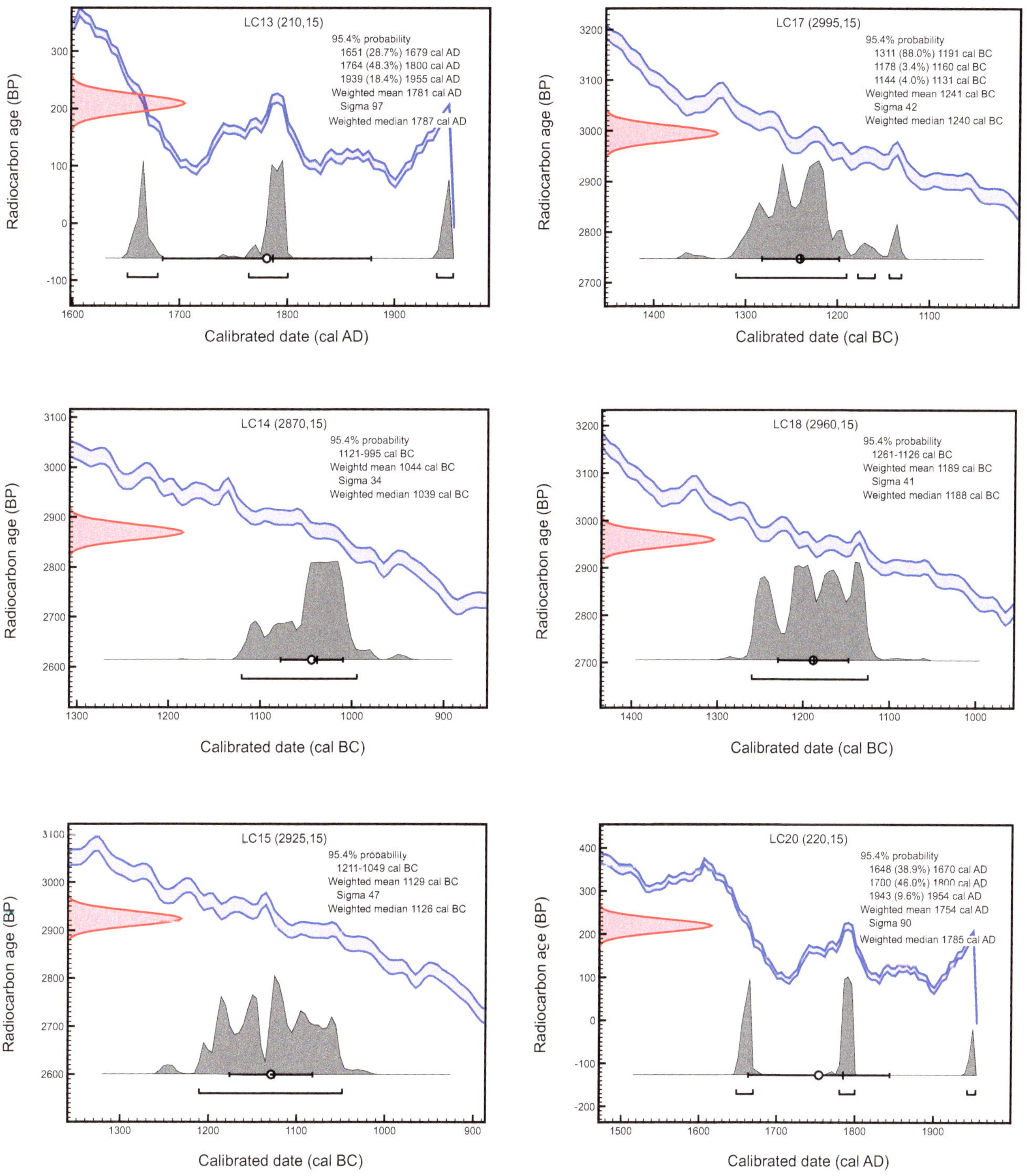

FIGURE 4.9 Calibrated radiocarbon results for samples LC13, LC14, LC15, LC17, LC18, and LC20 using OxCal 4.1.3 (Bronk Ramsey 2009). See caption for **Figure 4.5**.

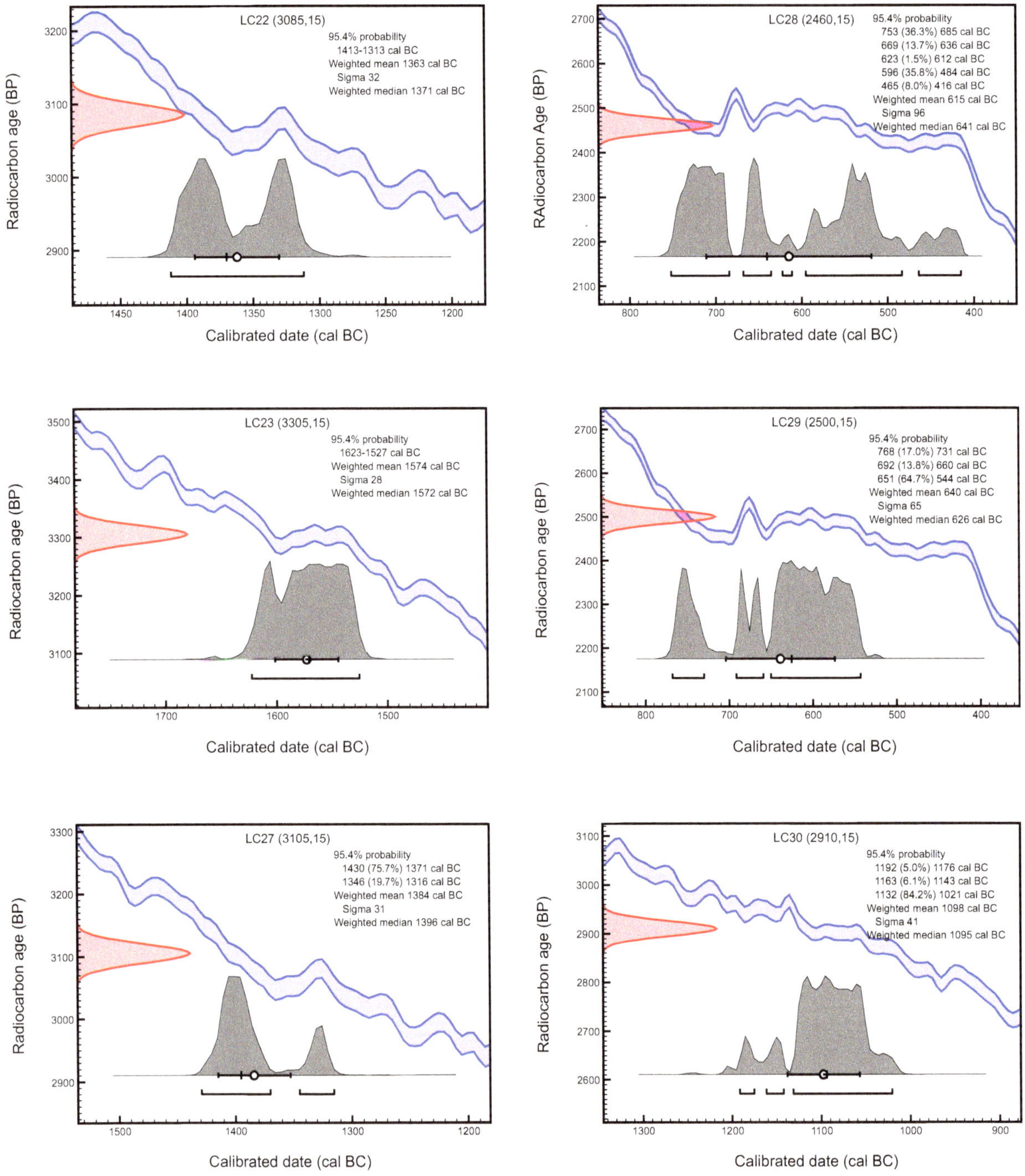

FIGURE 4.10 Calibrated radiocarbon results for samples LC22, LC23, LC27, LC28, LC29, and LC30 using OxCal 4.1.3 (Bronk Ramsey 2009). See caption for **Figure 4.5**.

FIGURE 4.11 Calibrated radiocarbon results for sample LC34 using OxCal 4.1.3 (Bronk Ramsey 2009). See caption for **Figure 4.5**.

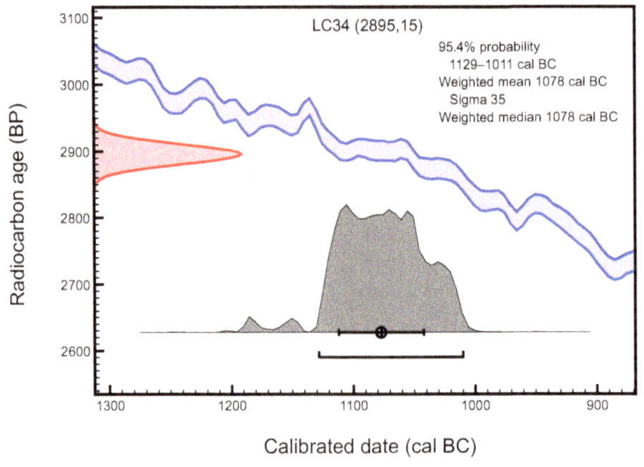

Chapter 5 Illustrations

FIGURE 5.1 The conservation laboratory of the Lofkënd Archaeological Project (taken during the 2006 field season)

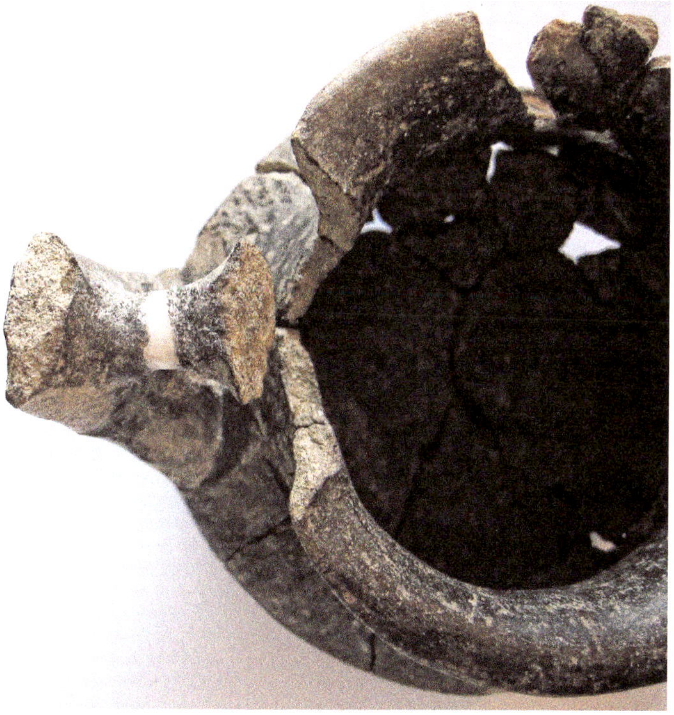

FIGURE 5.2 Detail of fill (white material) used to support an unstable join on the handle of vessel **9/94** (P322) (photo Allison Lewis)

a

b

FIGURE 5.3 (a) Calcareous accretions on the surface of a vessel **9/3** (P304), which are obscuring the matt-painted decoration (photo Alma Bardho); (b) Vessel **9/3** (P304) after cleaning, with the matt paint decoration revealed (photo Rich MacDonald)

Figure 5.4 Bronze fibula (**10/19** [SF 431a]) after cleaning

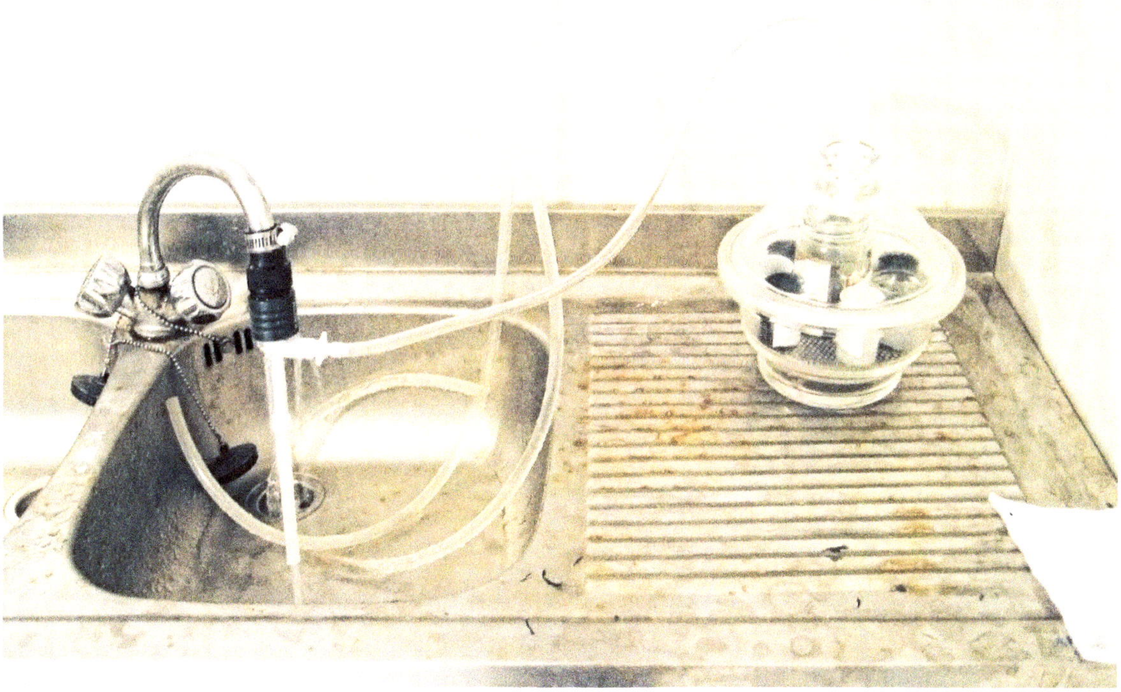

Figure 5.5 Treatment of copper alloy artifacts with BTA under vacuum using a small glass desiccator and running water from the sink faucet to draw a vacuum

Figure 5.6 Remnants of "shells" (spherical and hollow-looking corrosion) indicating active corrosion on one side of an iron boss (**10/64** [SF 092]) (photo Ian Coyle)

Figure 5.7 Pseudomorphs of a plain weave textile on a bimetallic pin (**10/50** [SF 111])

Figure 5.8 Conservators consolidate and face a bronze headband (**10/85** [SF 317]) on site in preparation for block lifting (photo Rich MacDonald)

Figure 5.9 Storage area for the Lofkënd finds in the old monastery at Apollonia

FIGURE 5.10 A ceramic vessel stored within a polypropylene container supported with Ethafoam® and acid-free tissue

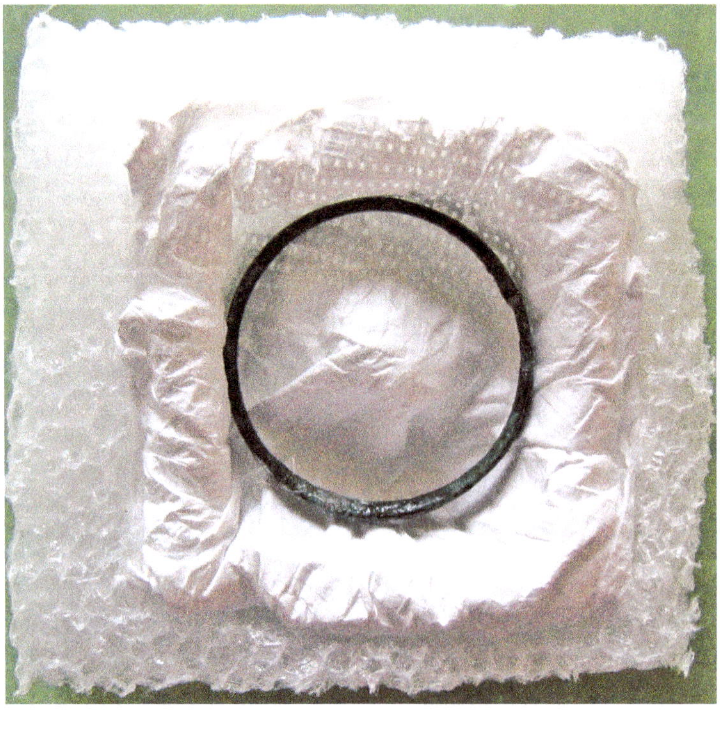

FIGURE 5.11 An example of a Tyvek®-lined Ethafoam® support for the storage of a bronze ring (**10/75** [SF 163e])

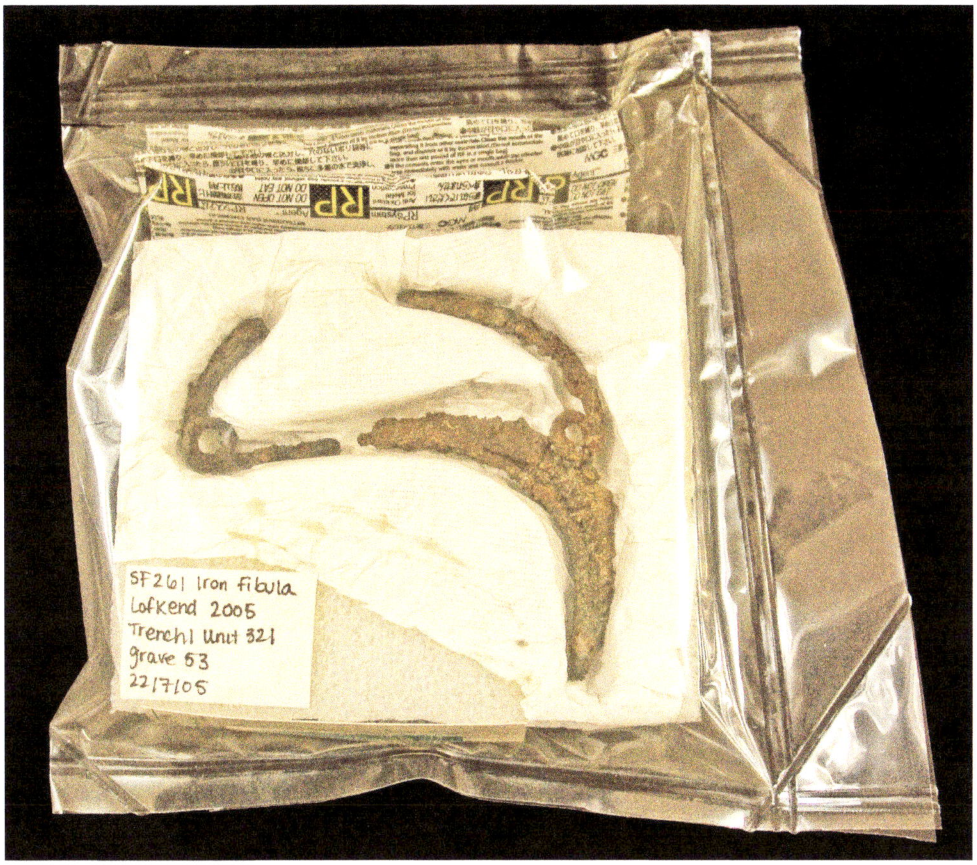

FIGURE 5.12 Storage of an iron fibula (**10/23** [SF 261]) using the RP System® and Escal™ barrier film. A humidity indicator strip was on the side of the Ethafoam® support inside the bag.

FIGURE 5.13 Labeling system employed for pottery using paper labels adhered to the artifact using Paraloid B-72

Chapter 6 Illustrations

FIGURE 6.1 Modern infant double burial en bloc, with the crania at the bottom. Tomb XC (Grave 14), Individuals 109 and 110.

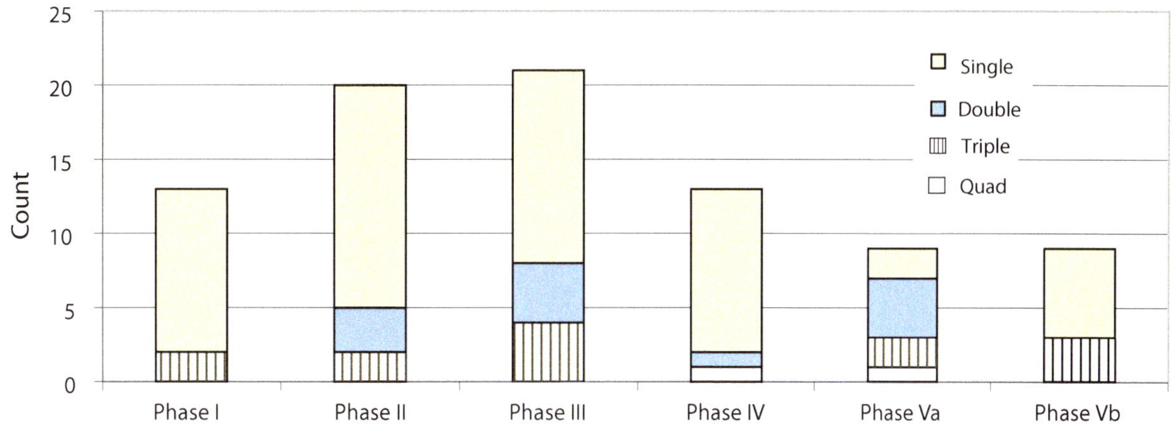

FIGURE 6.2 Temporal distribution of the prehistoric multiple burials

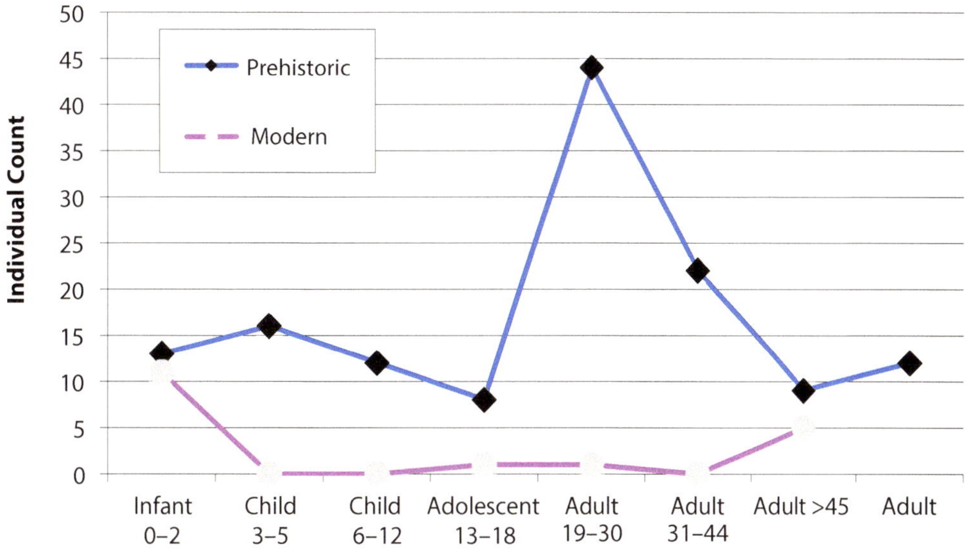

FIGURE 6.3 Mortality profiles for the prehistoric and modern subsamples

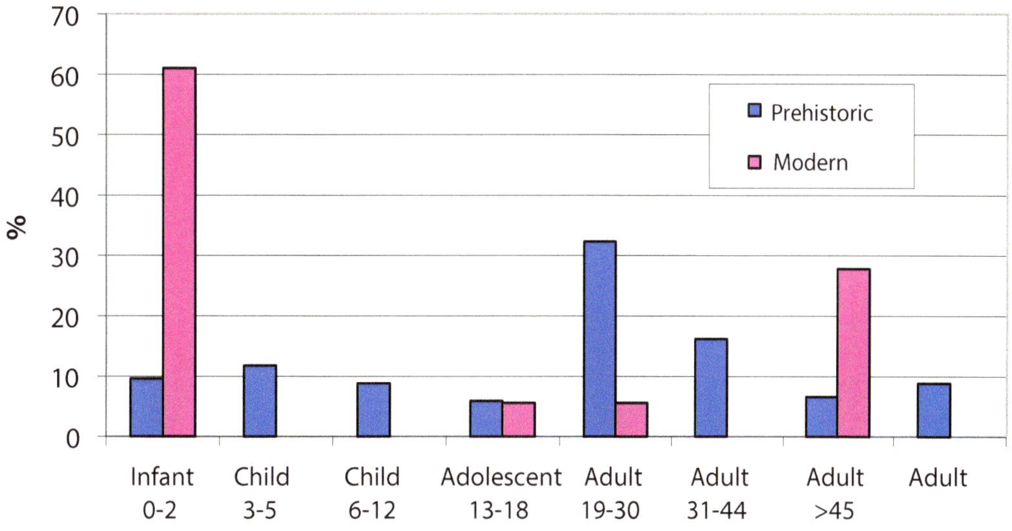

FIGURE 6.4 Proportional distribution of age

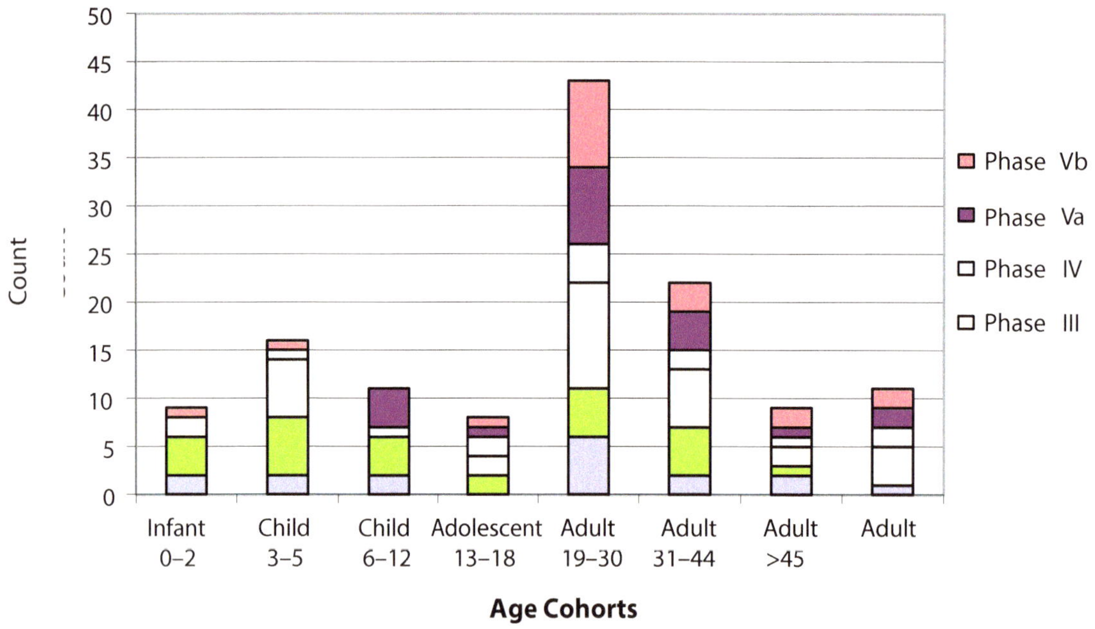

FIGURE 6.5 Representation of age cohorts by temporal phase

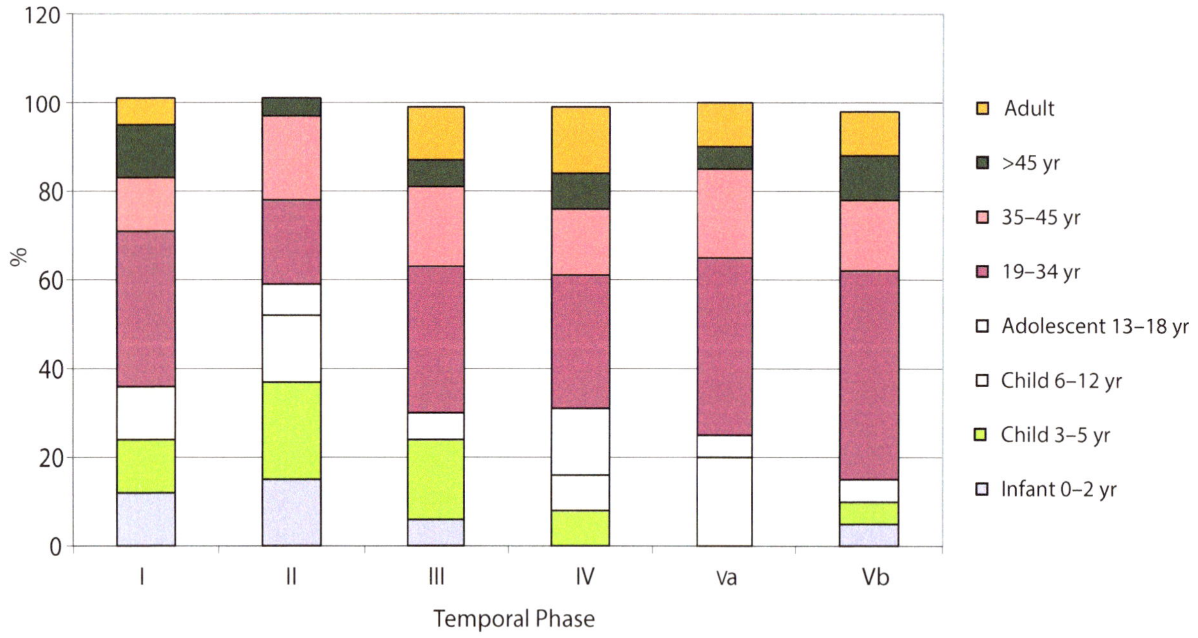

FIGURE 6.6 Age cohort proportional representation by temporal phase

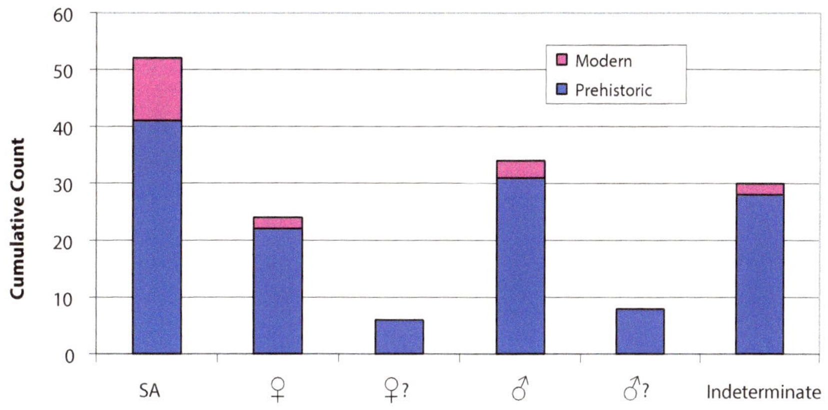

FIGURE 6.7 Sex distribution in the prehistoric and modern subsamples

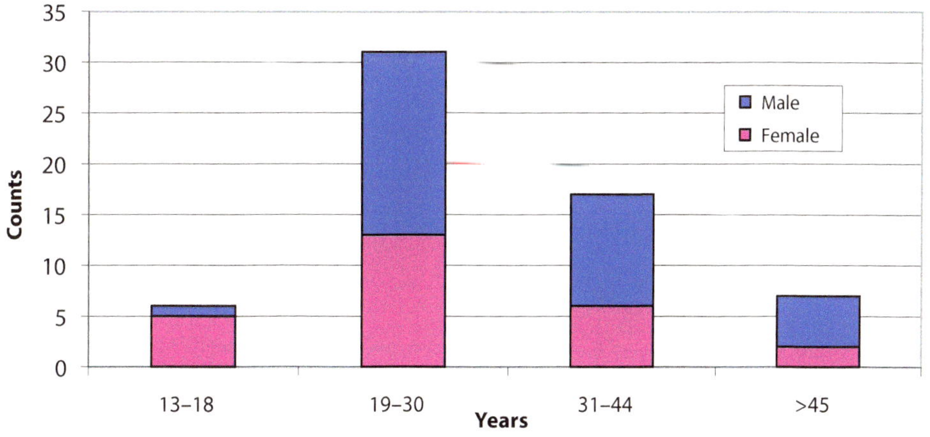

FIGURE 6.8 Sex distribution by age cohort for the prehistoric subsample

a

b

FIGURE 6.9 Prehistoric Lofkënd cranial form, Tomb X (Grave 95), Individual 554, adult female facing left: (a) lateral view; (b) superior view

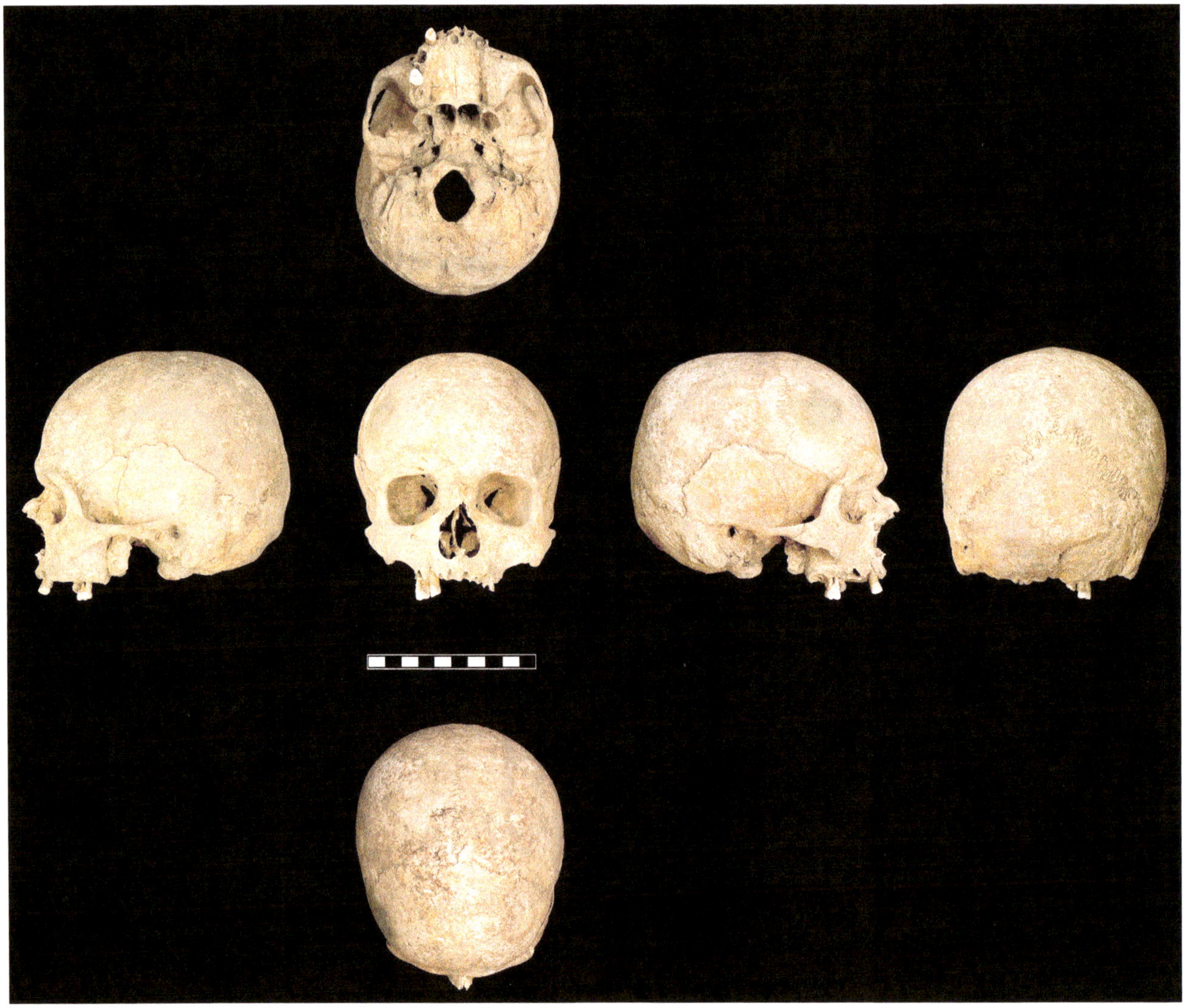

FIGURE 6.10 Modern Lofkënd cranial form, Tomb XCII (Grave 23), Individual 159

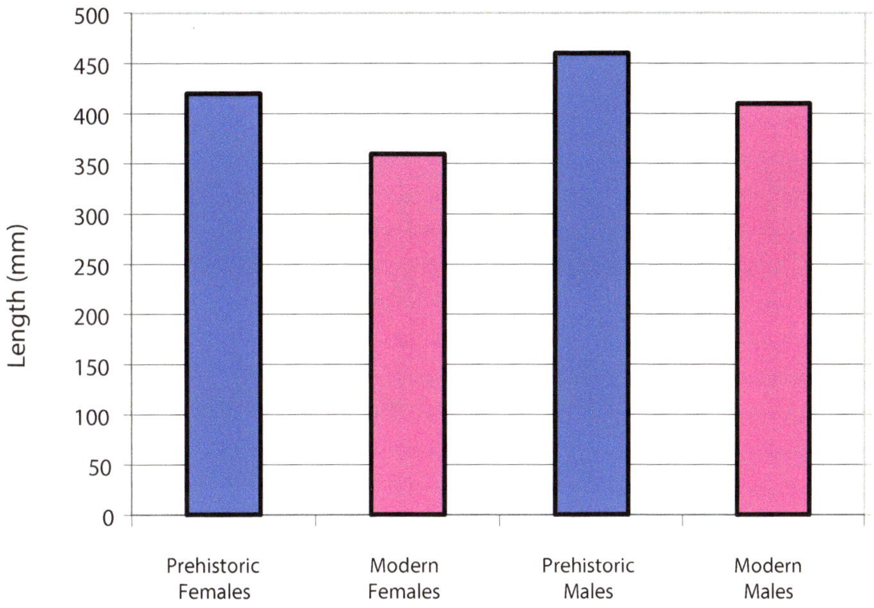

FIGURE 6.11 Mean femoral lengths (millimeters) for prehistoric and modern females and males

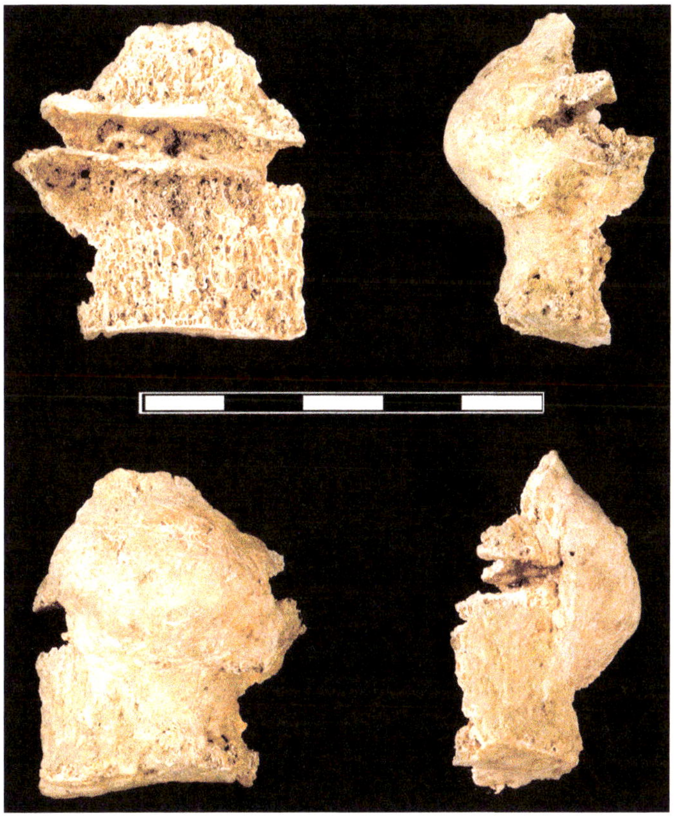

FIGURE 6.12 Thoracic vertebrae with fusion, possible case of DISH. Tomb XLV (Grave 60), Individual 345

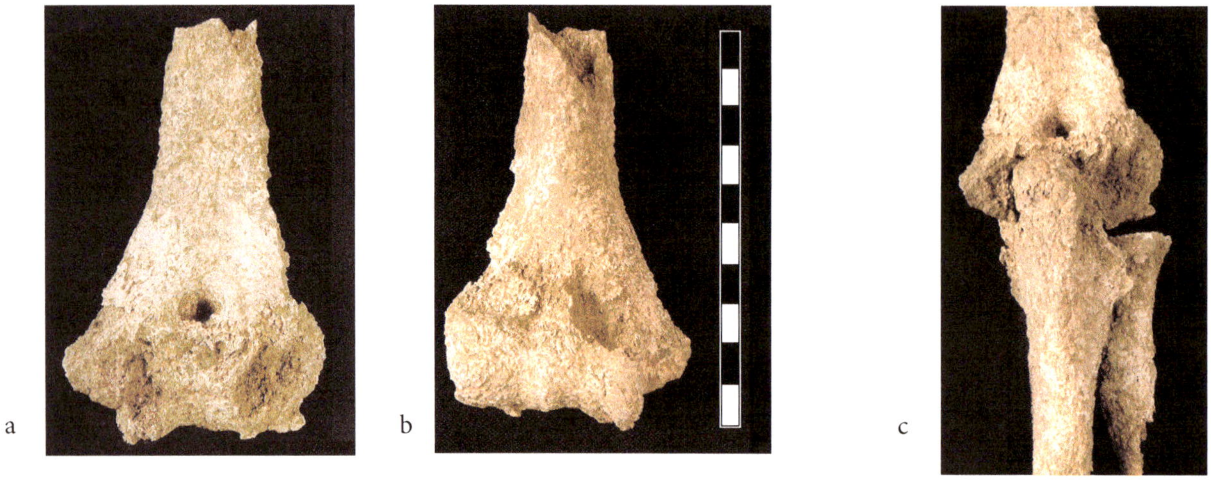

FIGURE 6.13 Elbow dislocation, Tomb LXXXIV (Grave 2), Individual 38, adult male: (a) posterior view of distal right humerus showing secondary olecranon fossa formation; (b) anterior view; (c) articulated elbow joint, posterior view

FIGURE 6.14 Healed cranial trauma, Tomb LXXVII (Grave 18), Individual 139: (a) exocranial view; (b) endocranial view

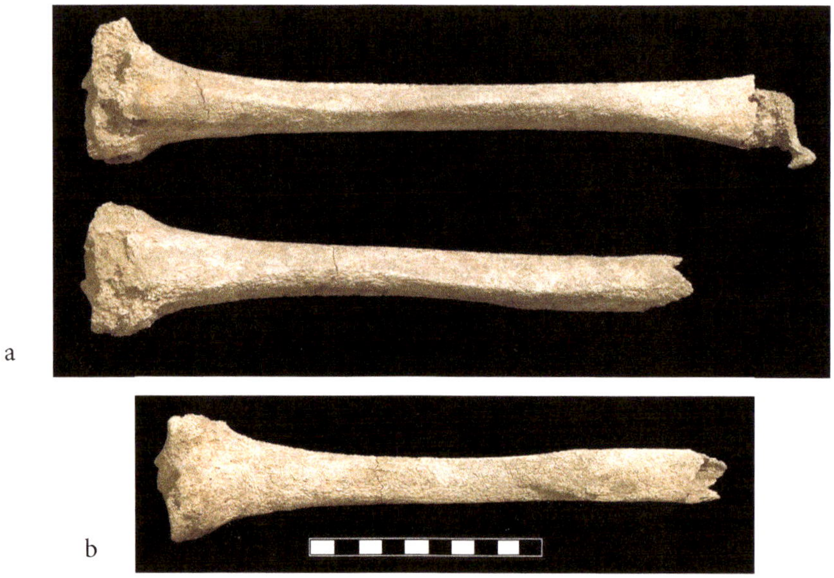

FIGURE 6.15 Right tibia trauma, Tomb XXXIX (Grave 66), Individual 369, adult female: (a) right and left tibiae, anterior view showing difference in the distal shafts; (b) right tibia, posterior view

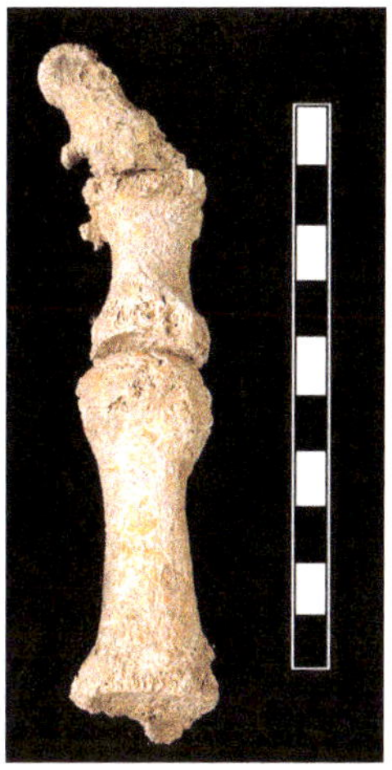

Figure 6.16 Hallux trauma, Tomb XXXVIII (Grave 79), Individual 439

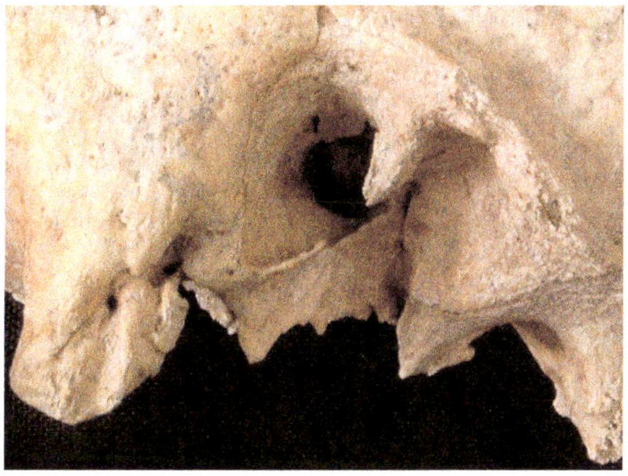

Figure 6.17 Mastoid and external auditory meatus infection, Tomb II (Grave 91), Individual 516

a

b

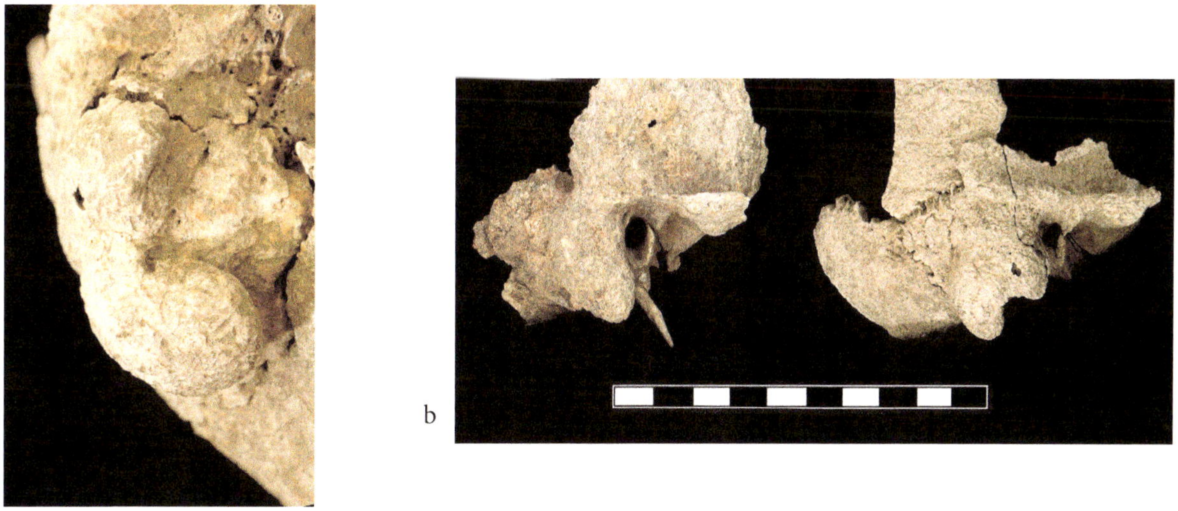

Figure 6.18 Possible mastoiditis: (a) double form of mastoid seen in basal view of Tomb XXXIX (Grave 66), Individual 369; (b) comparison with normal male mastoid morphology (left, Tomb XLIV [Grave 65]), Individual 365

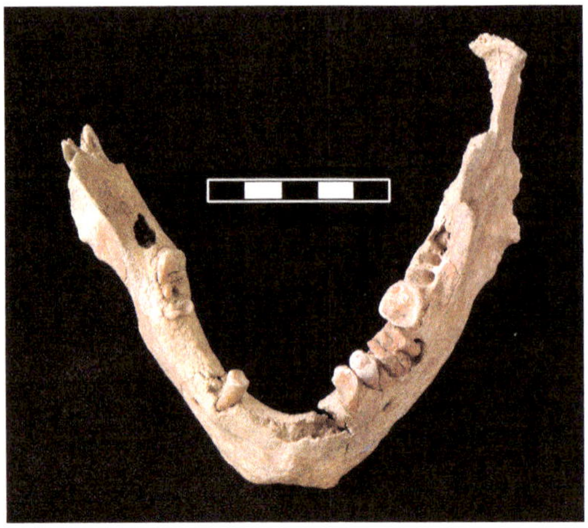

FIGURE 6.19 Antemortem tooth loss and caries, Tomb II (Grave 91), Individual 516, adult male

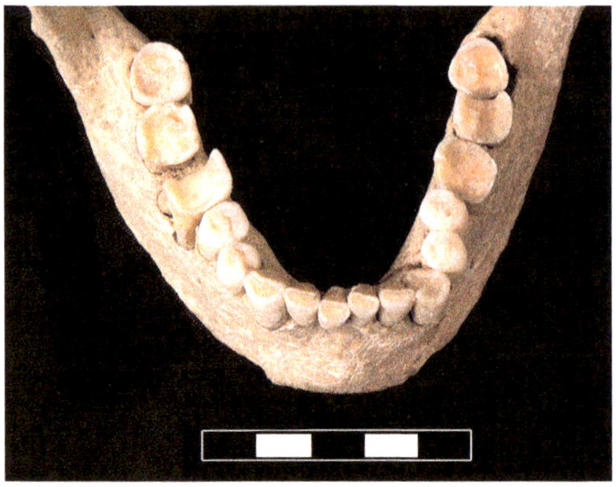

FIGURE 6.20 Dental crowding and rotation, Tomb XLV (Grave 60), Individual 351

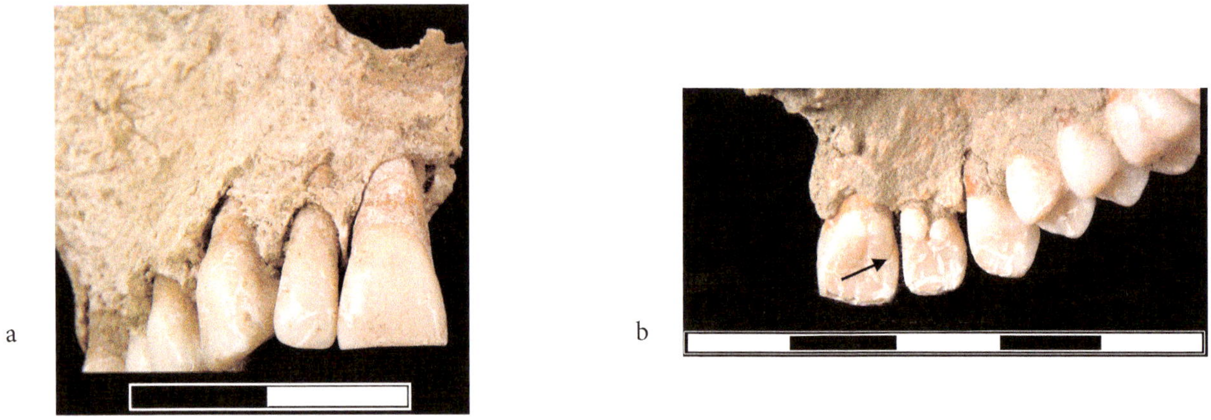

FIGURE 6.21 I2 Trait expressions: (a) reduced size, Tomb LXX (Grave 17), Individual 126; (b) form of the trait where the crown and root exhibit a groove, and the I1 has a faint groove on the distal aspect as indicated by the arrow; Tomb XLVIII (Grave 52), Individual 318

FIGURE 6.22 Foramen caecum molare indicated on dm2 and M1, Tomb XXIII (Grave 56), Individual 329

a
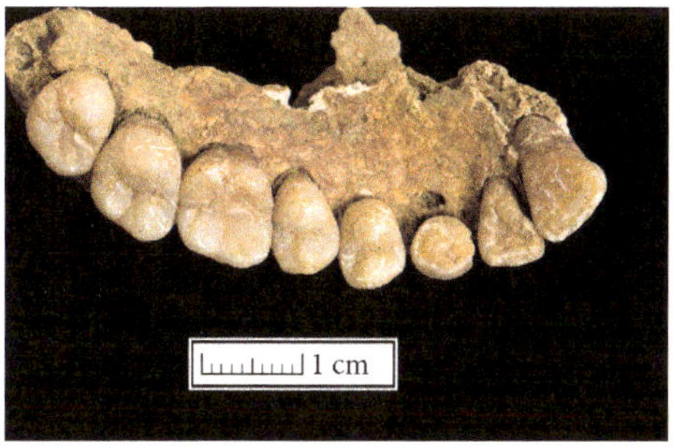

b
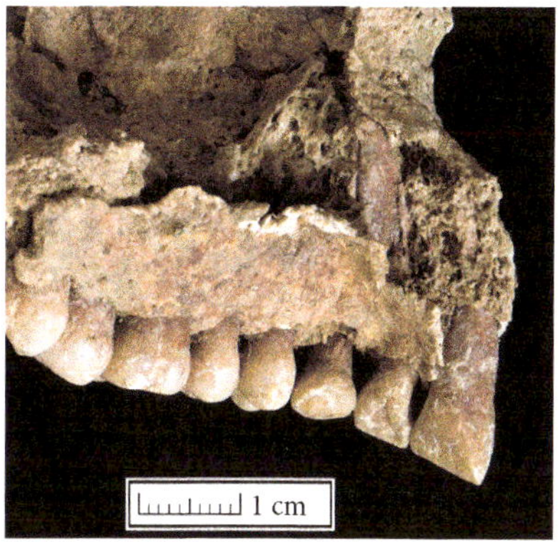

FIGURE 6.23 Retained left dc1, Tomb V (Grave 96), Individual 569: (a) lateral view; (b) palatal view with permanent canine root visible

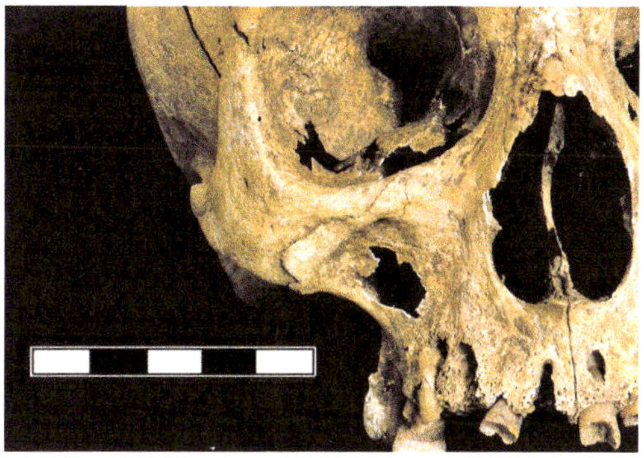

FIGURE 6.24 Infraorbital suture variant, Tomb C (Grave 48), Individual 302

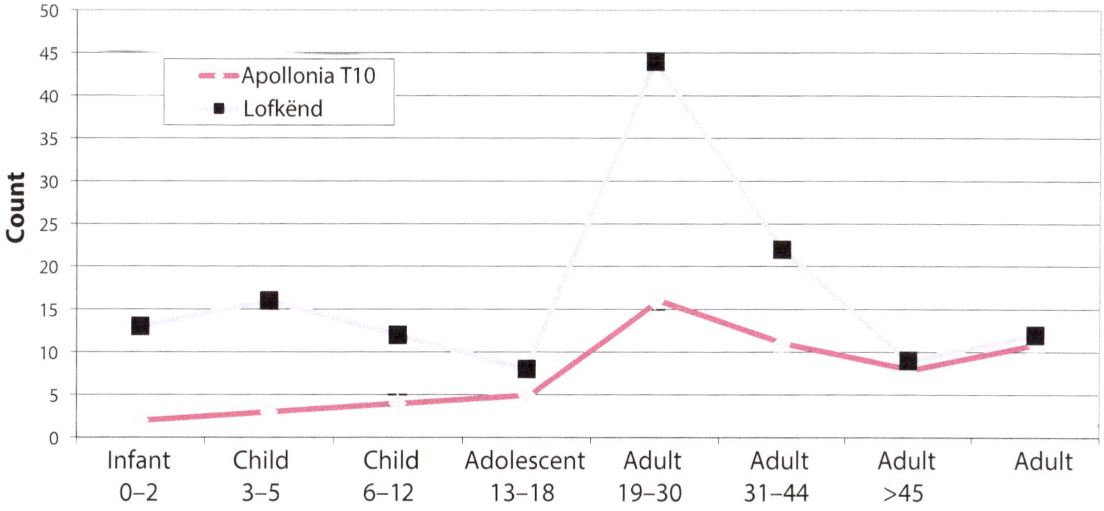

FIGURE 6.25 Comparative age profile of the Lofkënd and Apollonia prehistoric burials

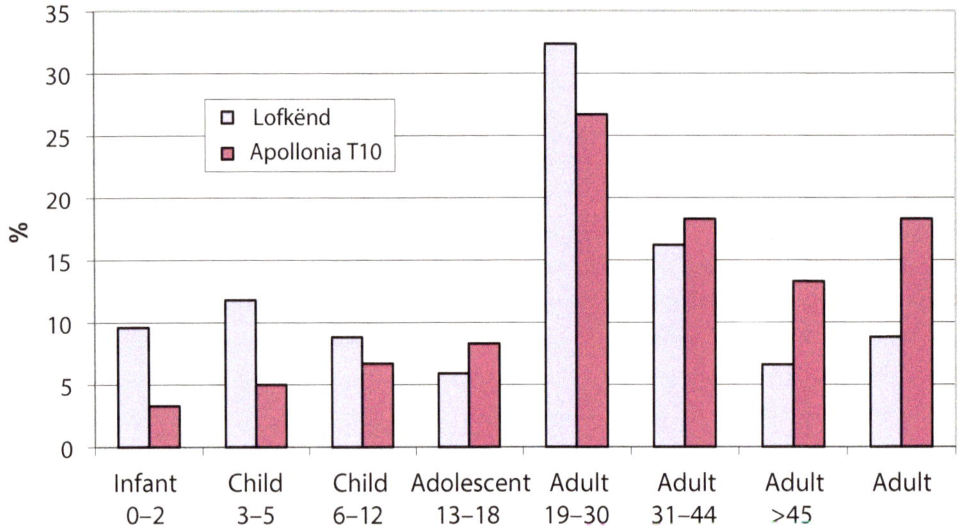

FIGURE 6.26 Proportional distribution of age for prehistoric individuals from Lofkënd and Apollonia Tumulus 10

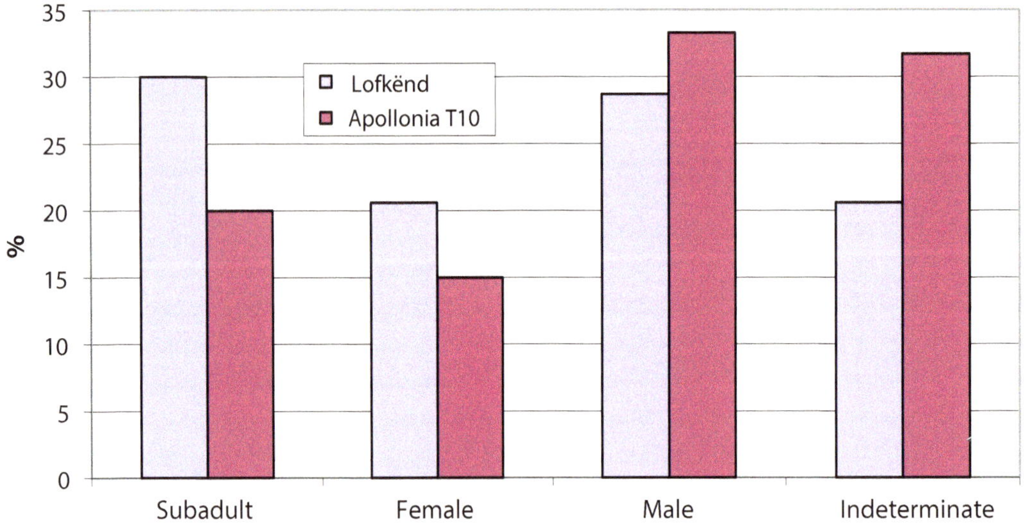

FIGURE 6.27 Proportional sex distribution of prehistoric burials from Lofkënd and Apollonia

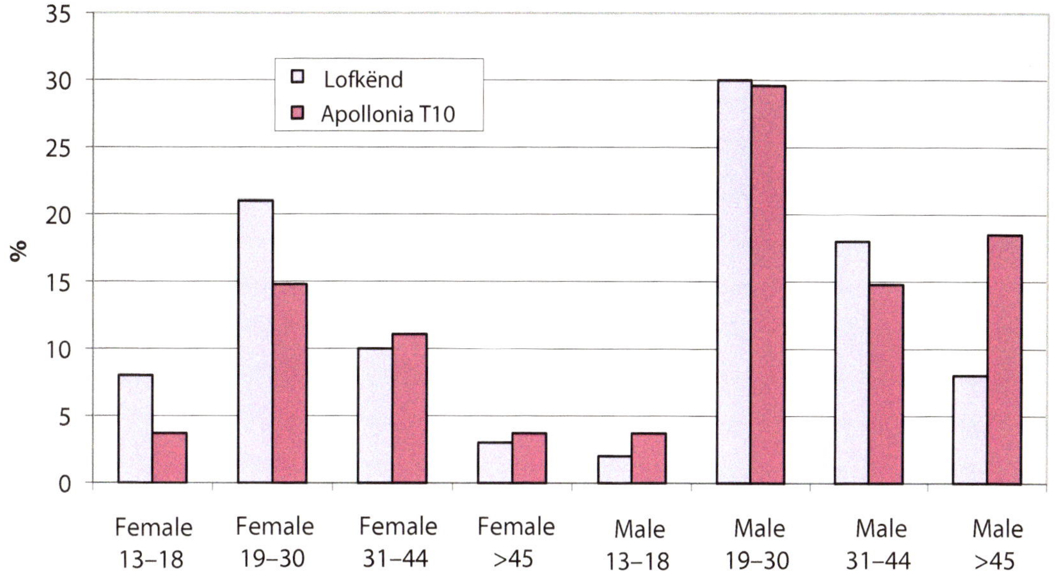

Figure 6.28 Age/sex proportional representation for the Lofkënd and Apollonia prehistoric subsamples

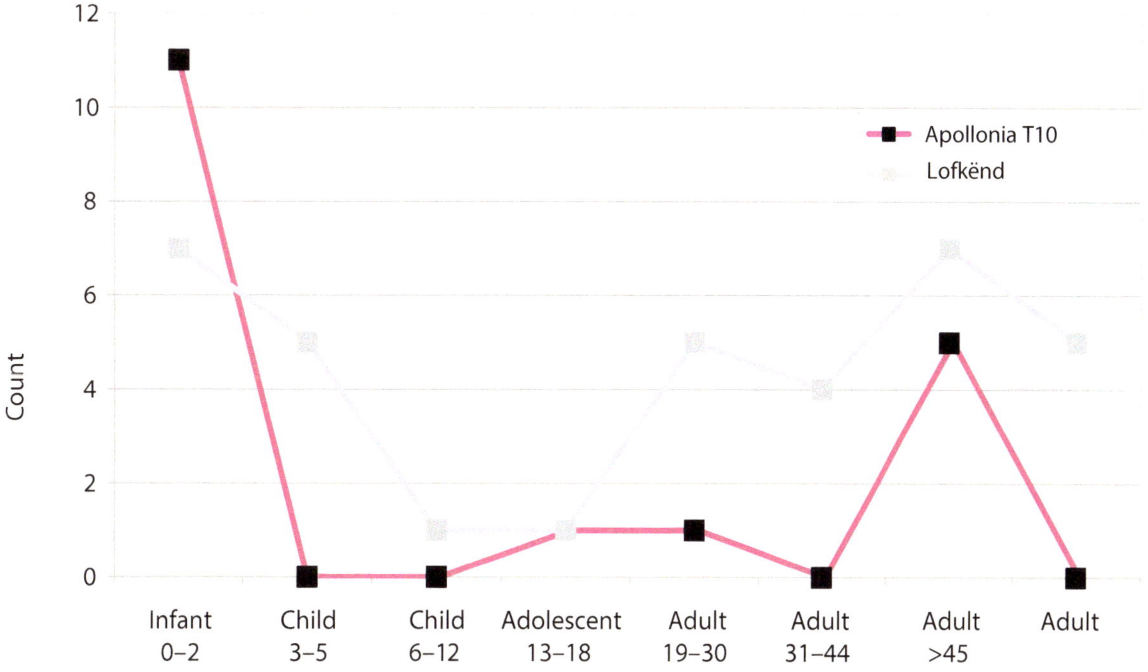

Figure 6.29 Age distribution of the post-medieval/modern burials

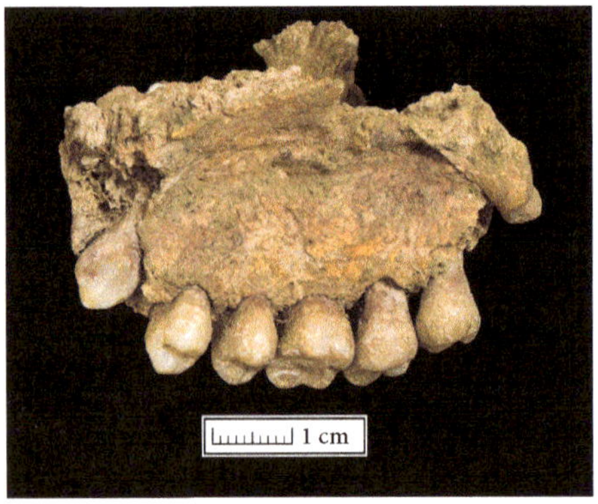

FIGURE 6.30 Tomb V (Grave 96), Individual 569, lingual view of right maxilla with unerupted C1 (photo 3583)

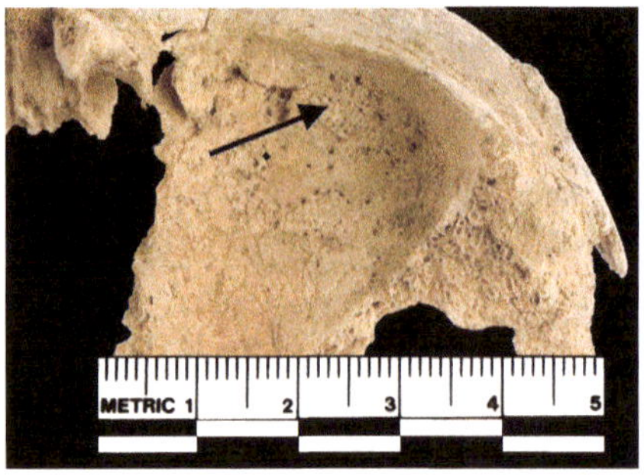

FIGURE 6.31 Tomb XII (Grave 88), left orbit of Individual 495. Arrow indicates porosity in region where cribra orbitalia occurs; other areas of porosity in the orbit suggest taphonomic damage to the surface (photo 3340)

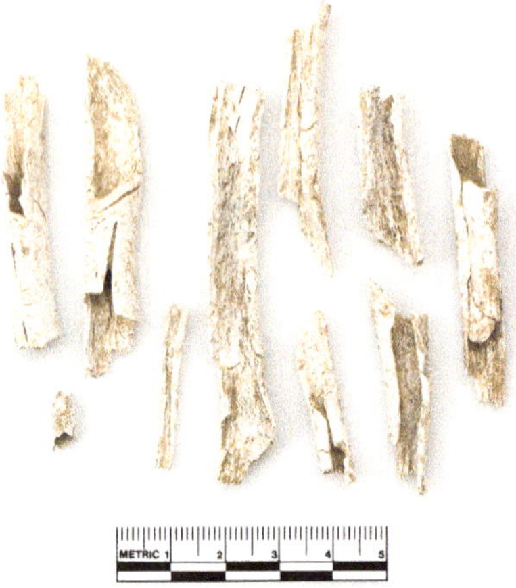

FIGURE 6.32 Long bone diaphyses of Tomb XXX (Grave 70), Individual 398 (cremation) (photo 2939)

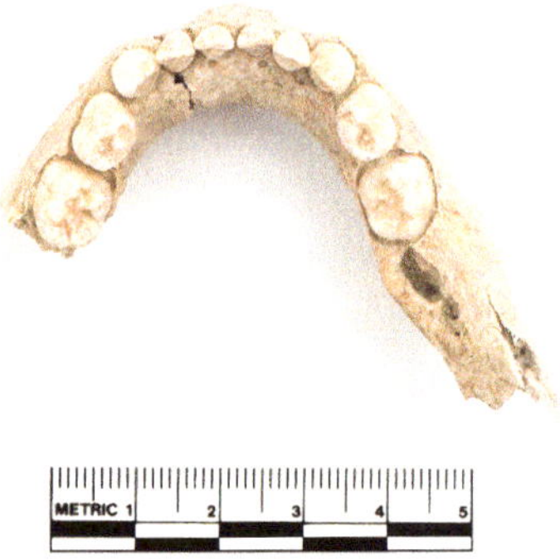

FIGURE 6.33 Occlusal view, Tomb XLII (Grave 59), Individual 340, mandible (photo 2874)

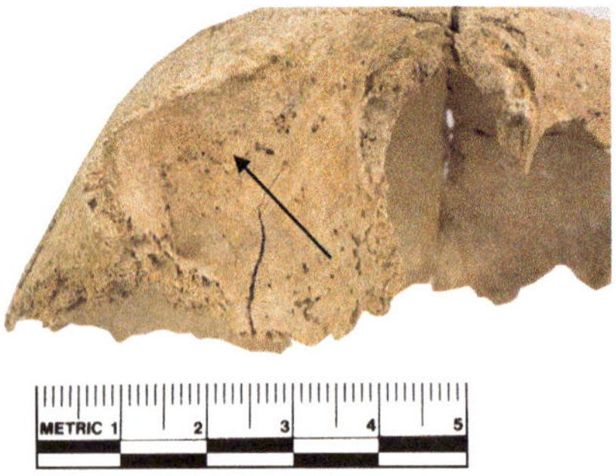

FIGURE 6.34 Tomb XLII (Grave 59), right orbit of Individual 340 with possible mild cribra orbitalia (photo 3346, Gr59340)

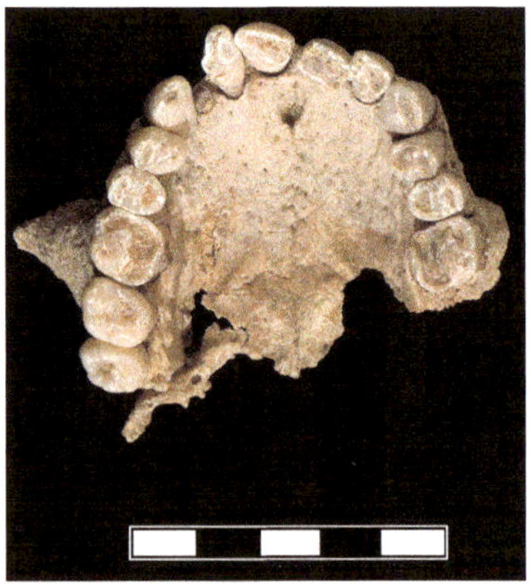

FIGURE 6.35 Occlusal view of Tomb XLIV (Grave 65), Individual 365: maxilla displaying crowding and rotation of the anterior dentition; I2 trait and a supernumerary tooth between the I2 and the C1 on the right side (photo 2936)

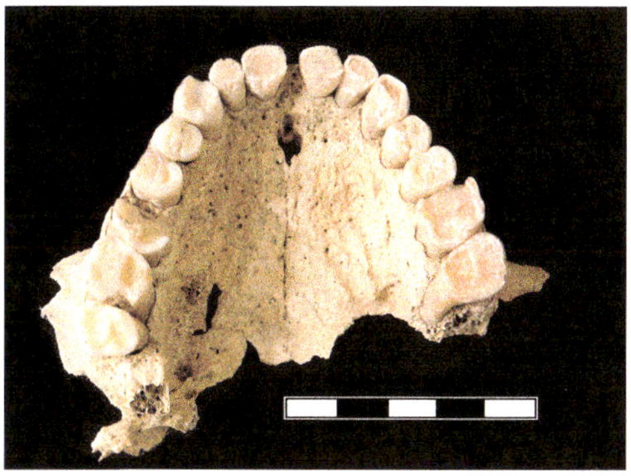

Figure 6.36 Occlusal view of Tomb XLV (Grave 60), Individual 345 maxilla. Note diastema between the central incisors, remnants of slight shoveling on the lateral incisors, right P3 rotation, caries, and M3 reduction (photo 3841)

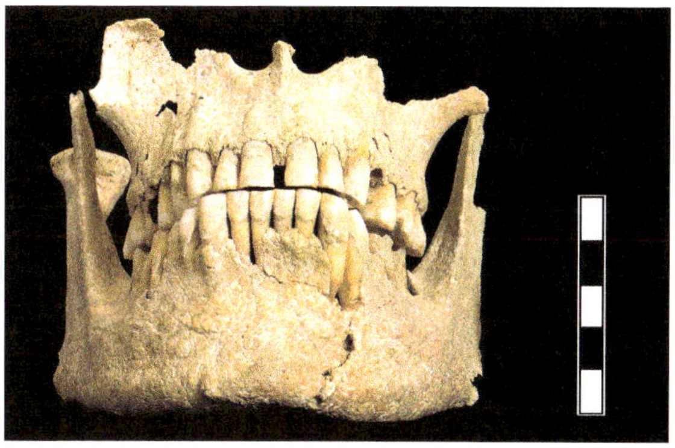

Figure 6.37 Articulated maxilla and mandible of Tomb XLV (Grave 60), Individual 345 (photo 3566)

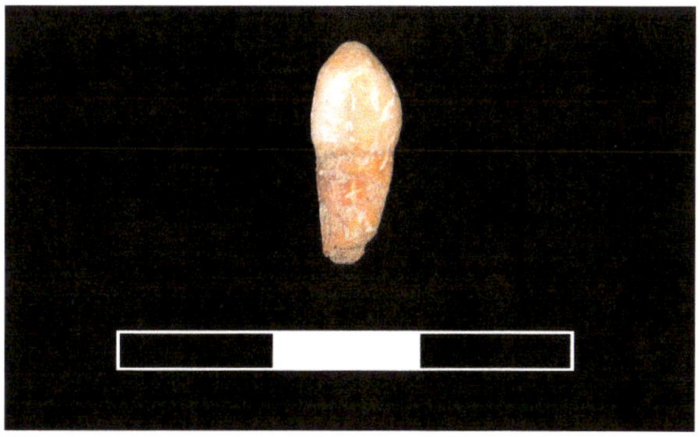

Figure 6.38 Peg tooth from Tomb XLVIII (Grave 52), Individual 366 (photo 3958)

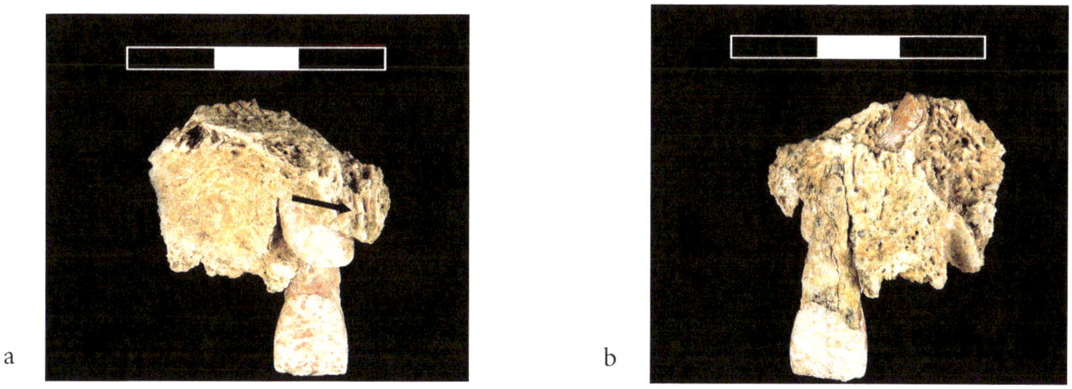

Figure 6.39 Tomb LII (Grave 69): (a) lingual view of Individual 390 maxilla with left I1 and displaced C1. A small portion of the anterior mid-palatal suture is indicated by the arrow (photo 3960); (b) labial view of Individual 390 maxilla with an arrow indicating the antemortem loss of the right I1 and the root of the displaced left C1 at the top of the image (photo 3962)

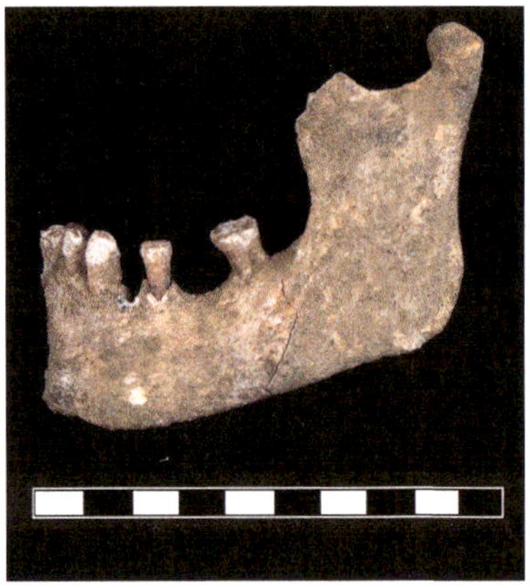

FIGURE 6.40 Tomb LXVII (Grave 12), left lateral view of Individual 100 mandible with antemortem loss of M1 and potential agenesis of M3 (photo 2094)

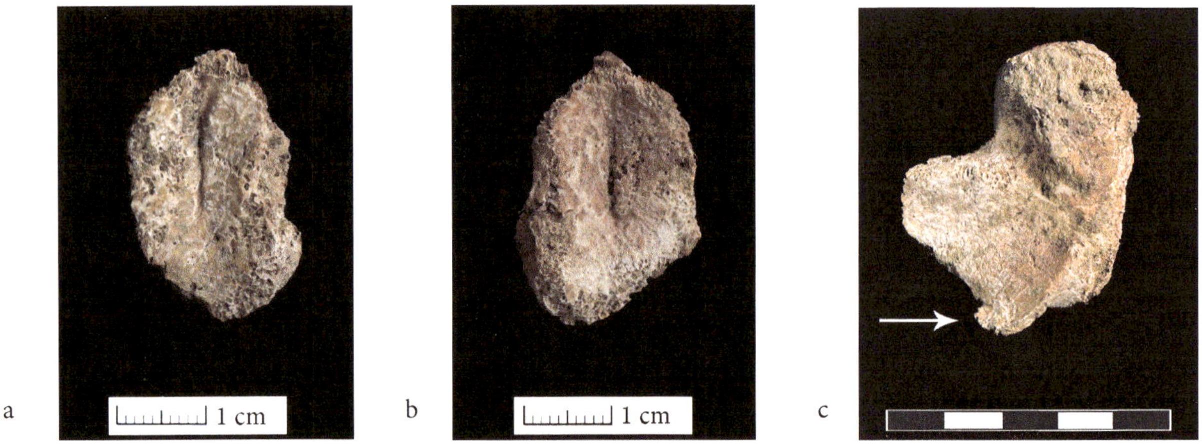

FIGURE 6.41 Foot variants, Tomb LXVII (Grave 12), Individual 171: (a) right hallux metatarsal with folded base (photo 2070); (b) left hallux metatarsal with folded base (photo 2071); (c) inferior view, left talus with extended facet indicated by arrow (photo 2073)

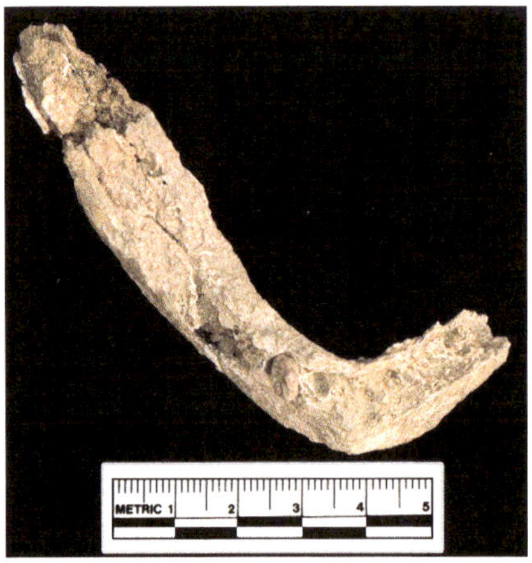

FIGURE 6.42 Tomb LXVIII (Grave 13): occlusal view of Individual 103 mandible with extensive antemortem losses (photo 343)

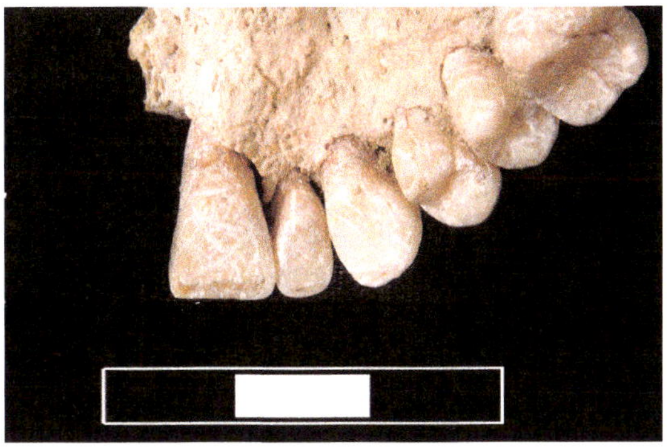

FIGURE 6.43 Tomb LXX (Grave 17): reduced I2, lingual view of the right maxilla of Individual 126 (photo 3839)

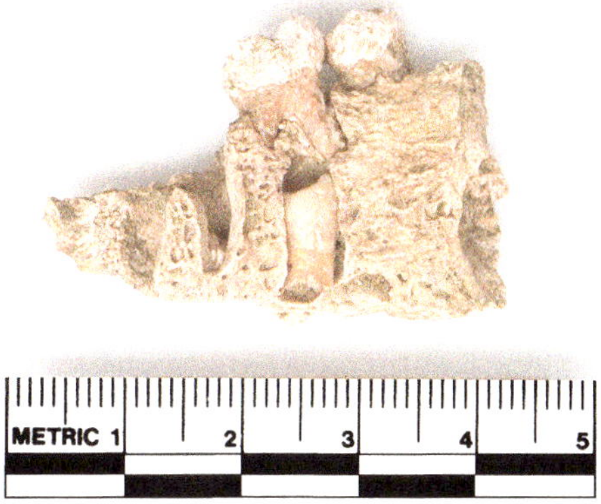

FIGURE 6.44 Tomb LXXI (Grave 28), Individual 177: left mandible, lingual view with dm1 and dm2, M1 socket and incomplete and unerupted P4 (photo 2757)

FIGURE 6.45 Tomb LXXX (Grave 4), Individual 32: left femur, posterior surface with healed periostitic reactive region (photo 352)

a

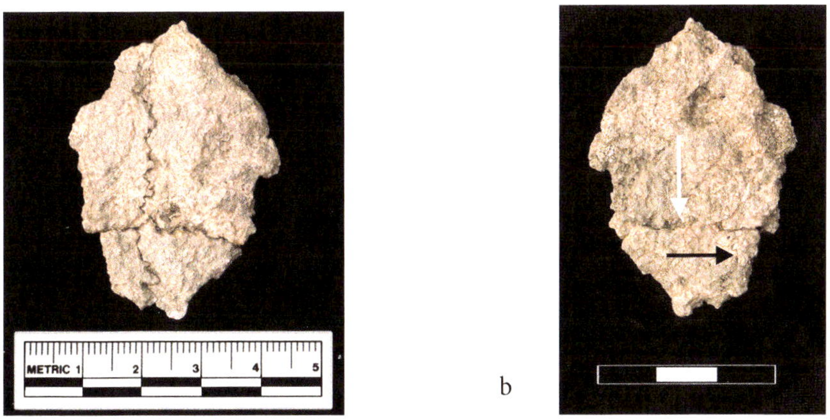

b

FIGURE 6.46 Tomb LXXXIII (Grave 7), Individual 60: (a) exocranial view of active metopic suture running toward the top of the image. The coronal and sagittal sutures are also active (photo 339); (b) endocranial view showing fusion of the metopic suture. Active coronal suture indicated by the white arrow; active sagittal suture indicated by the black arrow (photo 340)

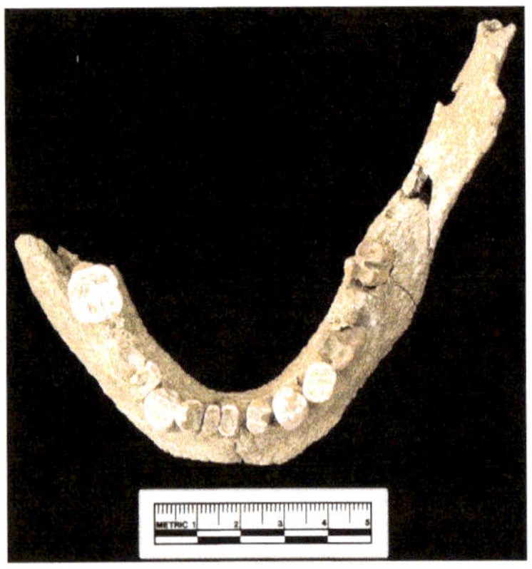

FIGURE 6.47 Tomb LXXXIV (Grave 2): occlusal view of Individual 38 mandible showing extreme wear on incisors, carious decay and loss of molar teeth or crowns on the left; premolar rotation (photo 346)

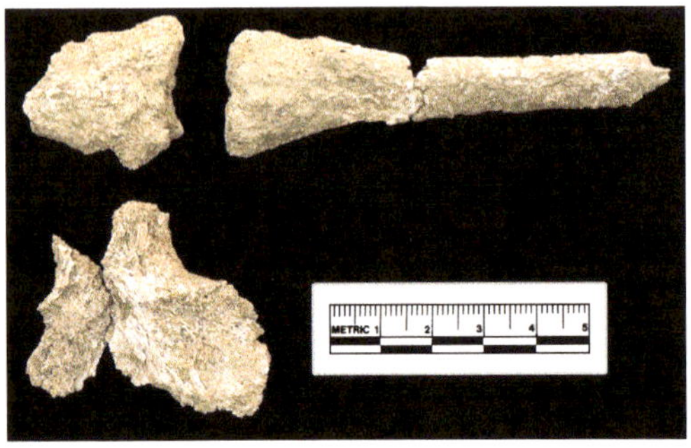

FIGURE 6.48 Tomb LXXXIV (Grave 2): fragments of child innominate, femur and tibia found above the knees of Individual 18 (photo 342)

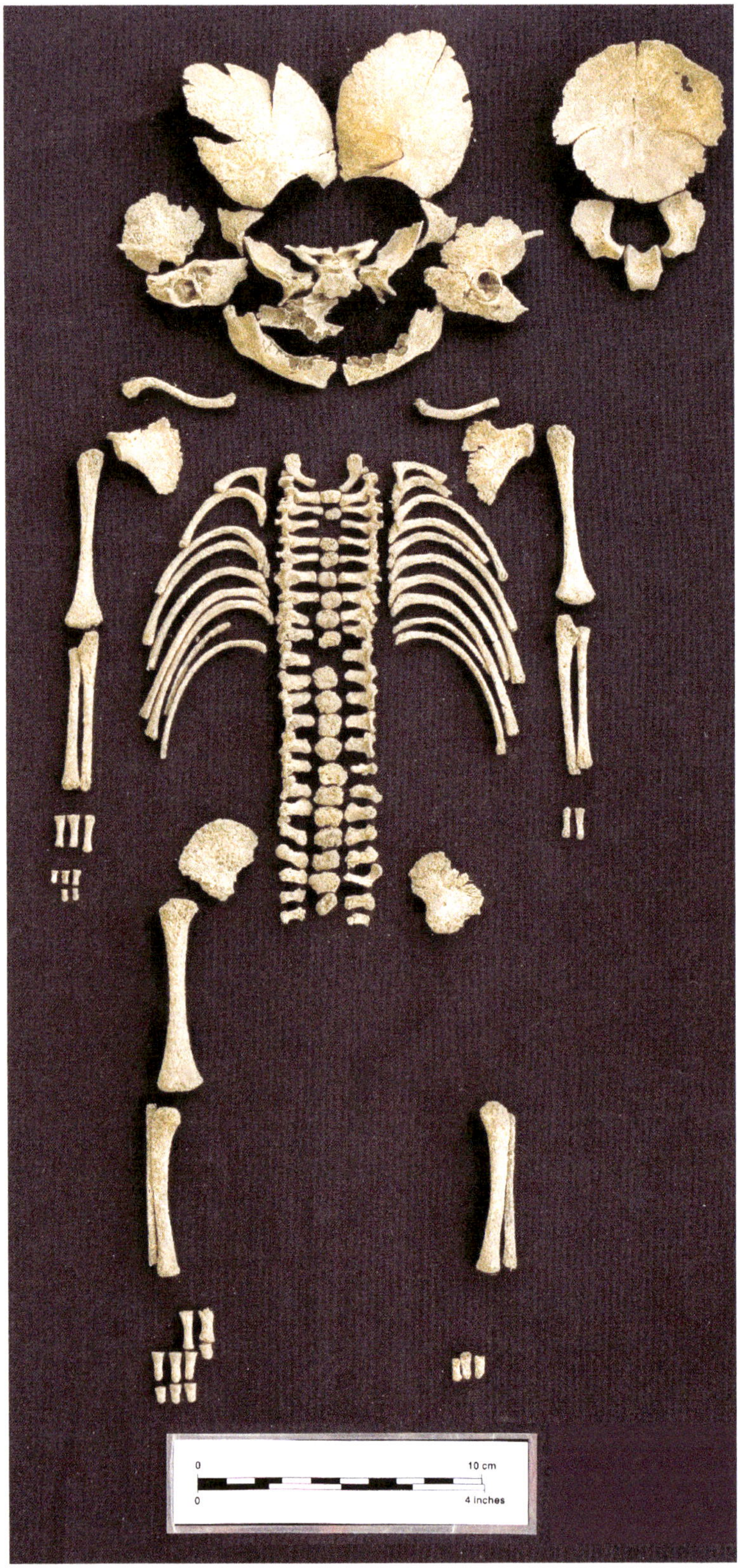

FIGURE 6.49 Tomb LXXXIX (Grave 11), Individual 96 as laid out, with occipital (photo 438)

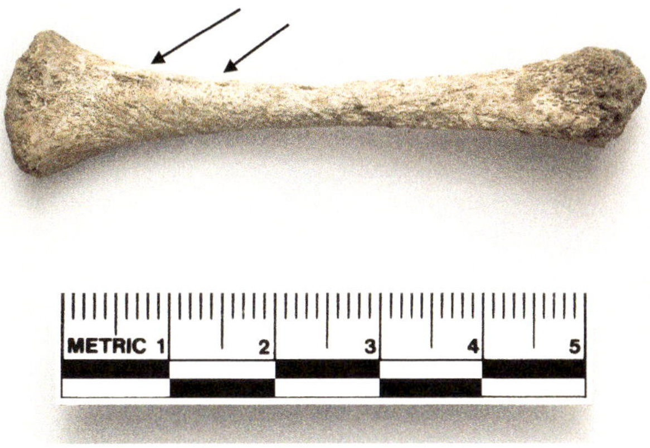

Figure 6.50 Tomb LXXXIX (Grave 11), Individual 96: left humeral diaphysis with rugosity along the medial supracondylar ridge (photo 2942)

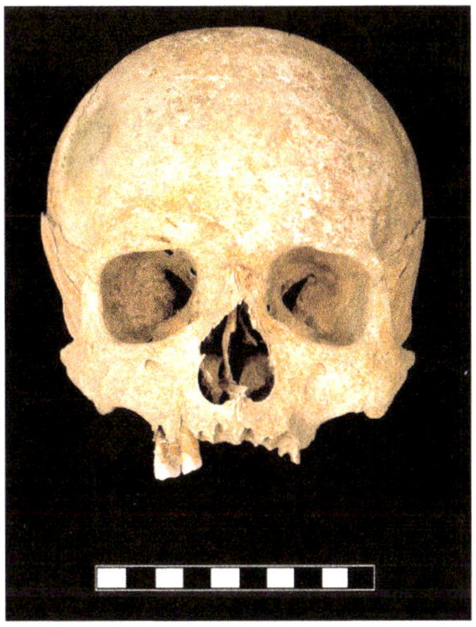

Figure 6.51 Tomb XCII (Grave 23), Individual 159, face with nasal asymmetry related to healed trauma (photo 2050)

Figure 6.52 Tomb XCIV (Grave 25): right frontal of Individual 165 with arrow indicating reactive area (photo 3343)

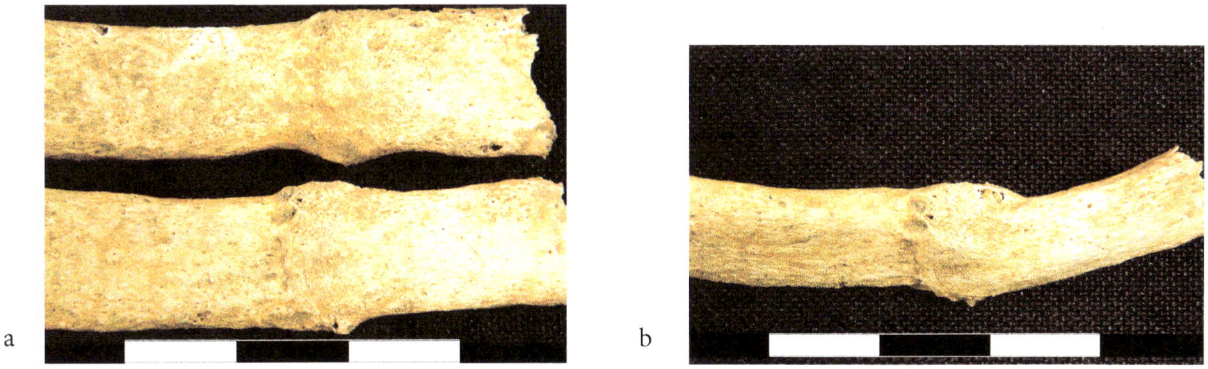

Figure 6.53 Tomb XCVII (Grave 39): (a) healed rib fractures, ventral view; Individual 250 (photo 3964); (b) healed rib fracture, lateral view; Individual 250 (photo 3965)

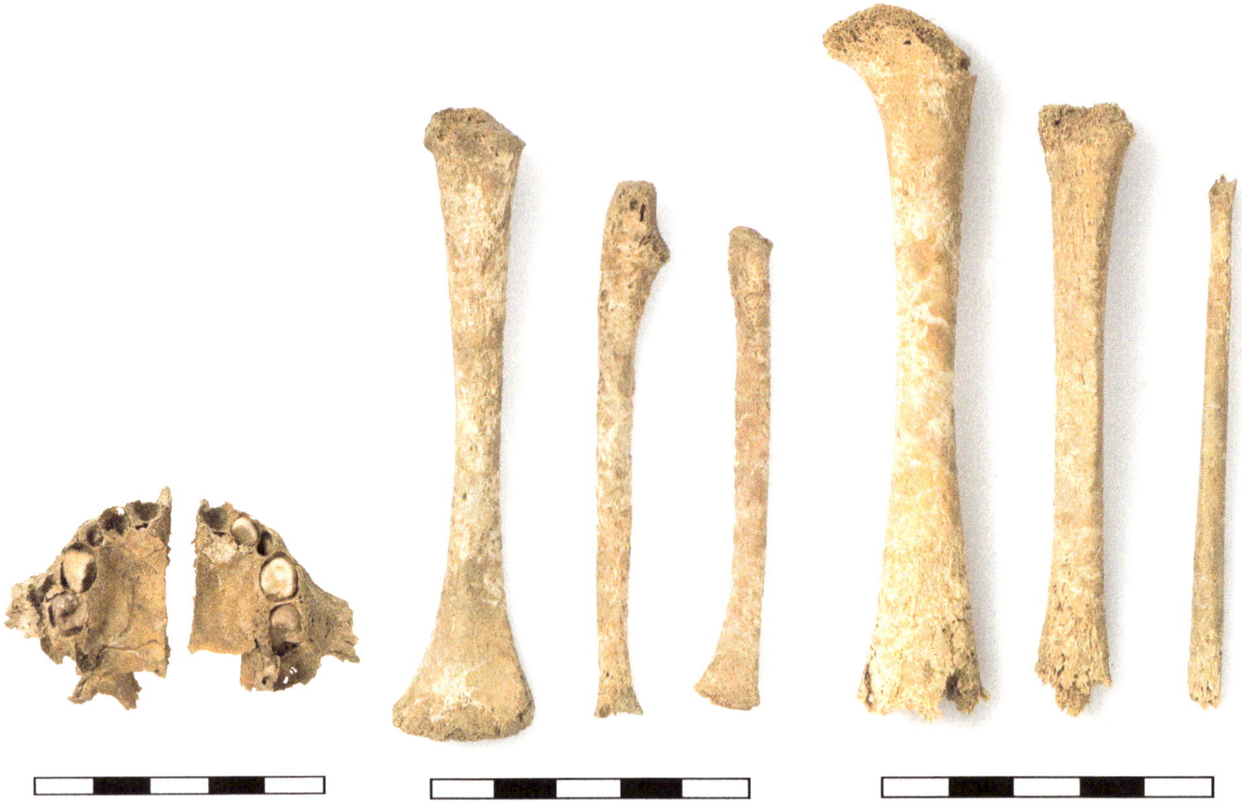

a b c

FIGURE 6.54 Tomb XCVIII (Grave 36), Individual 238: (a) maxillae, palatal view (photo 2324); (b) left humerus, radius, ulna (photo 2303); (c) left femur, tibia, fibula (photo 2309)

a b

FIGURE 6.55 Tomb C (Grave 48): (a) posterior view of Individual 302 cranium (photo 3577); (b) posterior-superior view of Individual 302 cranium highlighting the bilateral parietal depressions (photo 3576)

Appendix 1 Illustration

FIGURE A1.1 Seth Pevnick sampling for level 2 sterility of DNA

Chapter 7 Illustration

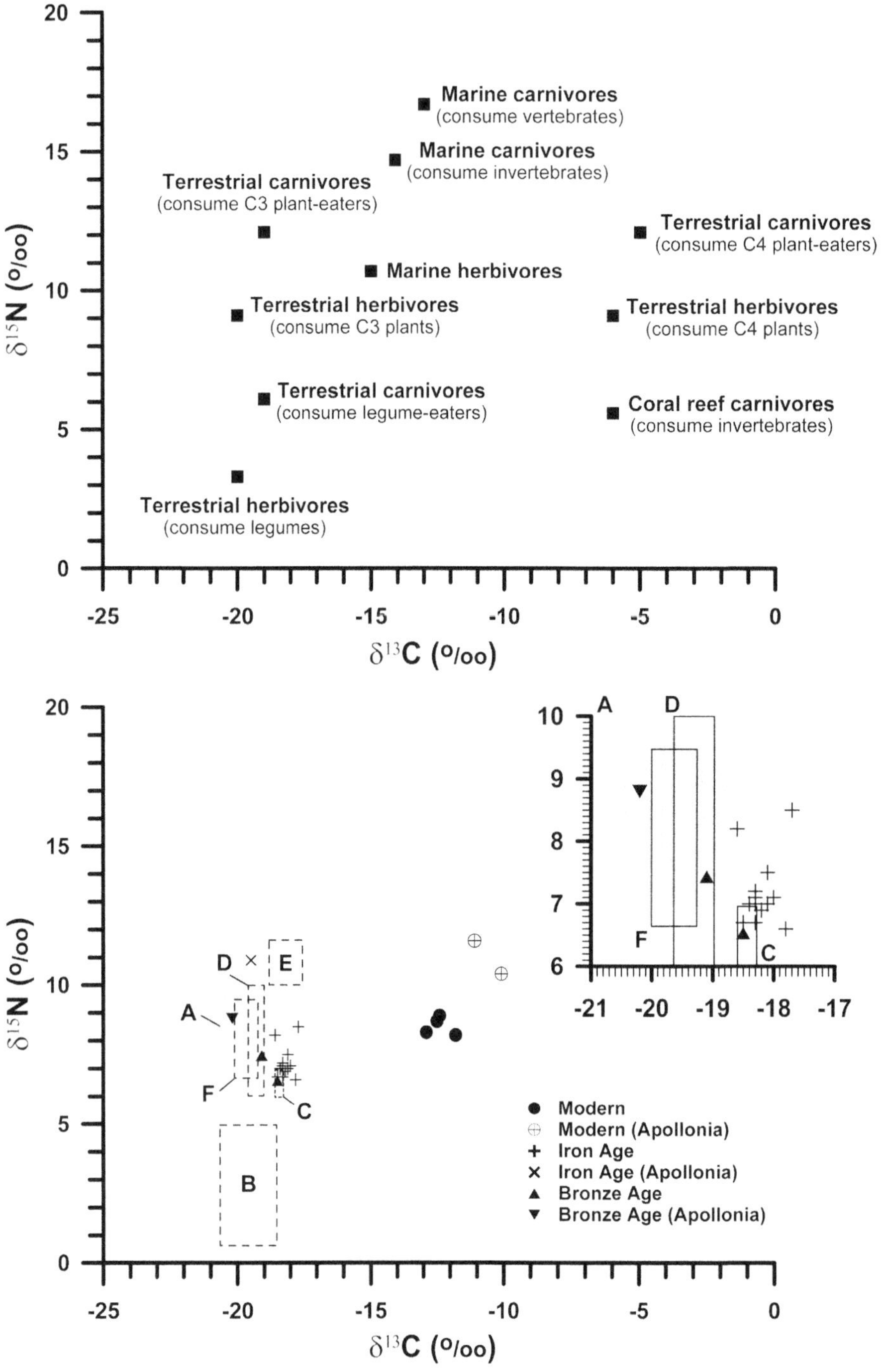

FIGURE 7.1 *Upper*: Generalized food web showing the relative spatial relationships between major groups (using average values) for bone samples (collagen) from prehistoric individuals feeding on a single type of food (modified from DeNiro 1987). *Lower*: Stable-isotope data for human bone samples from the Lofkënd tumulus and the nearby necropolis at Apollonia (Damiata and Southon 2010). Solid triangle denotes Bronze Age sample, cross denotes Iron Age sample, and solid circles are modern age. Stippled area and boxes denote results from Papathanasiou (2003) and Richards and Hedges (2008) for sites in Greece used for comparison. A, human bone (collagen) from Neolithic sites; B and C, faunal bone (collagen) from herbivores and carnivores, respectively, from the Minoan site at Chamalevri (Crete, 2160–1070 BC); D and E, human bone (collagen) from Mycenae (1600–1200 BC); and F, human bone (collagen) from the Minoan Cemetery at Armenoi (Crete, 1390–1190 BC).

Chapter 8 Illustrations

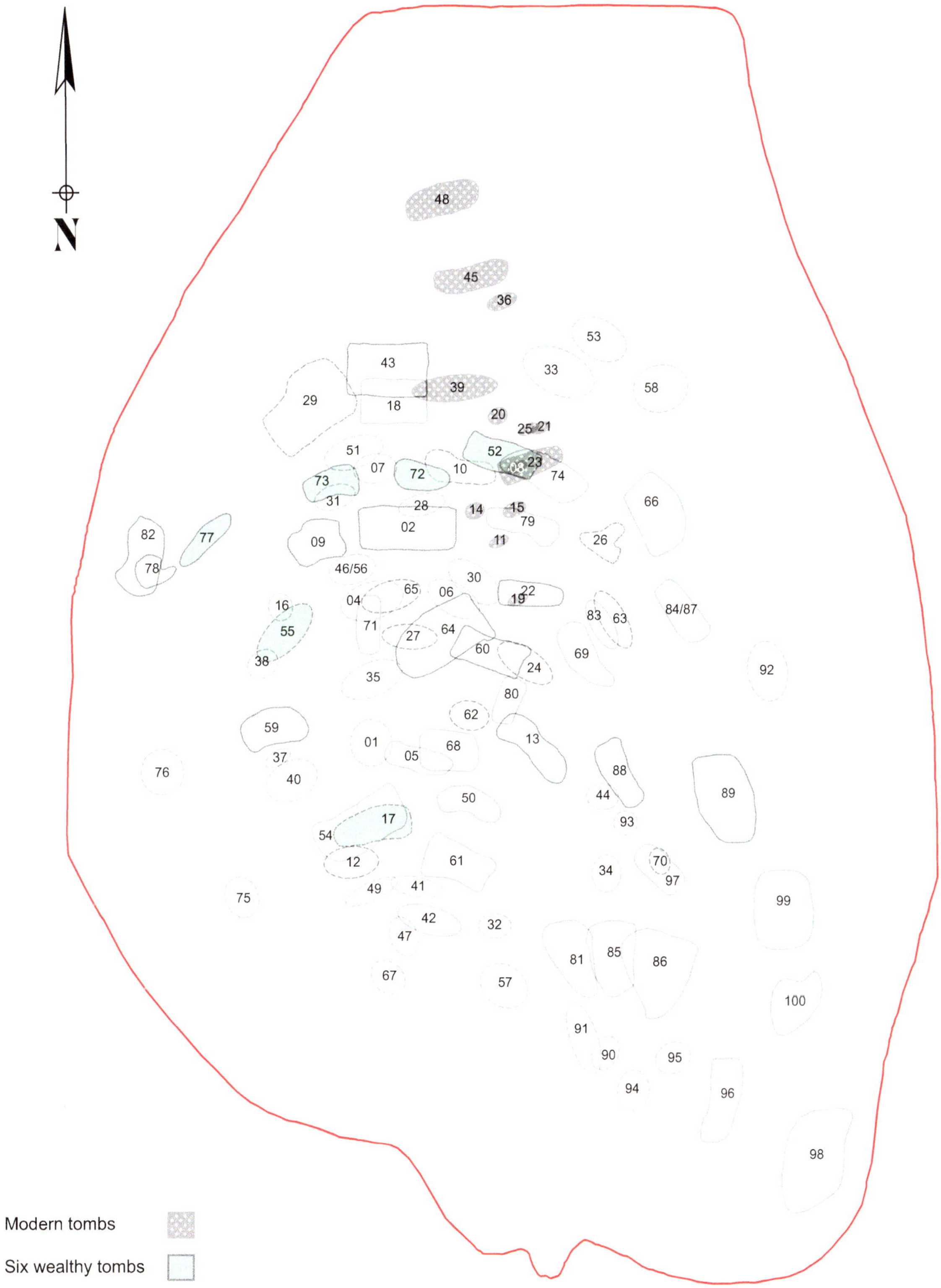

FIGURE 8.1 Plan of the tumulus showing all tombs and highlighting modern tombs as well as the focus group. Tombs with a solid outline are those where extent of tomb was clear or reasonably clear; those with a dashed outline are those where extent of tomb was either partially preserved or not clear.

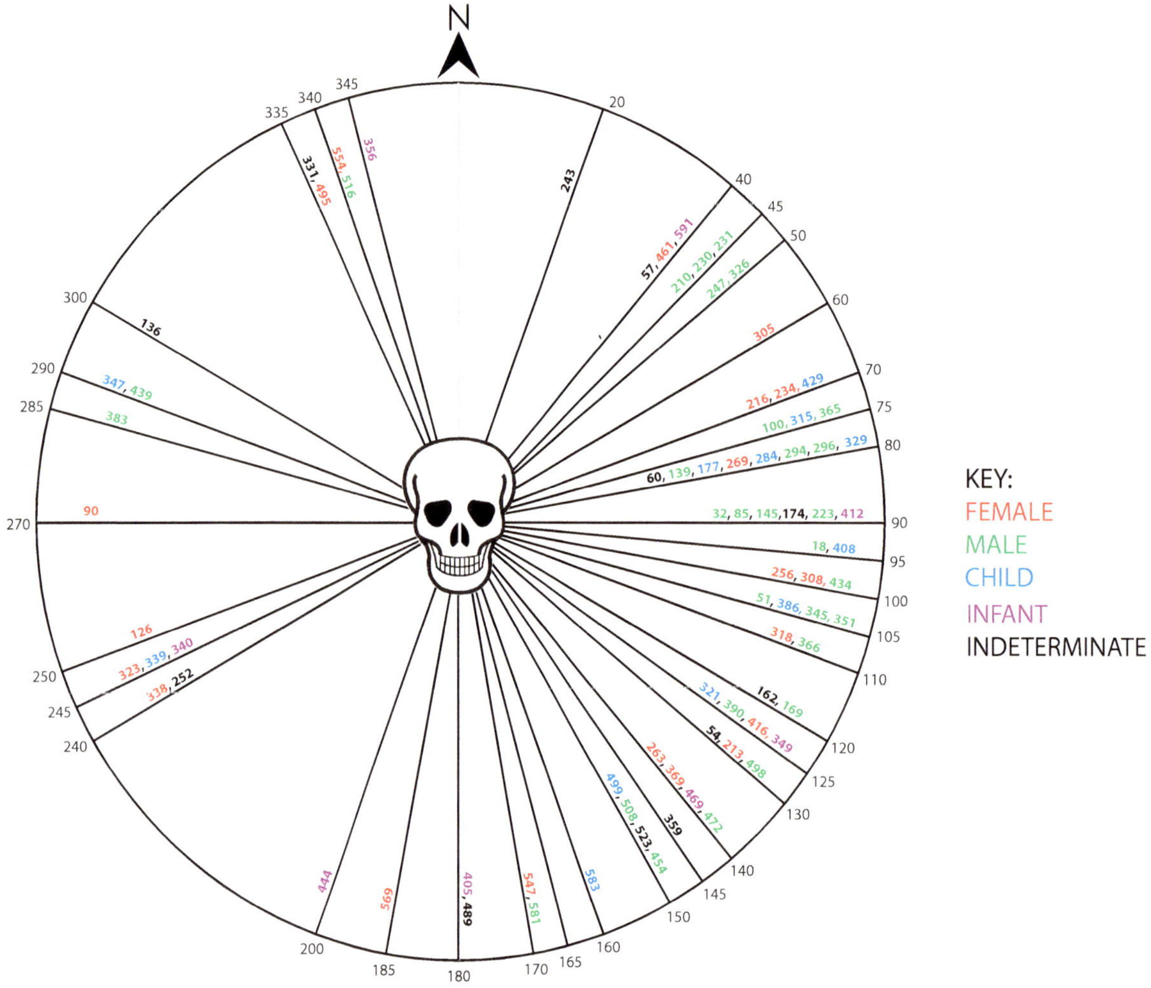

FIGURE 8.2 Schematic diagram showing orientation of skeletons in the tumulus

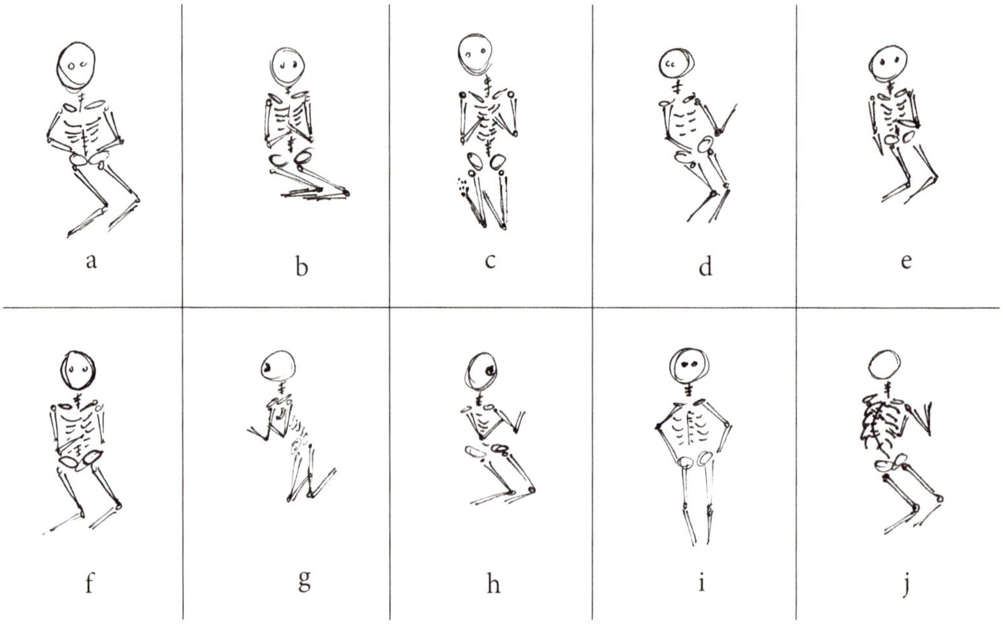

FIGURE 8.3 Sketches of the range of positions of prehistoric skeletons at Lofkënd

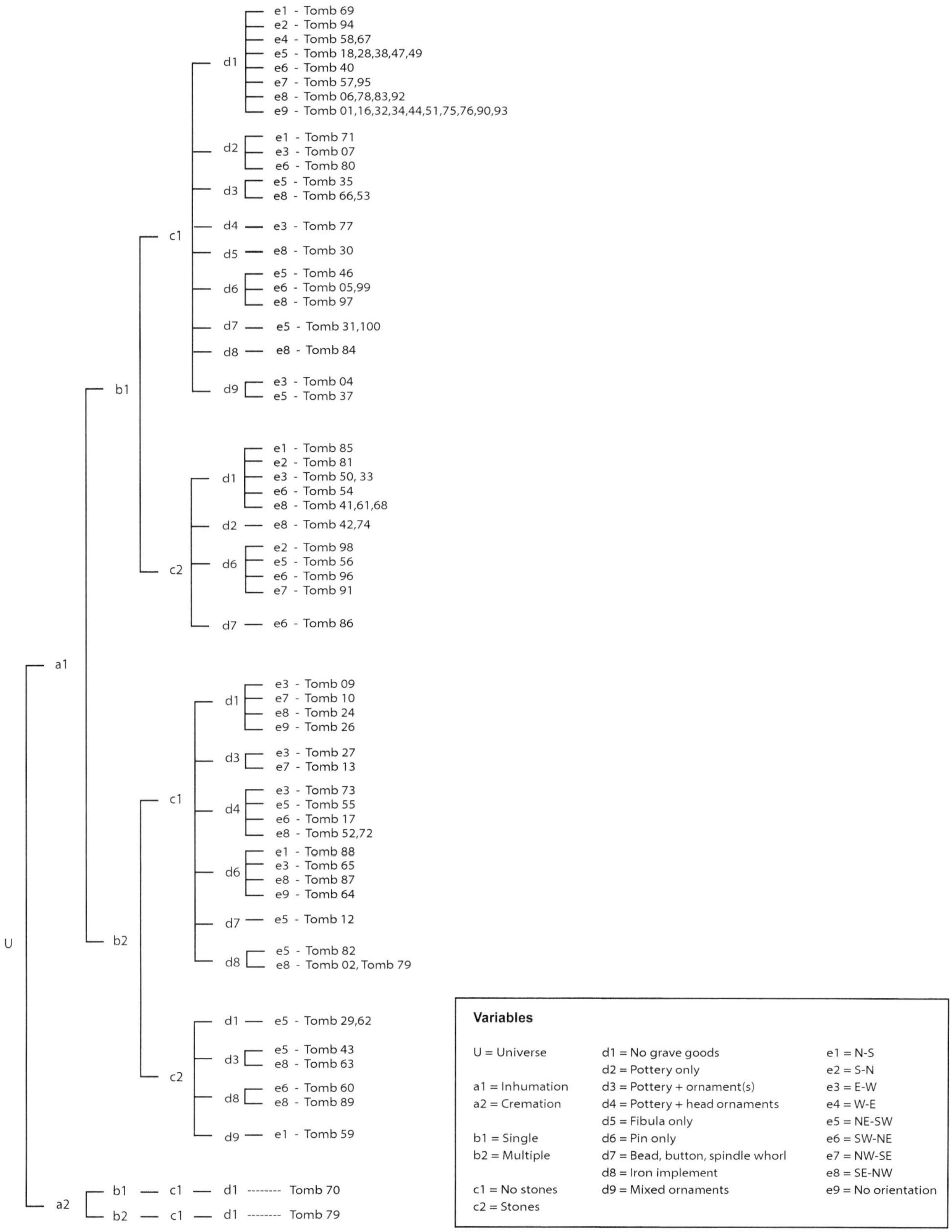

FIGURE 8.4 Key diagram showing primary attributes of prehistoric burials

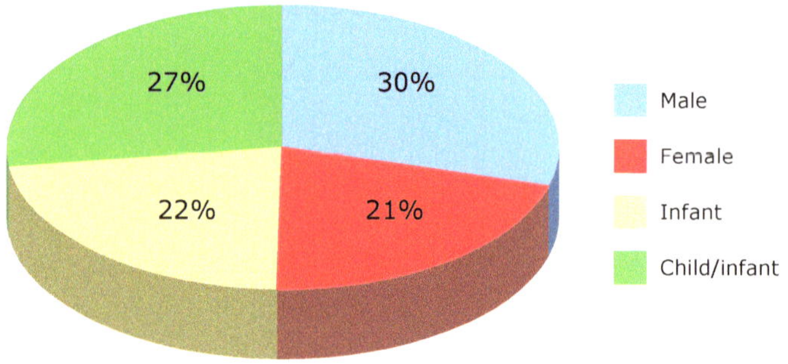

Figure 8.5 Distribution of prehistoric skeletons by sex and age (including infants and children)

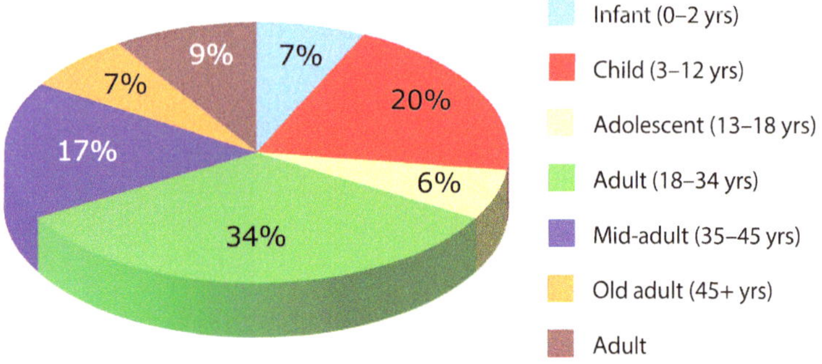

Figure 8.6 Distribution of prehistoric skeletons by biological age

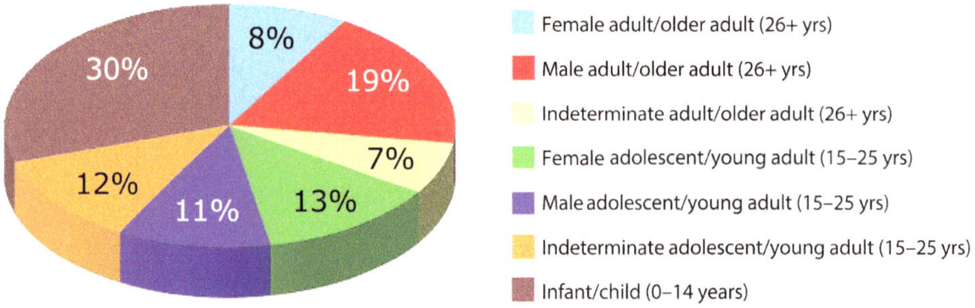

Figure 8.7 Distribution of prehistoric skeletons by social age

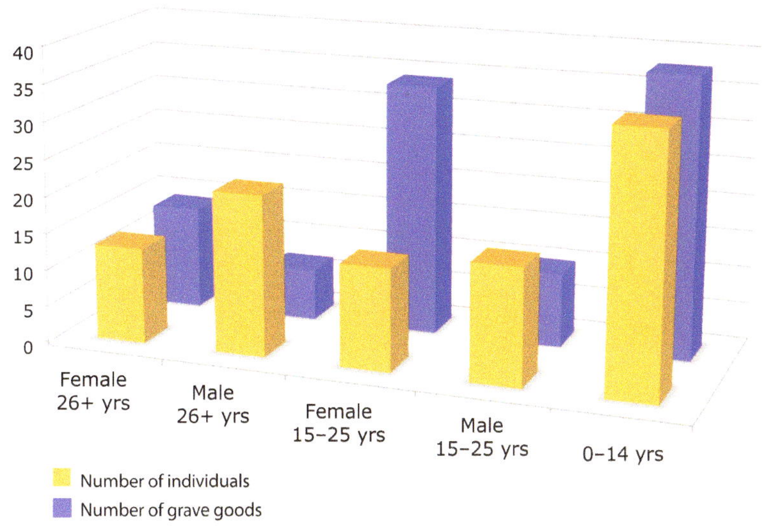

FIGURE 8.8 Total grave goods sorted by biological age group (excluding those of indeterminate sex)

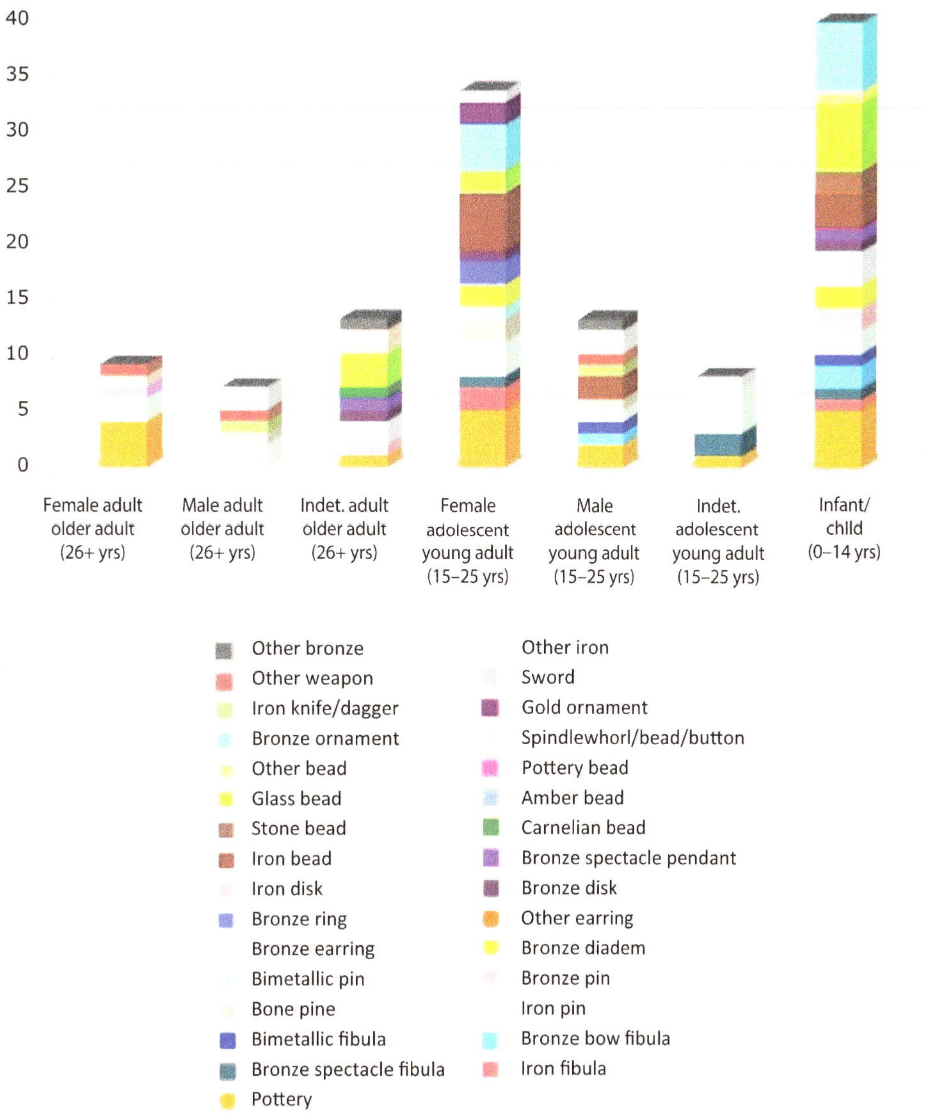

FIGURE 8.9 Total grave goods sorted by social age and sex groups

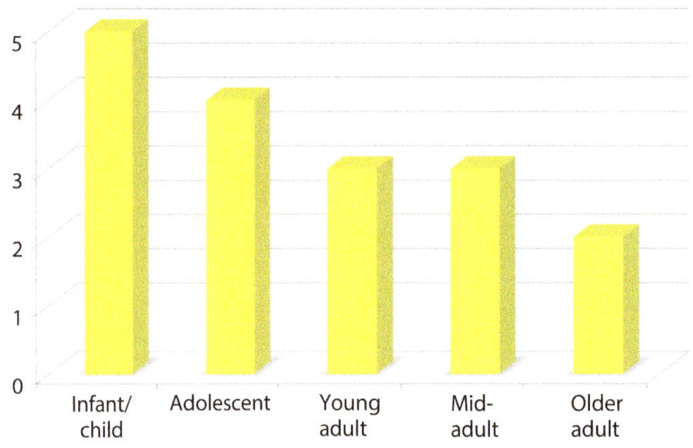

FIGURE 8.10 Total pottery (whole vessels in burials) sorted by biological age

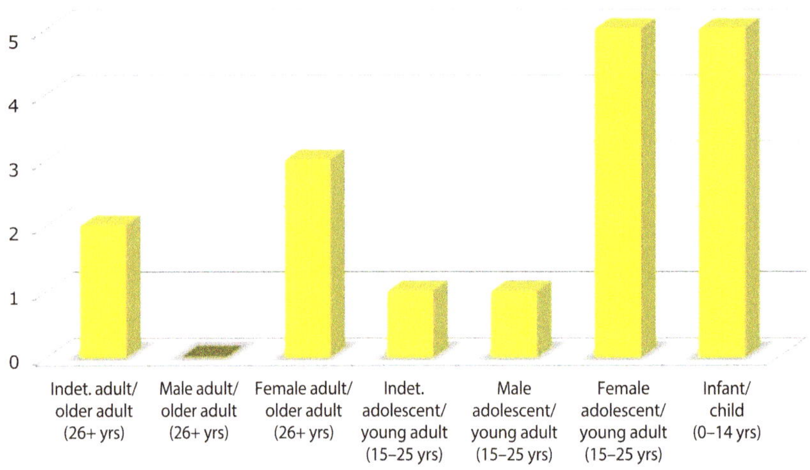

FIGURE 8.11 Total pottery (whole vessels in burials) sorted by social age

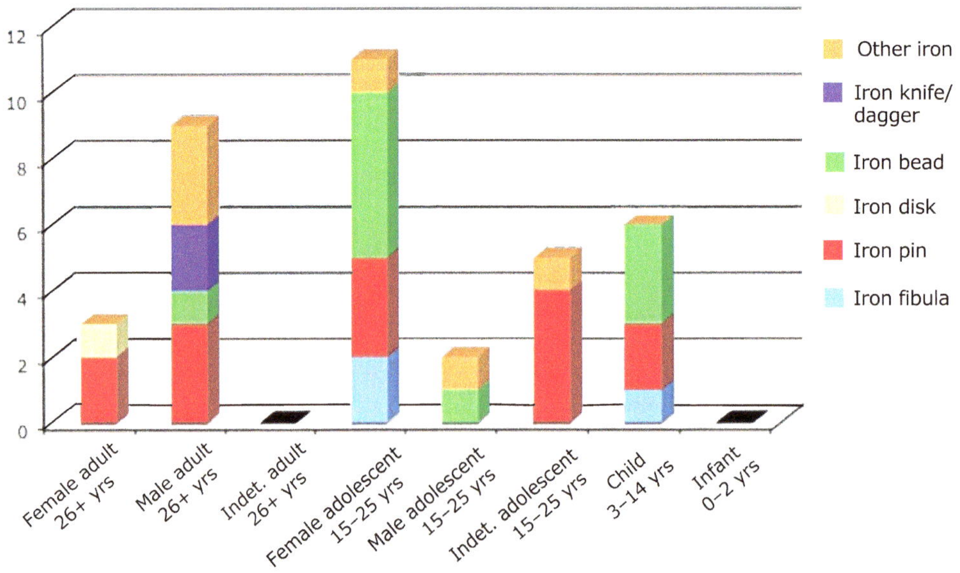

FIGURE 8.12 Iron finds sorted by social age and sex

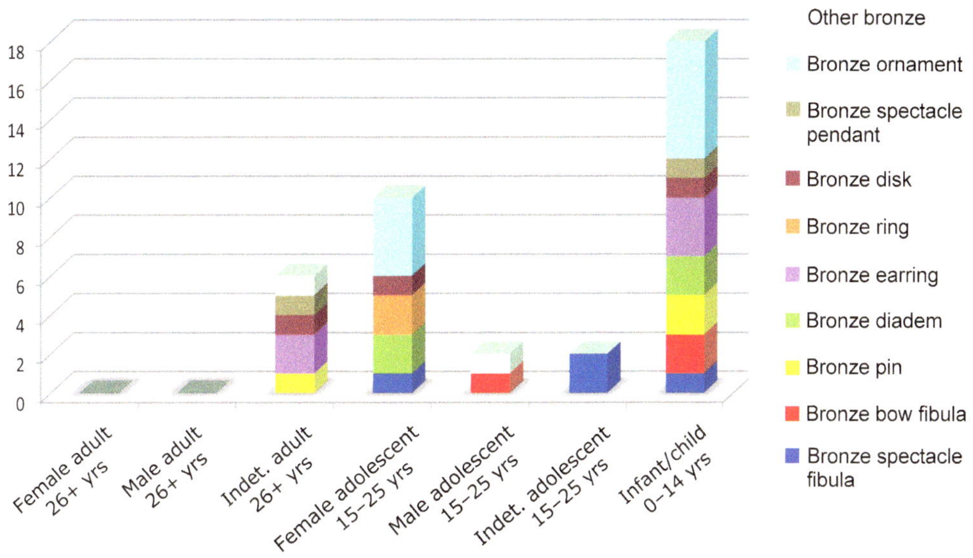

Figure 8.13 Bronze finds sorted by social age

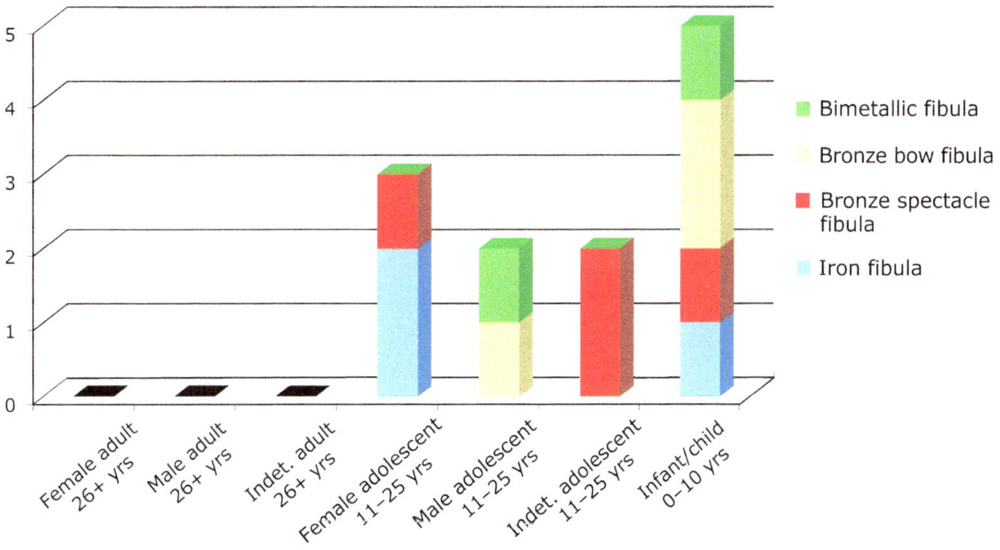

Figure 8.14 All fibulae sorted by social age and sex

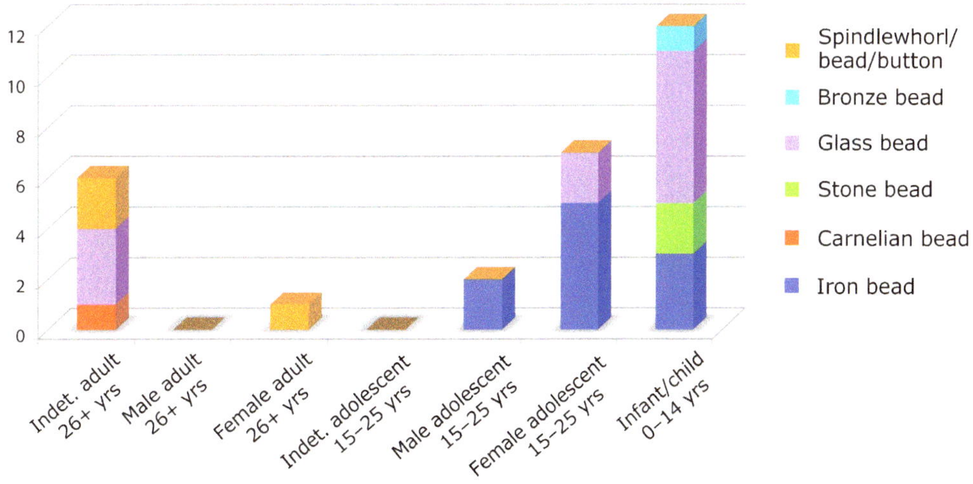

Figure 8.15 All beads sorted by social age and sex

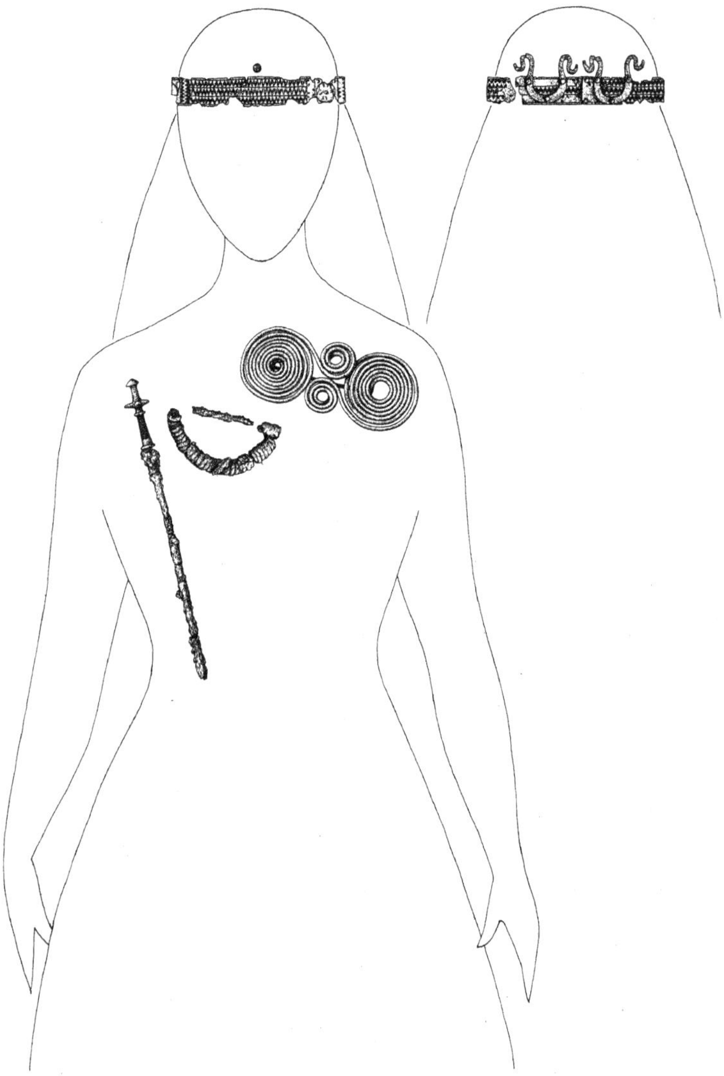

FIGURE 8.16 Reconstruction of bronze and iron ornaments (headdress, fibulae, pin, and bead of semi-precious stone) worn by the young female in Tomb LXX (17) (Ilir Zaloshnja)

FIGURE 8.17 Drawing of female heads from frescoes in Xeste 3, Akrotiri, Thera (Greece); Late Bronze Age (after Davis 1986)

Chapter 9 Illustrations

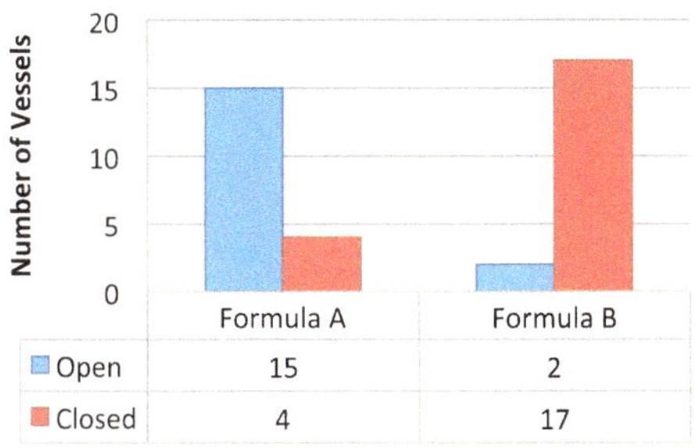

FIGURE 9.1 Open versus closed vessels as a function of definition

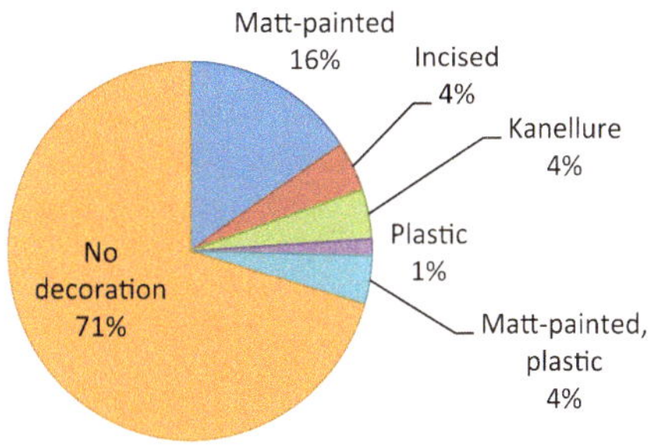

FIGURE 9.2 Decoration of fineware sherds (light and dark) from tumulus fill

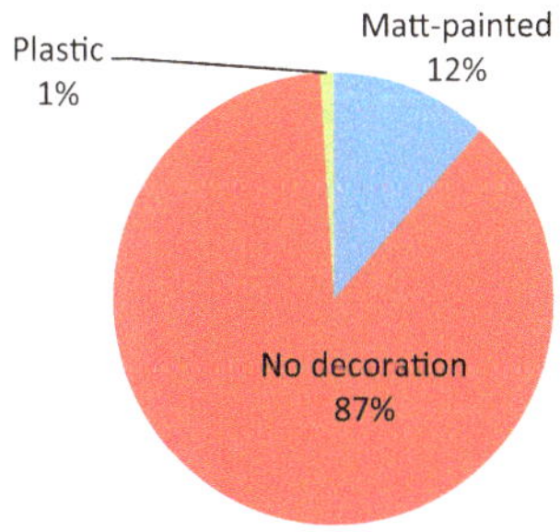

FIGURE 9.3 Decoration of semi-coarse sherds from tumulus fill

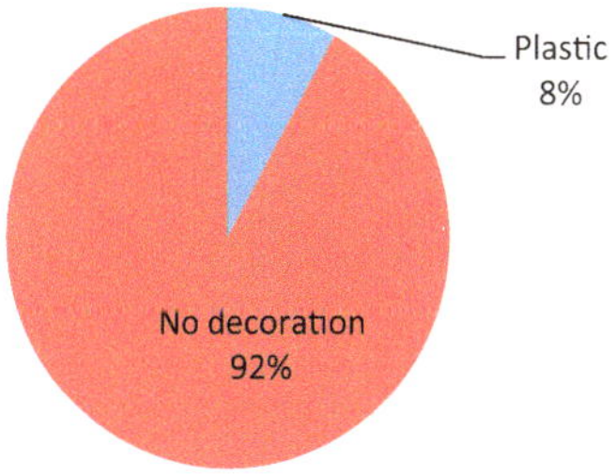

FIGURE 9.4 Decoration of coarseware sherds from tumulus fill

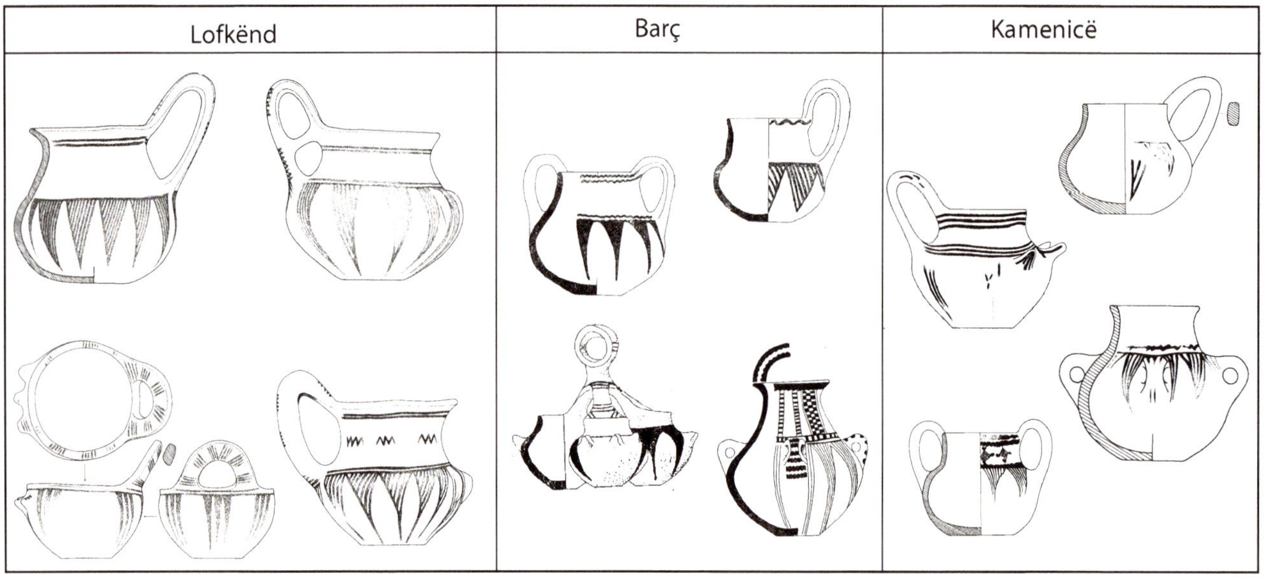

FIGURE 9.5 Similarities between the pottery from Lofkënd, Barç, and Kamenicë

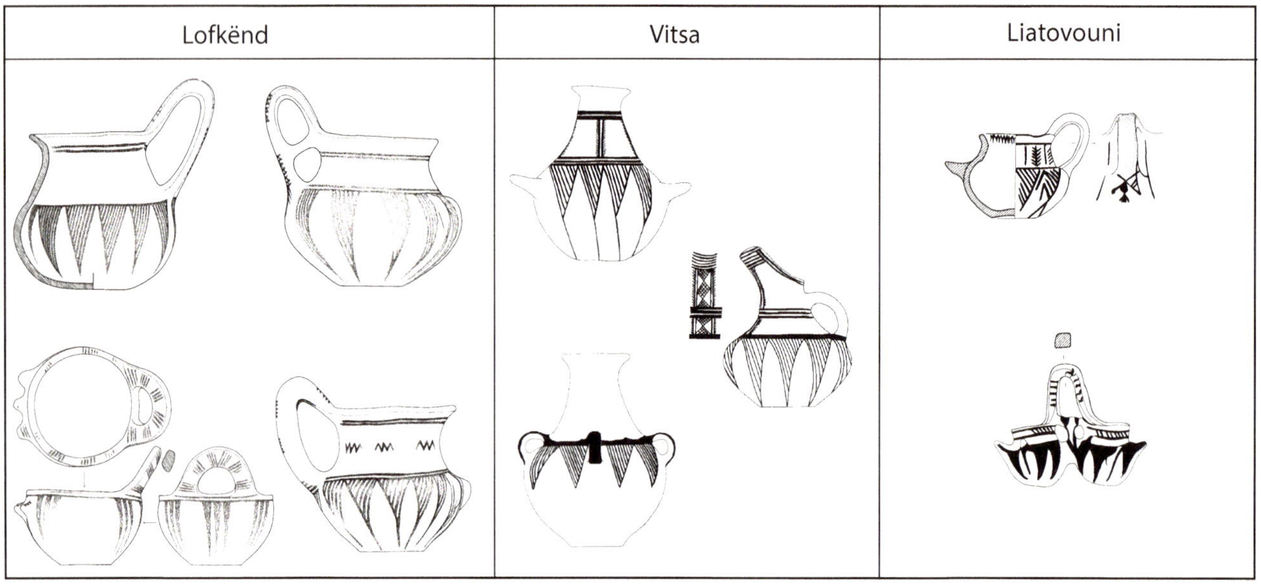

FIGURE 9.6 Similarities between the pottery from Lofkënd, Vitsa, and Liatovouni

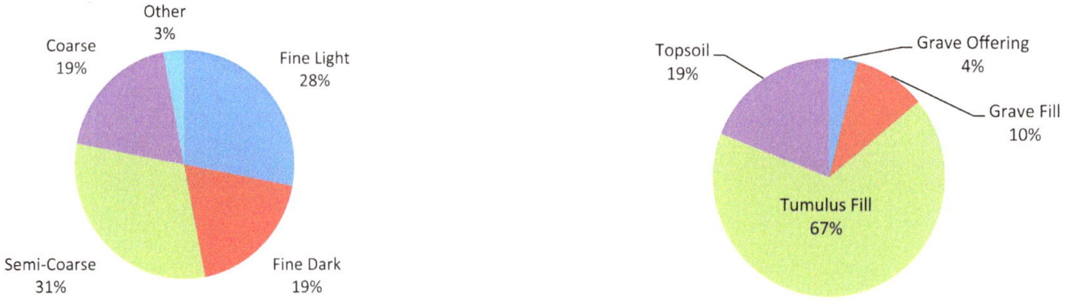

FIGURE 9.7 Pottery fabric distributions and pottery find context distributions

Type 1

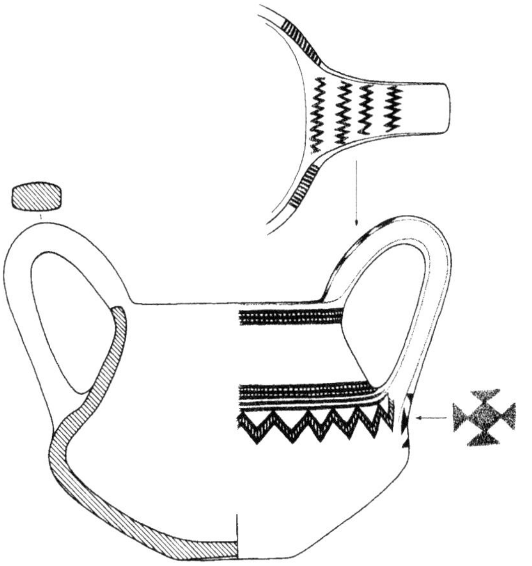

9/1 (P350)

Type 2a

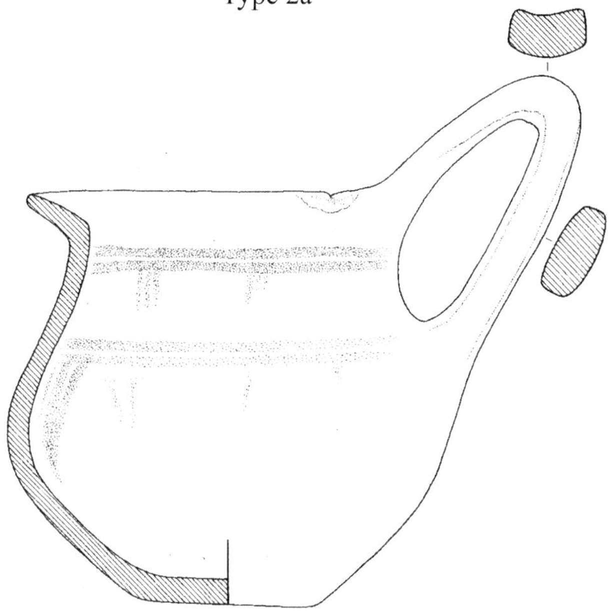

9/4 (P077)

Type 2a

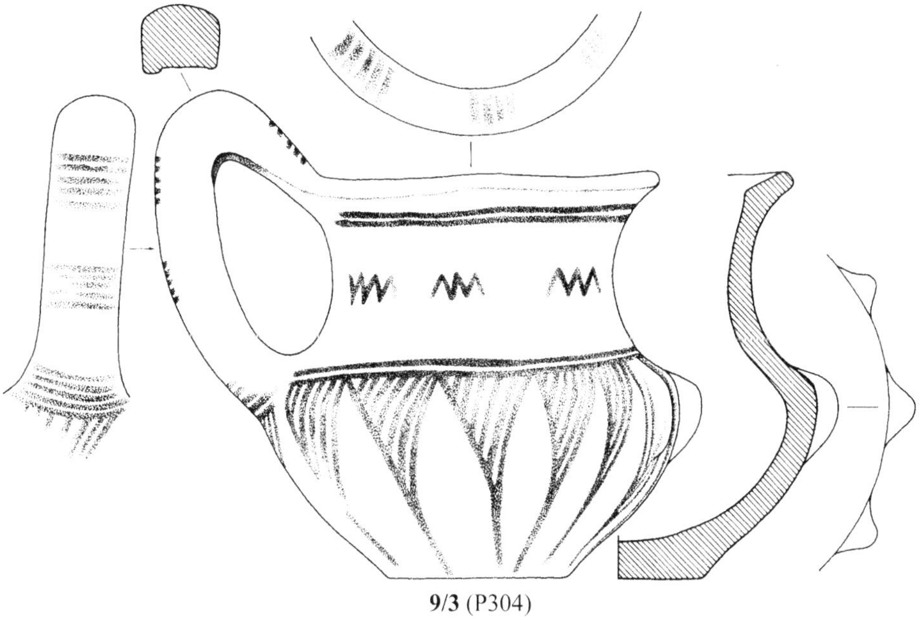

9/3 (P304)

FIGURE 9.8 **9/1, 9/3–9/4**

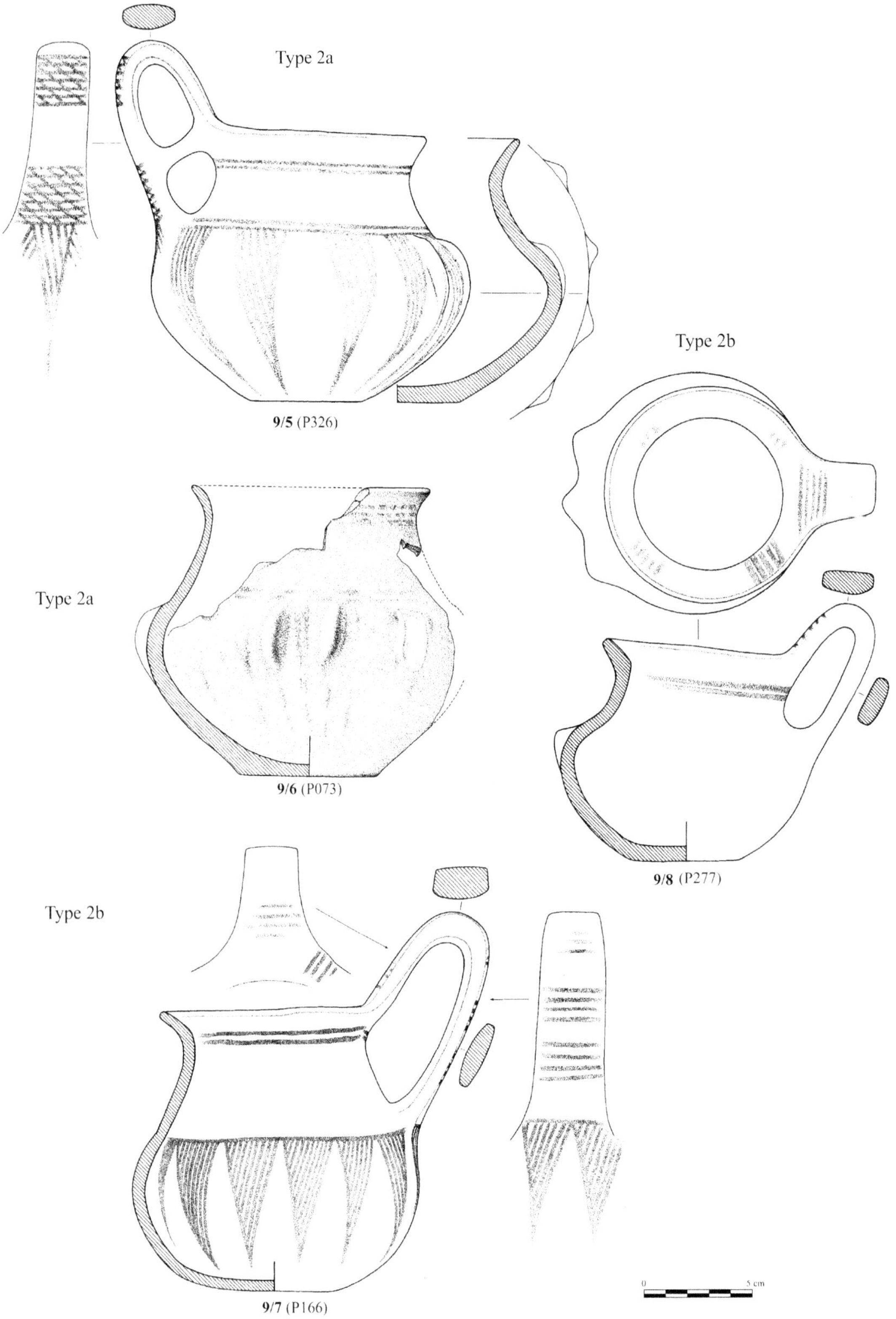

FIGURE 9.9 **9/5–9/8**

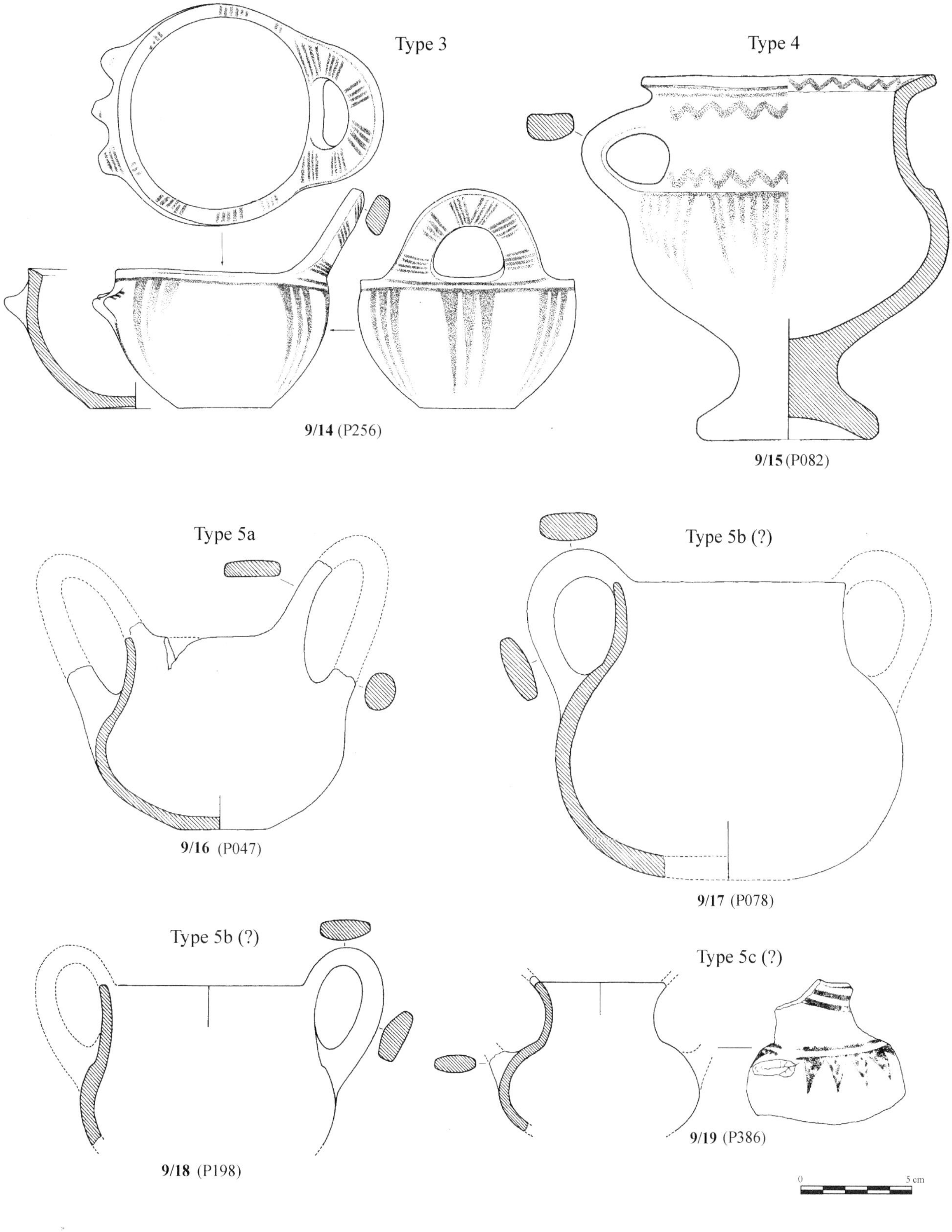

FIGURE 9.10 **9/14–9/19**

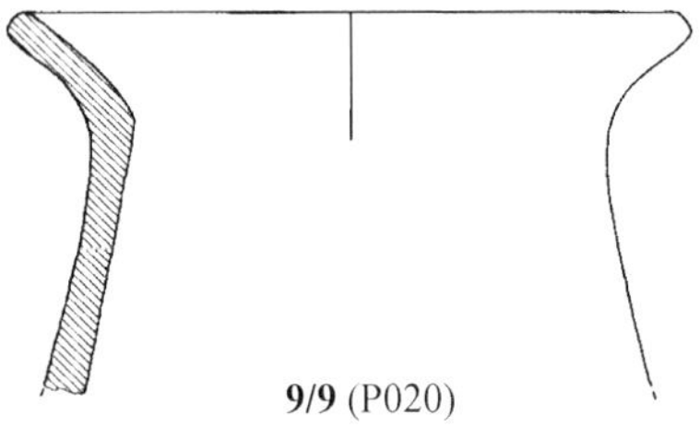

9/9 (P020)

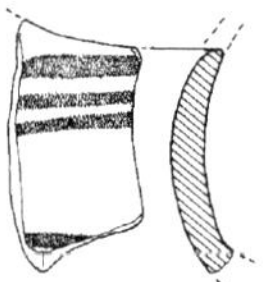

9/10 (P213)

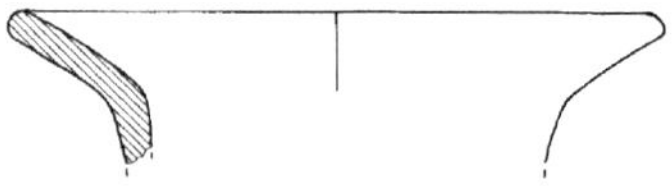

9/11 (P004)

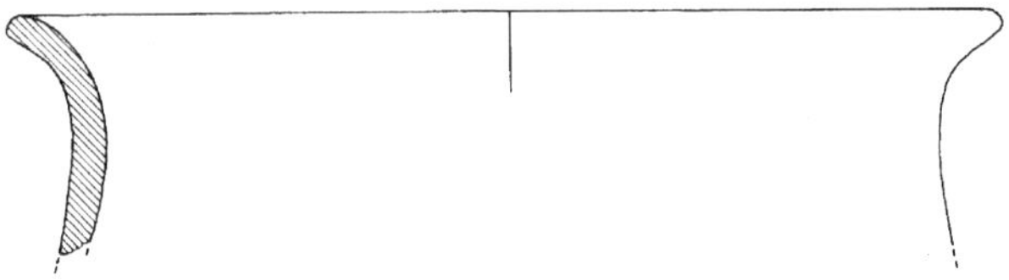

9/12 (P131)

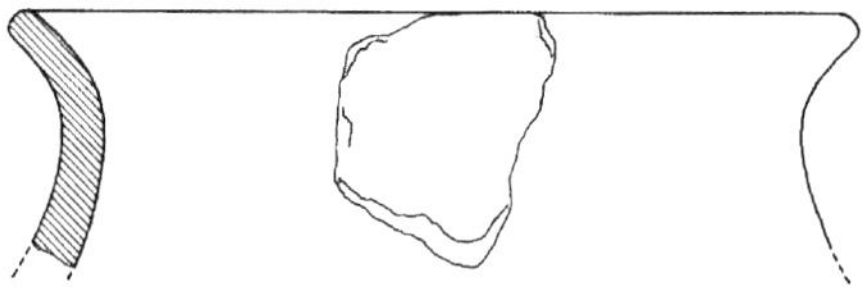

9/13 (P202)

Type 5c (?)

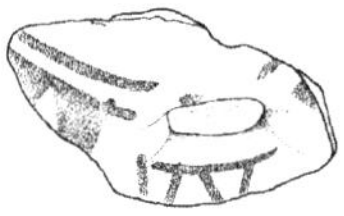

9/20 (P290 + P221)

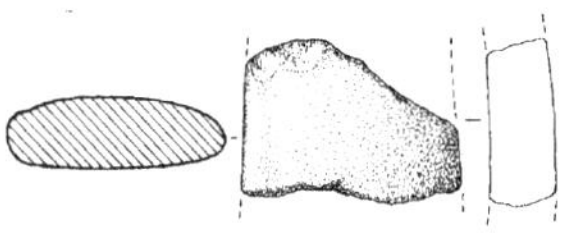

9/21 (P113)

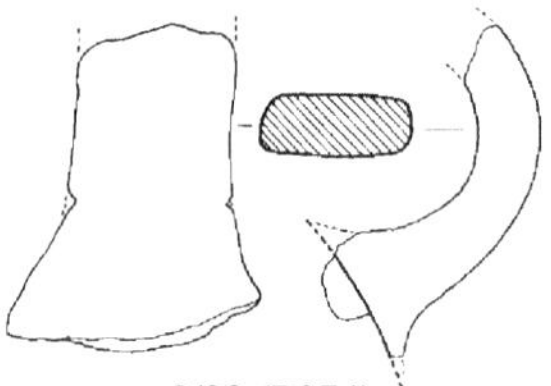

9/22 (P074)

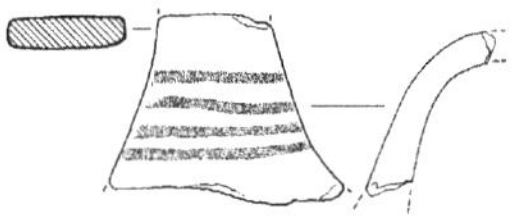

9/25 (P373)

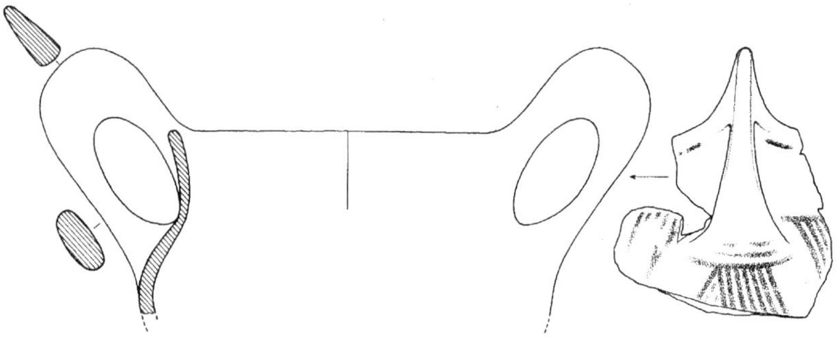

9/26 (P107)

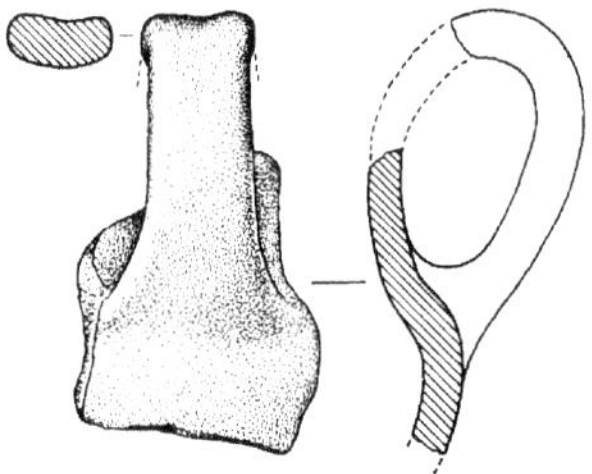

9/27 (P240 + P251)

FIGURE 9.11 **9/9–9/13, 9/20–9/22, 9/25–9/27**

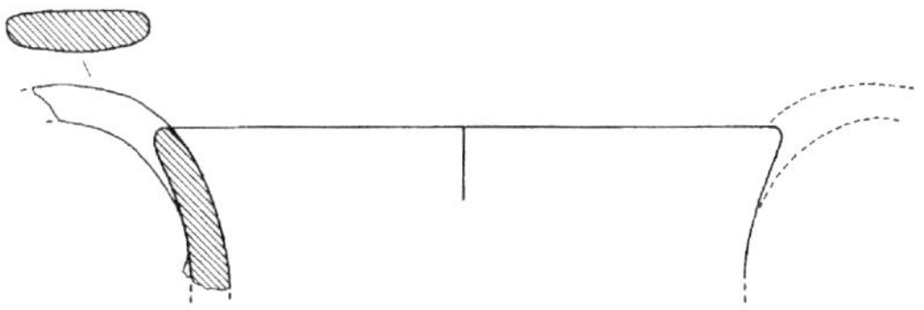

9/28 (P005)

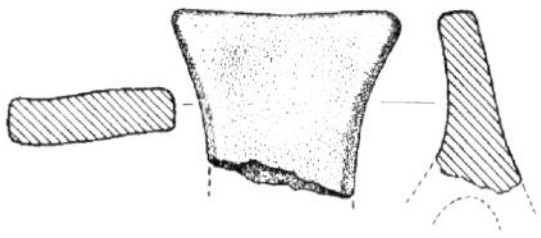

9/29 (P006)

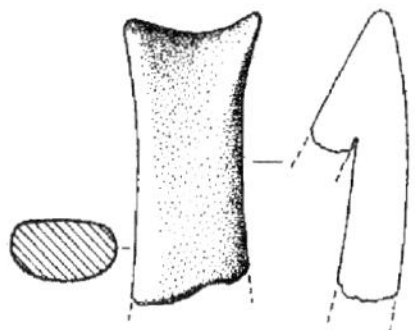

9/30 (P064)

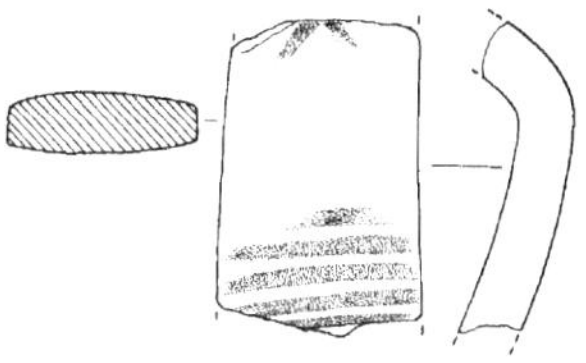

9/31 (P114)

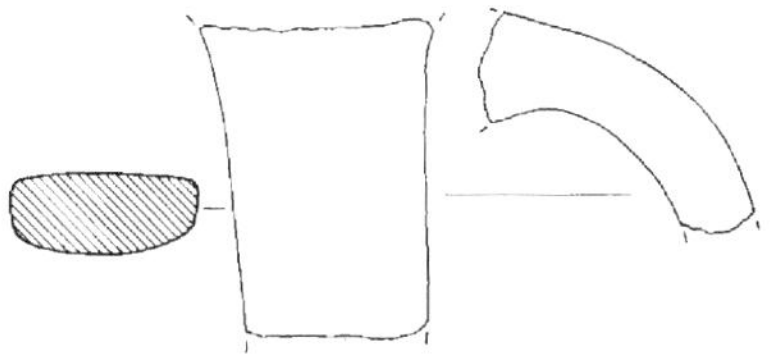

9/32 (P031)

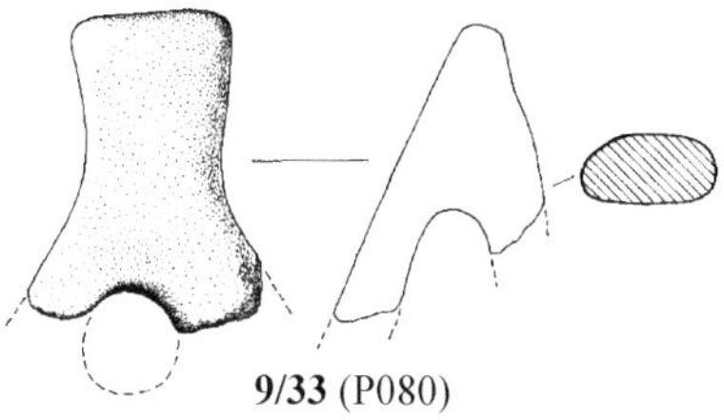

9/33 (P080)

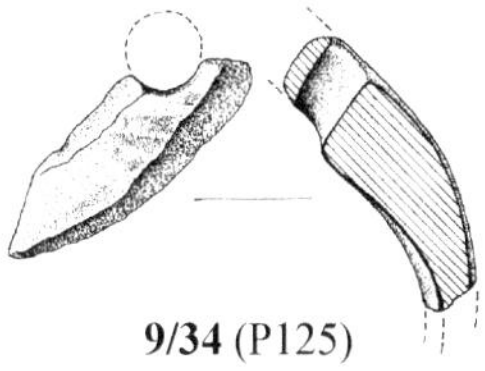

9/34 (P125)

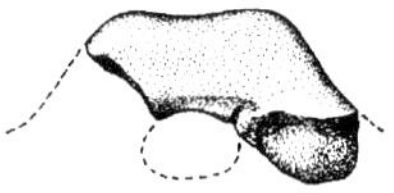

9/35 (P287)

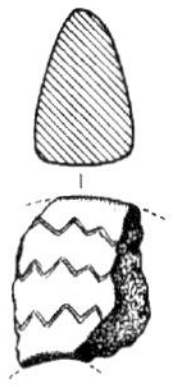

9/37 (P342)

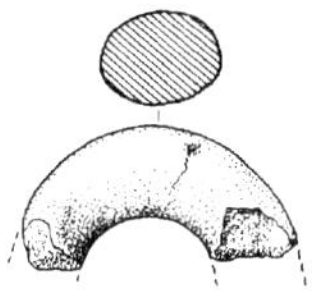

9/39 (P172)

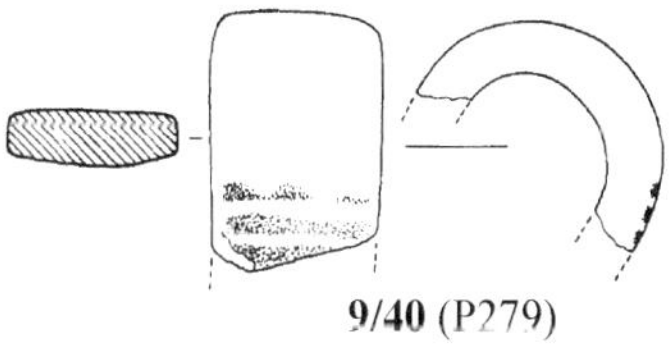

9/40 (P279)

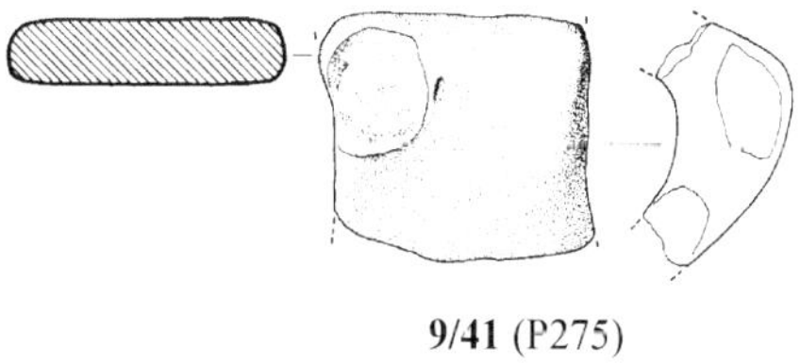

9/41 (P275)

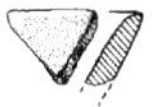

9/42 (P159)

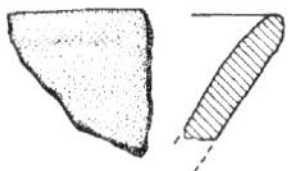

9/43 (P220)

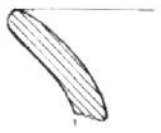

9/44 (P013)

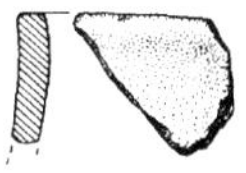

9/45 (P173)

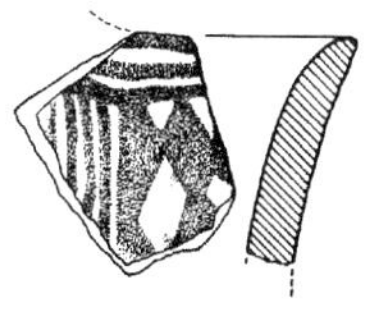

9/47 (P258)

0 5 cm

FIGURE 9.12 **9/28–9/35, 9/37, 9/39–9/45, 9/47**

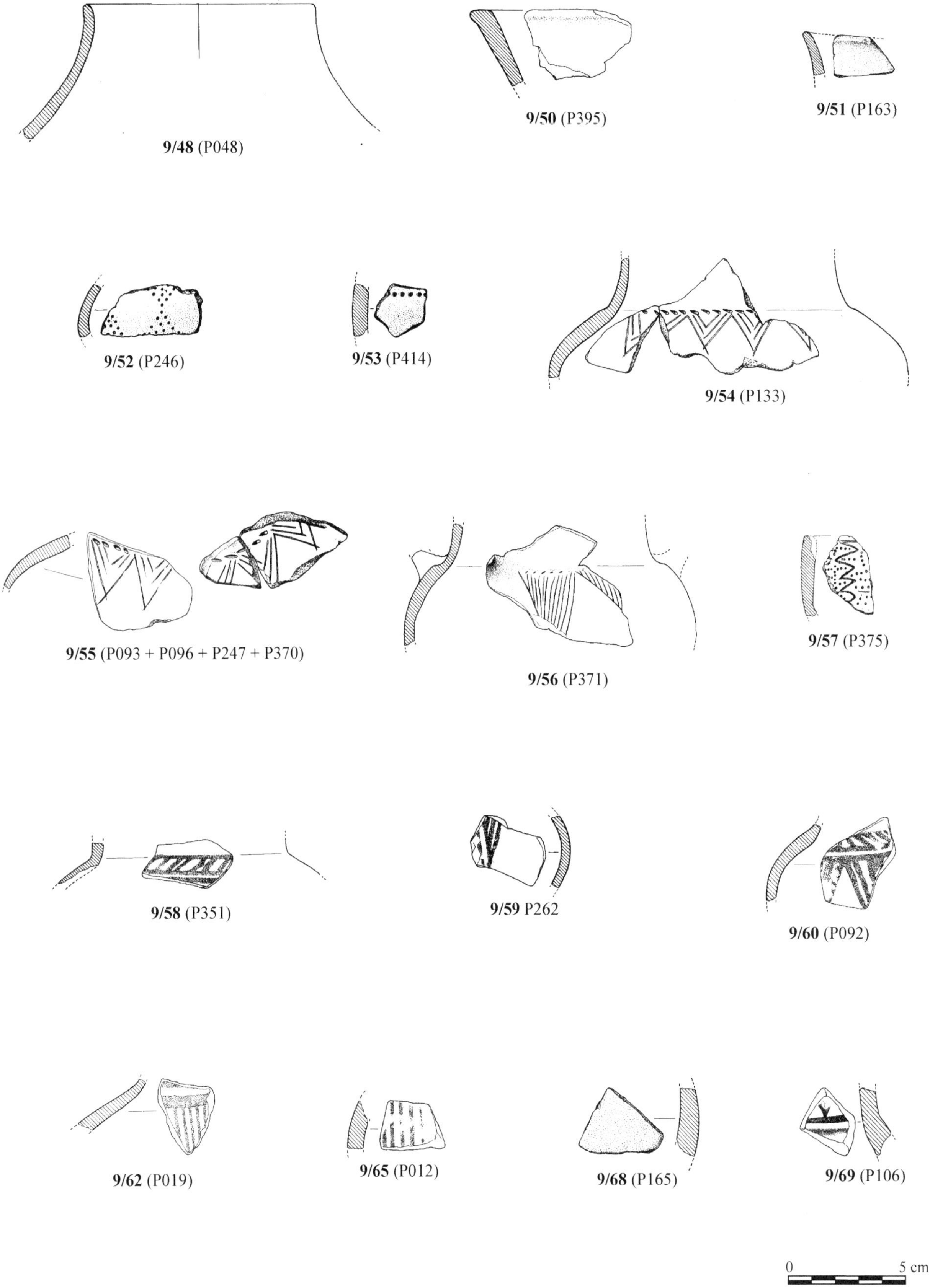

FIGURE 9.13 **9/48, 9/50–9/60, 9/62, 9/65, 9/68–9/69**. In the case of **9/55** the fragments have been digitally joined from two or more earlier drawings of individual fragments.

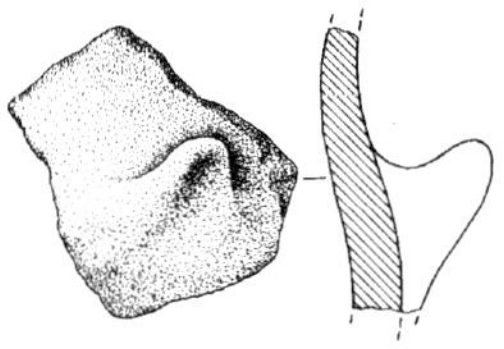

9/71 (P127)

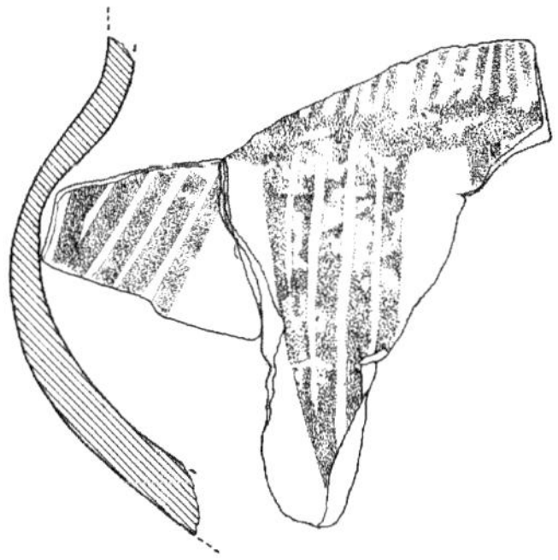

9/72 (P058 + P391)

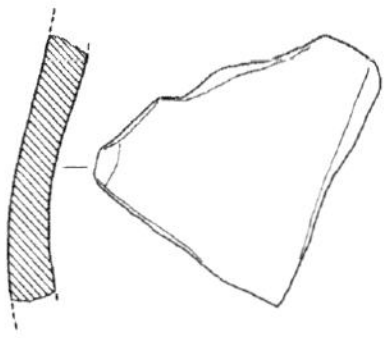

9/73 (P356)

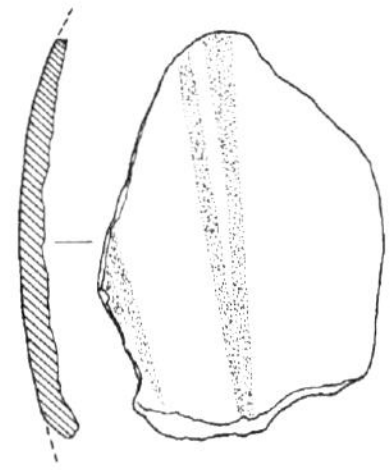

9/74 (P252)

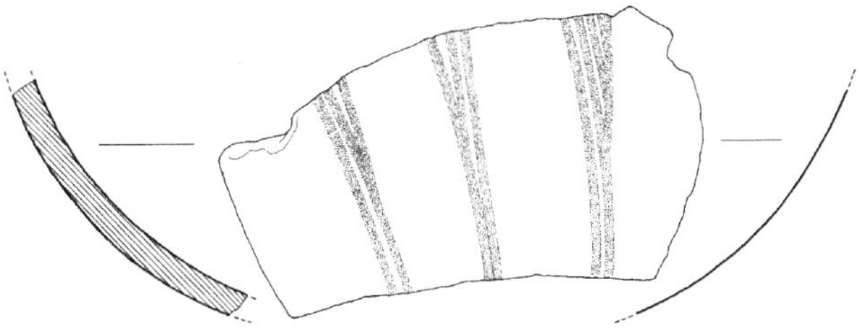

9/75 (P132 + P324)

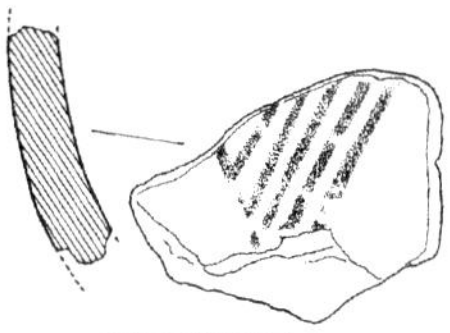

9/76 (P431)

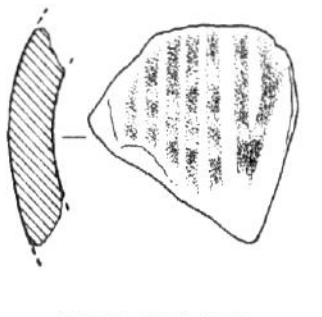

9/77 (P195)

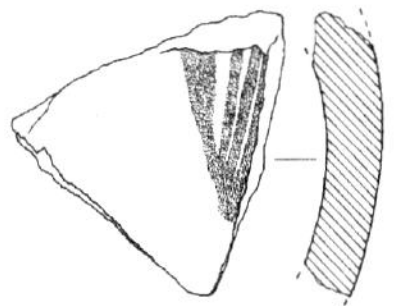

9/78 (P104)

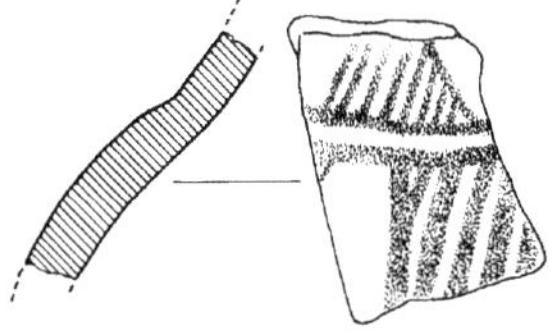

9/80 (P284)

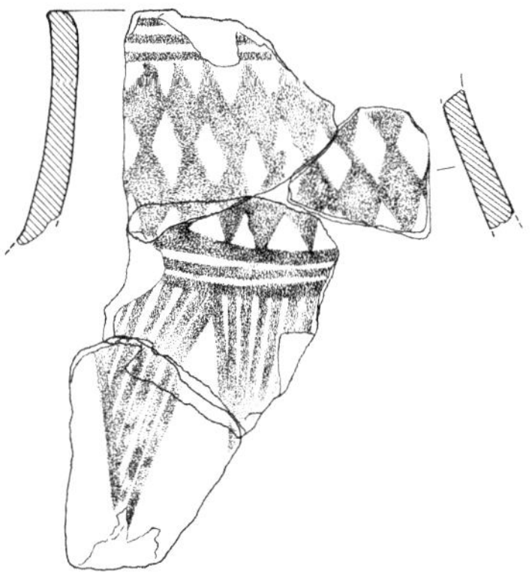

9/79 (P193 + P353 + P115 +P 167)

0 5 cm

Figure 9.14 **9/71–9/80**. Fragments of **9/72** and **9/79** digitally joined from two or more earlier drawings of individual fragments.

9/84 (P030)

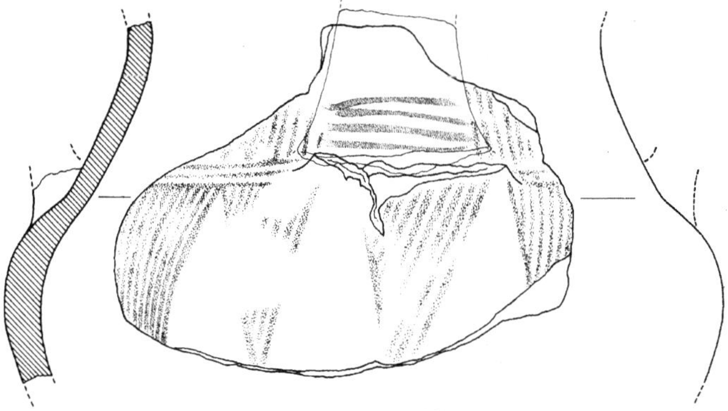

9/83 (P255 + P112)

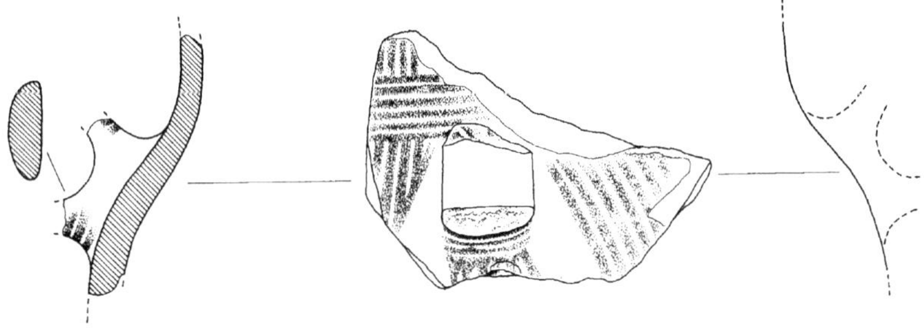

9/82 (P305)

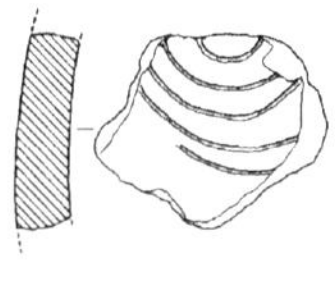

9/85 (P346)

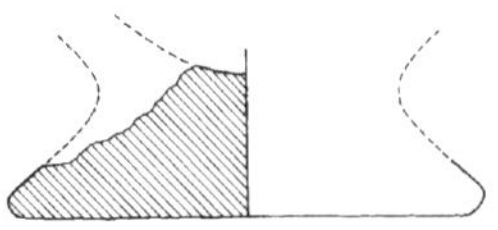

9/86 (P108+P388)

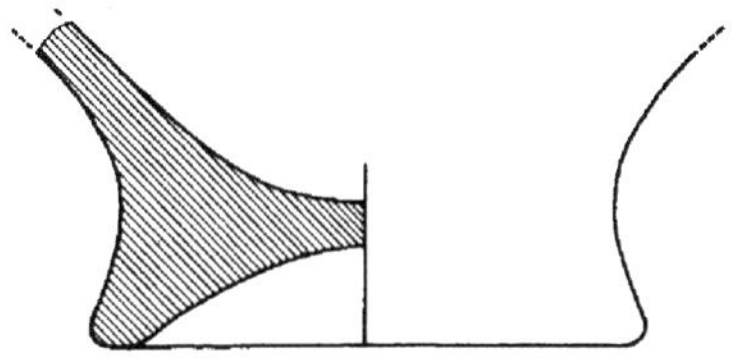

9/87 (P097 + P237 + P285)

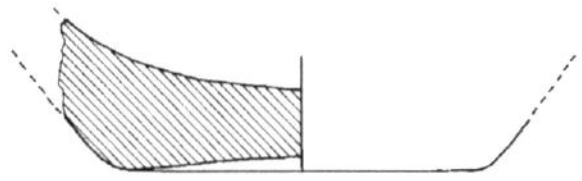

9/88 (P001)

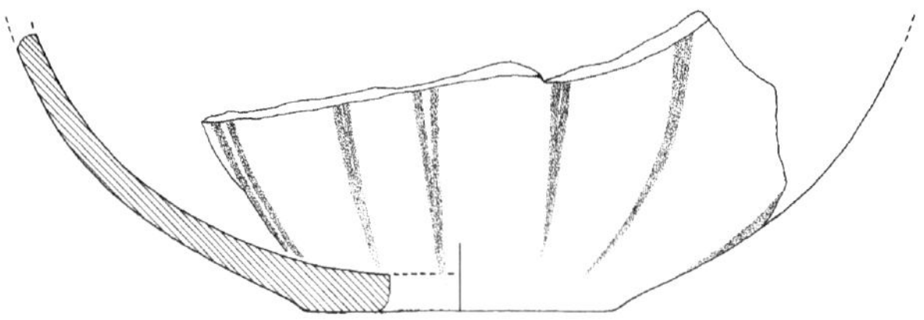

9/89 (P232)

Figure 9.15 **9/82–9/89**. Fragments of **9/83** digitally joined from two or more earlier drawings of individual fragments.

Type 1

9/90 (P422)

Type 2a

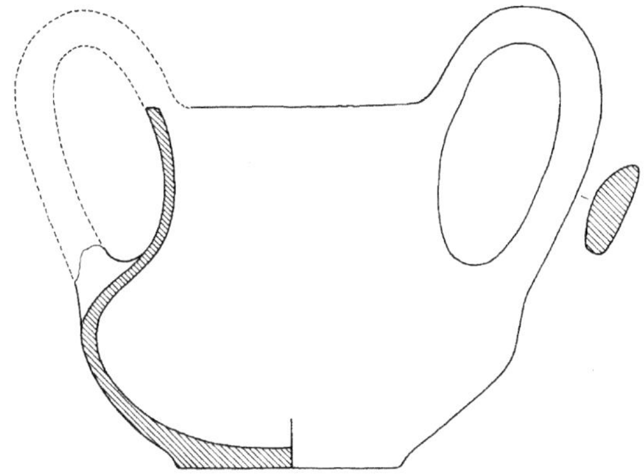

9/91 (P253)

Type 2b

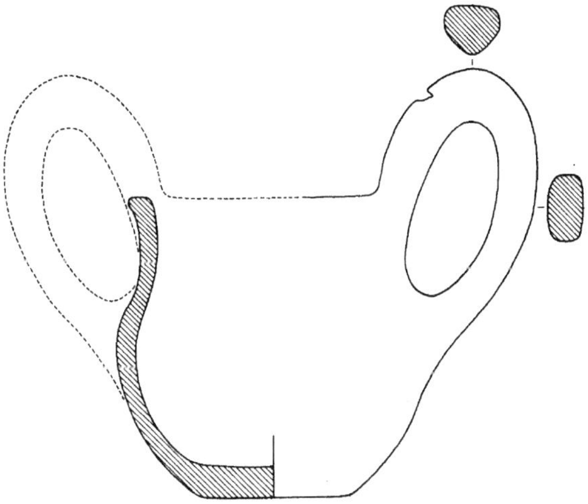

9/92 (P364)

Type 3

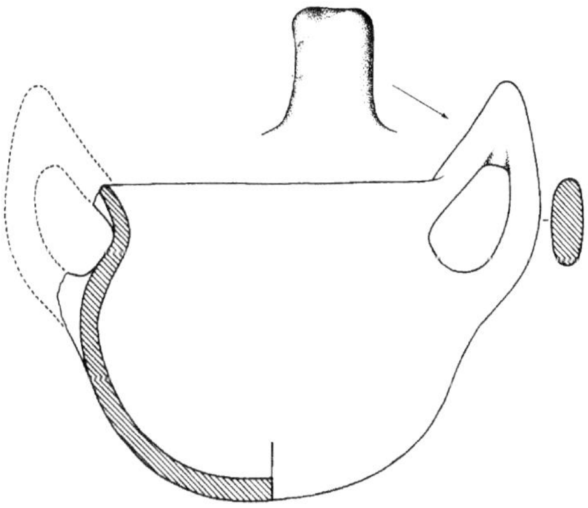

9/93 (P223)

FIGURE 9.16 **9/90–9/93**

Type 4

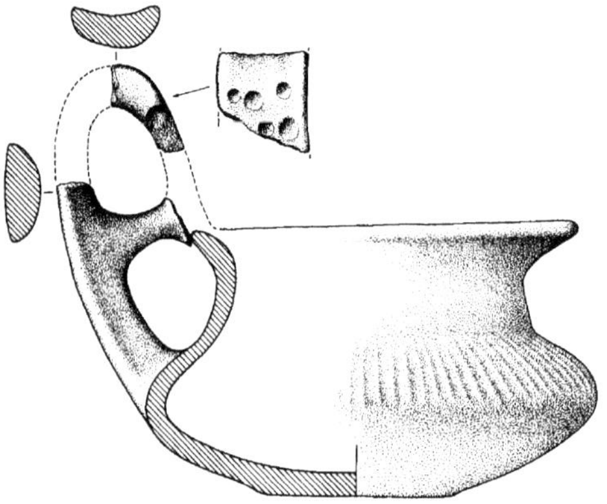

9/94 (P322)

Type 5

9/95 (P228)

Type 5

9/96 (P276)

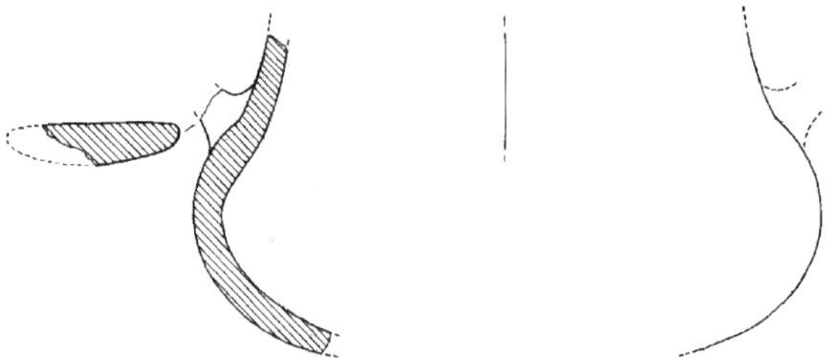

9/100 (P049 + P063)

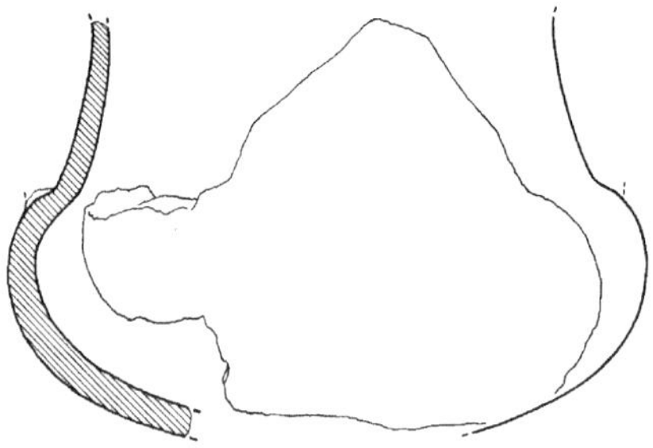

9/101 (P137)

FIGURE 9.17 **9/94–9/96, 9/100–9/101**

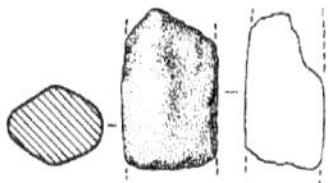

9/97 (P123)

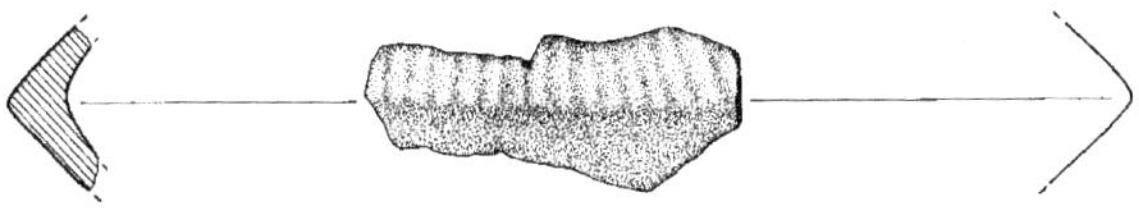

9/98 (P032)

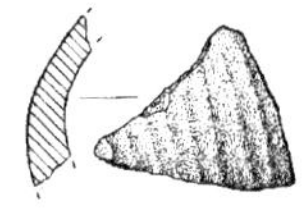

9/99 (P050)

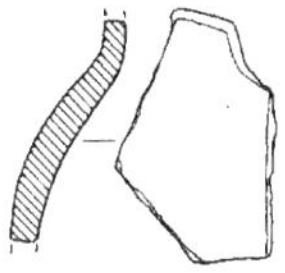

9/102 (P051)

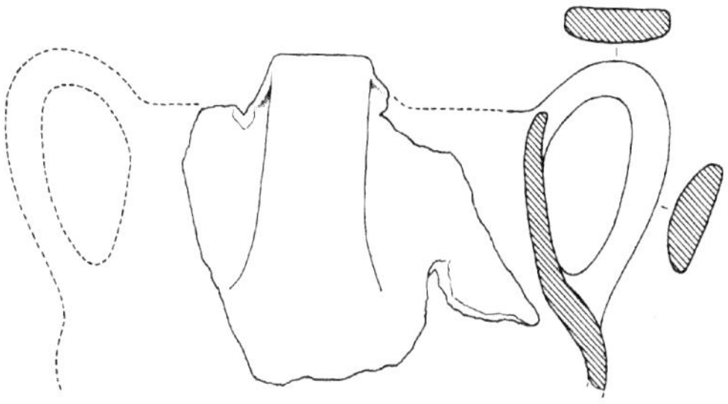

9/103 (P134)

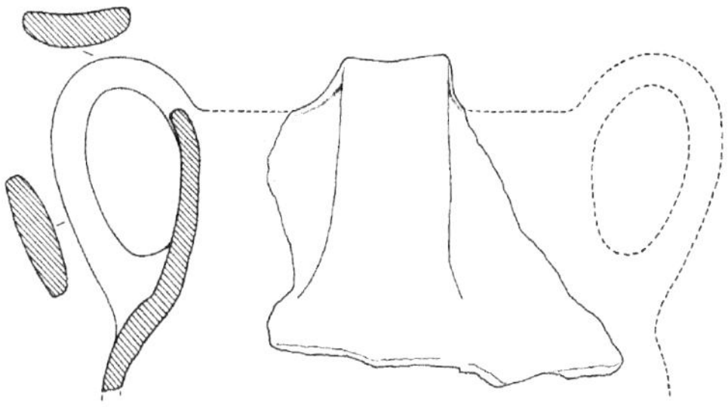

9/104 (P233)

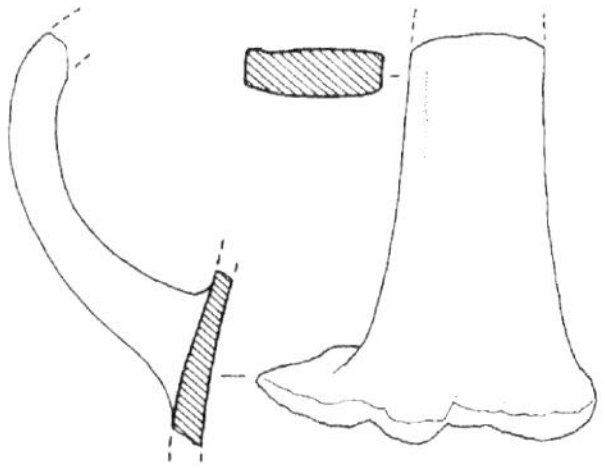

9/105 (P235)

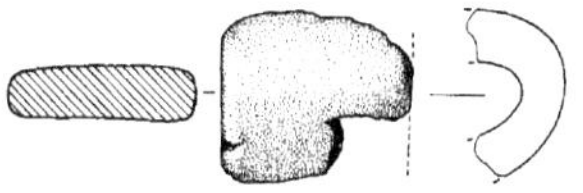

9/107 (P142)

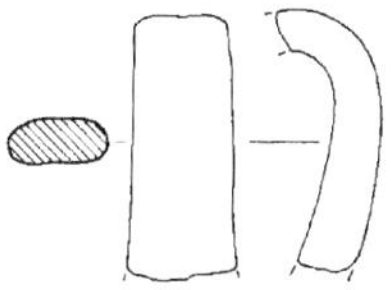

9/108 (P043)

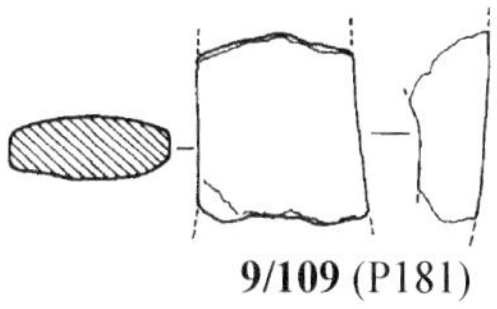

9/109 (P181)

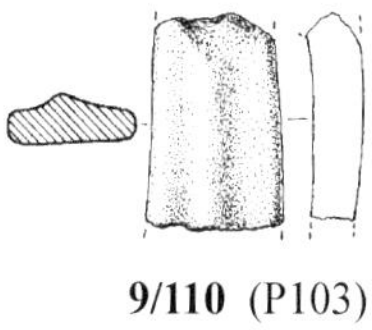

9/110 (P103)

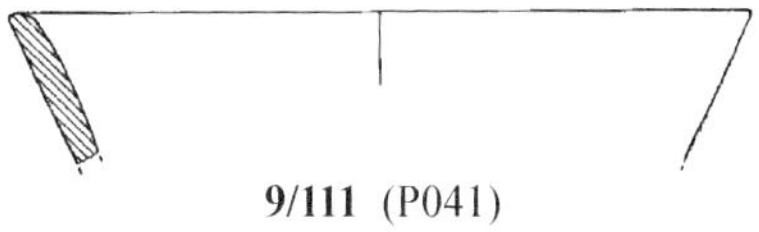

9/111 (P041)

0 5 cm

Figure 9.18 **9/97–9/99, 9/102–9/105, 9/107–9/111**

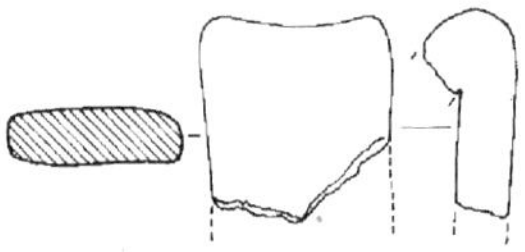

9/118 (P056)

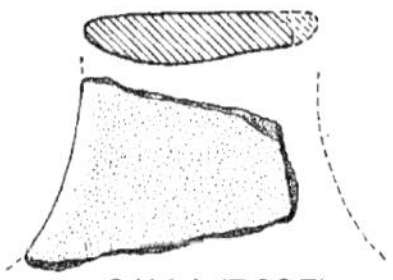

9/114 (P027)

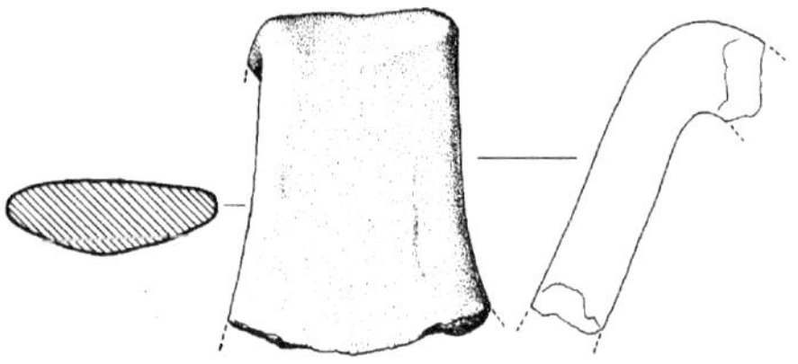

9/116 (P357A)

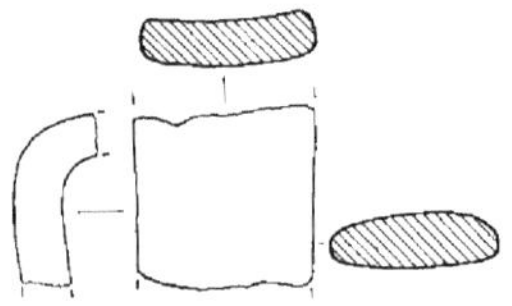

9/117 (P055)

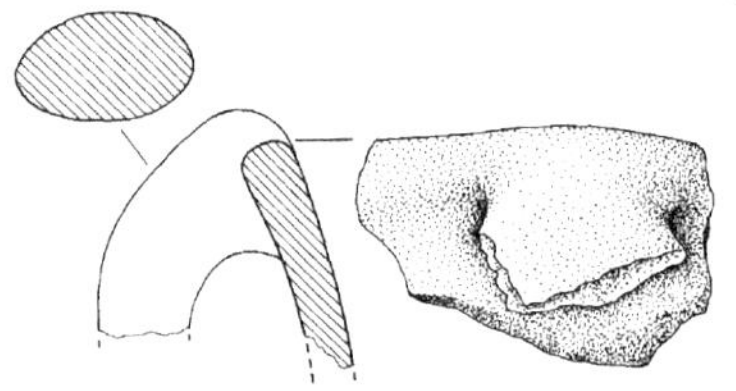

9/120 (P021)

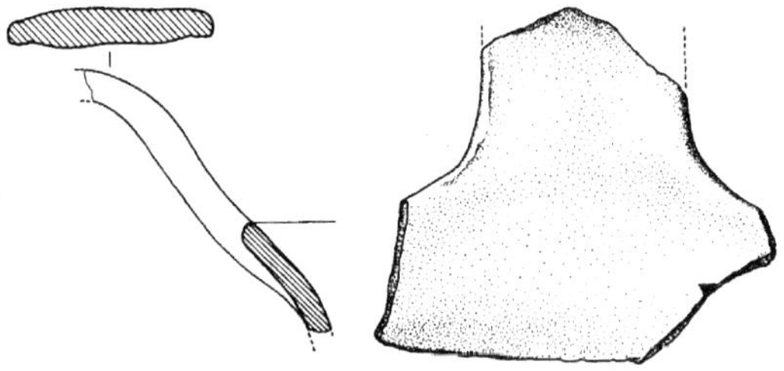

9/121 (P307)

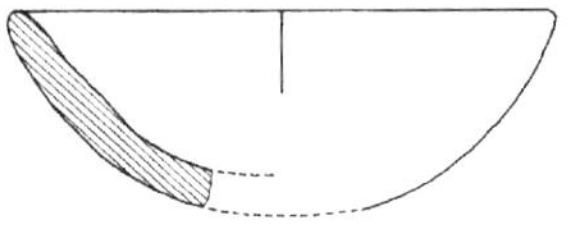

9/126 (P191)

9/127 (P089)

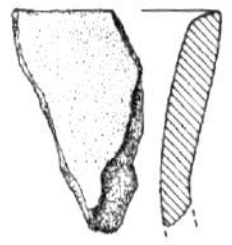

9/125 (P126)

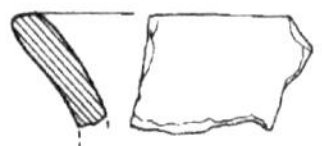

9/128 (P204)

9/129 (P160)

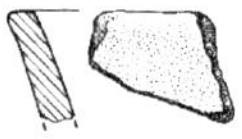

9/131 (P023)

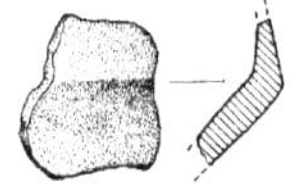

9/132 (P150)

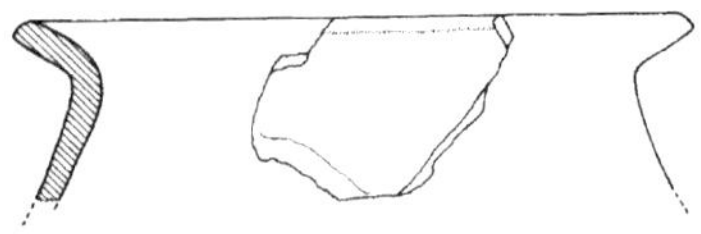

9/133 (P171)

9/134 (P068)

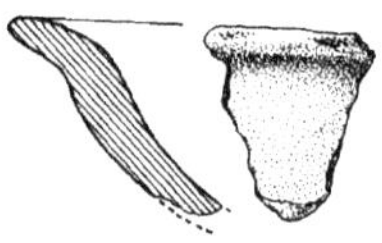

9/135 (P406)

FIGURE 9.19 **9/114, 9/116–9/118, 9/120–9/121, 9/125–9/129, 9/131–9/135**

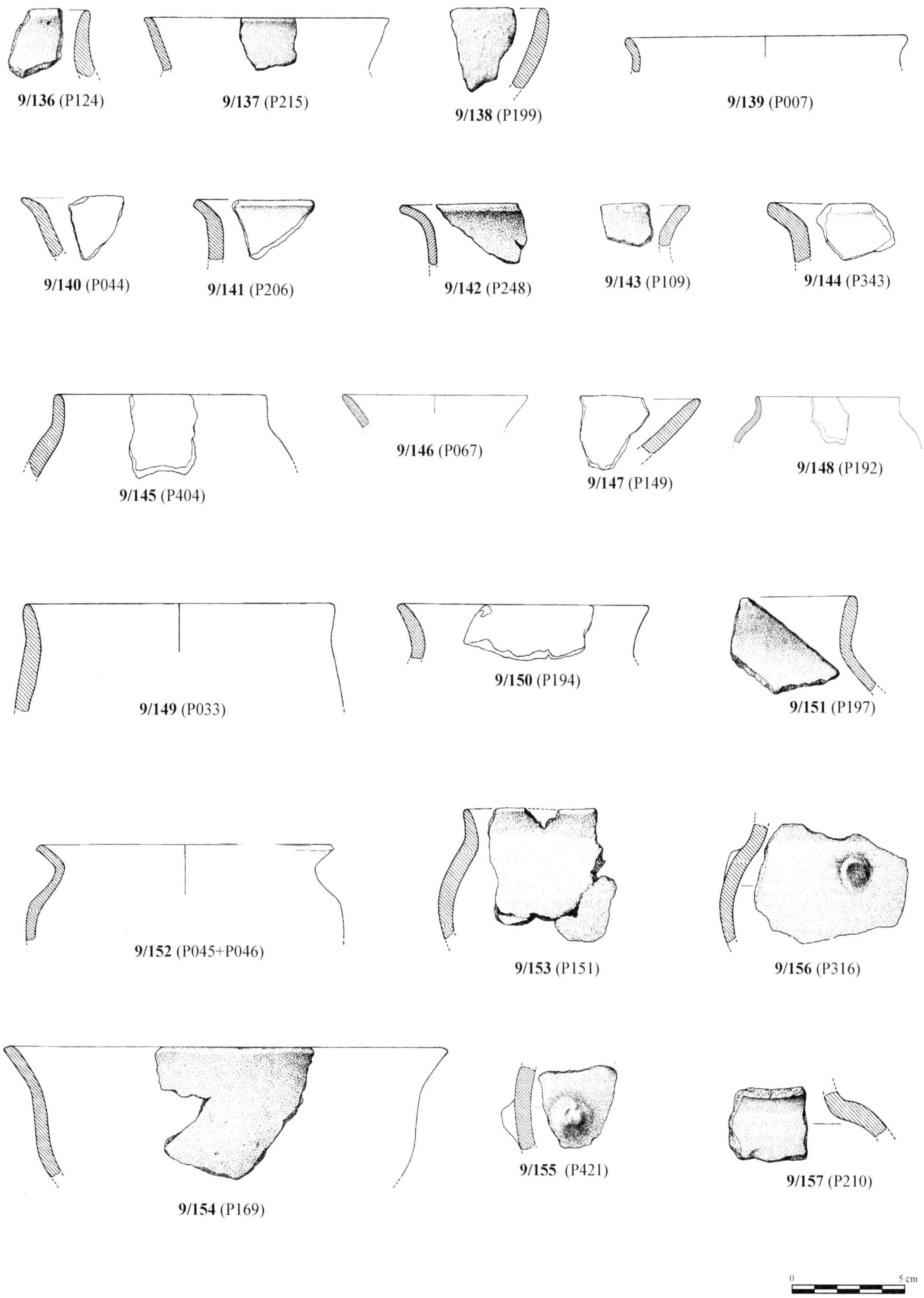

FIGURE 9.20 **9/136–9/157**

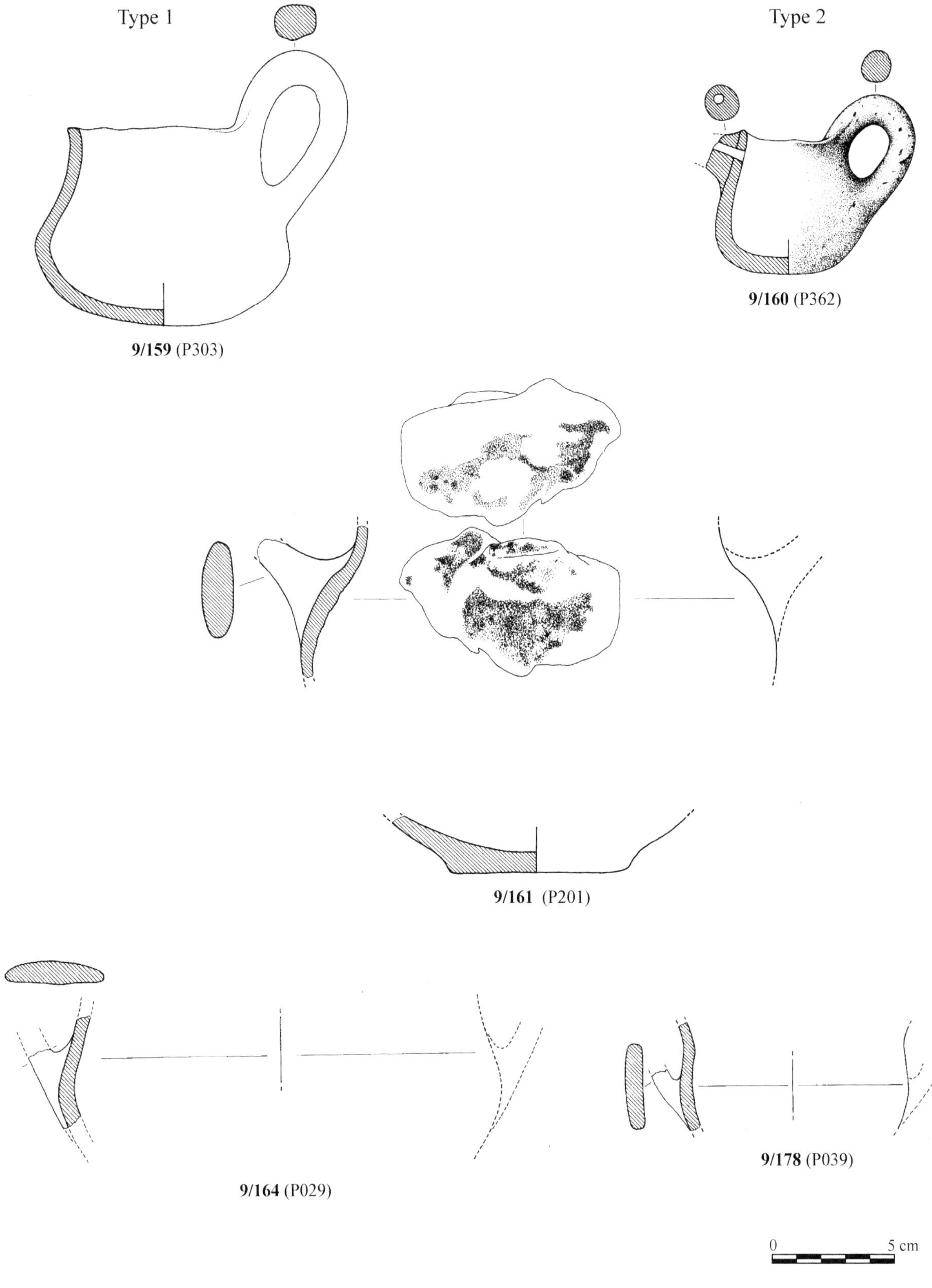

FIGURE 9.21 **9/159–9/161, 9/164, 9/178**

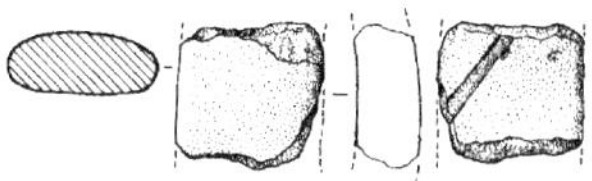

9/163 (P008)

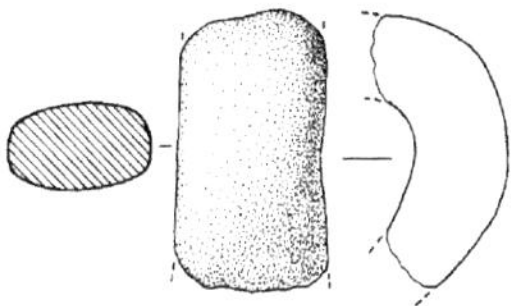

9/166 (P120)

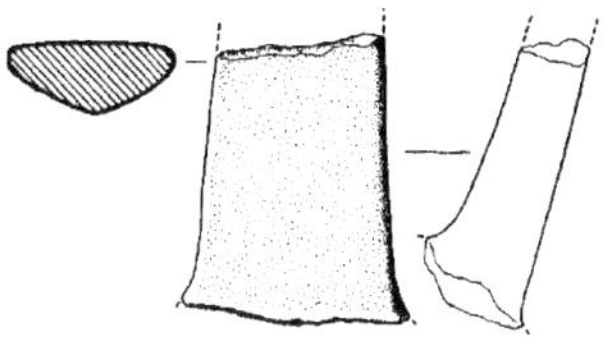

9/167 (P094B)

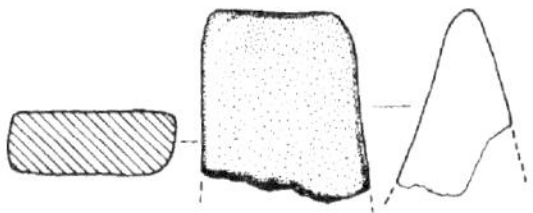

9/168 (P116)

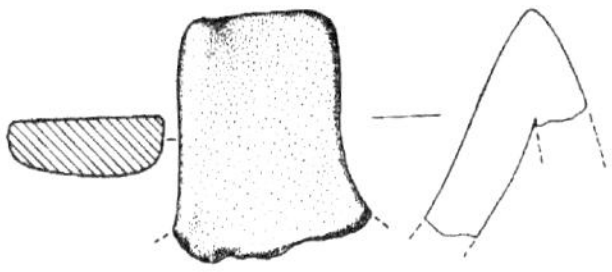

9/169 (P119)

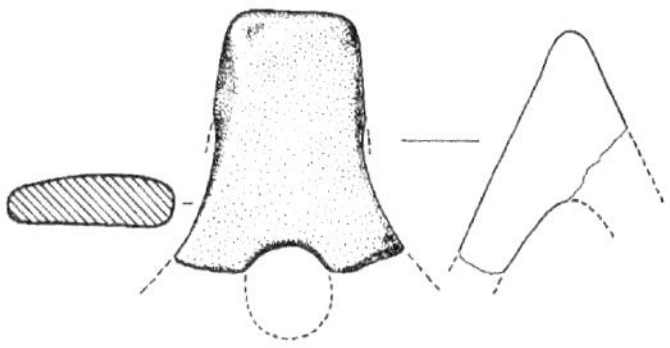

9/170 (P140)

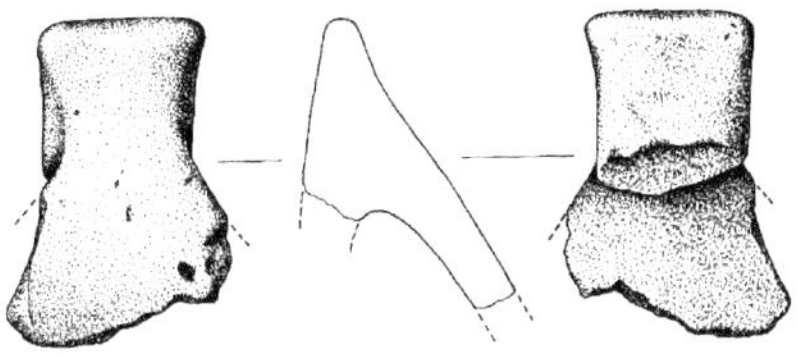

9/171 (P203)

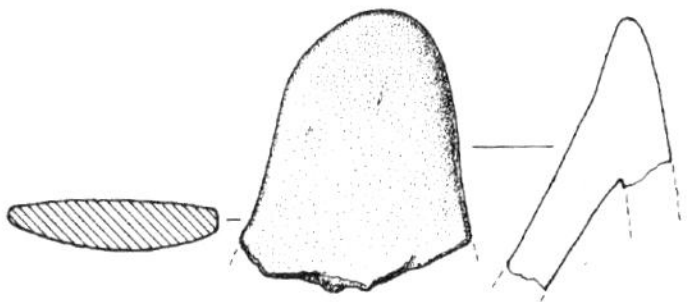

9/173 (P314)

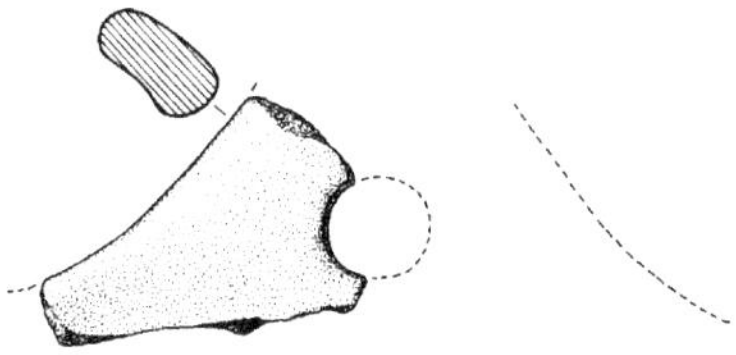

9/176 (P289)

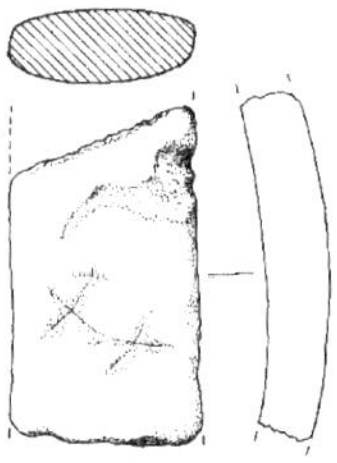

9/177 (P037)

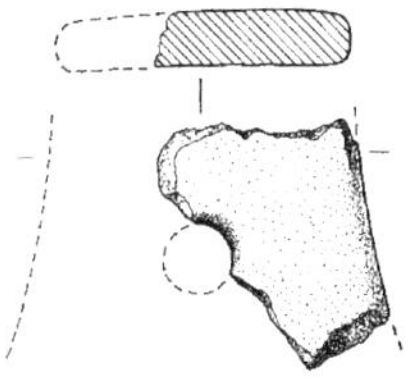

9/179 (P038)

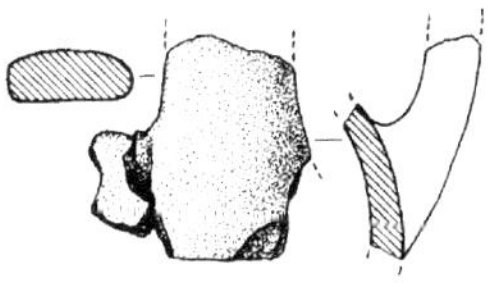

9/180 (P158)

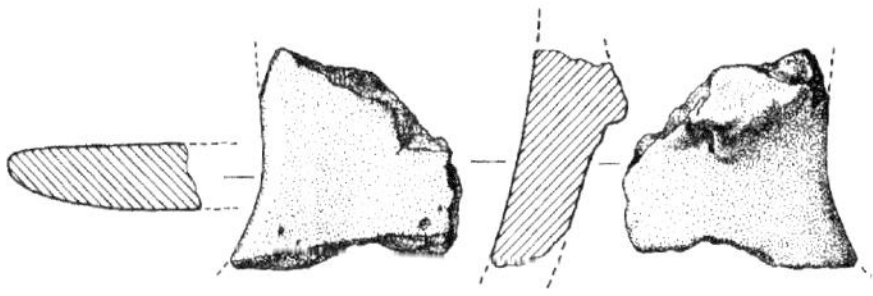

9/181 (P111)

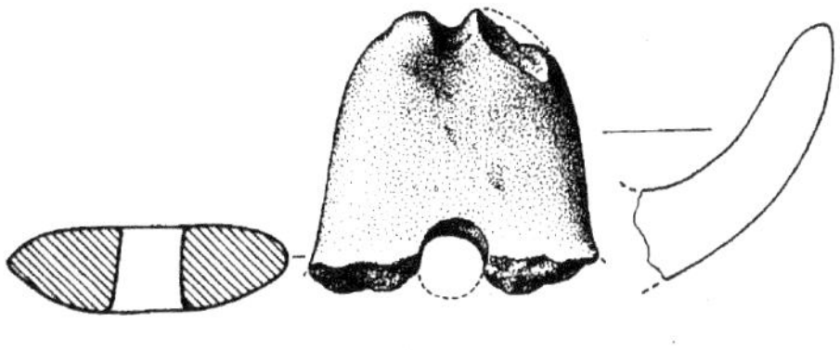

9/187 (P354)

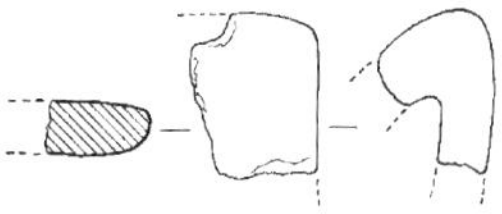

9/188 (P076)

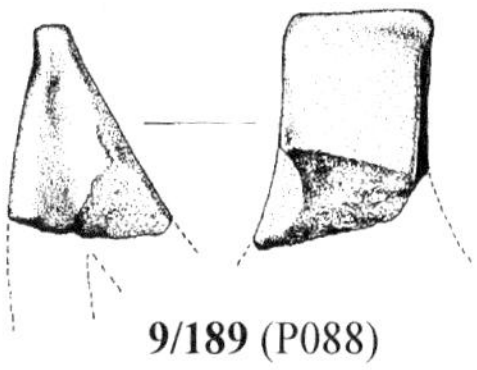

9/189 (P088)

0 5 cm

FIGURE 9.22 **9/163, 9/166–9/171, 9/173, 9/176–9/177, 9/179–9/181, 9/187–9/189**

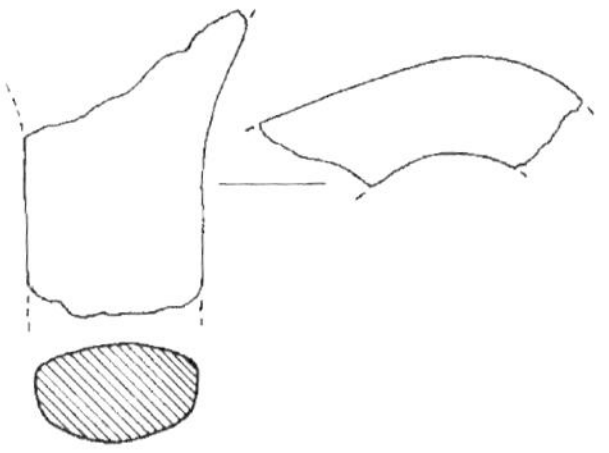

9/182 (P035)

9/183 (P062)

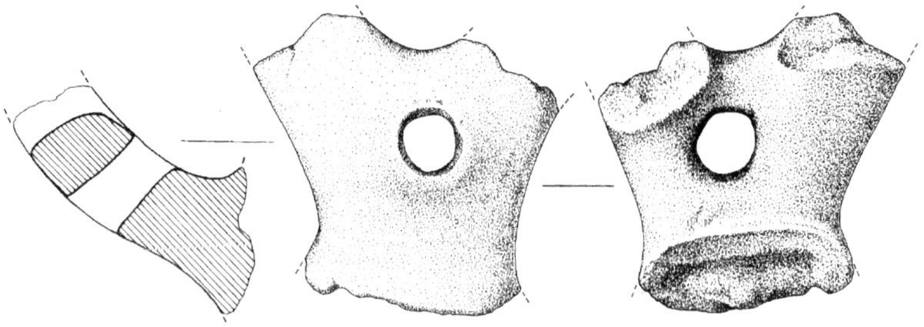

9/184 (P099)

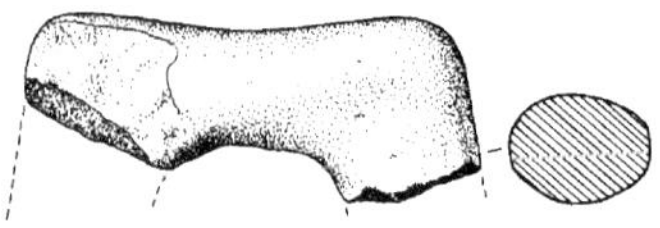

9/185 (P245)

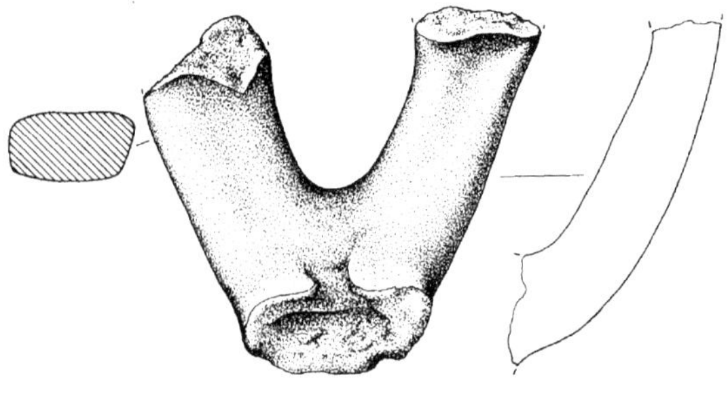

9/186 (P344)

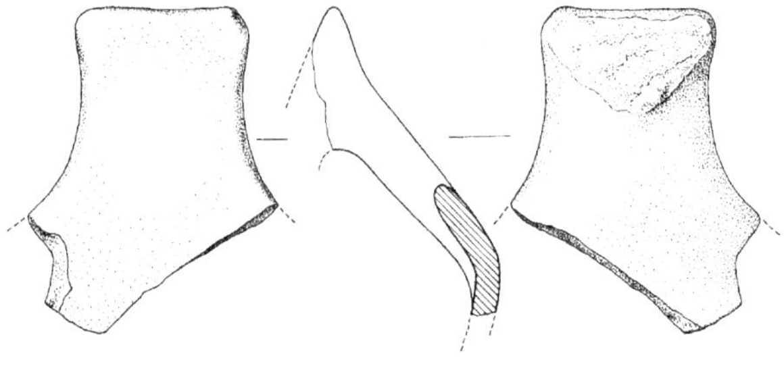

9/190 (P098)

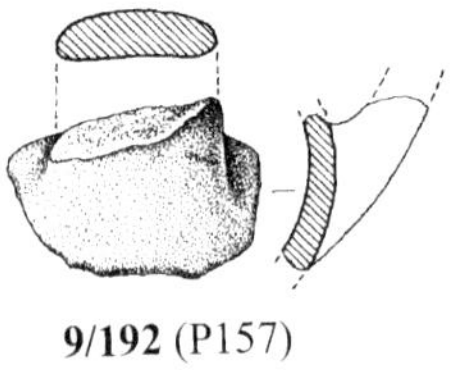

9/192 (P157)

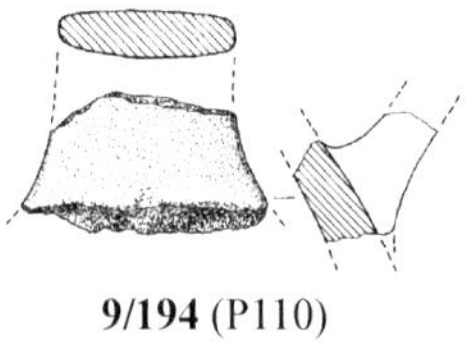

9/194 (P110)

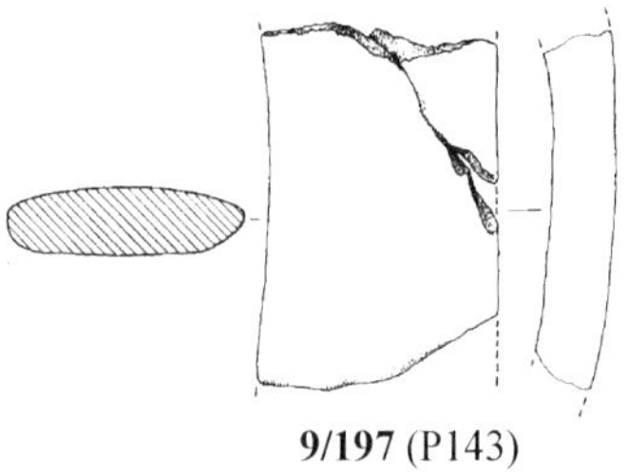

9/197 (P143)

FIGURE 9.23 **9/182–9/186, 9/190, 9/192, 9/194, 9/197**

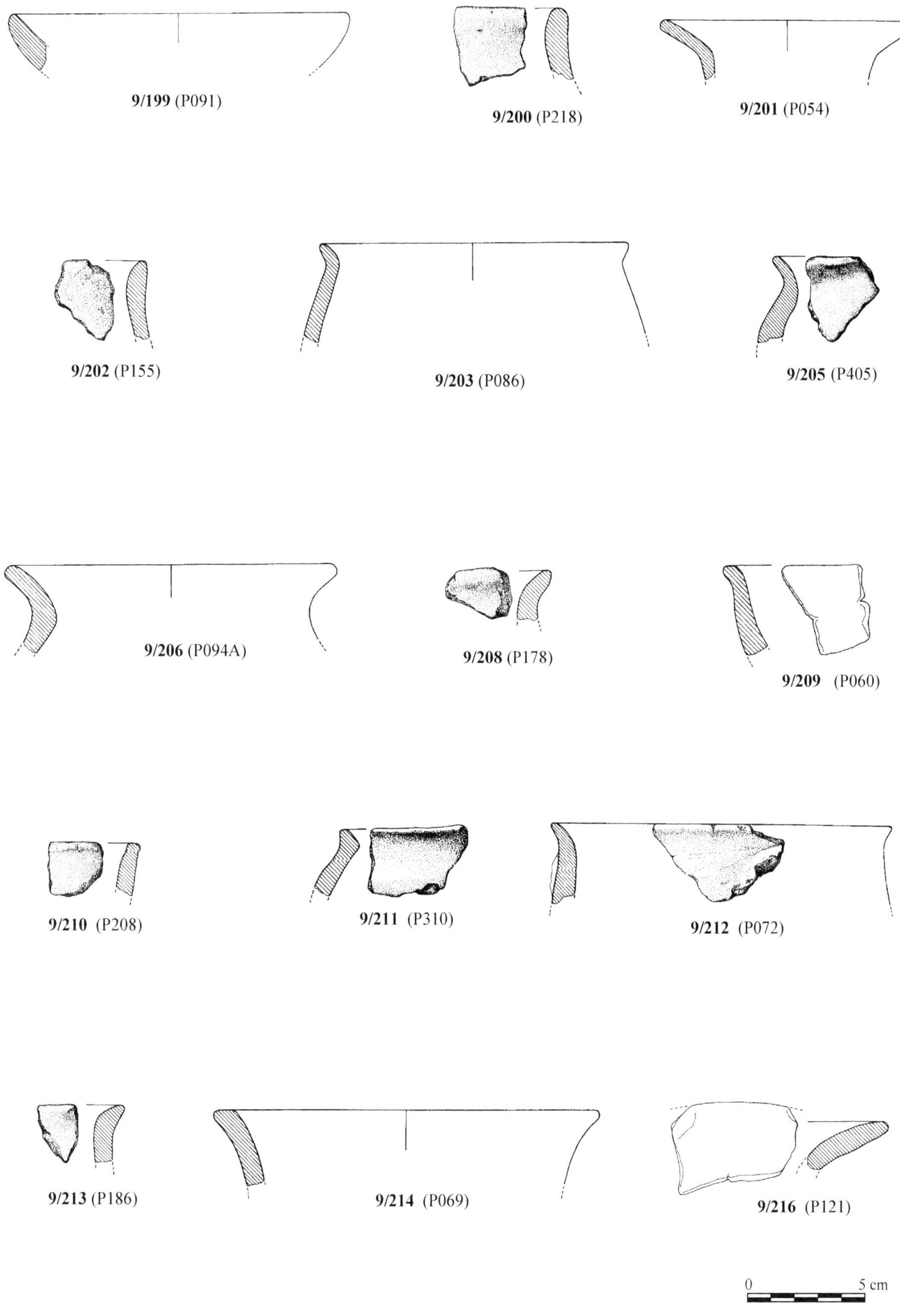

FIGURE 9.24 **9/199–9/203, 9/205–9/206, 9/208–9/214, 9/216**

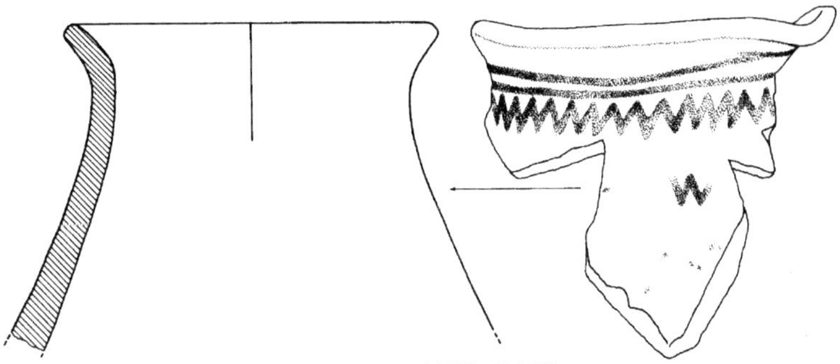

9/220 (P168)

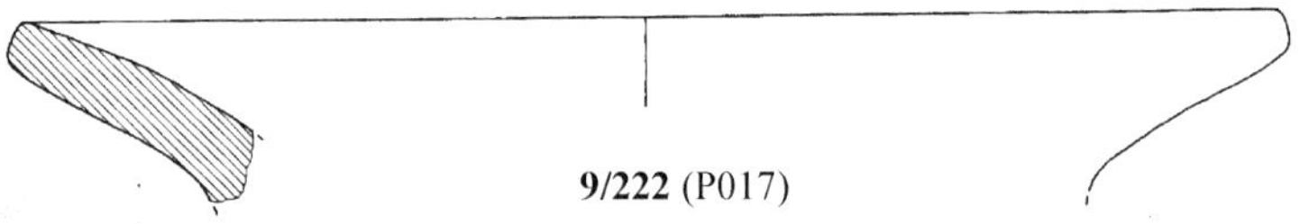

9/222 (P017)

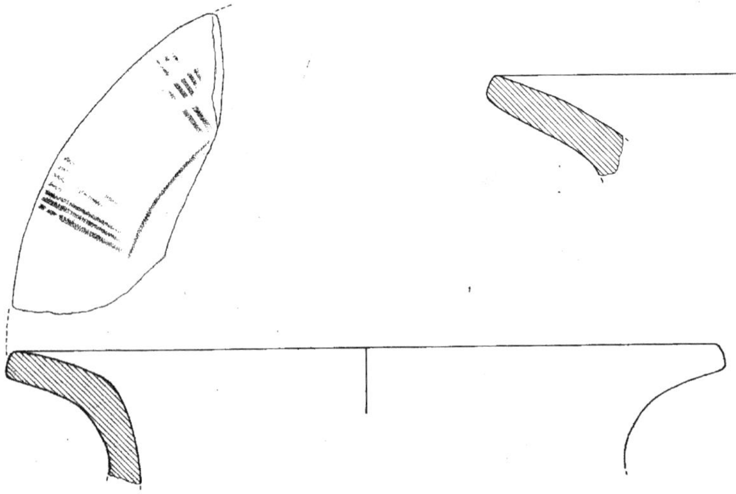

9/221 (P057)

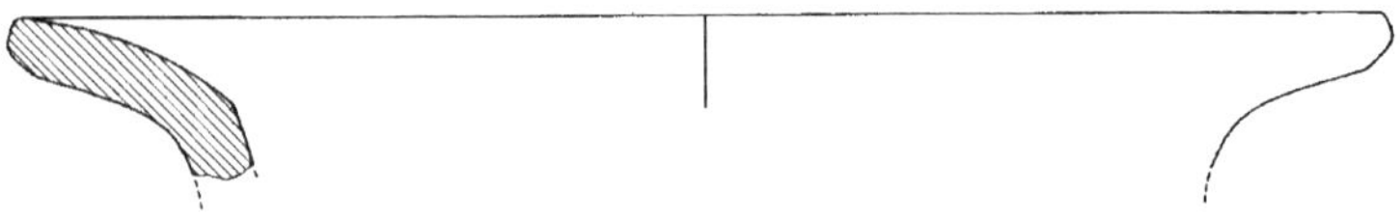

9/223 (P095)

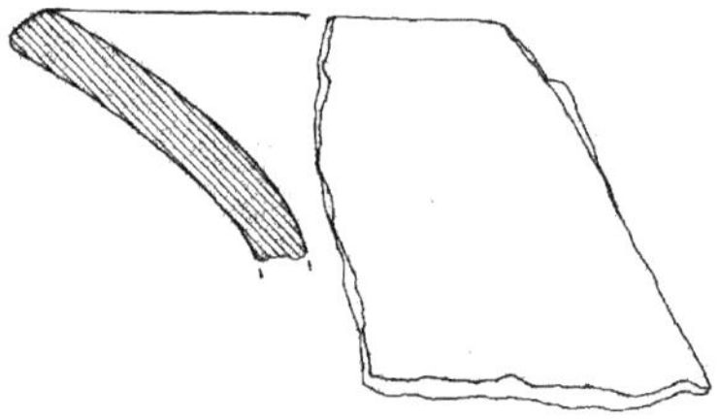

9/224 (P040)

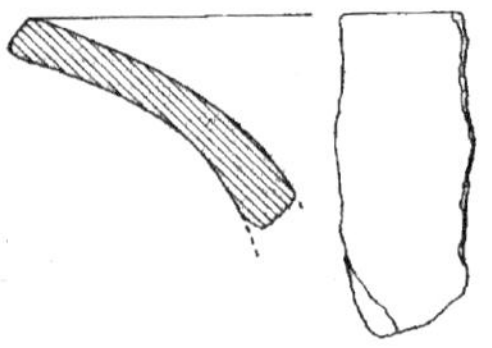

9/225 (P053)

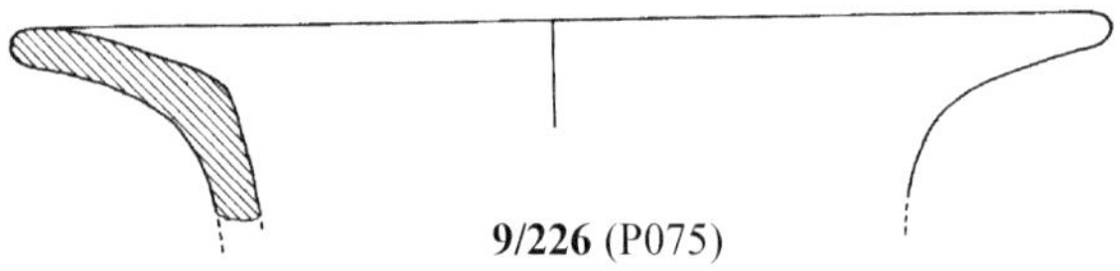

9/226 (P075)

0 5 cm

FIGURE 9.25 **9/220–9/226**

9/218 (P156)

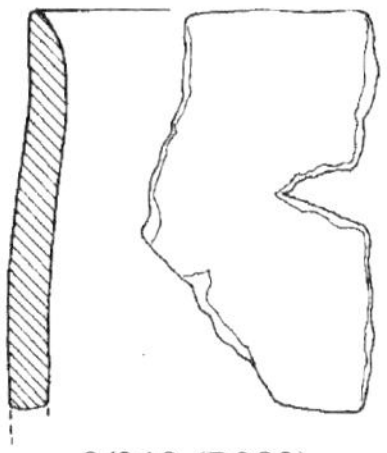

9/219 (P083)

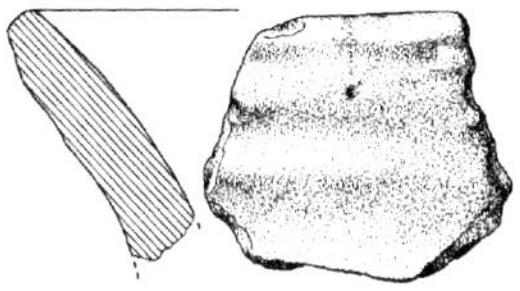

9/227 (P366)

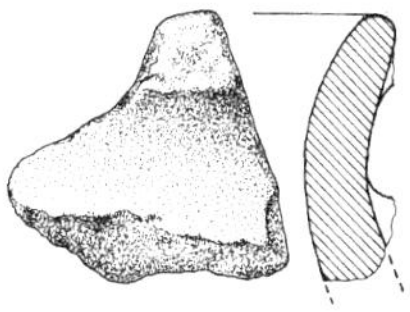

9/229 (P122)

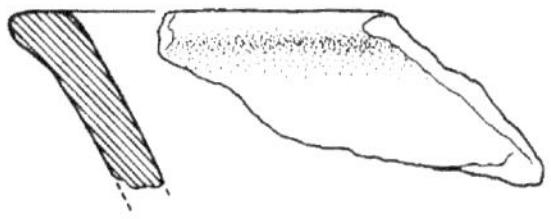

9/230 (P413)

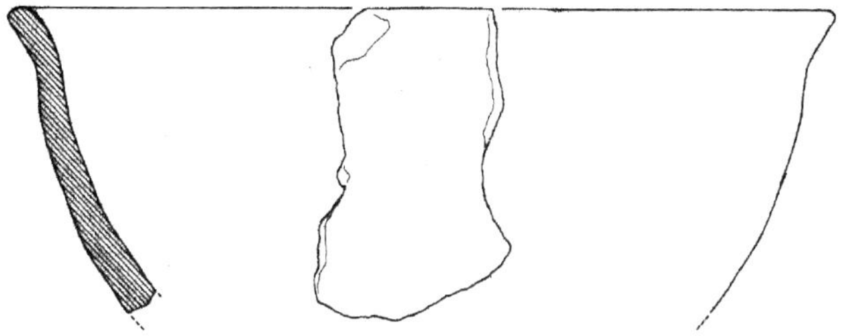

9/232 (P377)

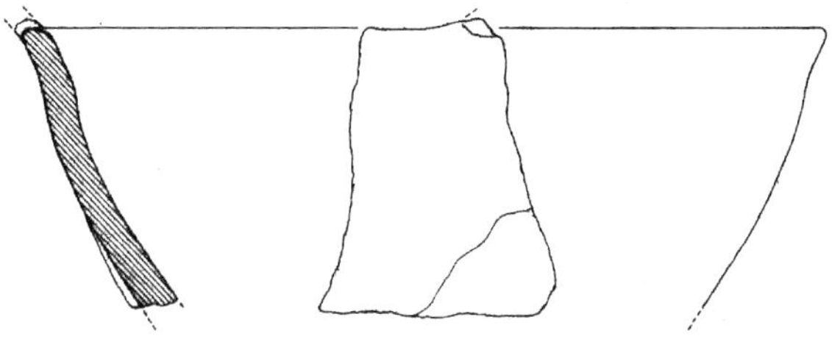

9/233 (P380)

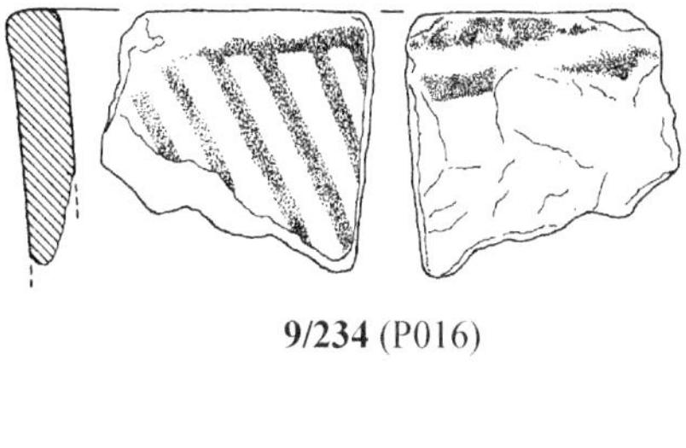

9/234 (P016)

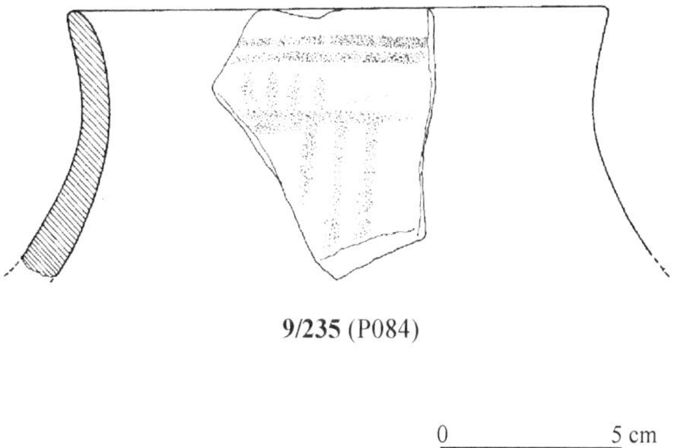

9/235 (P084)

0 5 cm

FIGURE 9.26 **9/218–9/219**, **9/227**, **9/229–9/230**, **9/232–9/235**

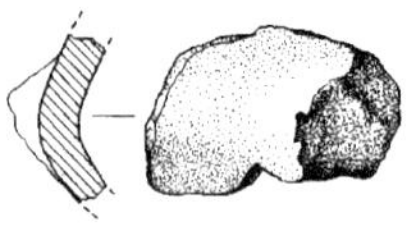

9/236 (P059)

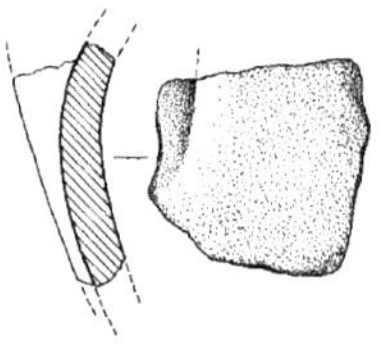

9/238 (P028)

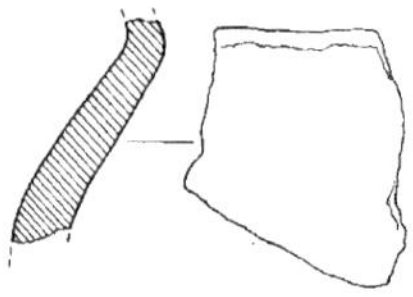

9/239 (P070)

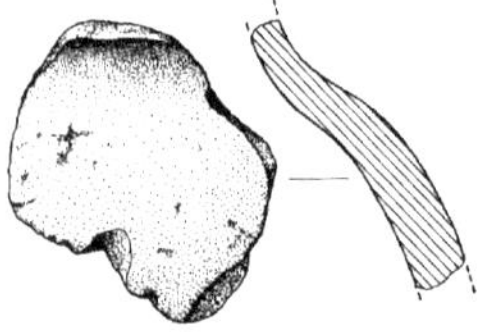

9/240 (P207)

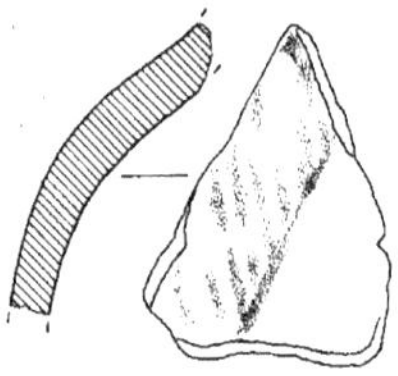

9/241 (P034)

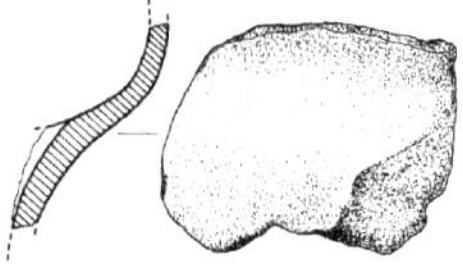

9/242 (P222)

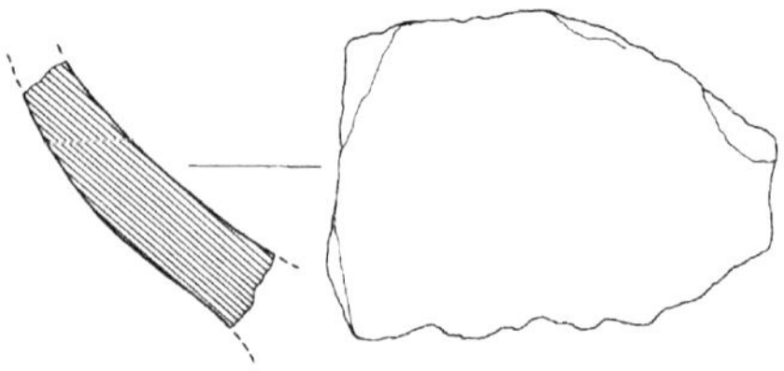

9/243 (P243)

9/244 (P288)

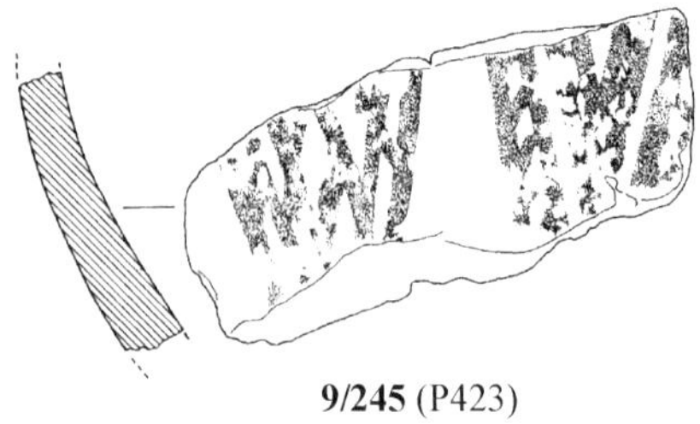

9/245 (P423)

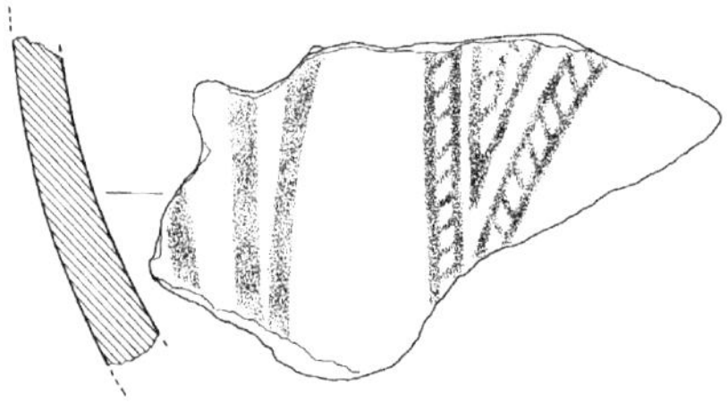

9/246 (P249)

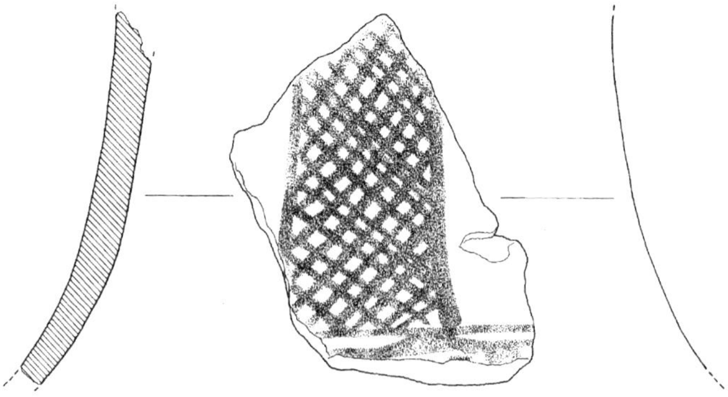

9/247 (P196)

0 5 cm

FIGURE 9.27 **9/236**, **9/238–9/247**

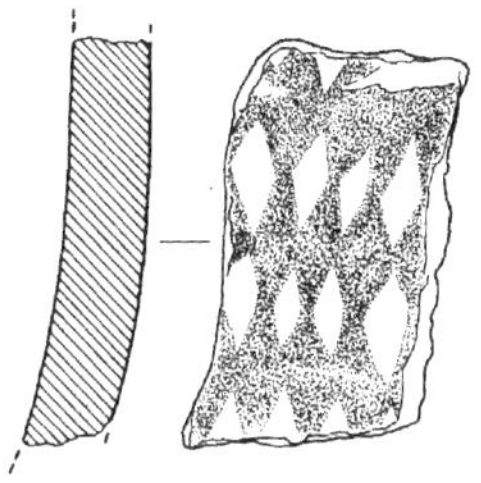

9/248 (P234)

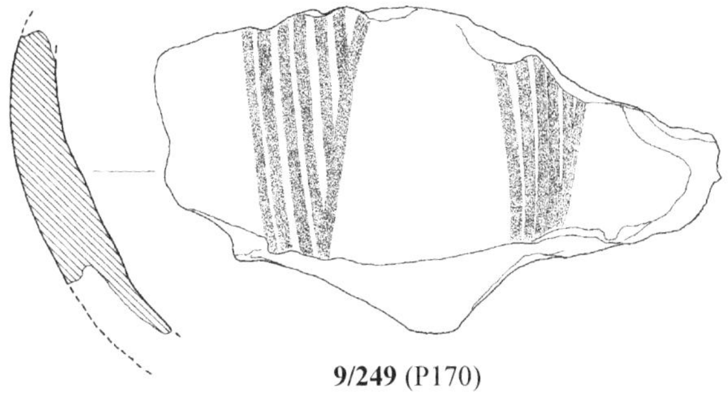

9/249 (P170)

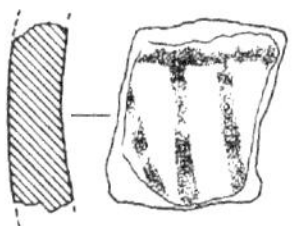

9/250 (P429)

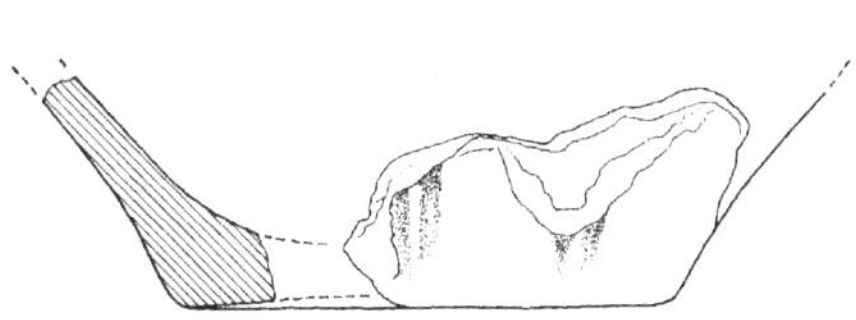

9/251 (P105)

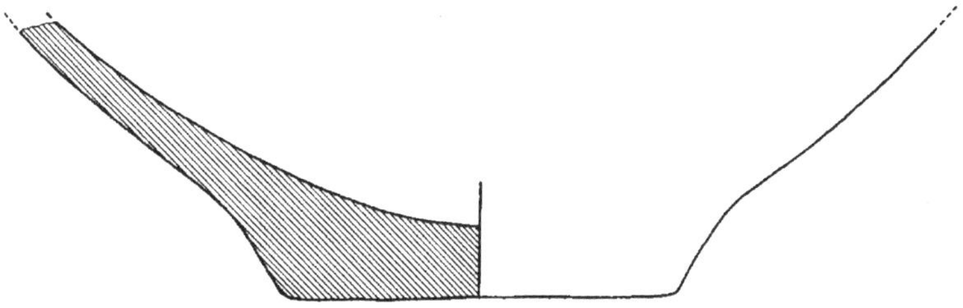

9/252 (P174)

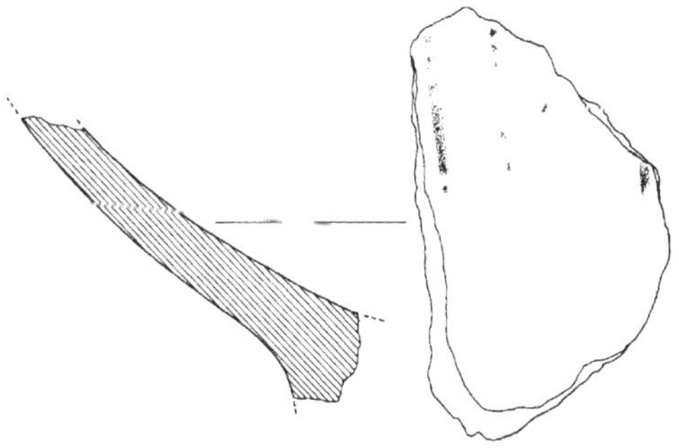

9/253 (P335)

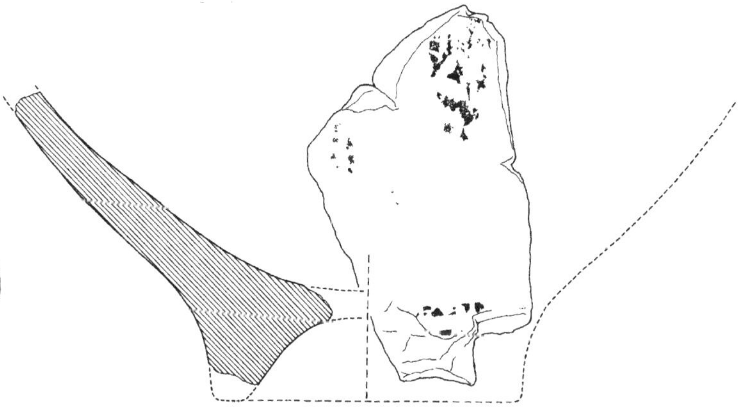

9/254 (P430)

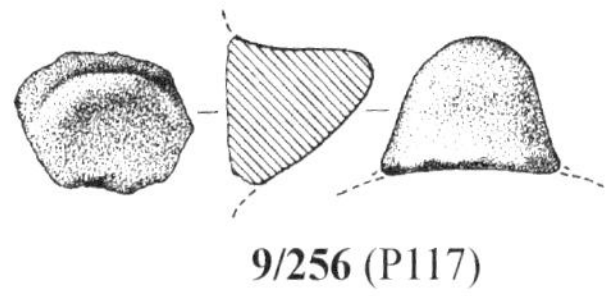

9/256 (P117)

FIGURE 9.28 **9/248–9/254, 9/256**

Type 1

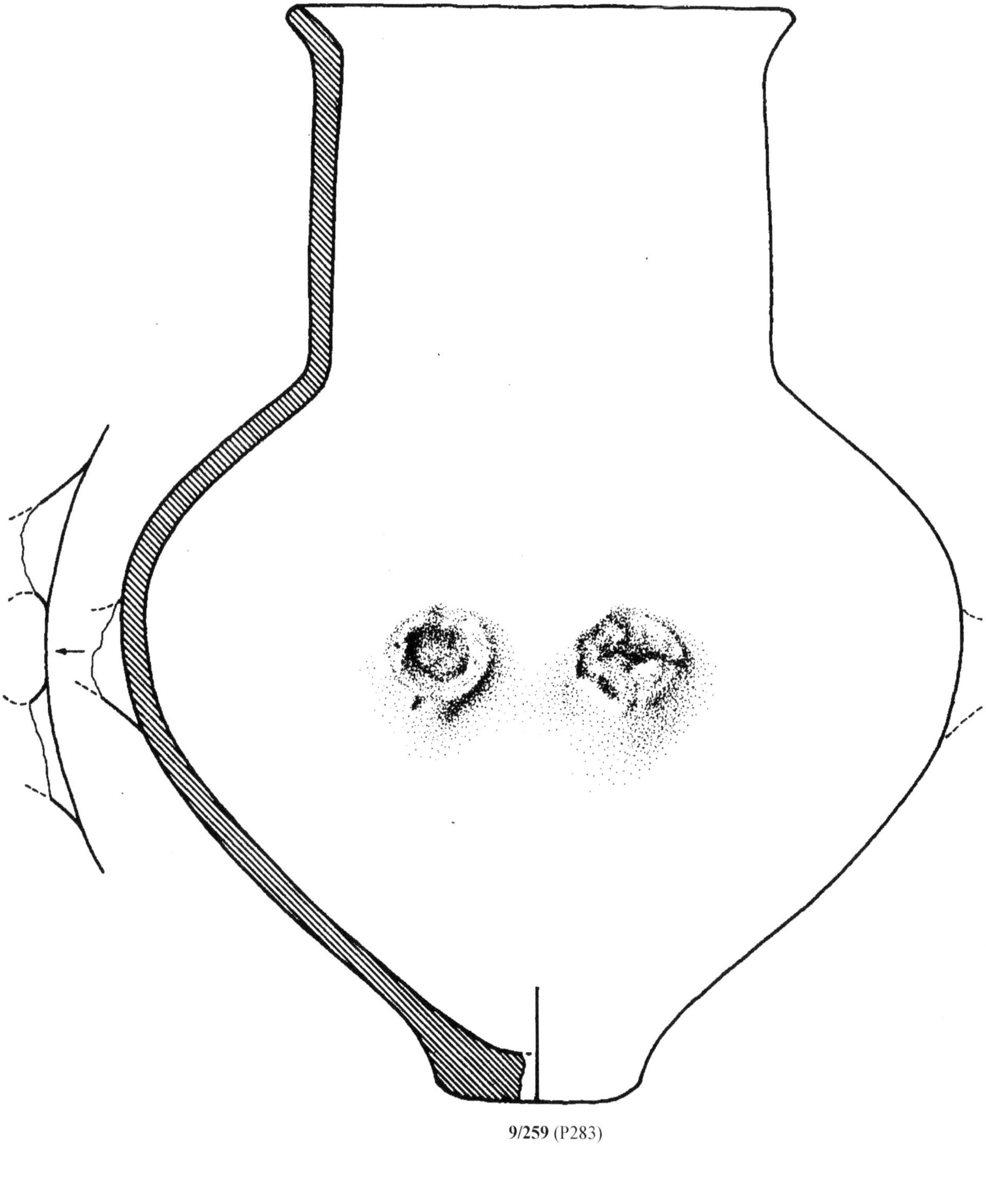

9/259 (P283)

FIGURE 9.29 **9/259**

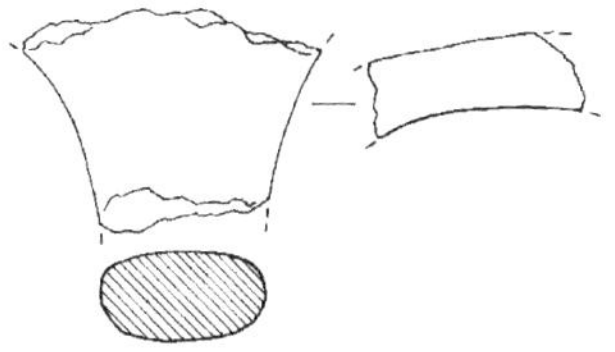

9/261 (P036)

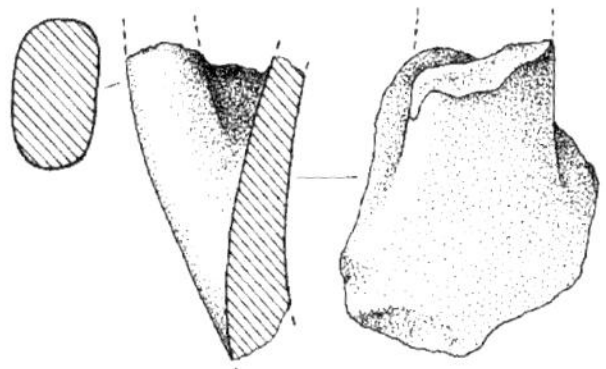

9/262 (P052)

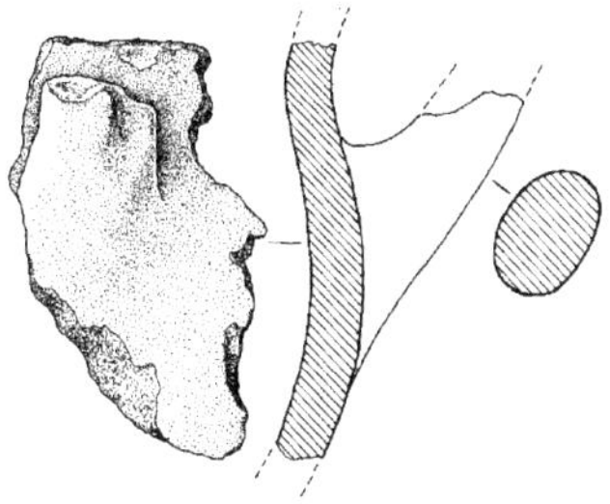

9/264 (P061)

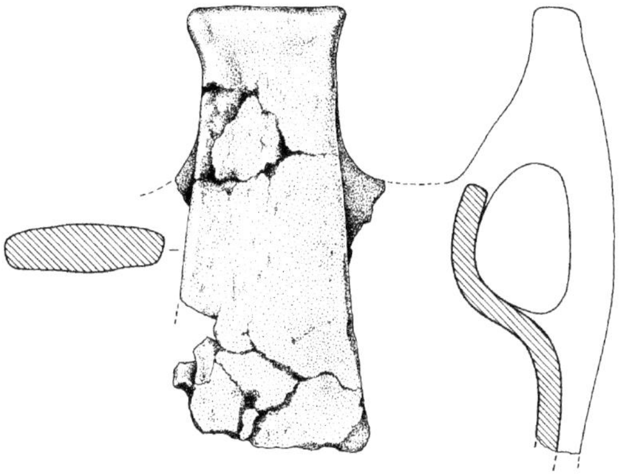

9/265 (P136)

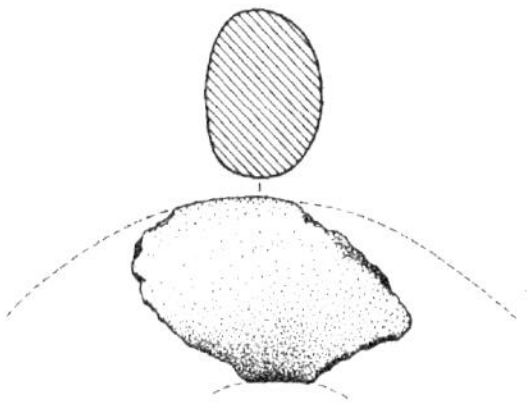

9/266 (P128)

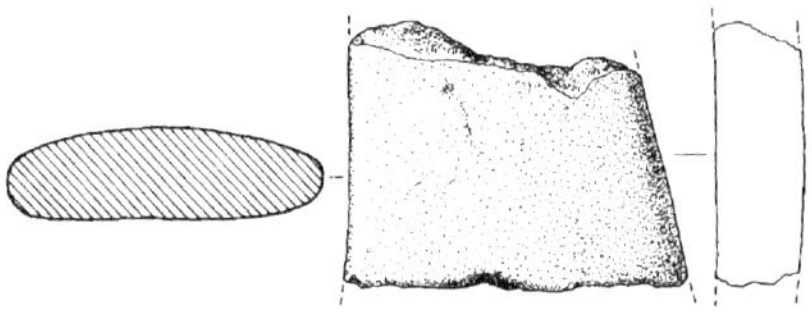

9/269 (P118)

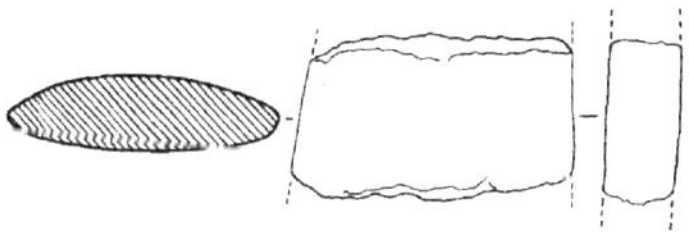

9/271 (P392)

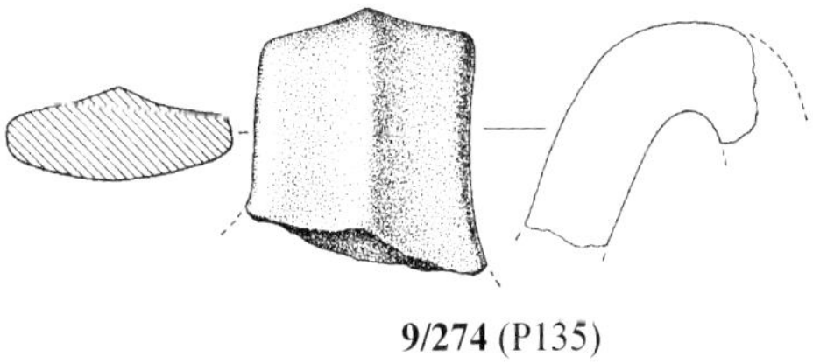

9/274 (P135)

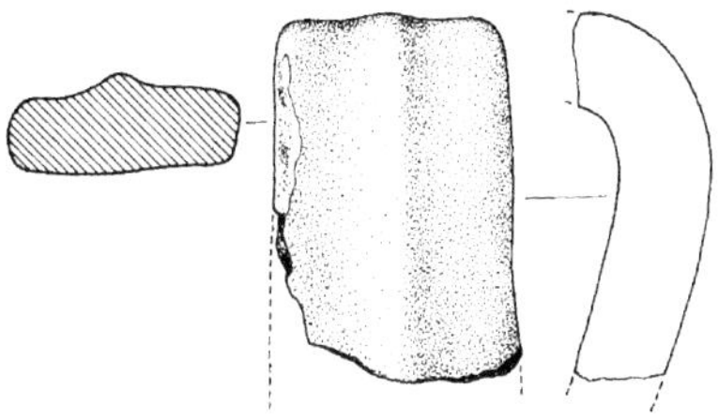

9/275 (P238)

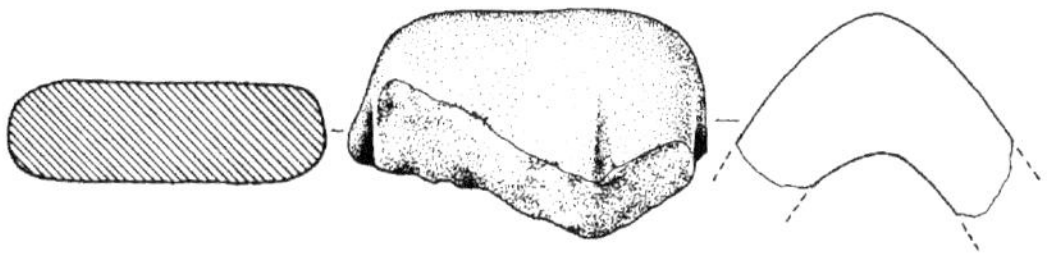

9/277 (P394)

0 5 cm

FIGURE 9.30 **9/261–9/262, 9/264–9/266, 9/269, 9/271, 9/274–9/275, 9/277**

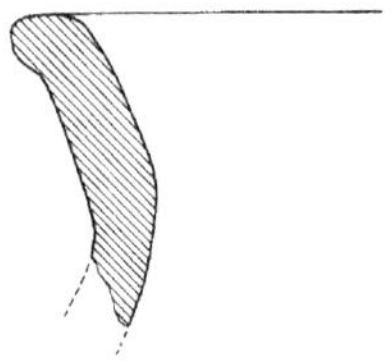

9/279 (P003)

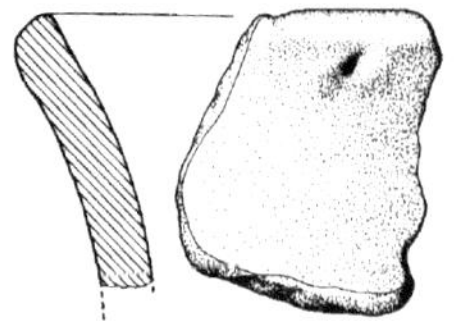

9/281 (P239)

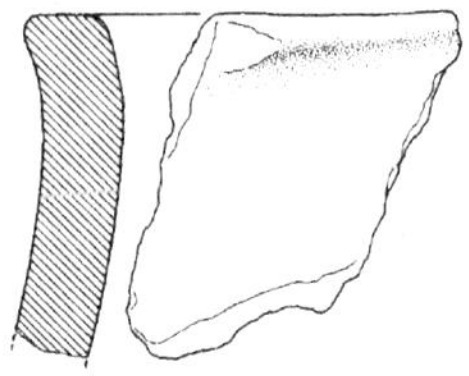

9/282 (P397)

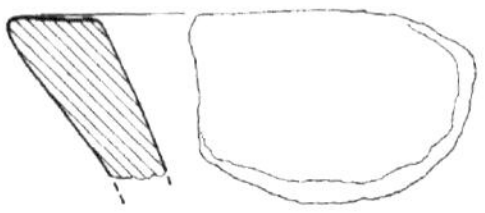

9/283 (P081)

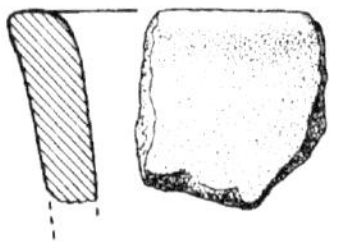

9/284 (P227)

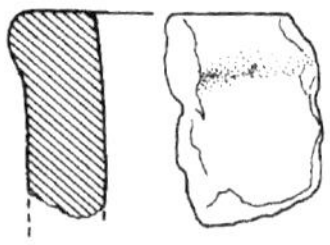

9/285 (P154)

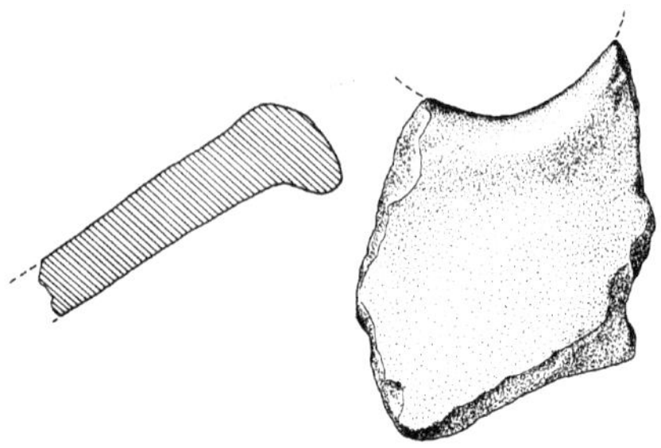

9/286 (P230)

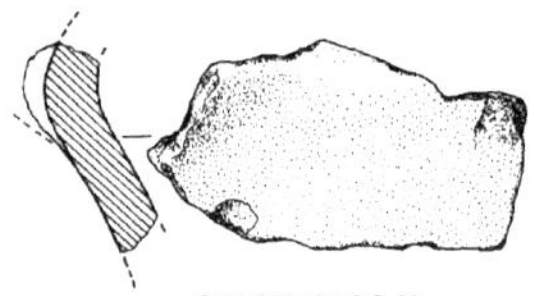

9/287 (P231)

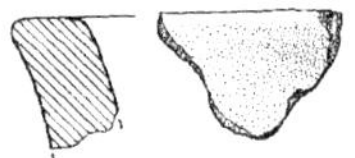

9/293 (P022)

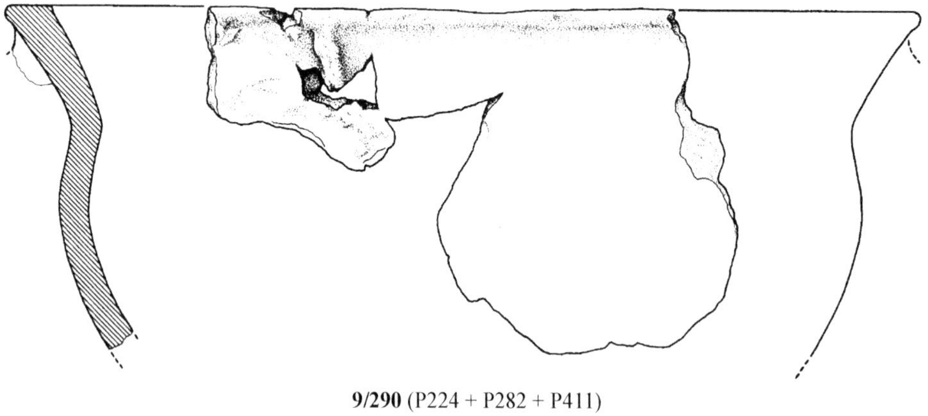

9/290 (P224 + P282 + P411)

9/294 (P026)

FIGURE 9.31 **9/279, 9/281–9/287, 9/290, 9/293–9/294**

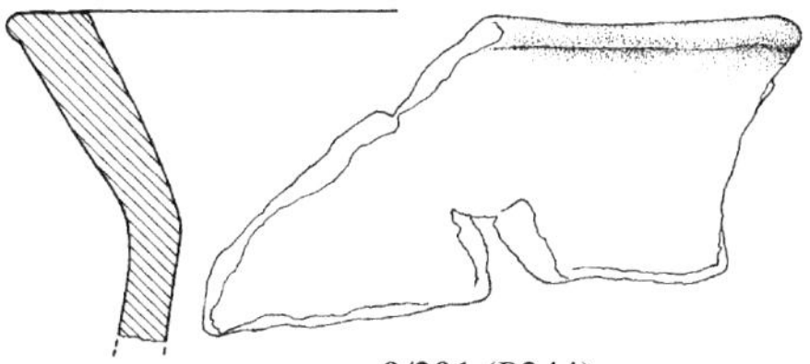

9/291 (P244)

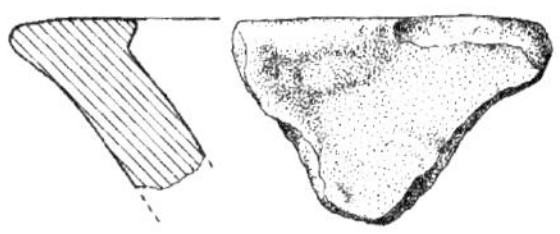

9/295 (P085)

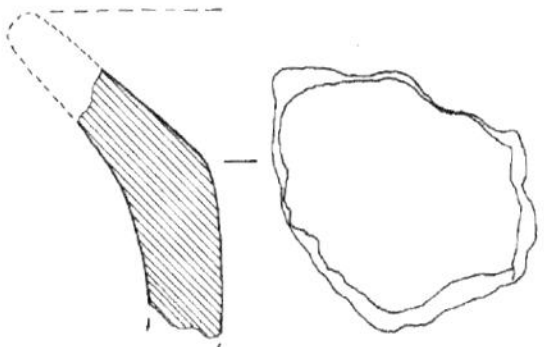

9/296 (P014)

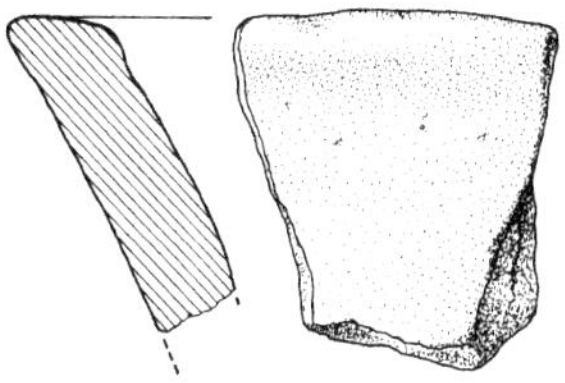

9/298 (P291)

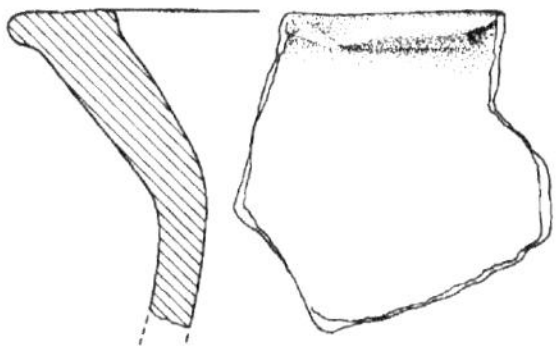

9/299 (P145)

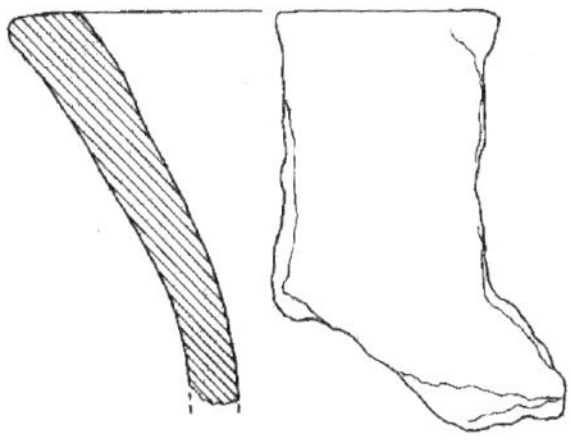

9/300 (P065)

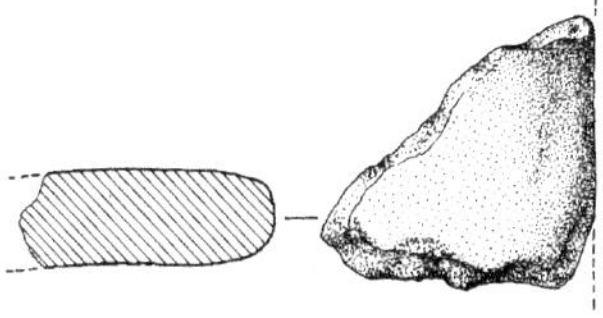

9/301 (P185)

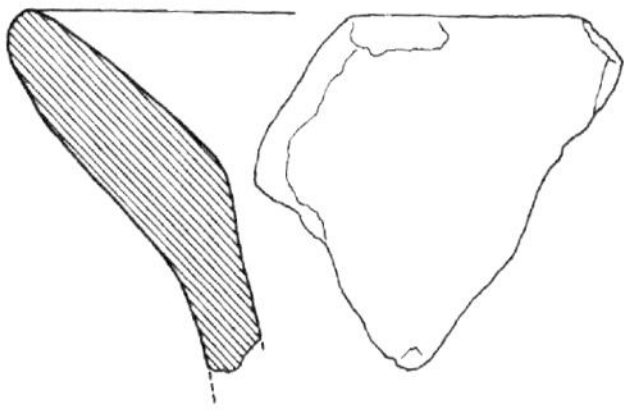

9/302 (P313)

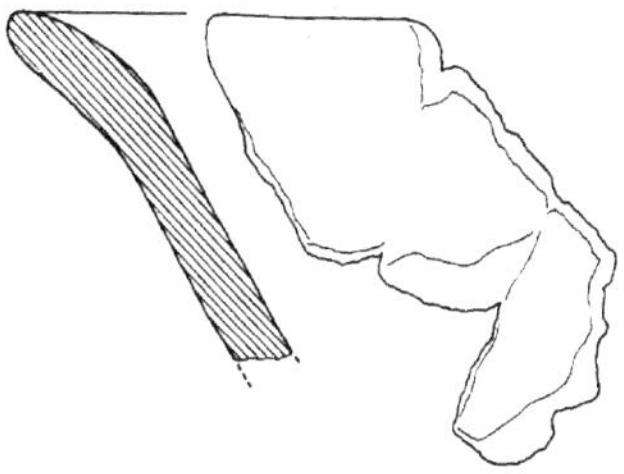

9/303 (P336)

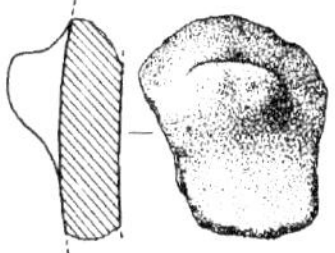

9/304 (P189)

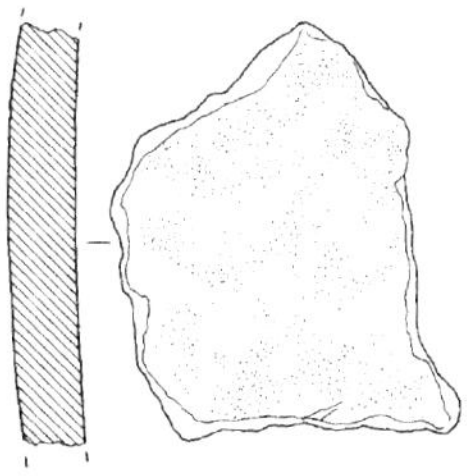

9/308 (P010)

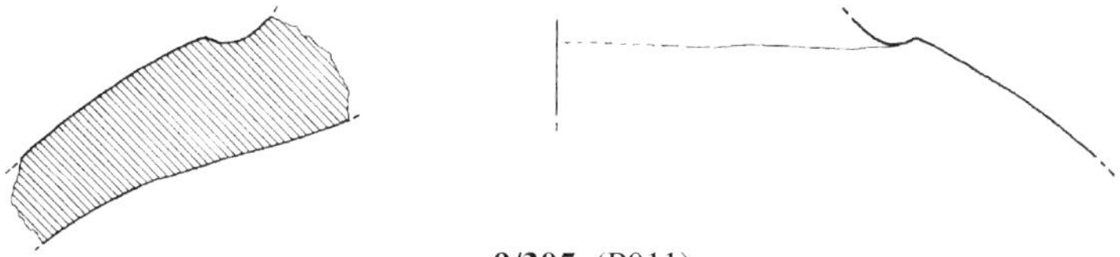

9/305 (P011)

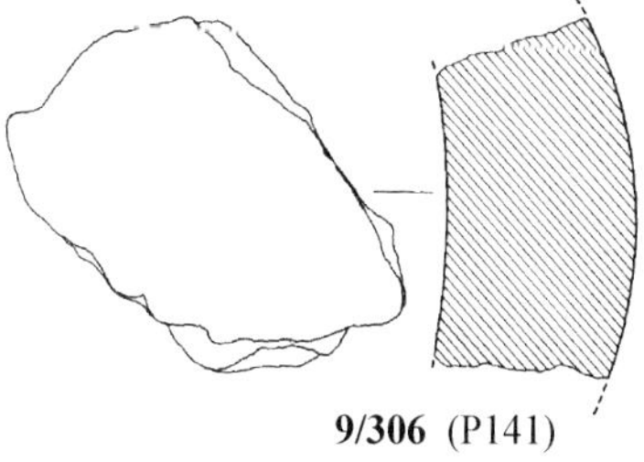

9/306 (P141)

FIGURE 9.32 **9/291, 9/295–9/296, 9/298–9/306, 9/308**

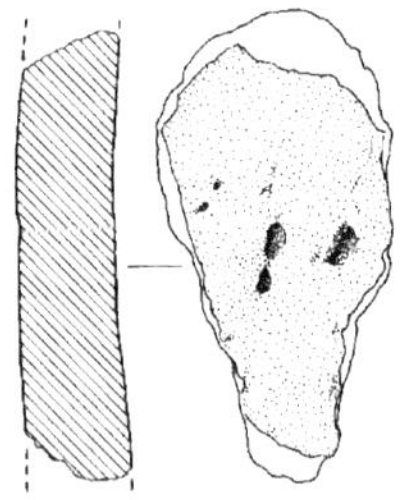

9/309 (P042)

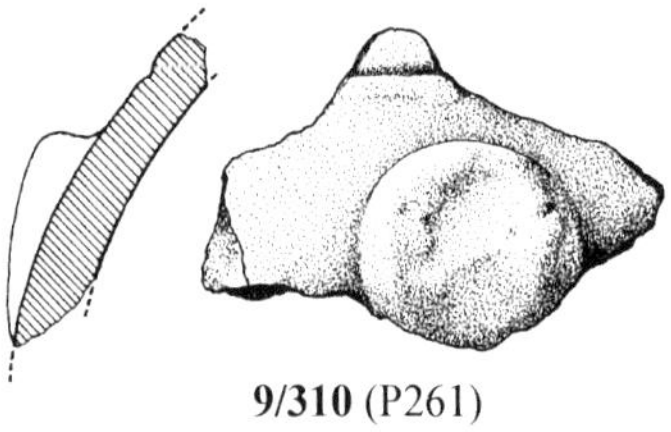

9/310 (P261)

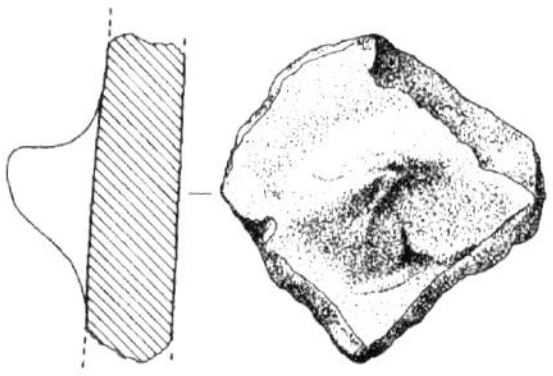

9/312 (P087)

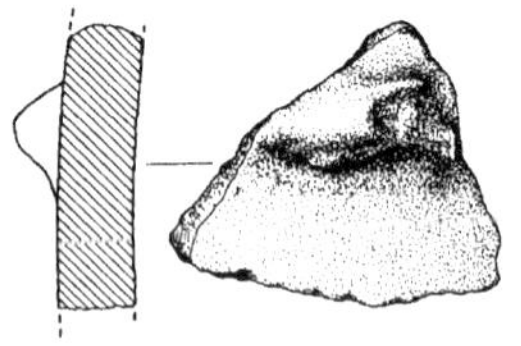

9/313 (P333)

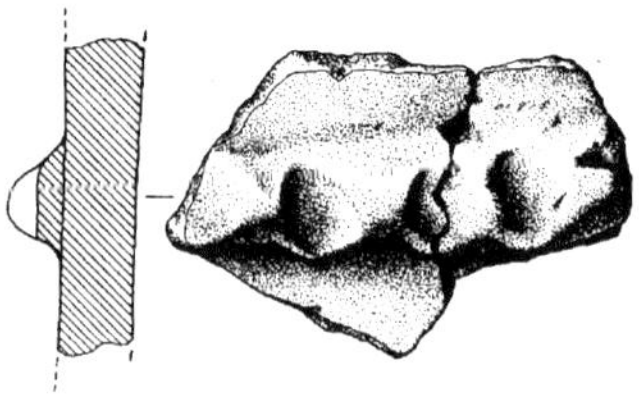

9/314 (P387 + P408)

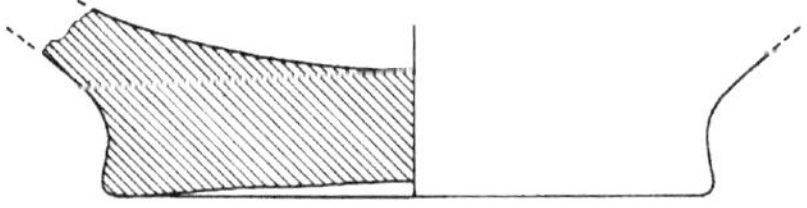

9/315 (P281)

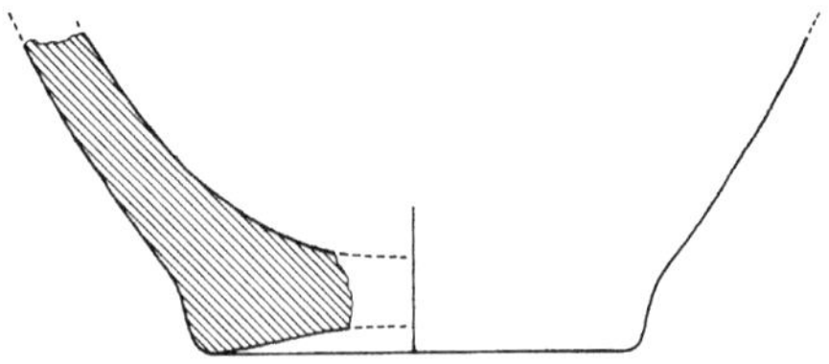

9/316 (P325)

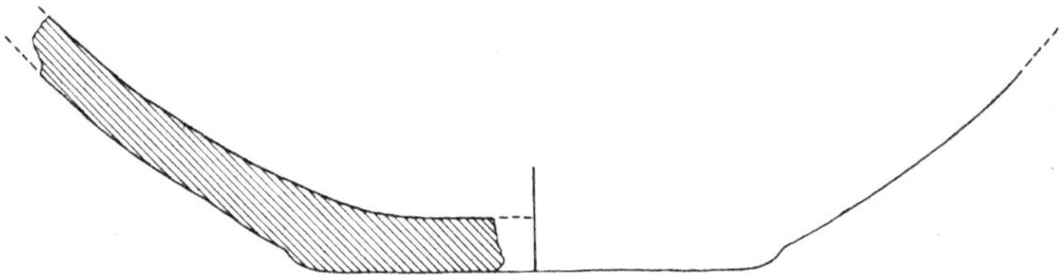

9/317 (P100)

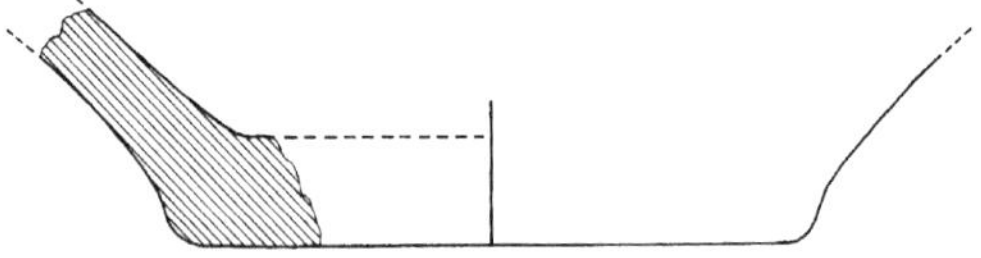

9/319 (P309)

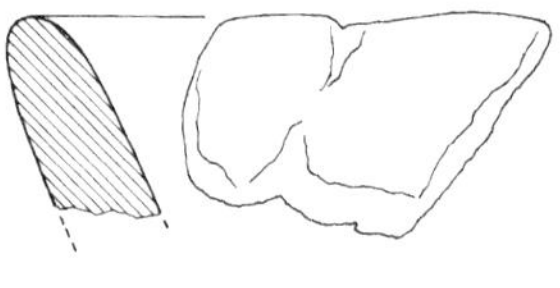

9/320 (P090)

FIGURE 9.33 **9/309–9/310**, **9/312–9/317**, **9/319–9/320**

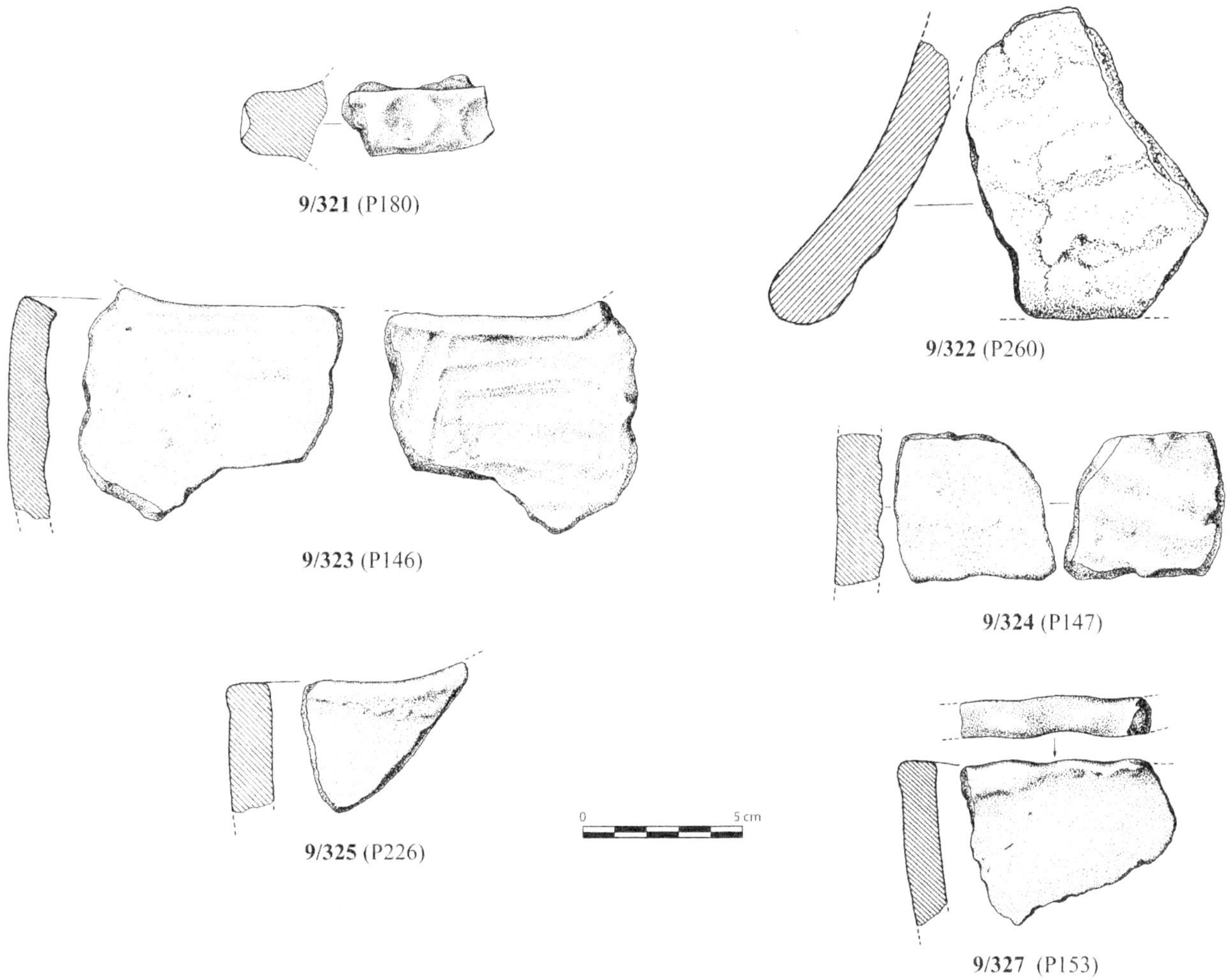

FIGURE 9.34 **9/321–9/325, 9/327**

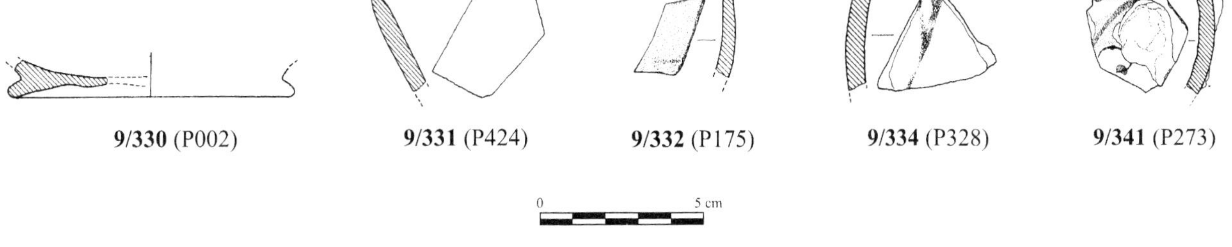

FIGURE 9.35 **9/330–9/332, 9/334, 9/341**

Type 1: **9/1** (P350)

Type 1: **9/2** (P350)

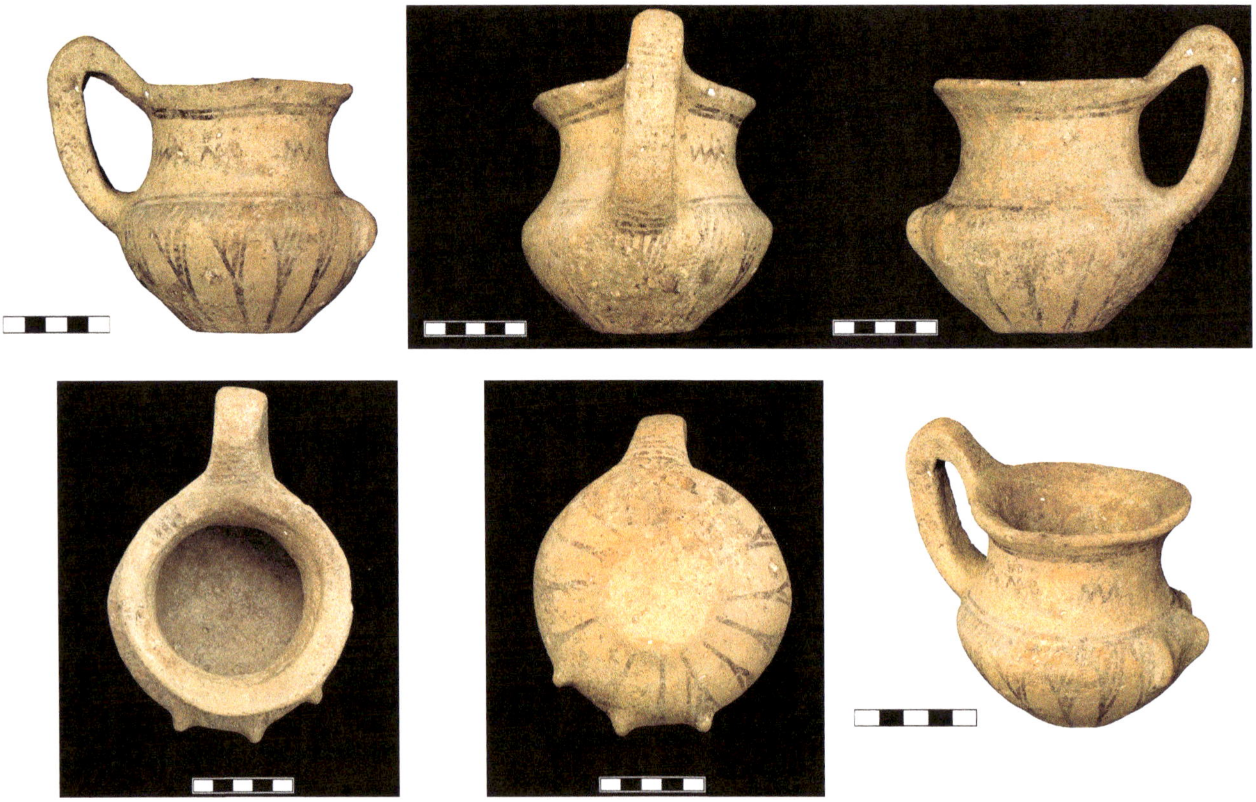

Type 2a: **9/3** (P304)

Figure 9.36 **9/1–9/3**

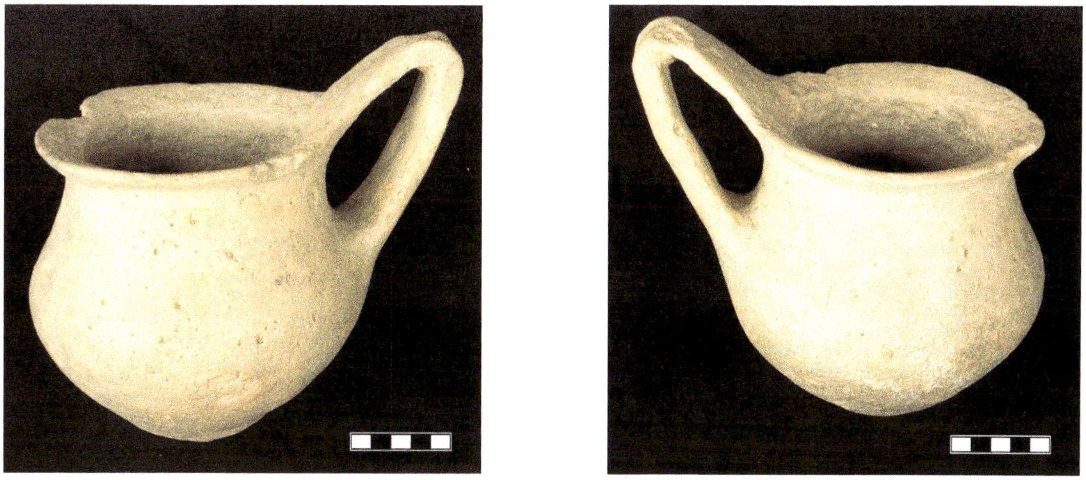

Type 2a: **9/4** (P077)

Type 2a: **9/5** (P326)

FIGURE 9.37 **9/4–9/5**

Type 2a: **9/6** (P073)

Type 2b: **9/7** (P166)

Type 2b: **9/8** (P277)

FIGURE 9.38 **9/6–9/8**

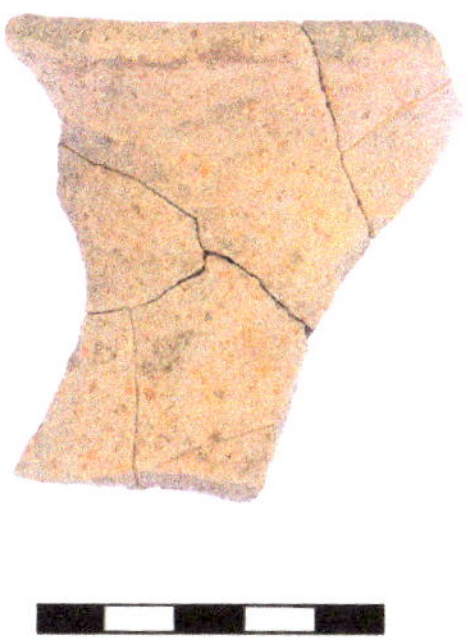

Type 2: **9/9** (P020)

Type 2: **9/10** (P213)

Type 2: **9/11** (P004)

Type 2: **9/12** (P131)

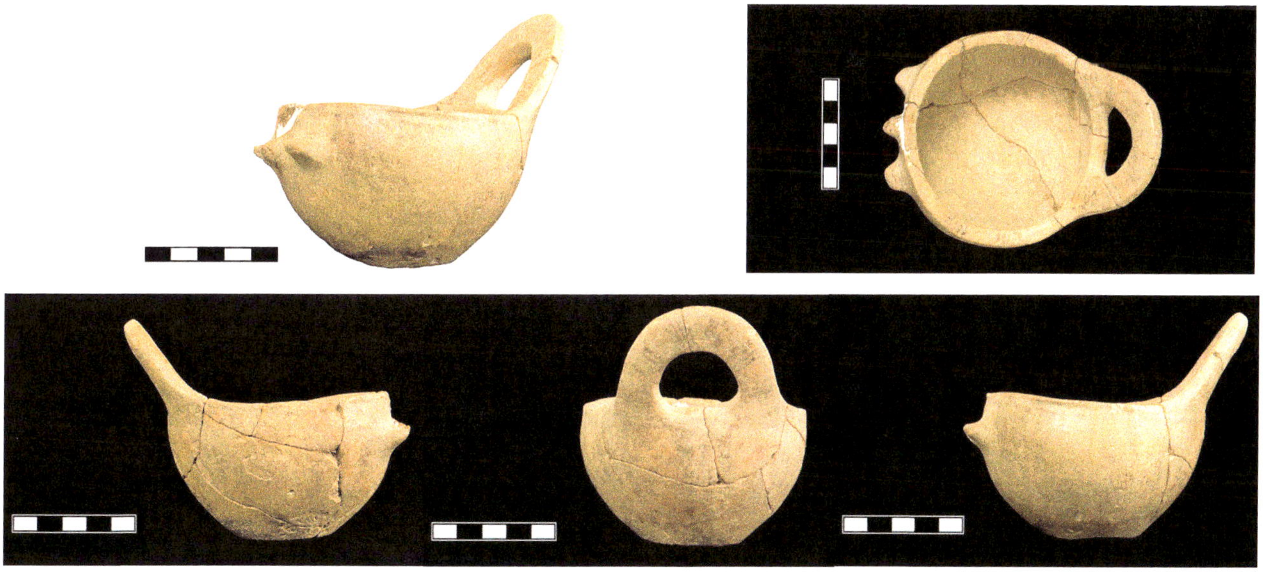

Type 3: **9/14** (P256)

Figure 9.39 **9/9–9/12**, **9/14**

Type 4: **9/15** (P082)

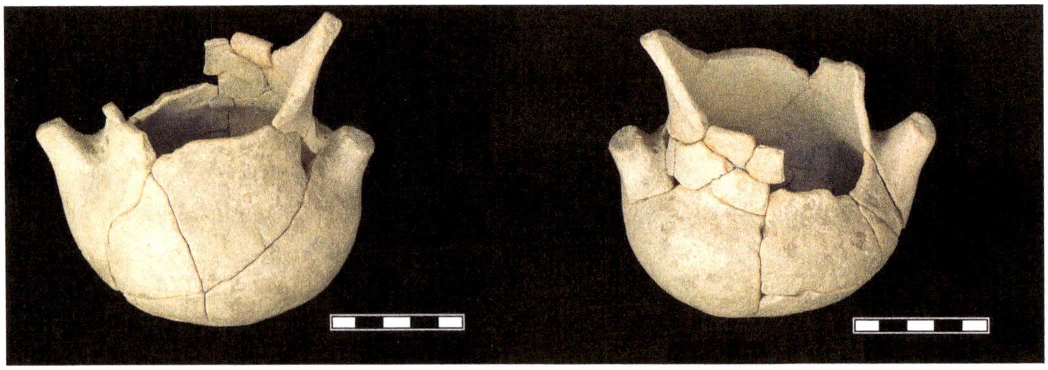

Type 5a: **9/16** (P047)

Type 5b: **9/17** (P078)

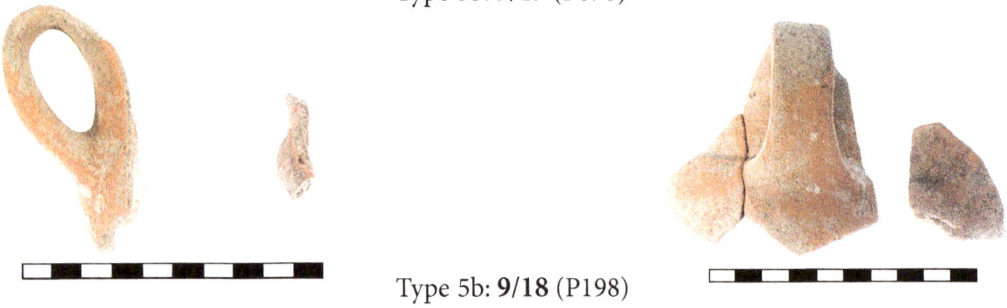

Type 5b: **9/18** (P198)

Figure 9.40 **9/15–9/18**

Type 5c(?): **9/19** (P386)

Type 5c(?): **9/20** (P290 + P221)

9/19 (P386) beside **9/20** (P290 + P221)

9/23 (P278)

9/24 (P418)

9/25 (P373)

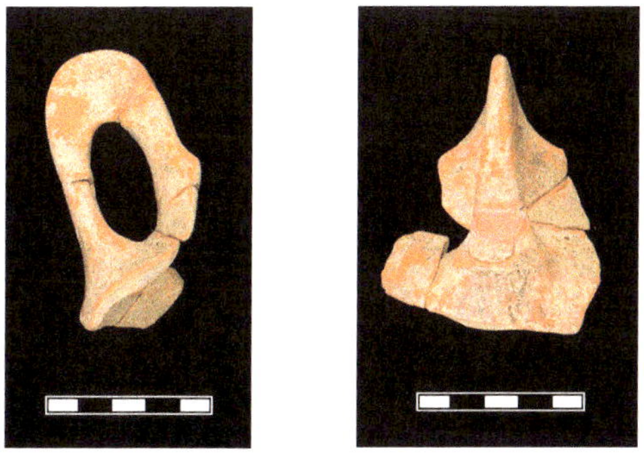

9/26 (P107)

9/27 (P240 + P251)

FIGURE 9.41 **9/19–9/20**, **9/23–9/27**

9/28 (P005)

9/29 (P006)

9/30 (P064)

9/31 (P114)

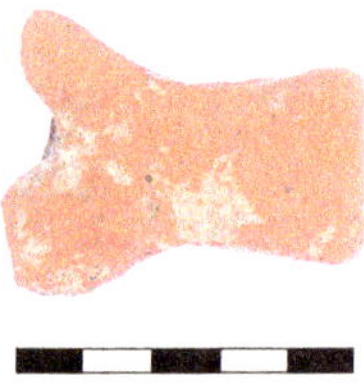

9/33 (P080)

9/34 (P125)

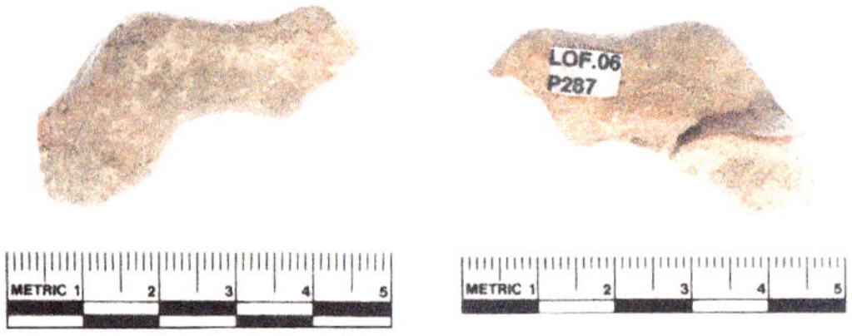

9/35 (P287)

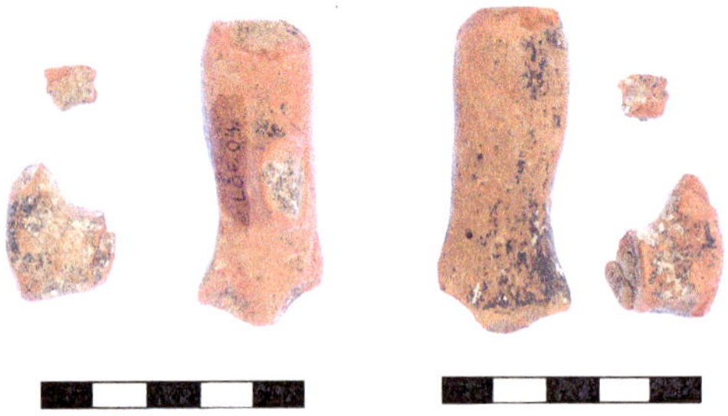

9/36 (P066)

9/37 (P342)

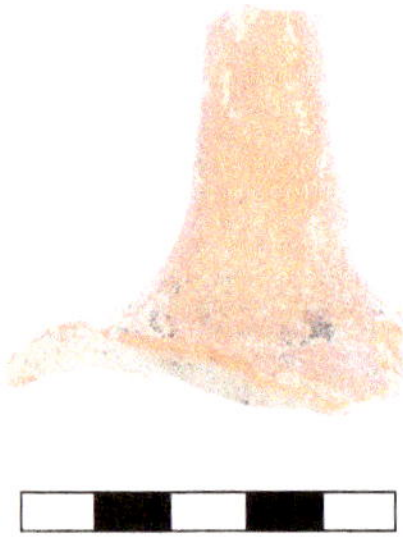

9/38 (P236)

FIGURE 9.42 **9/28–9/31**, **9/33–9/36**, **9/38**

9/39 (P172)

9/41 (P275)

9/46 (P268)

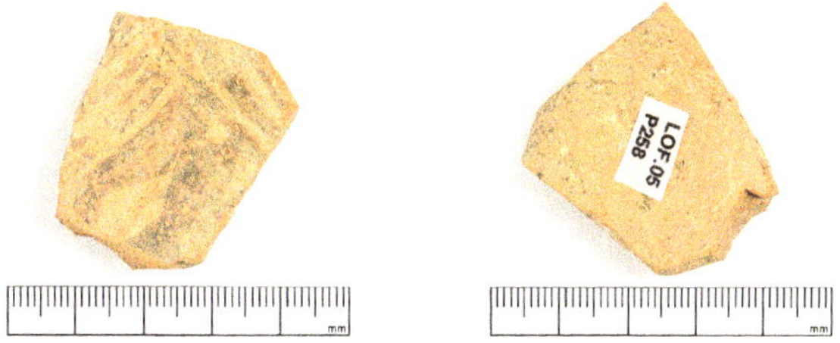

9/47 (P258)

9/48 (P048)

9/49 (P389)

9/50 (P395)

9/52 (P246)

9/53 (P414)

9/54 (P133)

Figure 9.43 **9/39**, **9/41**, **9/46–9/50**, **9/52–9/54**

9/55 (P093 + P096 + P247 + P370)

9/56 (P371)

9/57 (P375)

9/58 (P351)

9/59 (P262)

9/60 (P092)

9/61 (P071)

9/62 (P019)

9/63 (P416)

9/64 (P259)

9/65 (P012)

FIGURE 9.44 **9/55–9/65**

9/66 (P250)

9/67 (P296)

9/69 (P106)

9/70 (P272)

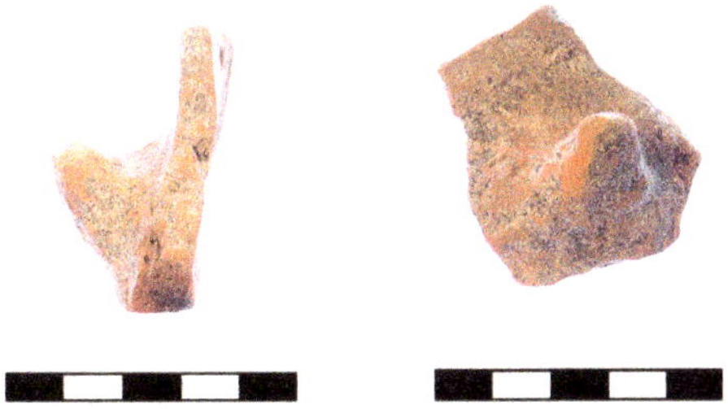

9/71 (P127)

9/72 (P058 + P391)

9/74 (P252)

9/75 (P132 + P324)

9/76 (P431)

FIGURE 9.45 **9/66–9/67**, **9/69–9/72**, **9/74–9/76**

9/77 (P195)

9/78 (P104)

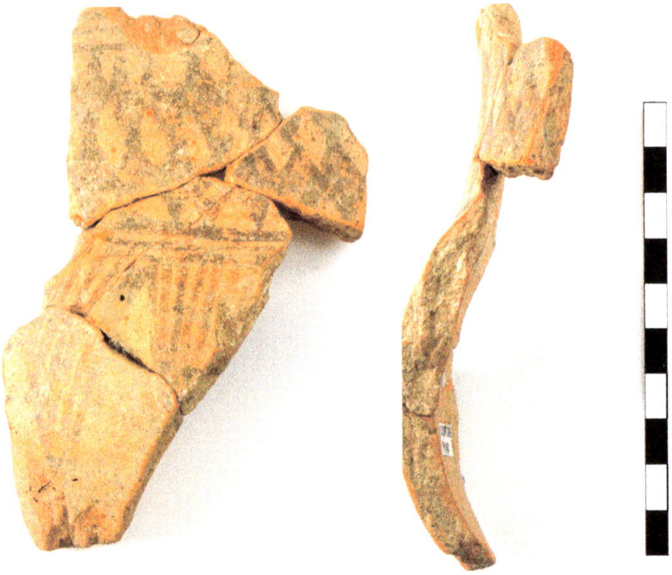

9/79 (P193 + P353 + P115 + P167)

9/80 (P284)

9/81 (P367)

9/82 (P305)

9/83 (P255 + P112)

9/84 (P030)

FIGURE 9.46 **9/77–9/84**

9/85 (P346)

9/86 (P108 + P388)

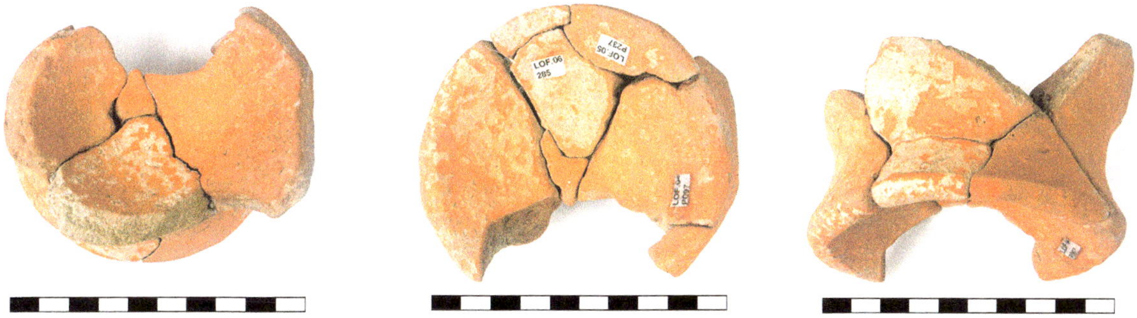

9/87 (P097 + P237 + P285)

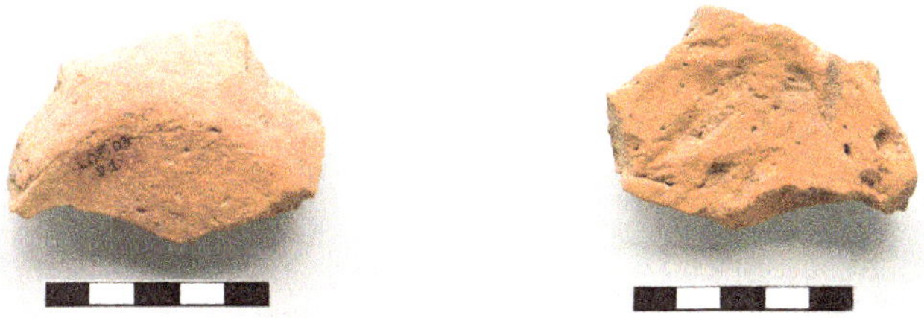

9/88 (P001)

9/89 (P232)

FIGURE 9.47 **9/85–9/89**

Type 1: **9/90** (P422)

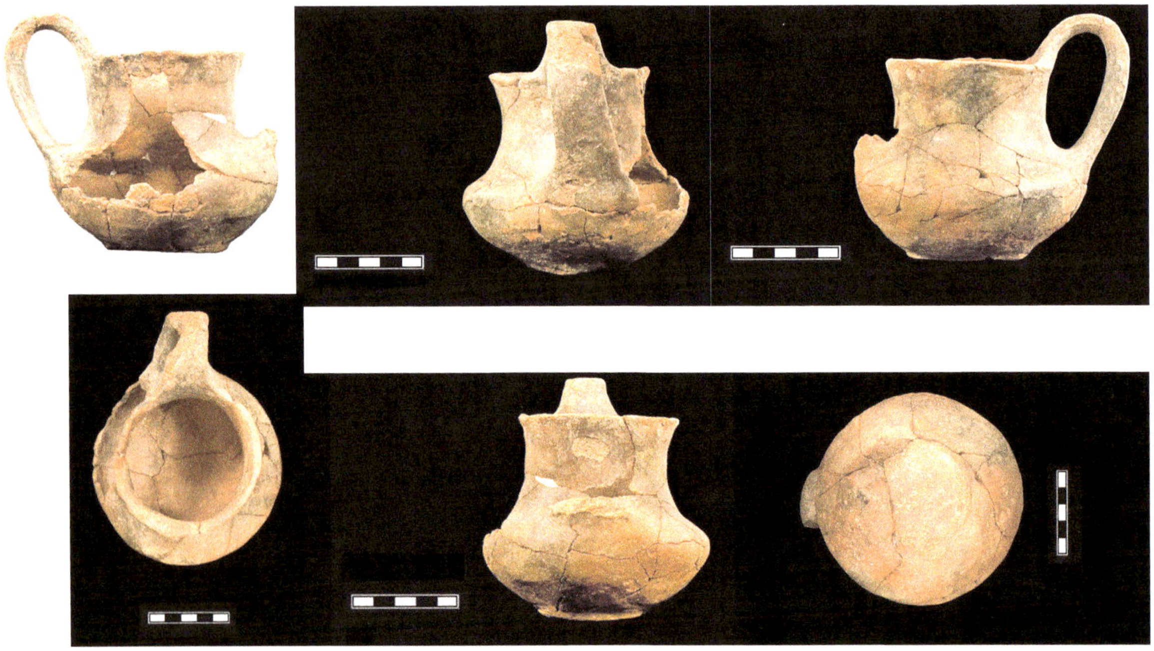

Type 2a: **9/91** (P253)

FIGURE 9.48 **9/90–9/91**

Type 2b: **9/92** (P364)

Type 3: **9/93** (P223)

FIGURE 9.49 **9/92–9/93**

Type 4: **9/94** (P322)

Type 4: **9/94** (P322)

Type 4: **9/94** (P322 non-joining handle frr)

Figure 9.50 **9/94**

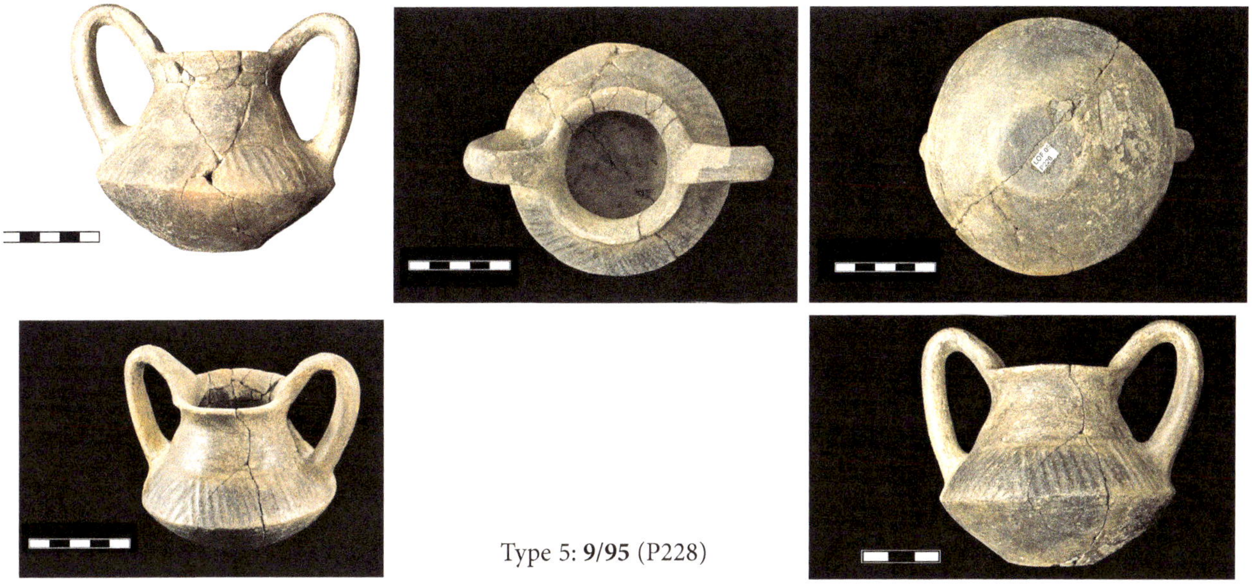

Type 5: **9/95** (P228)

Type 5: **9/96** (P276)

Type 5: **9/97** (P123)

Type 5: **9/98** (P032)

Type 5: **9/99** (P050)

Figure 9.51 **9/95–9/99**

9/100 (P049 + P063)

9/101 (P137)

9/102 (P051)

9/103 (P134)

9/104 (P233)

9/106 (P242)

9/111 (P041)

FIGURE 9.52 **9/100–9/104**, **9/106**, **9/111**

9/112 (P334)

9/113 (P211)

9/115 (P398)

9/119 (P229)

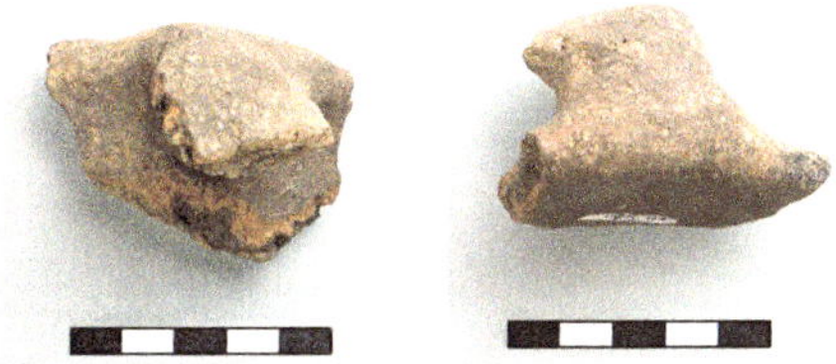

9/120 (P021)

9/122 (P361)

9/123 (P339)

9/124 (P299)

9/126 (P191)

9/127 (P089)

Figure 9.53 **9/112–9/113, 9/115, 9/119–9/120, 9/122–9/124, 9/126–9/127**

9/130 (P295)

9/134 (P068)

9/145 (P404)

9/152 (P045 + P046)

9/154 (P169)

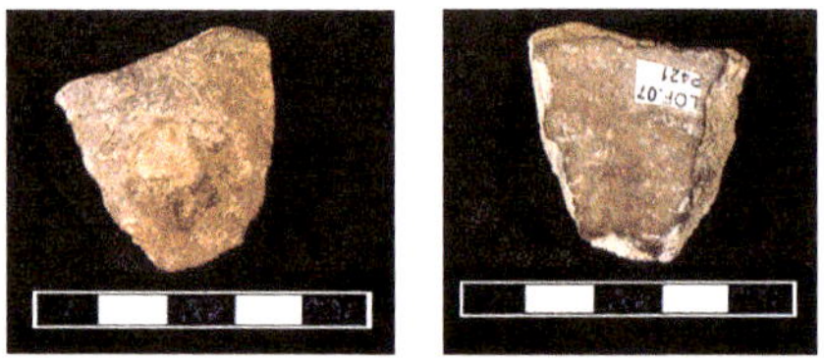

9/155 (P421)

9/156 (P316)

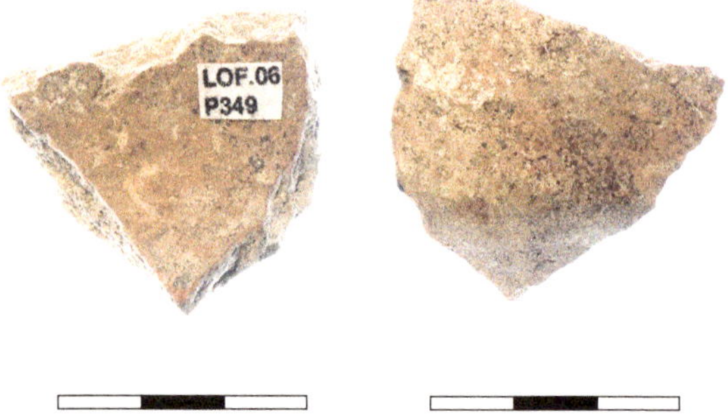

9/158 (P349)

FIGURE 9.54 **9/130, 9/134, 9/145, 9/152, 9/154–156, 9/158**

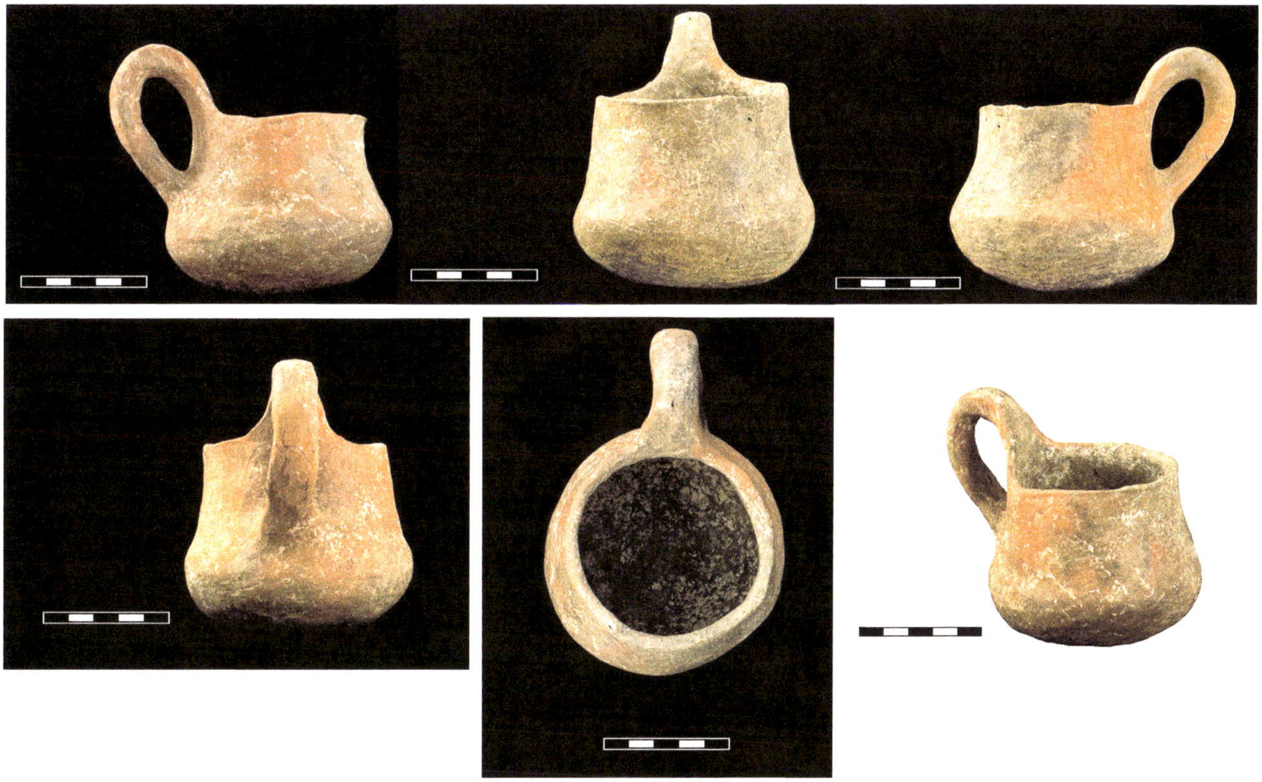

Type 1: **9/159** (P303)

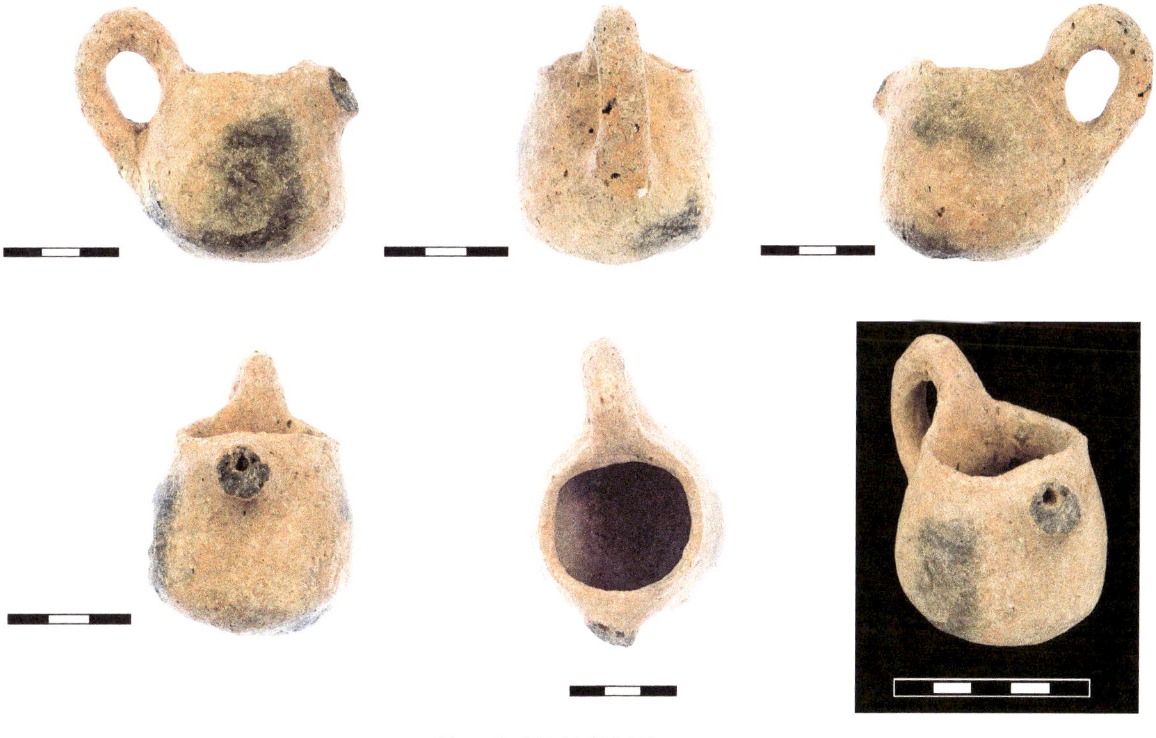

Type 2: **9/160** (P362)

Figure 9.55 **9/159–9/160**

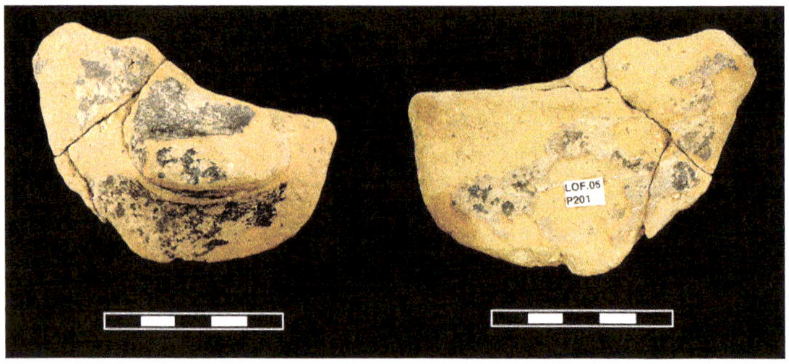

9/161 body frr (P201)

9/161 base frr (P201)

9/162 (P079)

9/163 (P008)

9/164 (P029)

Figure 9.56 **9/161–9/164**

9/165 (P347)

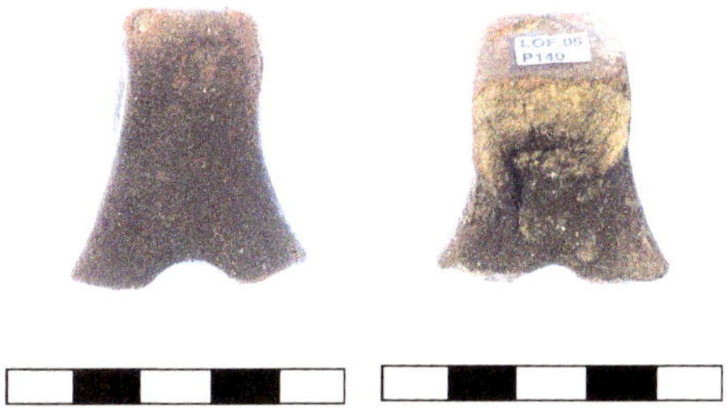

9/170 (P140)

9/172 (P401)

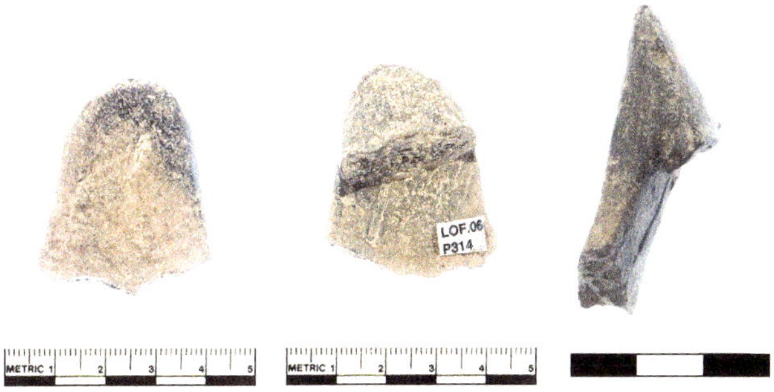

9/173 (P314)

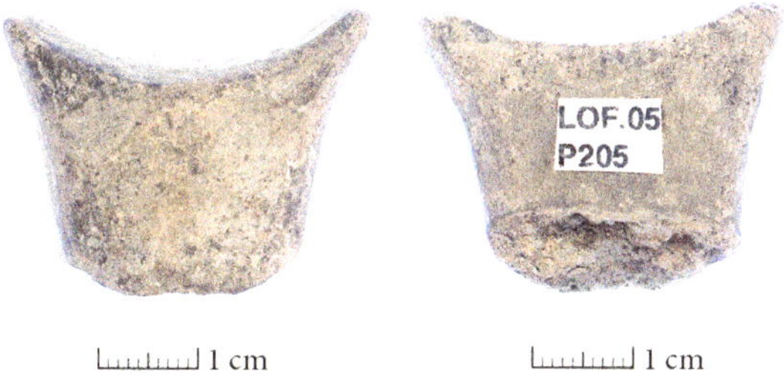

9/174 (P205)

9/175 (P293)

9/183 (P062)

FIGURE 9.57 **9/165**, **9/170**, **9/172–9/175**, **9/183**

9/184 (P099)

9/185 (P245)

9/186 (P344)

9/187 (P345)

9/190 (P098)

9/191 (P399)

9/193 (P182)

9/195 (P368)

9/196 (P216)

FIGURE 9.58 **9/184–9/187, 9/190–9/191, 9/193, 9/195–9/196**

9/198 (P320)

9/199 (P091)

9/201 (P054)

9/203 (P086)

9/204 (P417)

9/206 (P094A) and 9/167 (P094B)

9/207 (P321)

9/214 (P069)

9/215 (P337)

Figure 9.59 **9/198–9/199**, **9/201**, **9/203–9/204**, **9/206** (with **9/167**), **9/207**, **9/214–9/215**

9/217 (P318)

9/220 (P168)

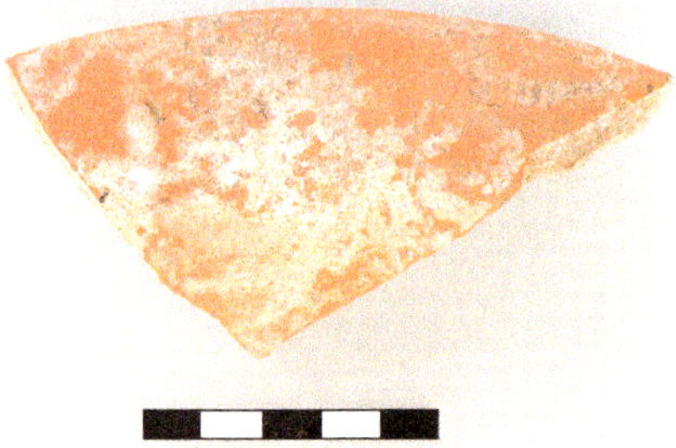

9/221 (P057)

9/223 (P095)

9/226 (P075)

9/228 (P267)

9/229 (P122)

9/231 (P323)

9/233 (P380)

FIGURE 9.60 **9/217**, **9/220–9/221**, **9/223**, **9/226**, **9/228–9/229**, **9/231**, **9/233**

9/234 (P016)

9/235 (P084)

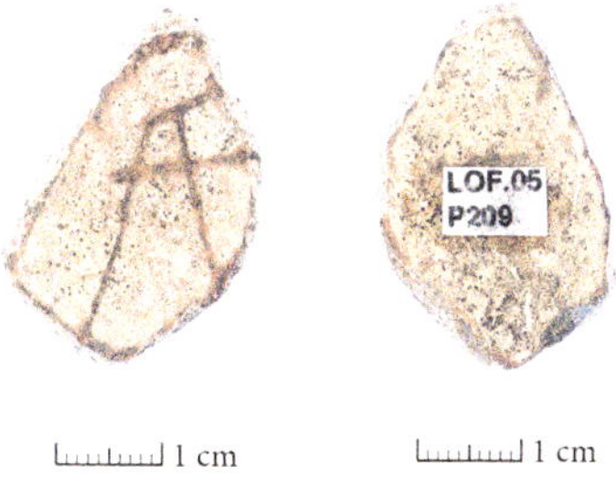

9/237 (P209)

9/241 (P034)

9/244 (P288)

9/245 (P423)

9/246 (P249)

Figure 9.61 **9/234–9/235**, **9/237**, **9/241**, **9/244–9/246**

9/247 (P196)

9/248 (P234)

9/249 (P170)

9/250 (P429)

9/254 (P430)

9/255 (P302)

9/257

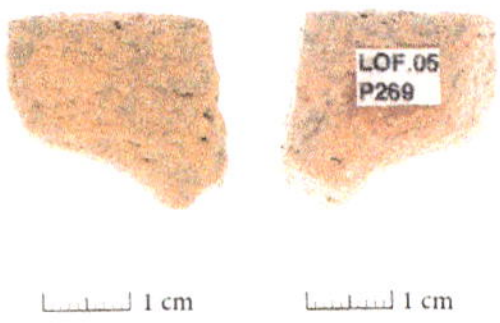

9/258 (P269)

FIGURE 9.62 **9/247–9/250**, **9/254–9/255**, **9/257–9/258**

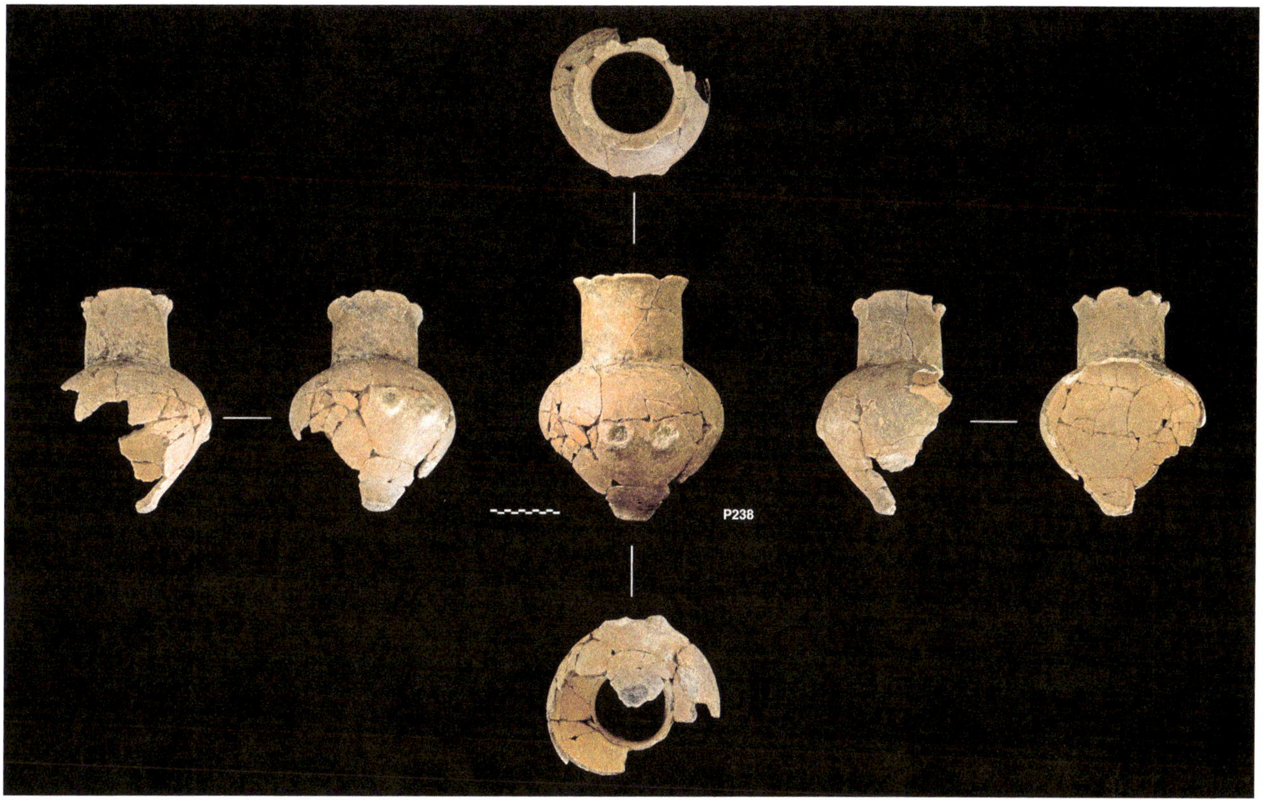

Type 1: **9/259** (P283)

9/260 (P024)

9/264 (P061)

9/265 (P136)

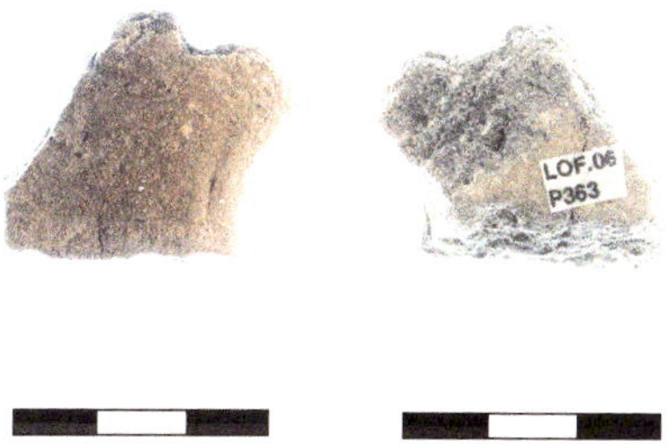

9/267 (P363)

FIGURE 9.63 **9/259–9/260, 9/264–20/265, 9/267**

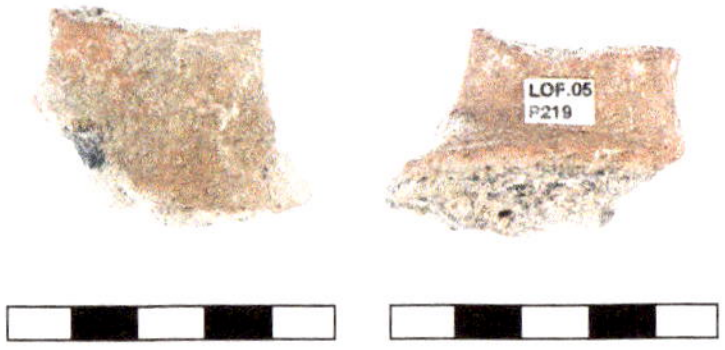

9/268 (P219)

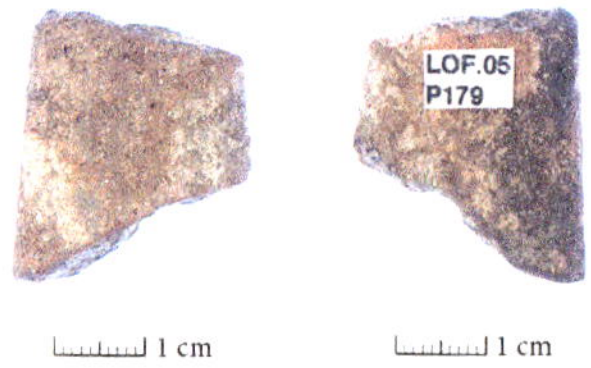

9/270 (P179)

9/272 (P400)

9/273 (P332)

9/274 (P135)

9/276 (P365)

9/278 (P376)

9/279 (P003)

9/280 (P217)

9/288 (P176)

FIGURE 9.64 **9/268**, **9/270**, **9/272**–**9/274**, **9/276**, **9/278**–**9/280**, **9/288**

9/289 (P161)

9/290 (P224 + P282 + P411)

9/292 (P317)

9/295 (P085)

9/296 (P014)

9/297 (P428)

9/298 (P291)

9/305 (P011)

9/307 (P306)

Figure 9.65 **9/289–9/290**, **9/292**, **9/295–9/298**, **9/305**, **9/307**

9/310 (P261)

9/311 (P360)

9/312 (P087)

9/314 (P387 + P408)

9/315 (P281)

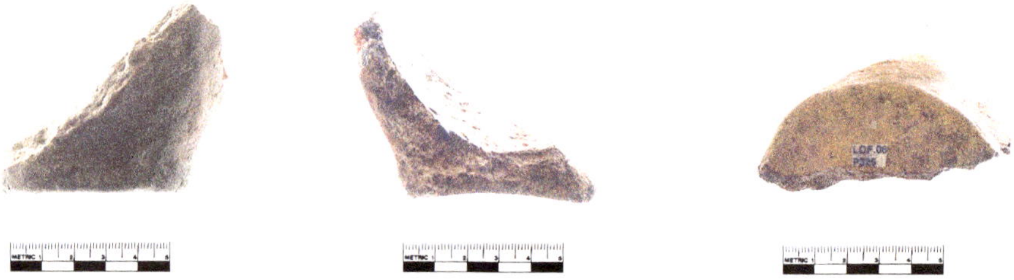

9/316 (P325)

9/317 (P100)

9/318 (P354)

FIGURE 9.66 **9/310–9/312**, **9/314–9/318**

9/319 (P309)

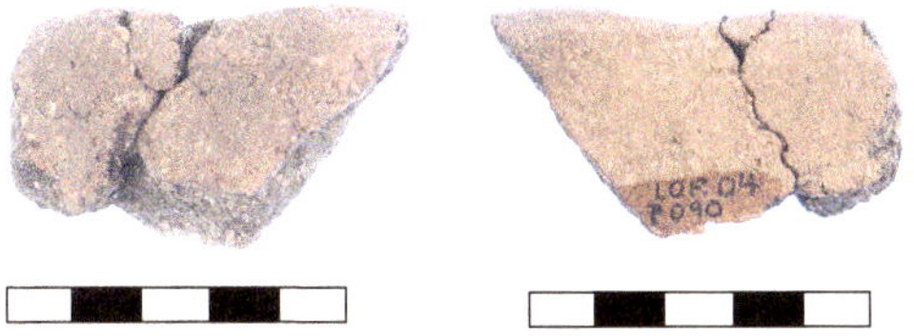

9/320 (P090)

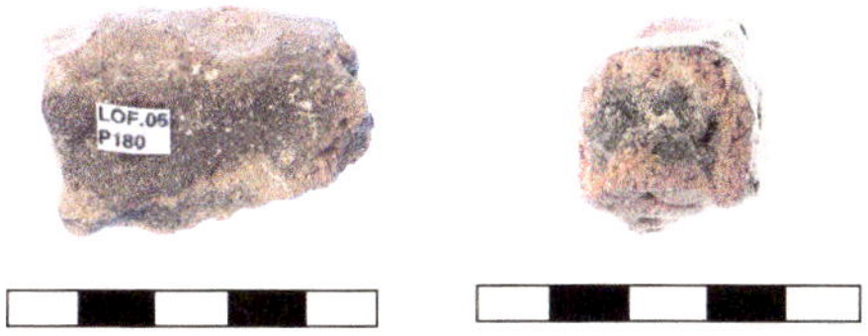

9/321 (P180)

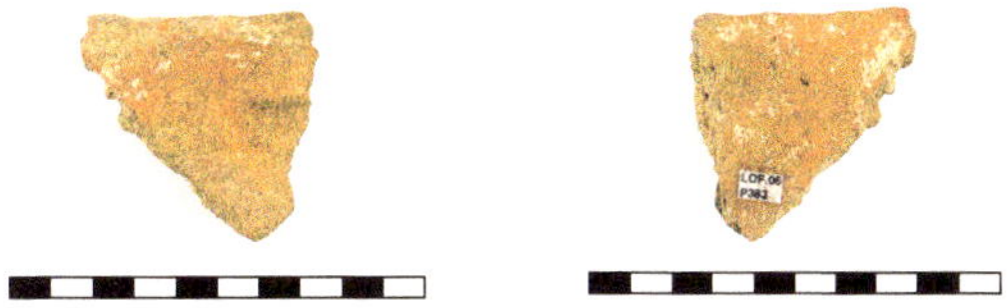

9/326 (P383)

9/328 (P327)

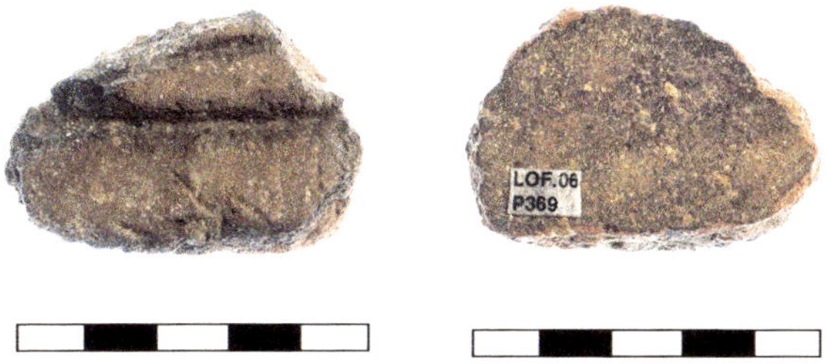

9/329 (P369)

Figure 9.67 **9/319–9/321**, **9/326**, **9/328–9/329**

9/330 (P002)

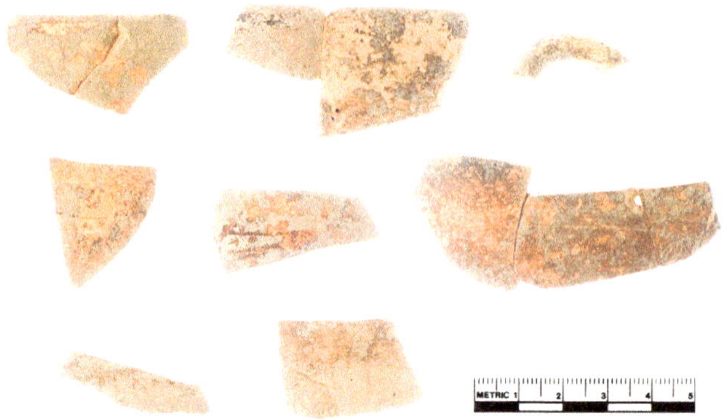

9/333 (P257)

9/335 (P372)

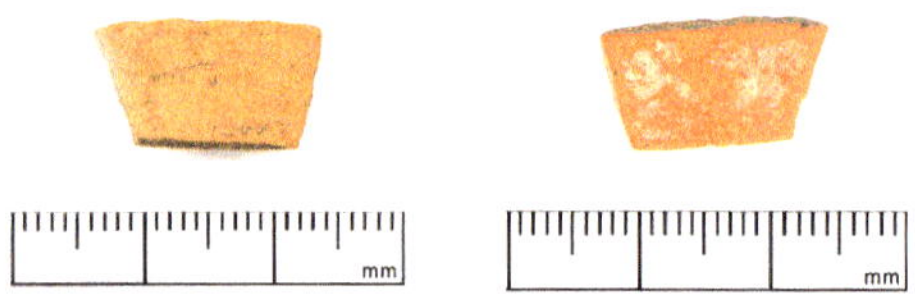

9/336 (P384)

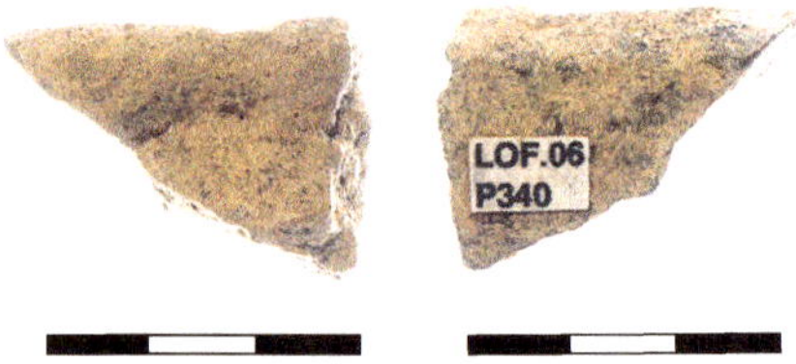

9/337 (P340)

FIGURE 9.68 **9/330**, **9/333**, **9/335–9/337**

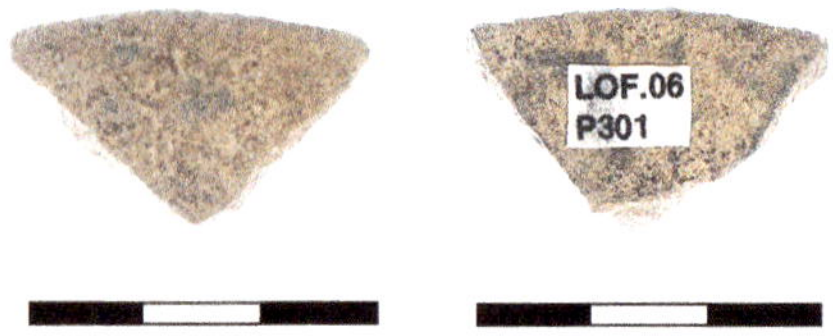

9/338 (P301)

9/339 (P338)

9/340 (P139)

9/341 (P273)

FIGURE 9.69 **9/338–9/341**

Appendix 2 Illustrations

FIGURE A2.1 Corinthian kotyle (rim, handle, and body sherds): **9/333** (P257), interior. See **Figure 9.68** for exterior views.

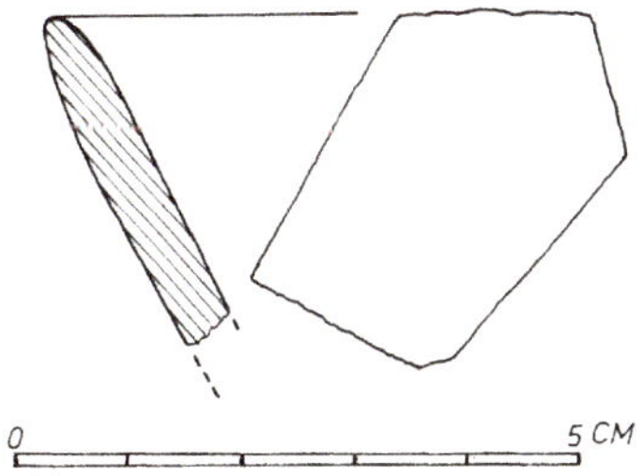

FIGURE A2.2 Rim fragment, wheelmade open shape (**9/331** [P424])

Chapter 10 Illustrations

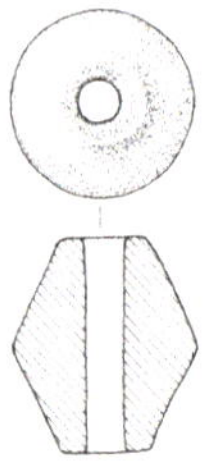

10/1 (TXXXI-2)

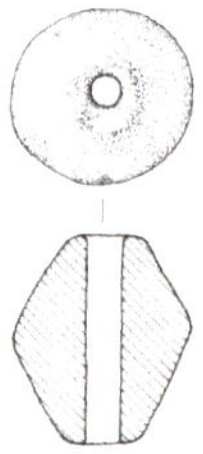

10/2 (TXXXI-1)

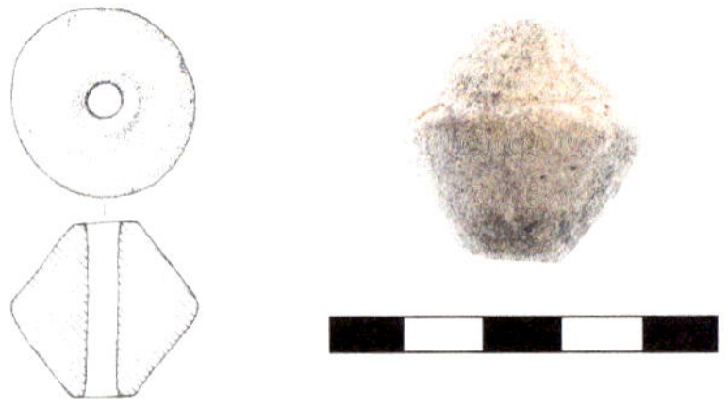

10/3 (SF 169)

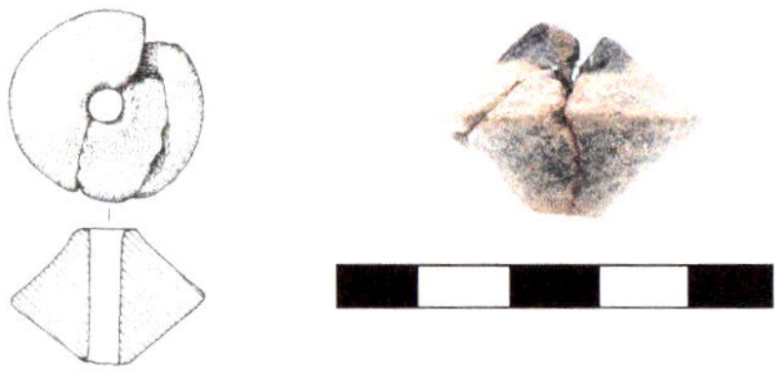

10/4 (SF 168)

10/5 (SF 065)

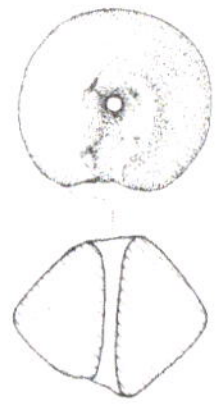

10/6 (TLXVII-1)

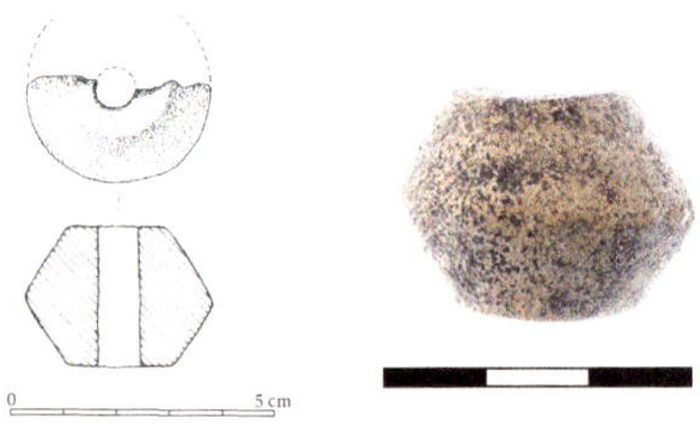

10/7 (SF 293)

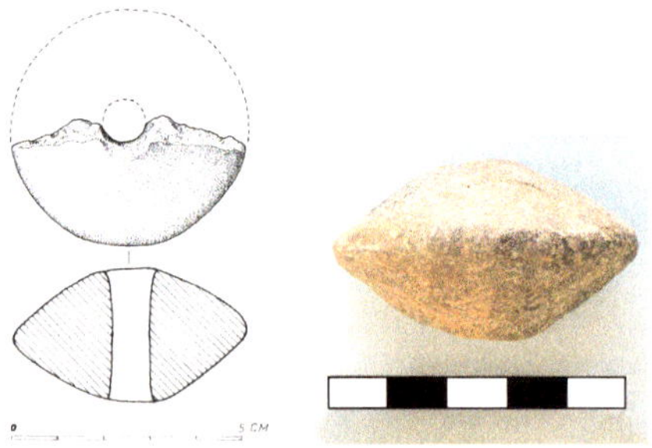

10/8 (SF 138)

FIGURE 10.1 Terracotta spindlewhorls, beads, or buttons (**10/1–10/8**)

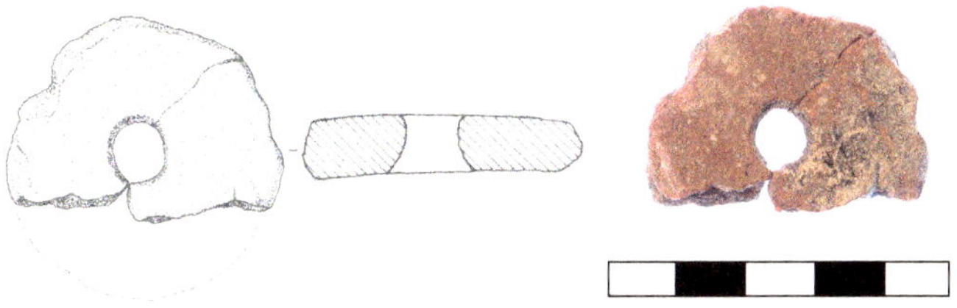

0/9 (SF 413)

Figure 10.2 Terracotta spindlewhorl, bead, or button formed from broken fragment of pottery (**10/9**)

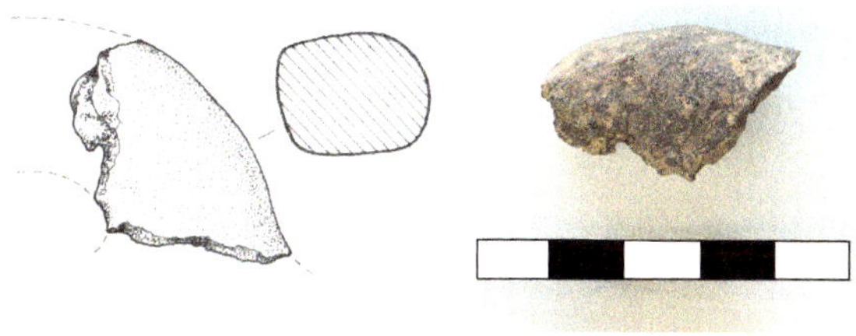

10/10 (SF 045)

Figure 10.3 Terracotta other (possible loomweight?) (**10/10**)

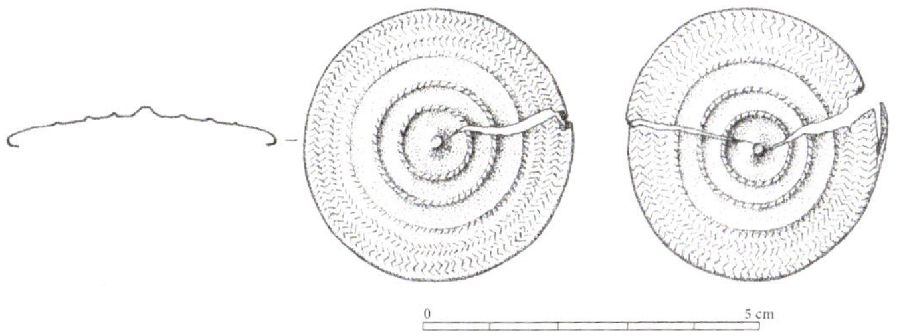

10/11, **10/12** (SF 290, 291)

Figure 10.4 Gold/electrum ear or head ornaments (**10/11** and **10/12**)

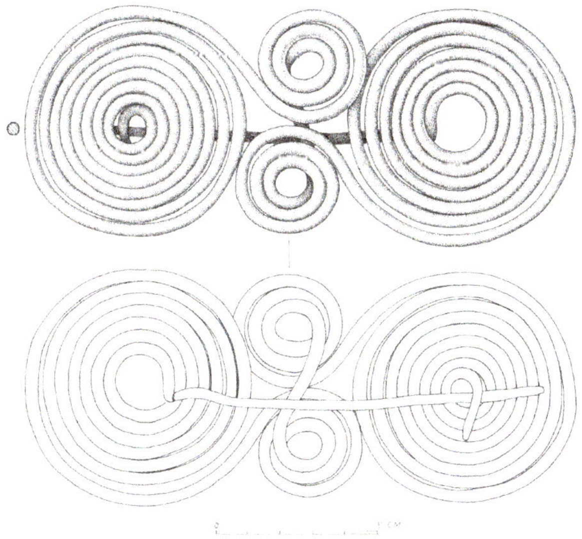

10/13 (SF 107)

Figure 10.5 Bronze fibula, Type I.1a (**10/13**)

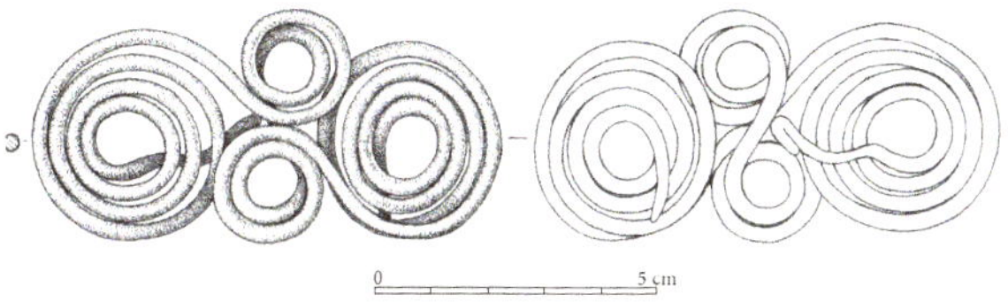

10/14 (SF 170)

Figure 10.6 Bronze fibula, Type I.1b (**10/14**)

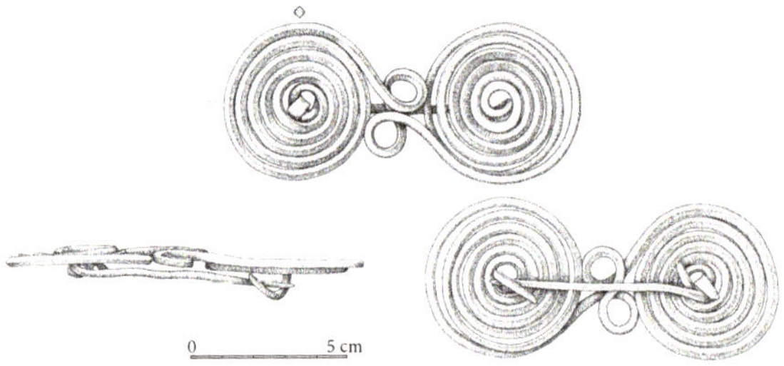

10/15 (SF 106)

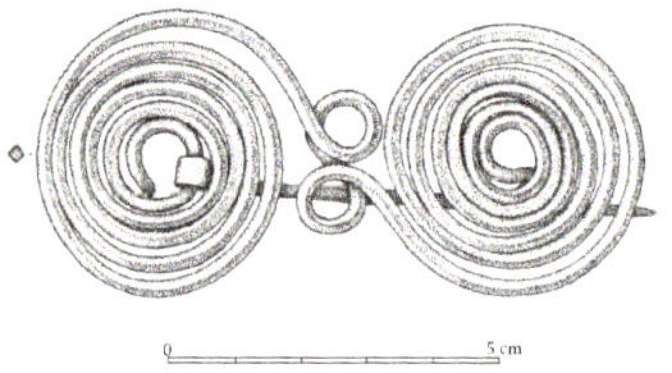

10/16 (SF 315)

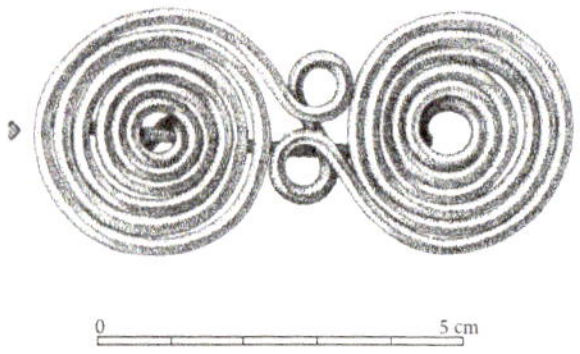

10/17 (SF 434)

Figure 10.7 Bronze fibulae, Type I.1c (**10/15**, **10/16**, **10/17**)

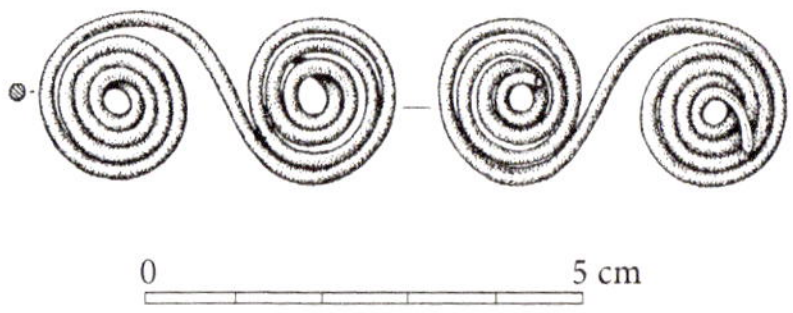

SF 157

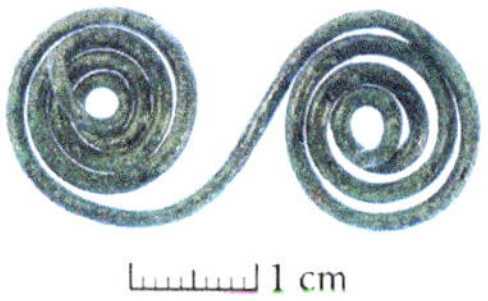

Figure 10.8 Bronze fibula, Type I.1d (**10/18**)

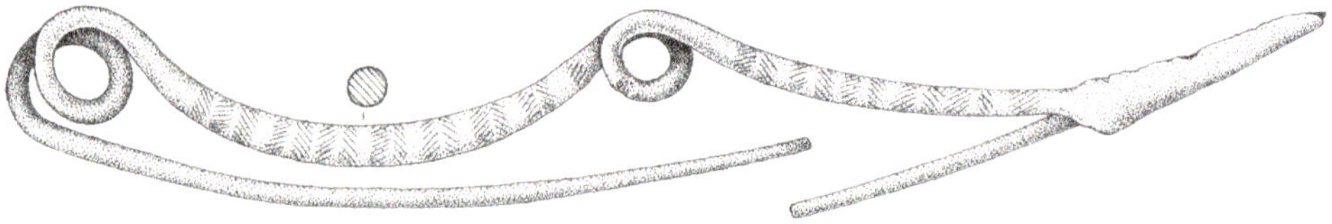

10/19 (SF 431a)

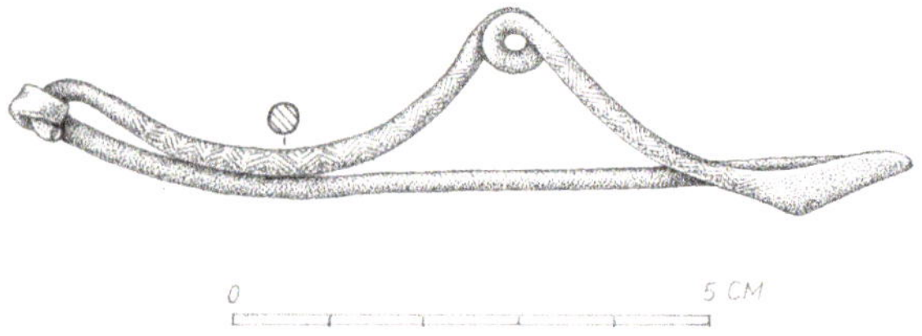

10/20 (SF 337)

Figure 10.9 Bronze fibulae, Type I.2a (**10/19**, **10/20**)

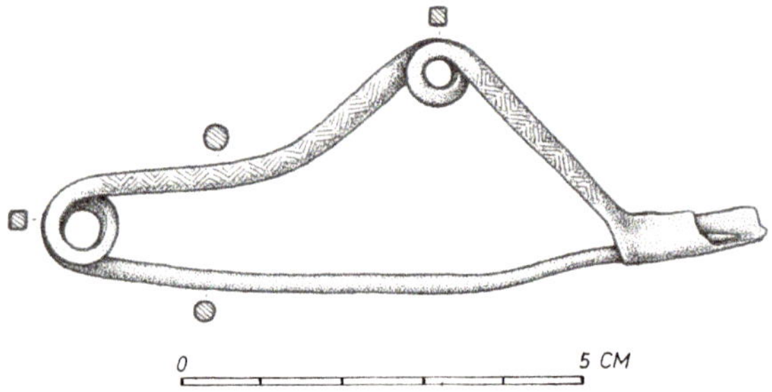

10/21 (SF 251)

Figure 10.10 Bronze fibula, Type I.2b (**10/21**)

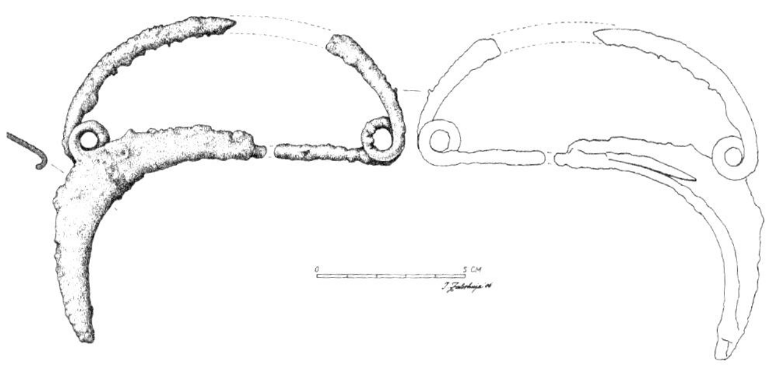

10/23 (SF 261)

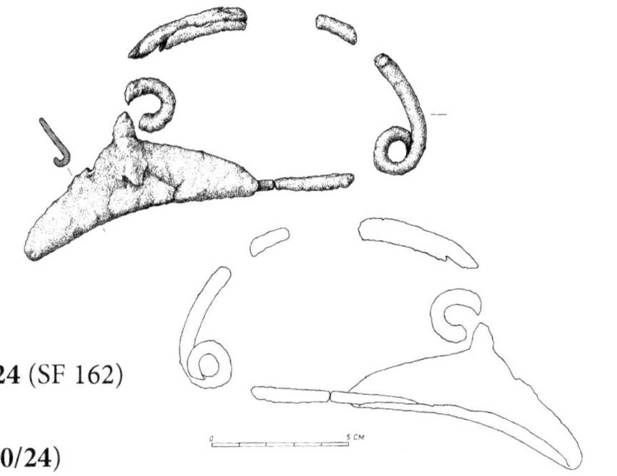

10/24 (SF 162)

Figure 10.12 Iron fibulae, Type II.2 (**10/23**, **10/24**)

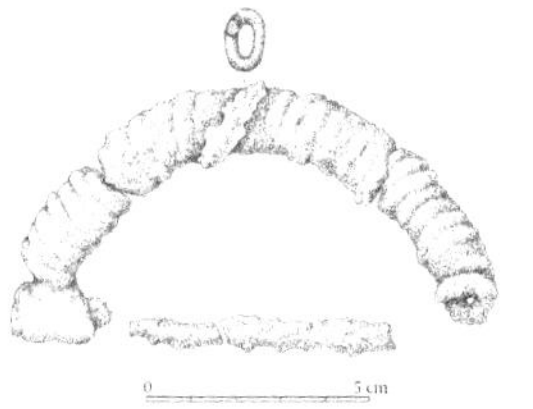

10/22 (SF 110)

Figure 10.11 Iron fibula, Type II.1 (**10/22**)

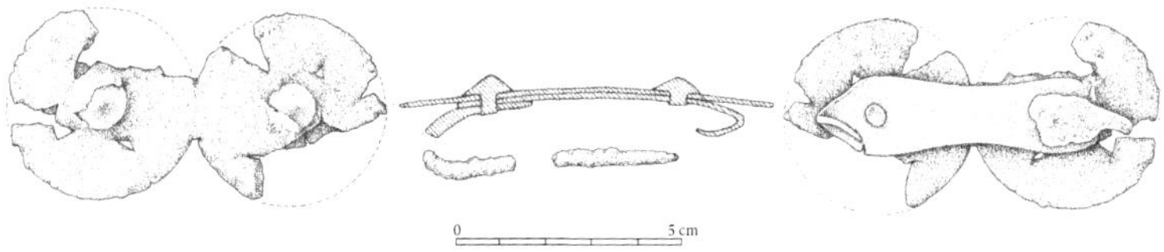

10/25 (**TLV-3**)

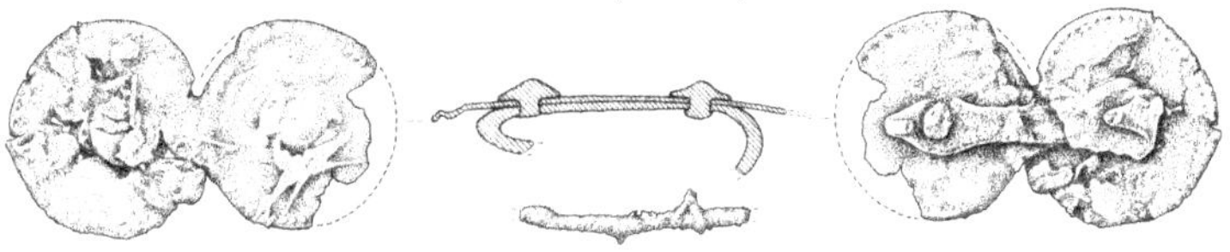

10/26 (**TLVI-2**)

Figure 10.13 Bimetallic fibulae, Type III.1 (**10/25**, **10/26**)

Type I.1 (**10/27**)

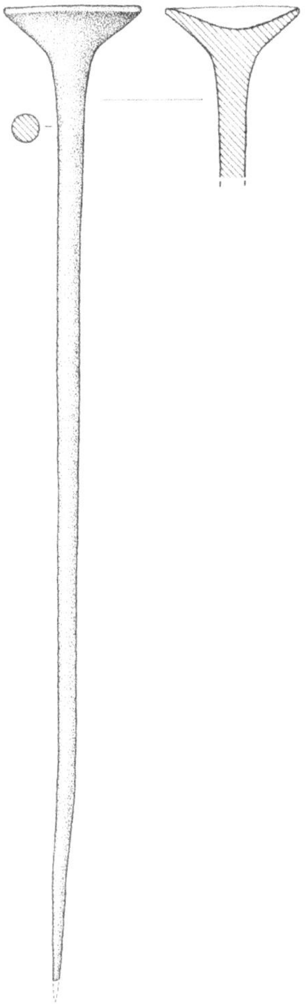

Type I.2 (**10/28**)

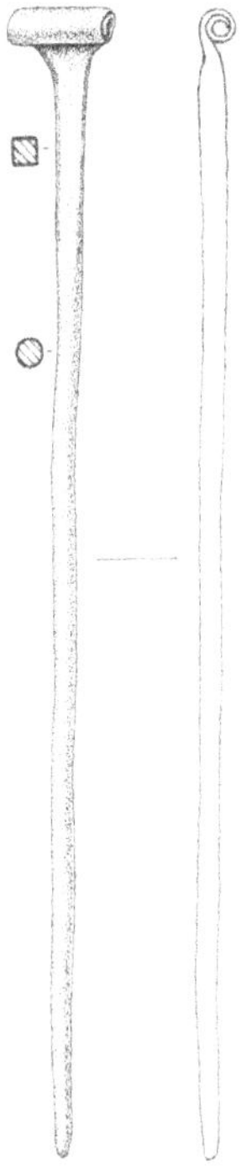

Type I.3 (**10/29**)

Figure 10.14 Bronze dress pins, Types I.1 (**10/27**), I.2 (**10/28**), I.3 (**10/29**)

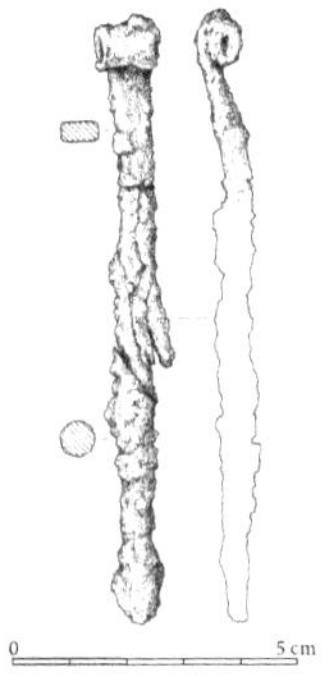

10/33

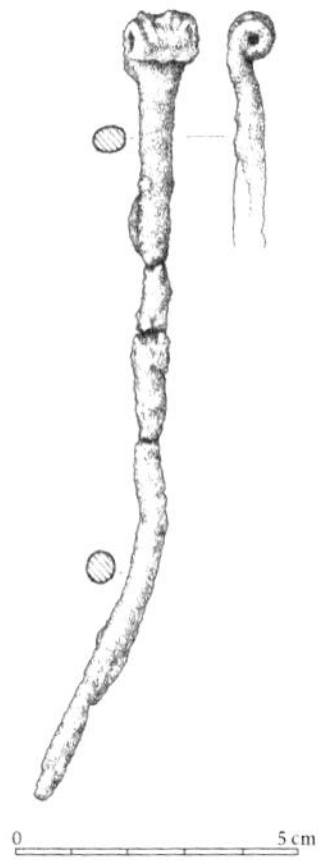

10/34

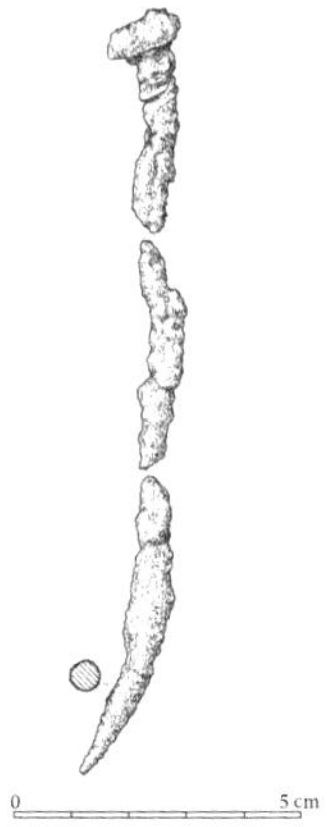

10/35

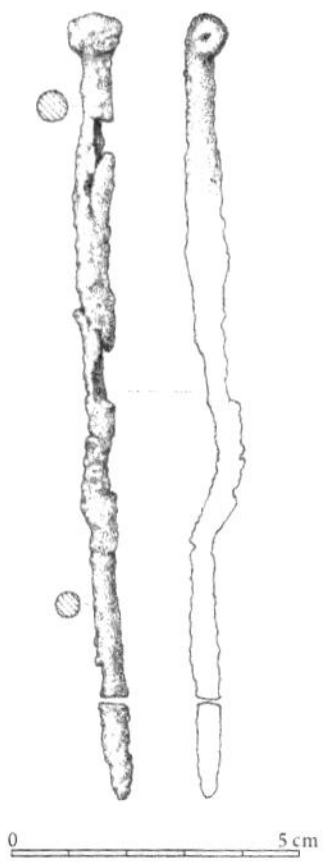

10/36

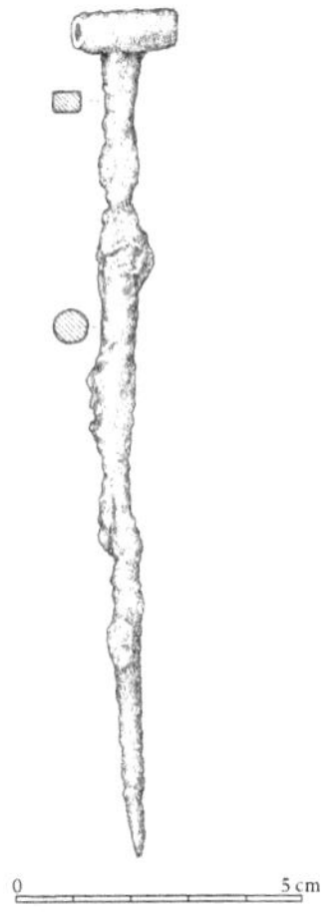

10/37

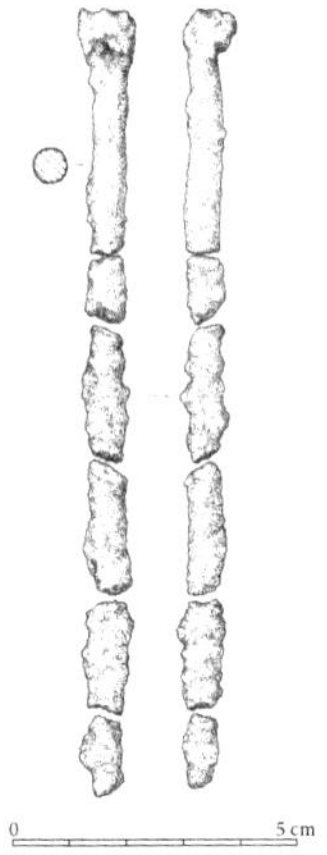

10/38

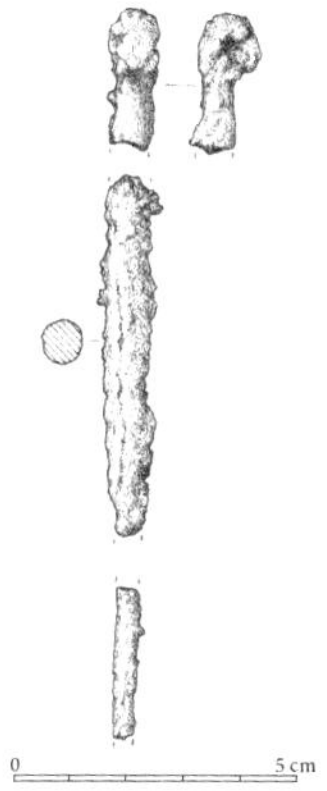

10/39

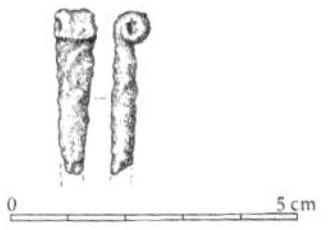

10/40

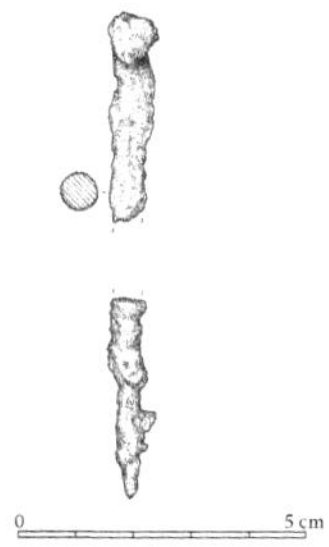

10/41

FIGURE 10.15 Iron pins, Type II.1 (**10/33-10/41**)

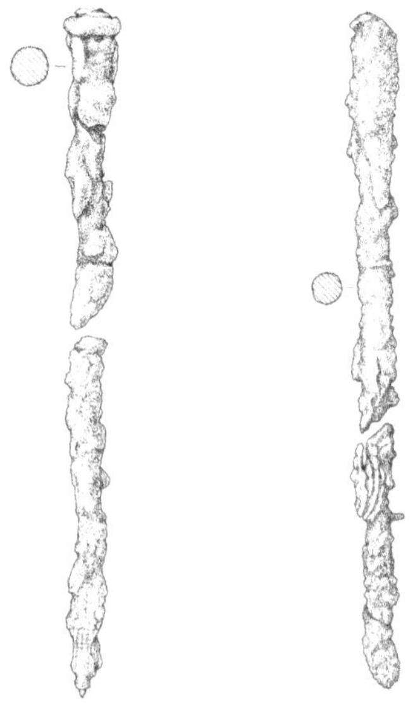

FIGURE 10.16 Iron pins, Type II.2 (**10/42**, **10/43**)

FIGURE 10.17 Iron pin, Type II.3 (**10/44**)

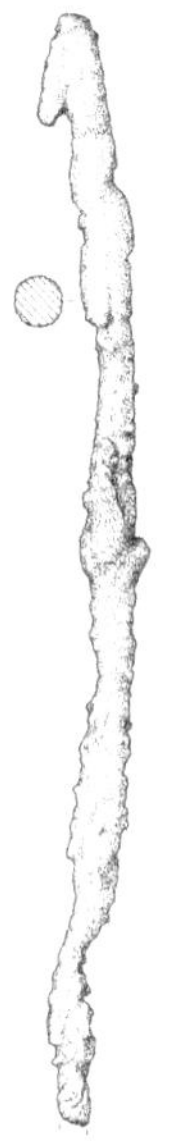

FIGURE 10.18 Iron pin, Type II.4 (**10/45**)

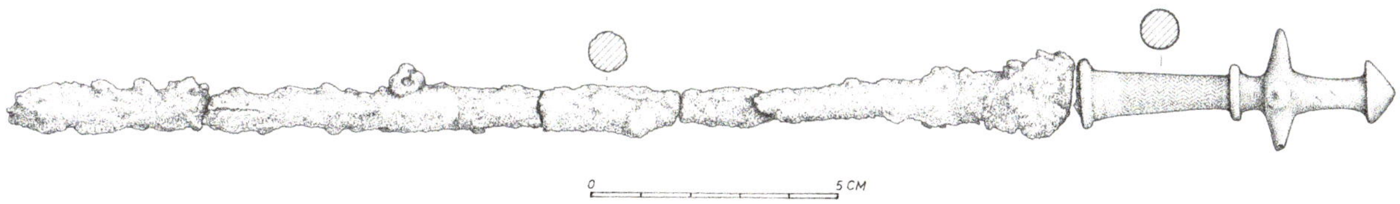

Figure 10.19 Bimetallic pin, Type III.1 (**10/50**)

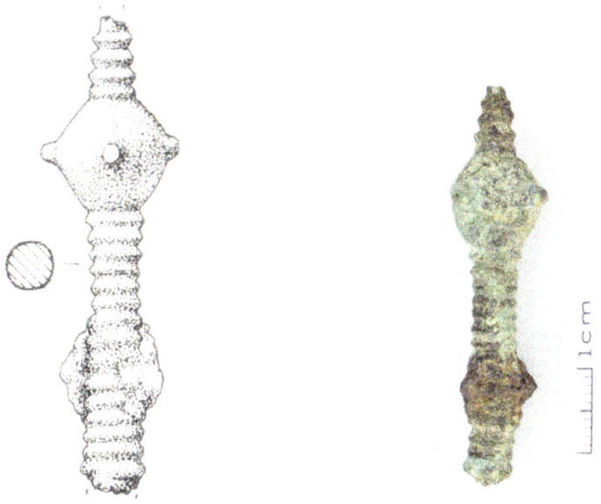

Figure 10.20 Bimetallic pin, Type III.2 (**10/51**)

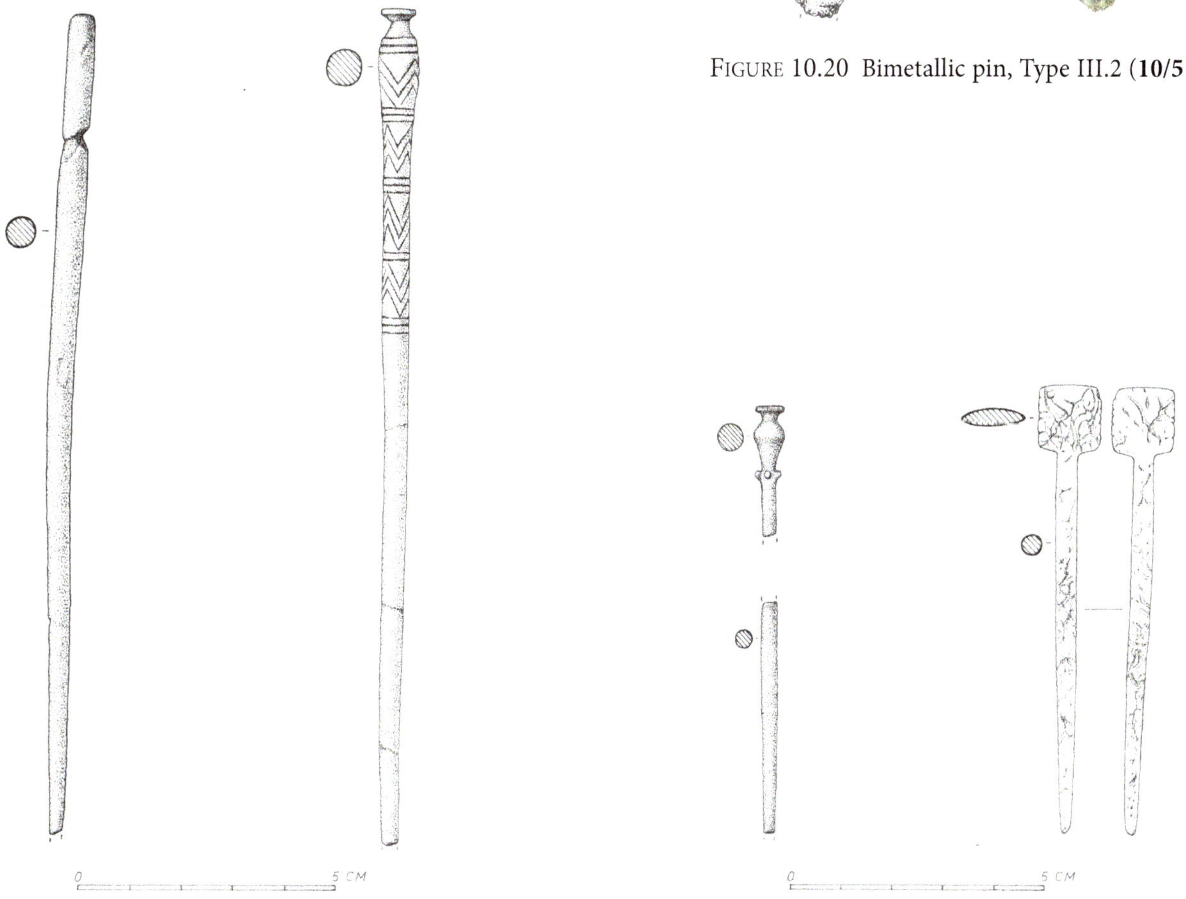

Type IV.1 (**10/52**) Type IV.2 (**10/53**) Type IV.3 (**10/54**) Type IV.4 (**10/55**)

Figure 10.21 Bone hair or dress pins, Types IV.1 (**10/52**), IV.2 (**10/53**), IV.3 (**10/54**), IV.4 (**10/55**)

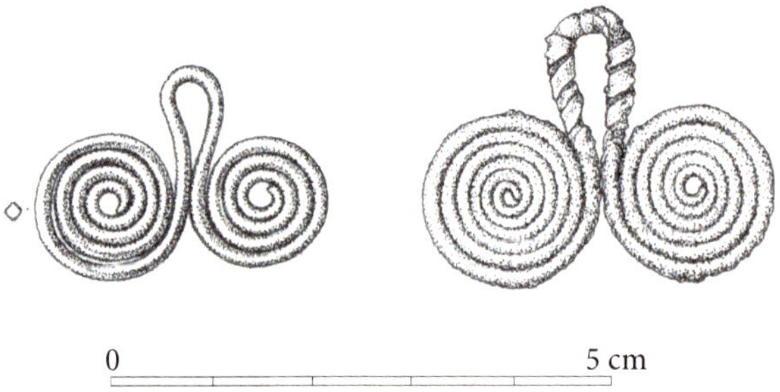

FIGURE 10.22 Small bronze double spiral ("spectacle") pendants (**10/57**, **10/58**)

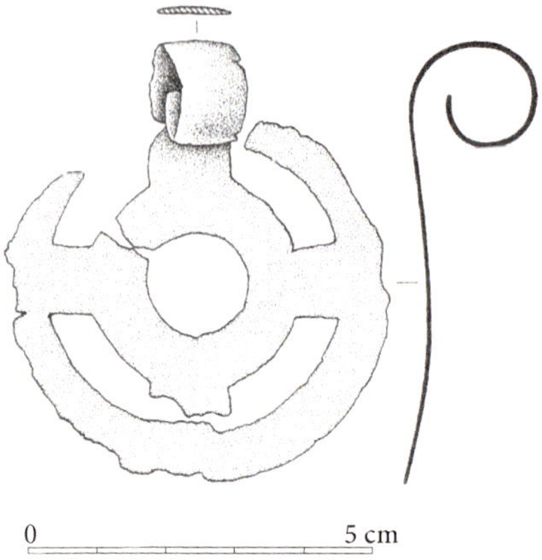

FIGURE 10.23 Wheel pendant (**10/59**)

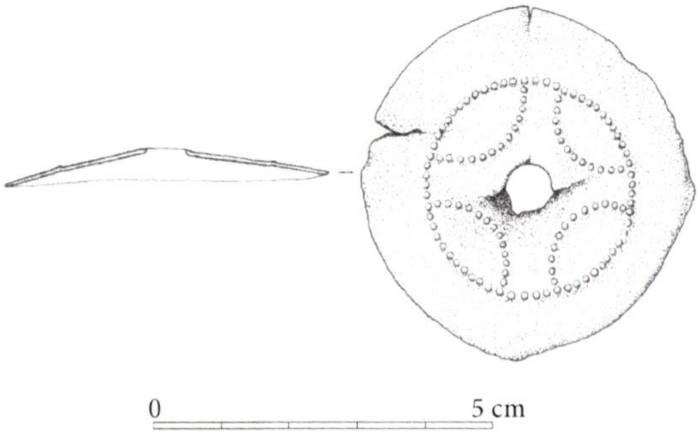

10/60 (**TXVIII-2**)

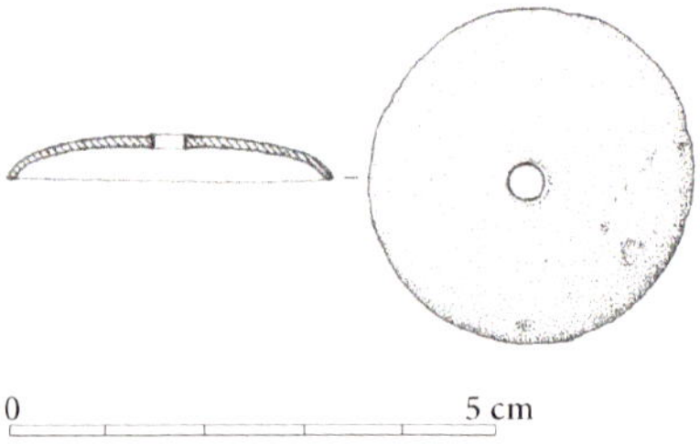

10/62 (**TLIII-2**)

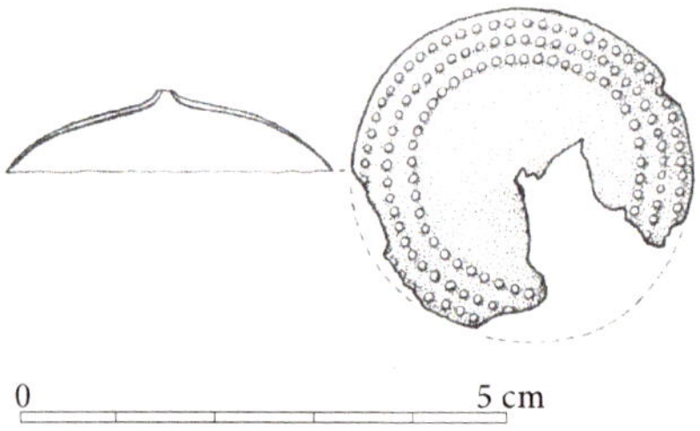

10/61 (SF 163a-d)

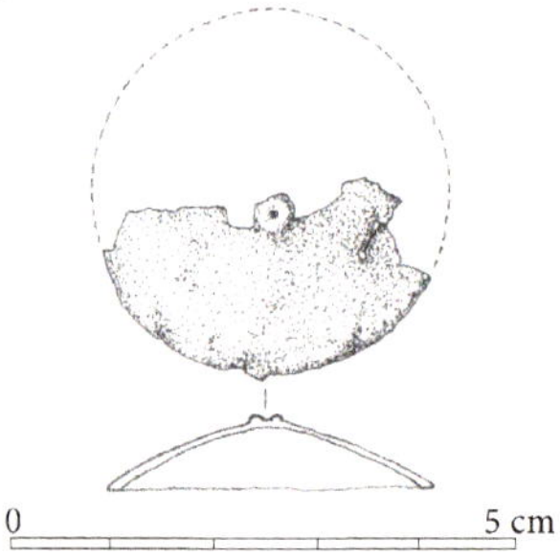

10/63 (**TXXI-5**)

FIGURE 10.24 Bronze disks/bosses (**10/60-10/63**)

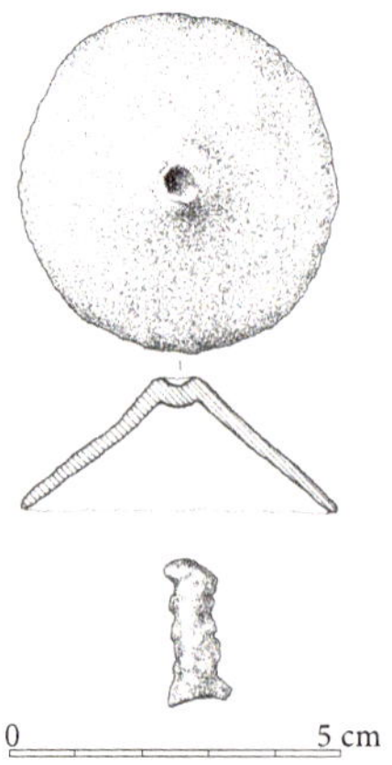

TLXVIII-2 (Iron)

FIGURE 10.25 Iron disk/boss (**10/64**)

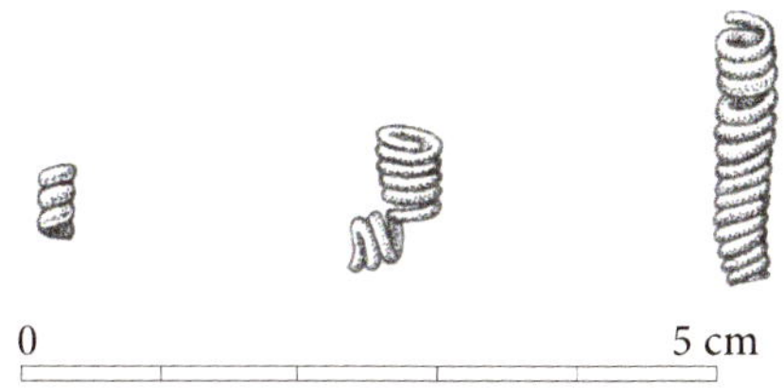

10/65 (TVIII-1) **10/66** (TXVII-5) **10/67** (TXVII-6)

Figure 10.26 Small bronze spiral coils (beads) (**10/65**, **10/66**, **10/67**)

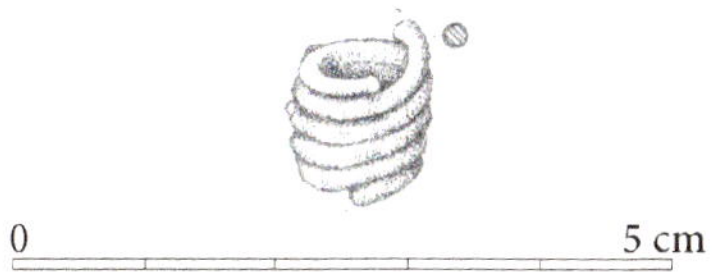

10/68 (TXLII-1)

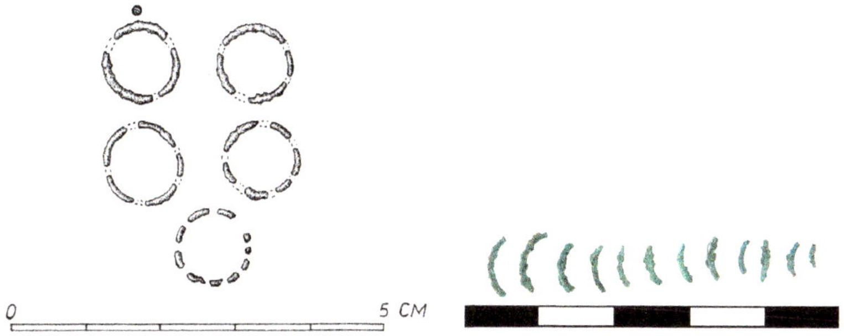

10/69 (TXLII-2)

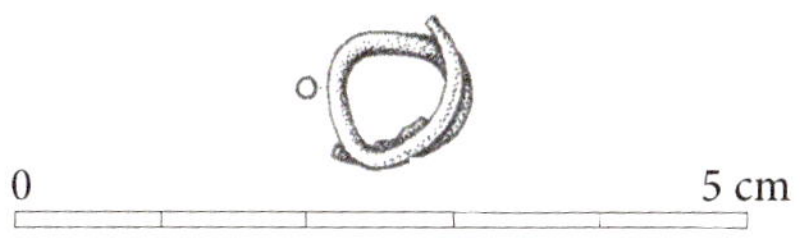

10/70 (TLIII-5)

Figure 10.27 Small bronze spirals (earrings or hair rings?) (**10/68**, **10/69**, **10/70**)

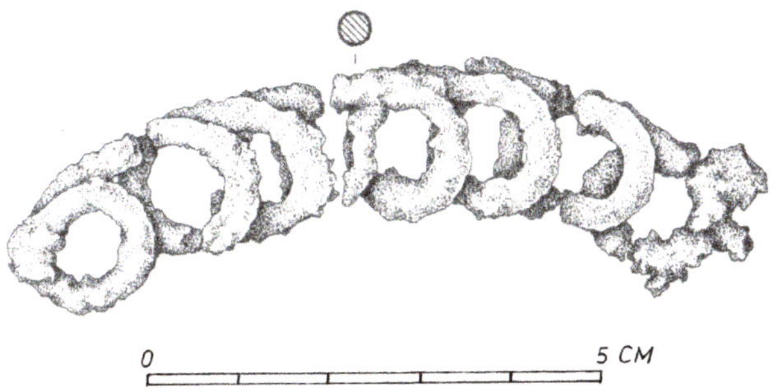

10/71 (TLVIII-4)

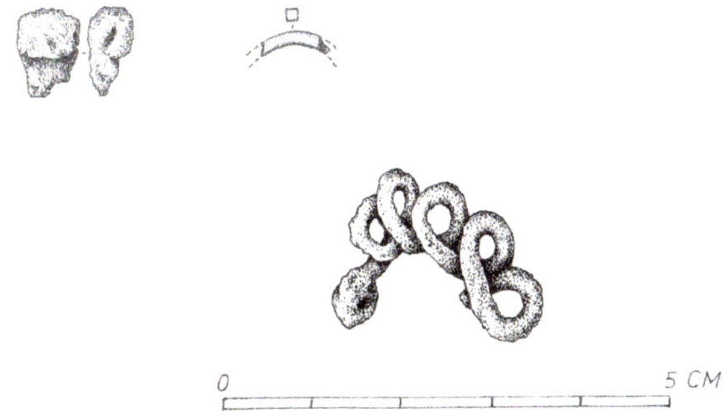

10/72 (TXXI-9)

Figure 10.28 Iron spiral coils (**10/71**, **10/72**)

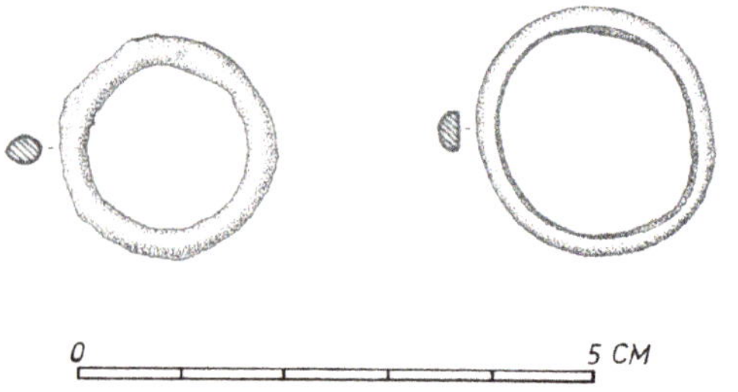

10/73 (TLXX-2a) **10/74** (TLXX-2b)

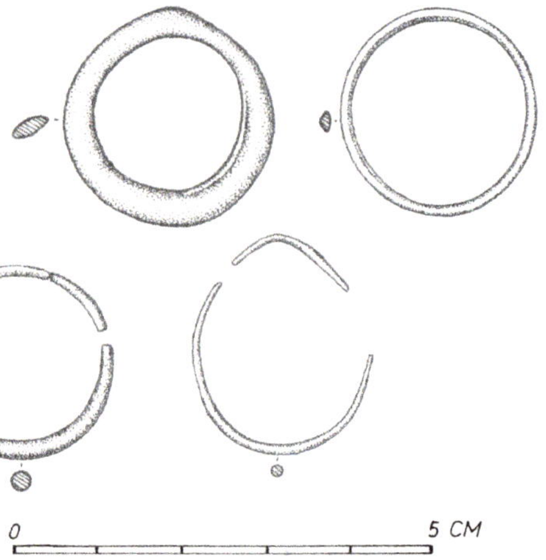

10/75 (SF 163e, top right), **10/76** (SF 163f, top left),
10/77 (SF 163g, bottom left), **10/78** (163h, bottom right)

10/75

10/76

10/77

10/78

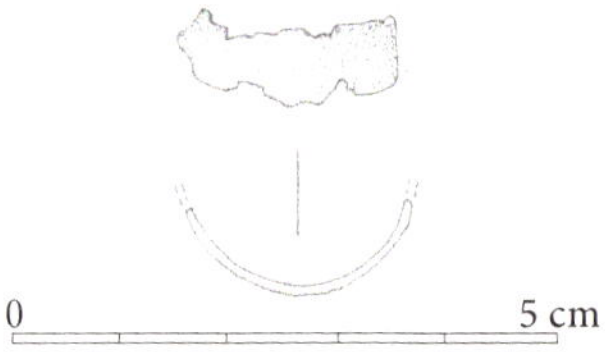

10/79 (**TXXI-10**)

FIGURE 10.29 Bronze rings (**10/73–10/79**)

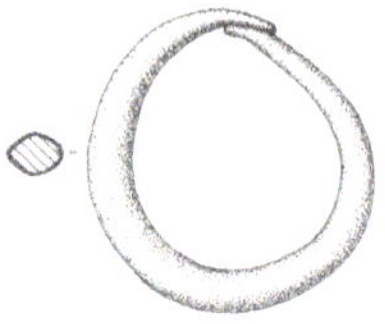

10/80 (**TXVII-3**)

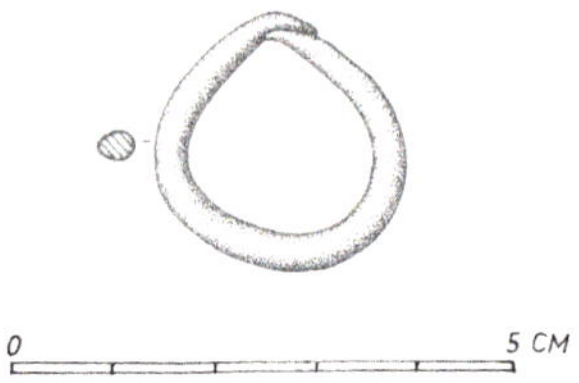

10/81 (**TXVII-4**)

10/82 (**TLIII-3**)

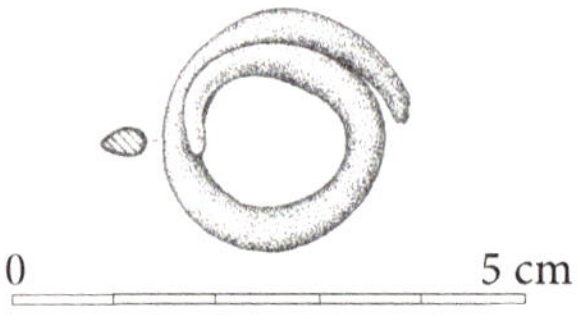

10/83 (**TLXIX-3**)

FIGURE 10.30 Bronze earrings (**10/80-10/83**)

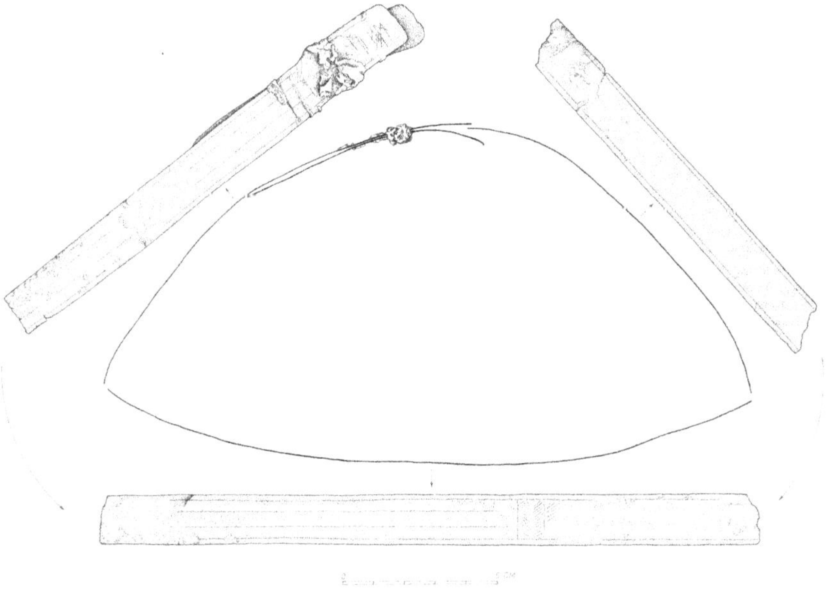

Figure 10.31 Bronze headband, **TXVII-2** (**10/84**)

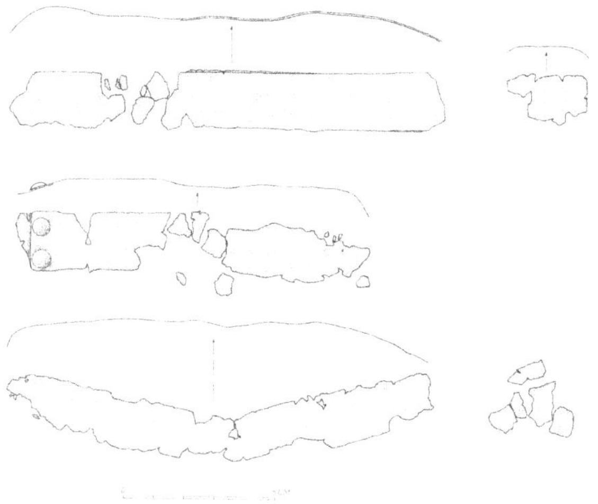

Figure 10.32 Bronze headband, **TXVIII-1** (**10/85**)

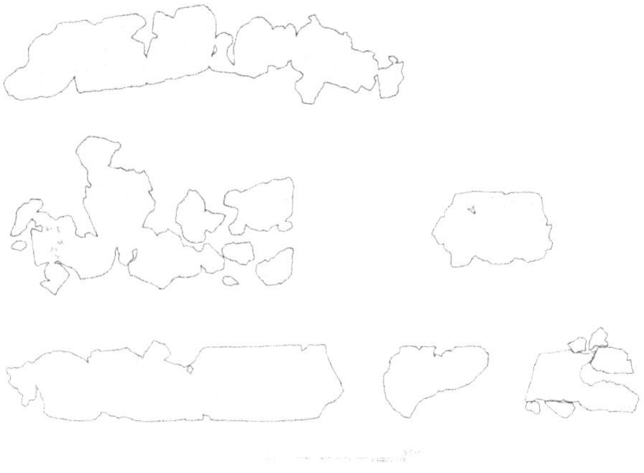

Figure 10.33 Bronze headband, **TXXI-4** (**10/86**)

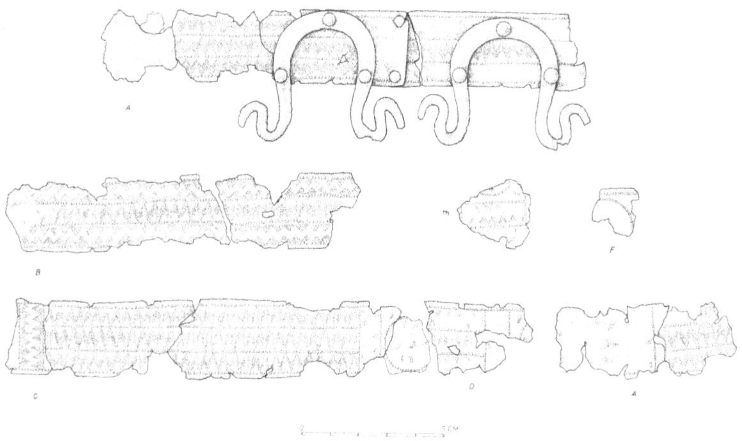

Figure 10.34 Bronze headband, **TLXX-2** (**10/87**)

Figure 10.35 Length of copper/copper alloy wire (**10/88**)

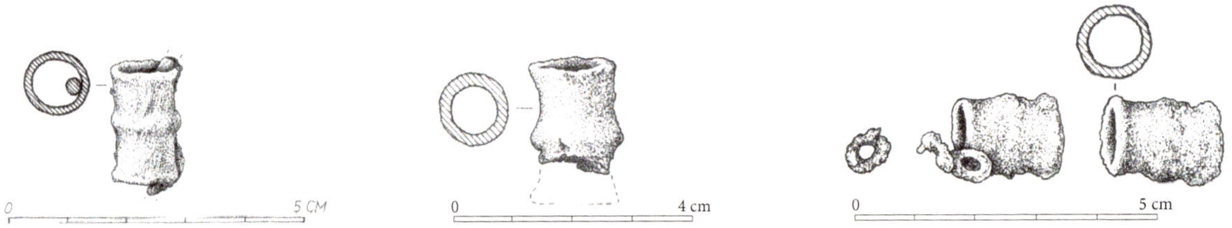

10/89 (**TLXVI-2**) **10/90** (**TLXXX-3**) **10/91** (**TXXI-7**) **10/92** (**TXXI-6**)

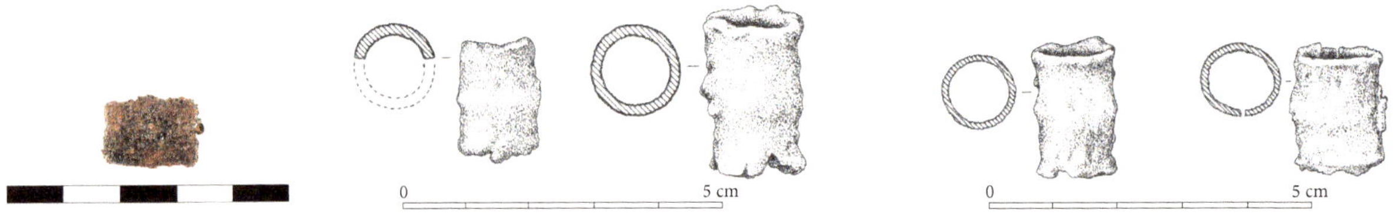

10/93 **10/93** (SF 126) **10/94** (**TXLVIII-5**) **10/95** (**TLV-6**) **10/96** (**TLV-7**)

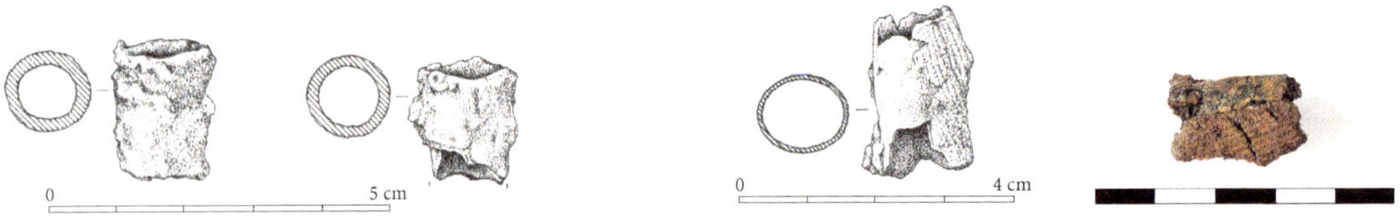

10/97 (**TLV-5**) **10/98** (**TLV-4**) **10/99** (SF 124)

FIGURE 10.36 Tubular iron beads, Type I (**10/89**–**10/99**)

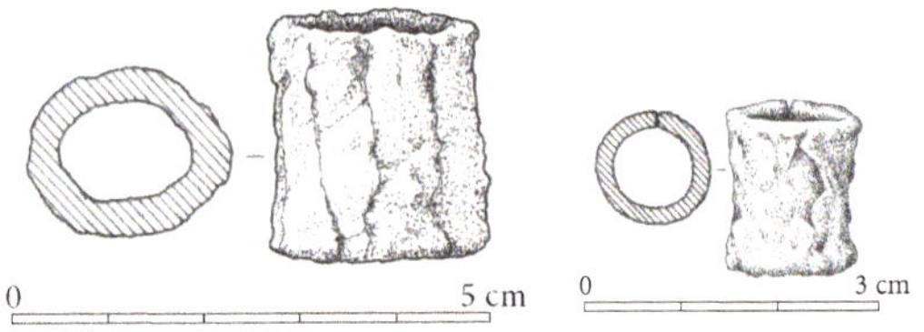

10/100 (**TLXXX-2**) **10/101** (**TLXVI-1**)

FIGURE 10.37 Tubular iron beads, Type II (**10/100**, **10/101**)

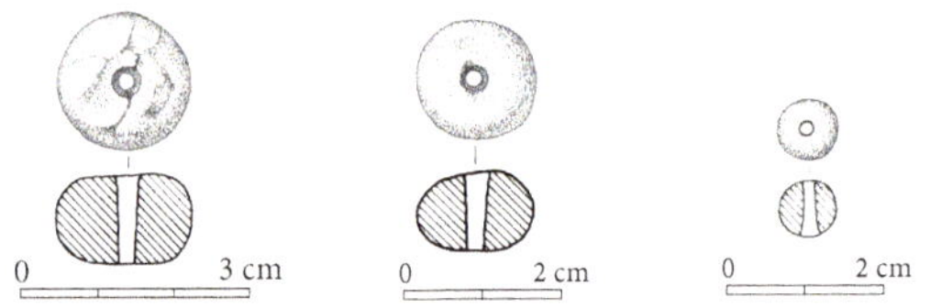

10/102 (**TXXVIII-6**) **10/103** (**TLIII-7**) **10/104** (**TLXX-6**)

FIGURE 10.38 Stone beads (**10/102**, **10/103**, **10/104**)

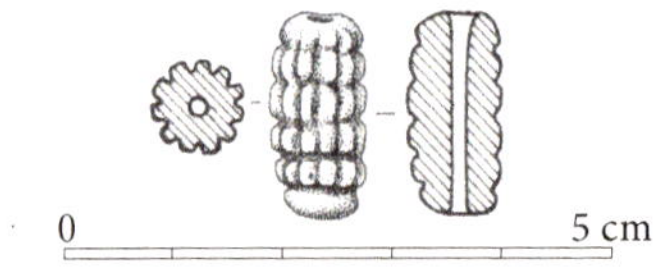

10/105 (**TLIII-9**)

FIGURE 10.39 Faience bead (**10/105**)

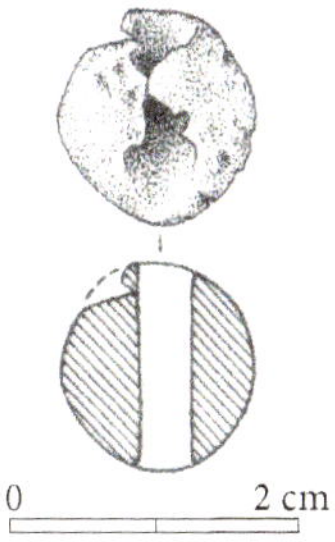

10/106
(**TXXVIII-8**)

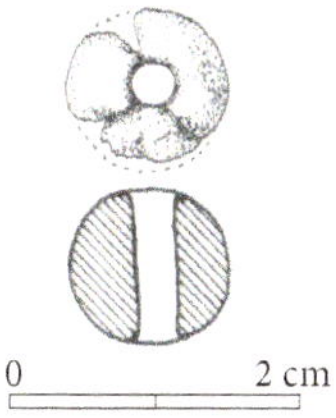

10/107
(**TLIII-8**)

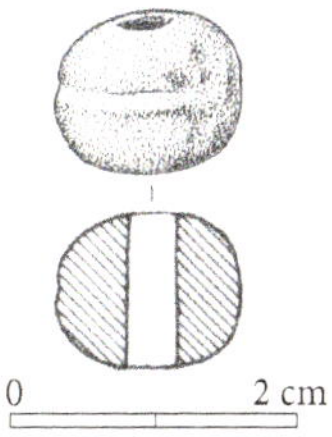

10/108
(**TXXVIII-7**)

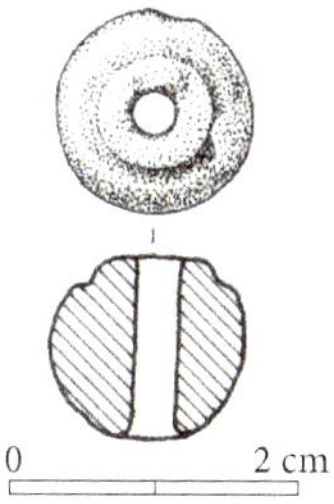

10/109
(**TLIII-6**)

FIGURE 10.40 Glass beads, Beck Type I.C.1.a and related (**10/106–10/109**)

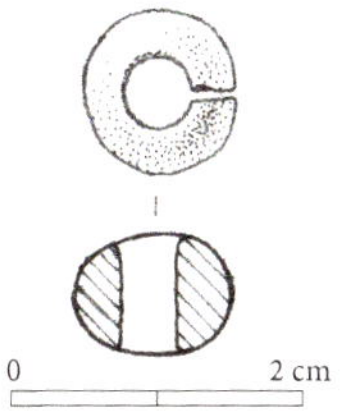

10/110 (SF 125)

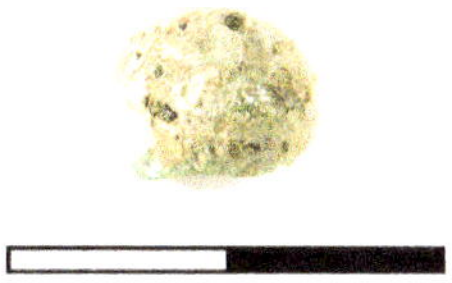

10/111 (SF 450)

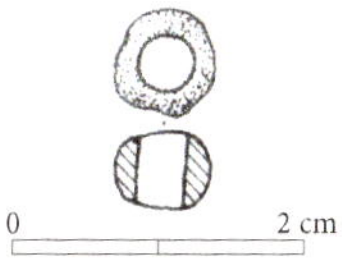

10/112
(**TXXVIII-4**)

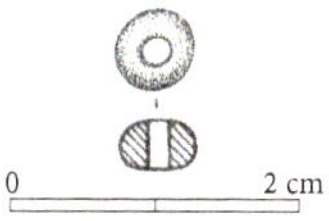

10/113
(**TLV-8**)

FIGURE 10.41 Glass beads, Beck Type I.B.1.a and related (**10/110–10/113**)

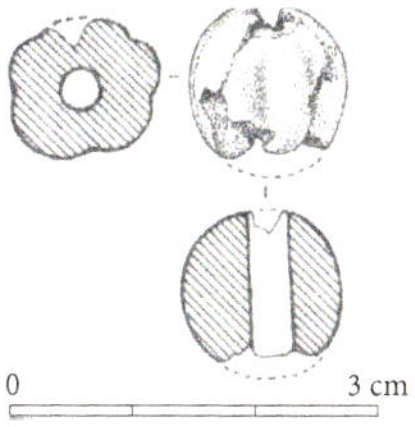

10/114
(**TXXI-8**)

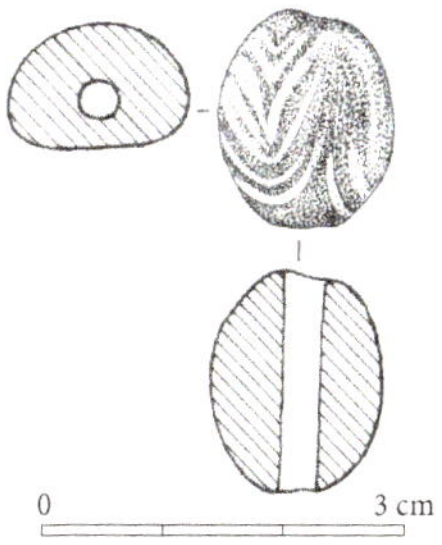

10/115
(**TXXVIII-5**)

FIGURE 10.42 Glass beads of various types (**10/114, 10/115**)

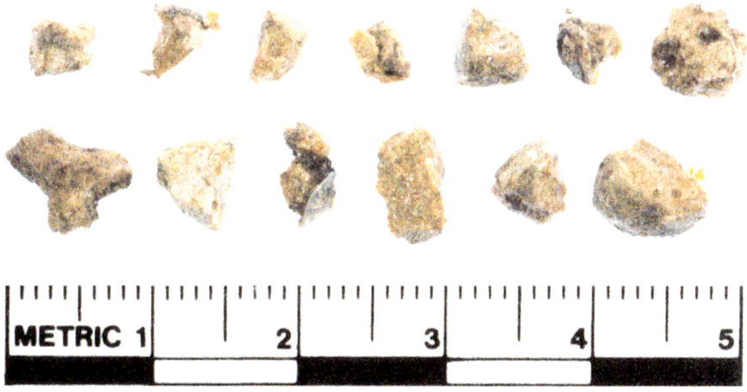

10/116 (**TXXVIII-9**)

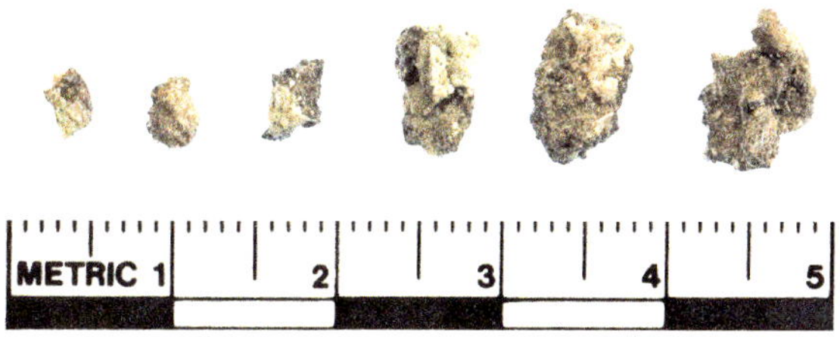

10/117 (**TXXVIII-10**)

FIGURE 10.43 Fragmentary glass beads of uncertain type (**10/116, 10/117**)

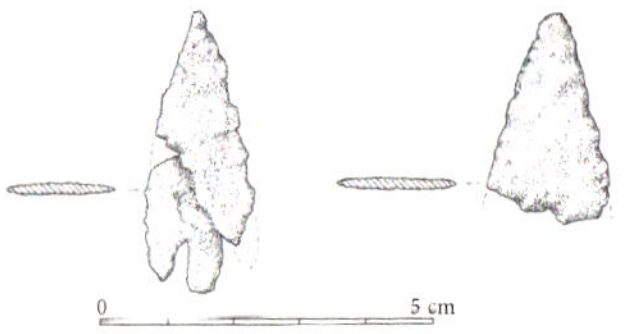

10/119 (TXXXII-1) **10/120** (TXXXII-2)

FIGURE 10.45 Iron arrowheads (**10/119**, **10/120**)

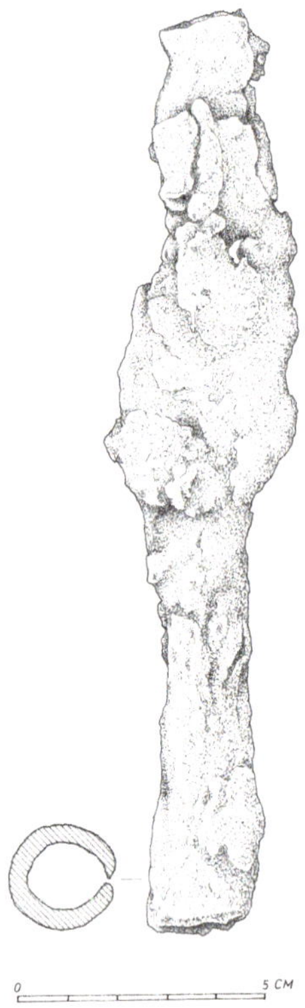

10/118 (TXLV-1) (SF 263)

FIGURE 10.44 Iron spearhead (**10/118**)

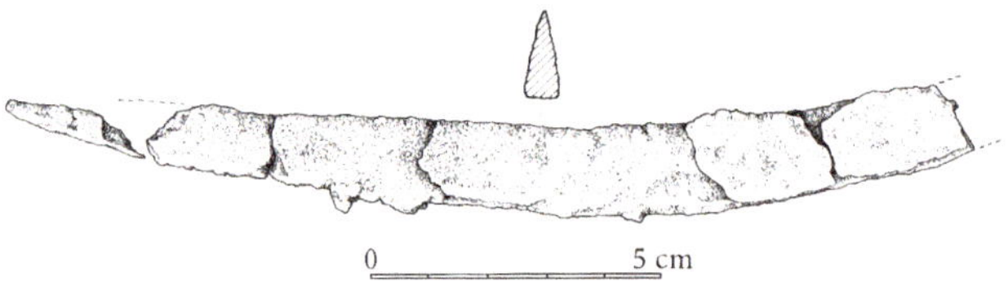

10/121 (TLXXXIV-1)

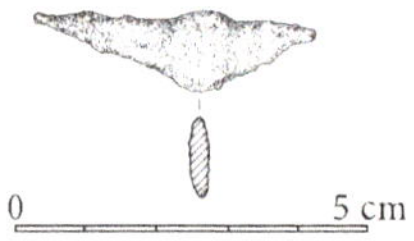

10/122 (TLXXXVIII-1)

10/123 (SF 230)

FIGURE 10.46 Iron knives (**10/121**, **10/122**, **10/123**)

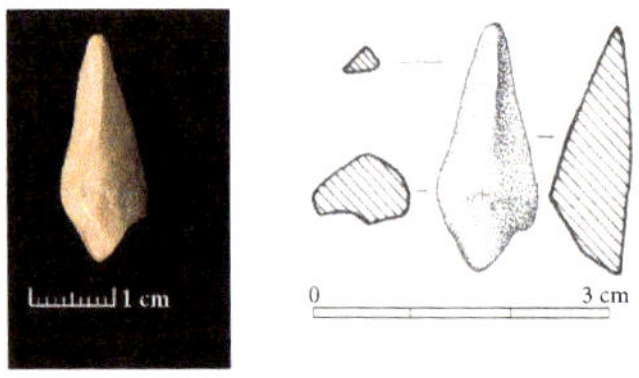

10/124 (SF 347)

FIGURE 10.47 Worked bone (**10/124**)

10/125 (SF 044) **10/126** **10/128** **10/127** **10/129** **10/130** **10/131**

FIGURE 10.48 Unidentified objects of copper/copper alloy (**10/125–10/126**, **10/128–10/131**)

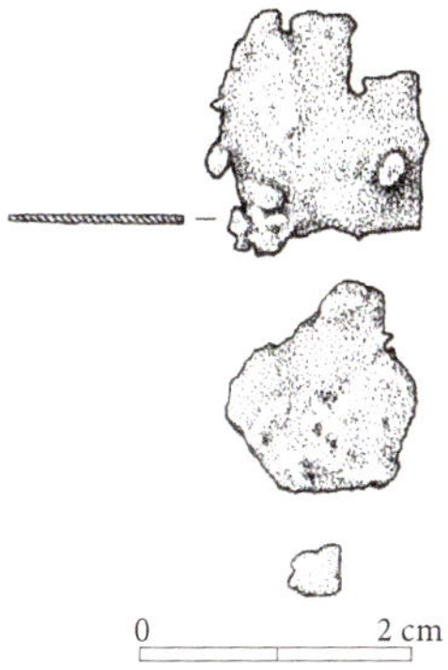

10/132 (**TXXI-3**) (SF 256)

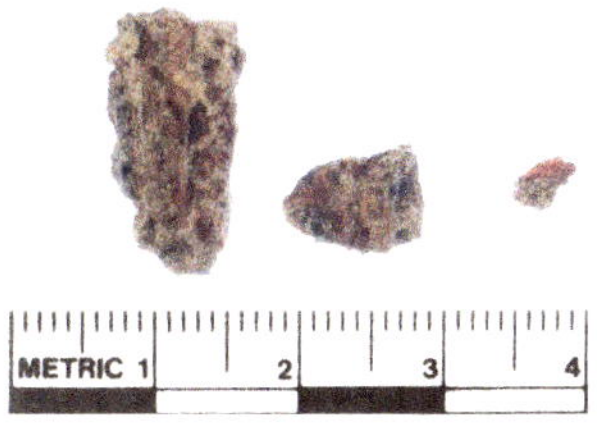

10/133 (**TXXXV-1**) (SF 384)

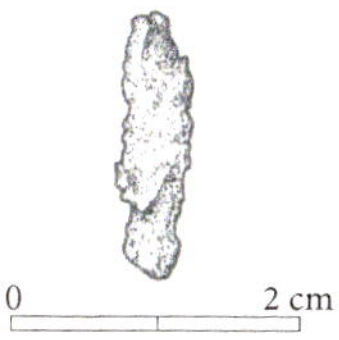

10/134 (SF 217)

Figure 10.49 Fragments of unidentified iron objects (**10/132**, **10/133**, **10/134**)

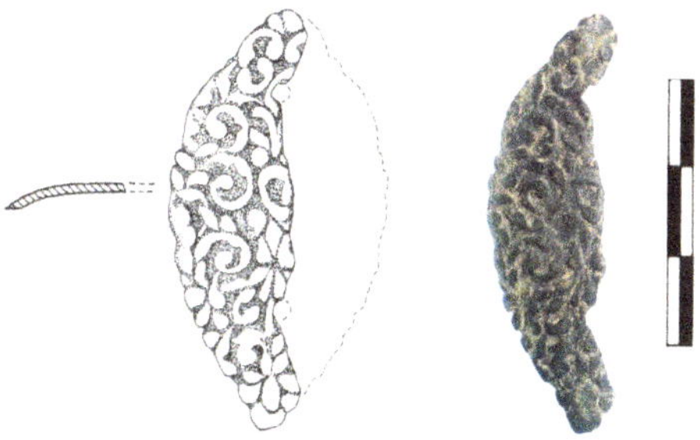

10/135 (SF 333)

Figure 10.50 Medieval/modern ornament made of copper, zinc, tin, and lead alloy (**10/135**)

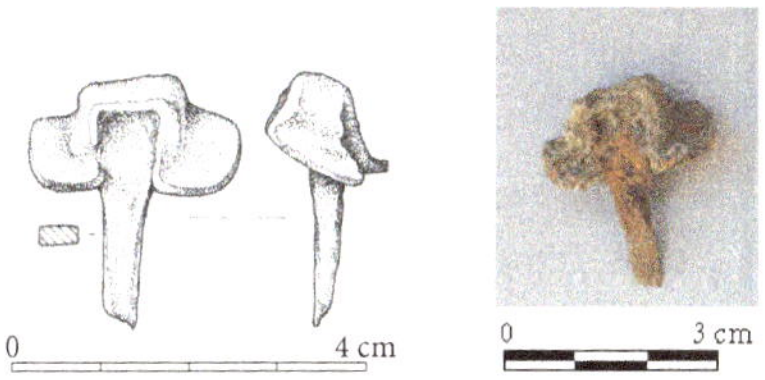

10/136 (SF 98)

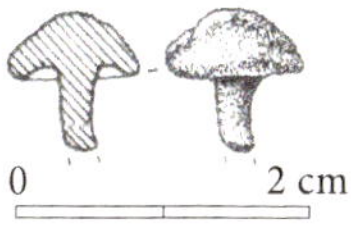

10/137 (SF 10)

10/138 (SF 289)

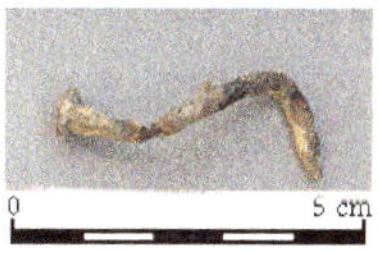

10/139 (SF 103)

10/140 (SF 86)

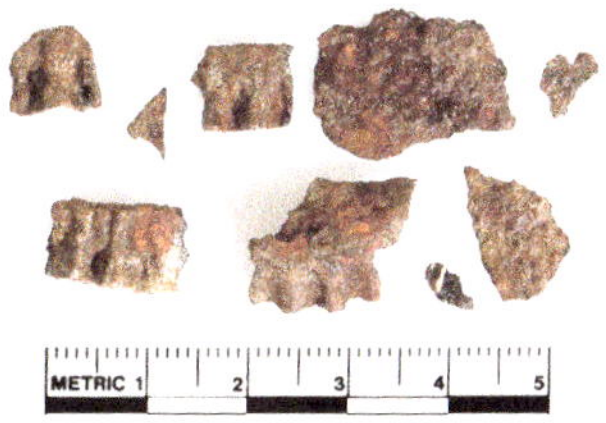

10/141 (SF 306)

10/142 (SF 114)

10/143 (SF 228)

10/144 (SF 301)

Figure 10.51 Modern iron objects and fragments of iron objects (**10/136**–**10/144**)

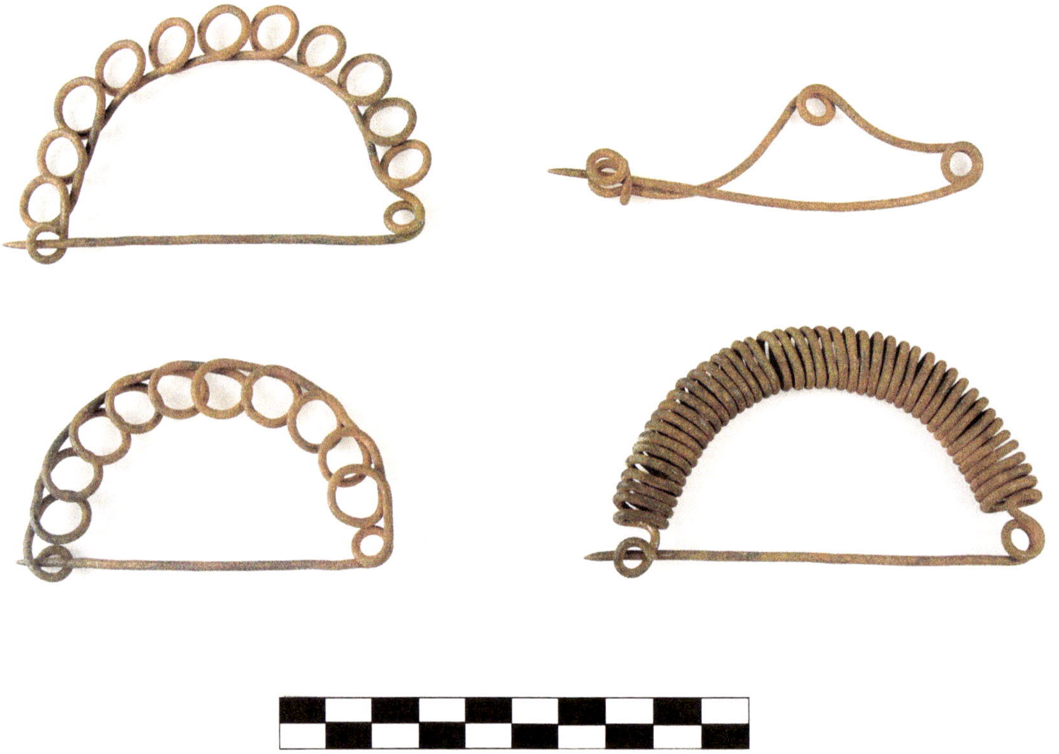

Figure 10.52 Modern fibulae made after some of the Lofkënd prehistoric types

Appendix 3 Illustrations

a

b

FIGURE A3.1 (a) Greek infantry advancing into Albania in November 1940; (b) the Greek army in the Albanian mountains, March 1941 (photos courtesy Yannis Mylonas)

FIGURE A3.2 Mannlicher-Schönauer M 1903, 6.5-mm caliber rifle used by the Greek Army from 1907 to 1940 and standard issue for the Greco-Italian War in Albania and Epirus

A3/1 (**AF 405**) **A3/2** (SF 072) **A3/5** (SF 057) **A3/8** (SF 381)

FIGURE A3.3 Cartridge shells, Greek series: **A3/1**, **A3/2**, **A3/5**, **A3/8**

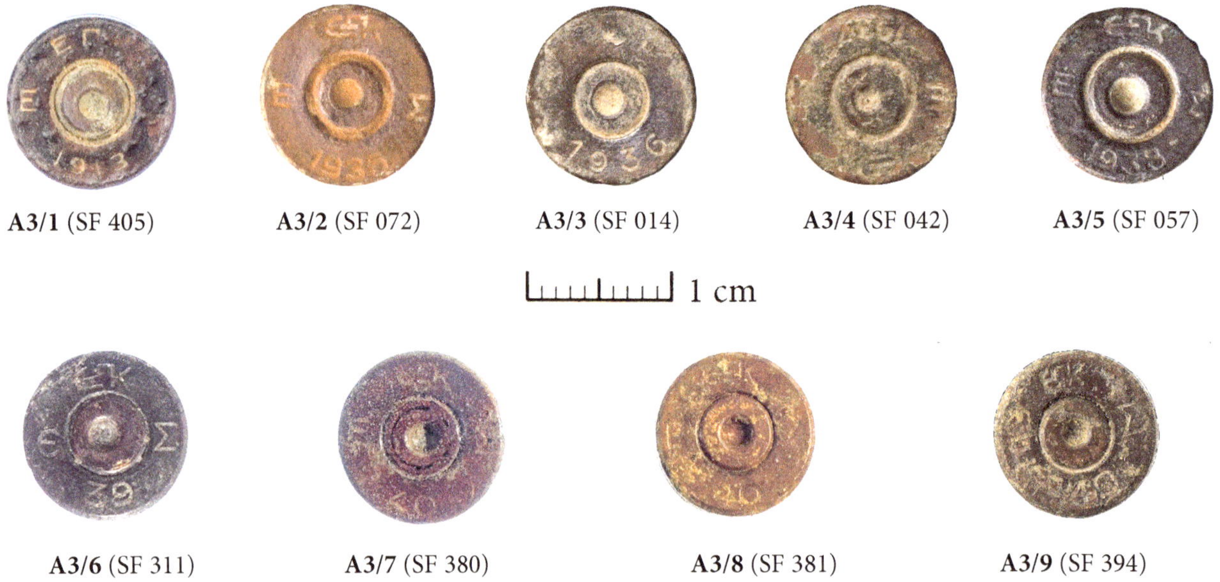

FIGURE A3.4 Cartridge shells, Greek series, head stamps: **A3/1–A3/9**

Figure A3.5 Mannlicher-Carcano M 1891, 6.5-mm caliber rifle produced from 1892 to 1945 and used by Italian troops during World War I and War World II

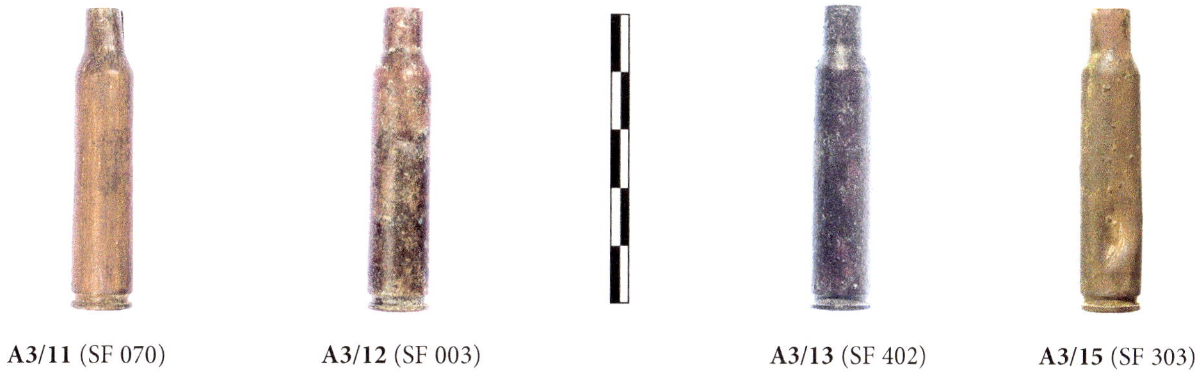

Figure A3.6 Cartridge shells, Italian SMI series: **A3/11**, **A3/12**, **A3/13**, **A3/15**

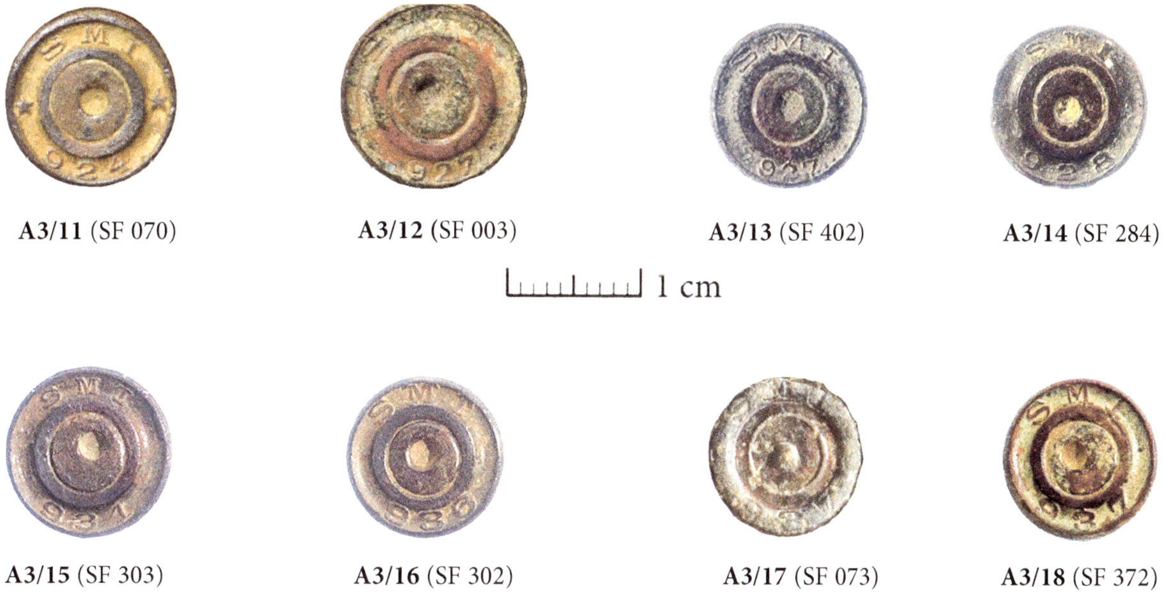

Figure A3.7 Cartridge shells, Italian SMI series, head stamps: **A3/11–A3/18**

A3/19 (SF 386) **A3/20** (SF 015) **A3/23** (SF 308)

Figure A3.8 Cartridge shells, Italian A.A series: **A3/19**, **A3/20**, **A3/23**

A3/19 (SF 386) **A3/20** (SF 015) **A3/21** (SF 160)

A3/22 (SF 250) **A3/23** (SF 308) **A3/24** (SF 391)

Figure A3.9 Cartridge shells, Italian A.A series, head stamps: **A3/19–A3/24**

A3/25
(SF 108)

Figure A3.10 Cartridge shell, Yugoslavian: **A3/25**

A3/26 (SF 422) **A3/27** (SF 229)

Figure A3.11 Cartridge shells, unidentified (illegible head stamps): **A3/26–A3/27**

A3/29 (SF 211) **A3/30** (SF 350)

Figure A3.12 Bullets: **A3/29–A3/30**

Chapter 11 Illustrations

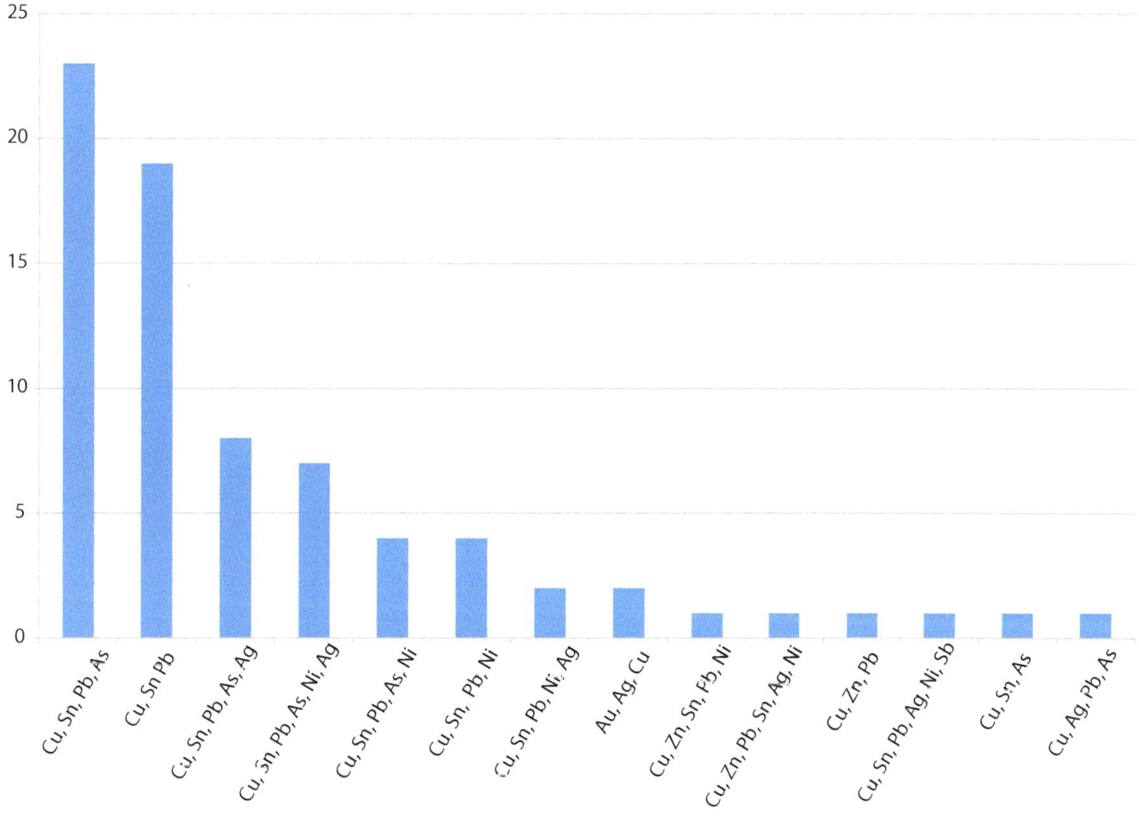

FIGURE 11.1 Types of alloys represented at Lofkënd

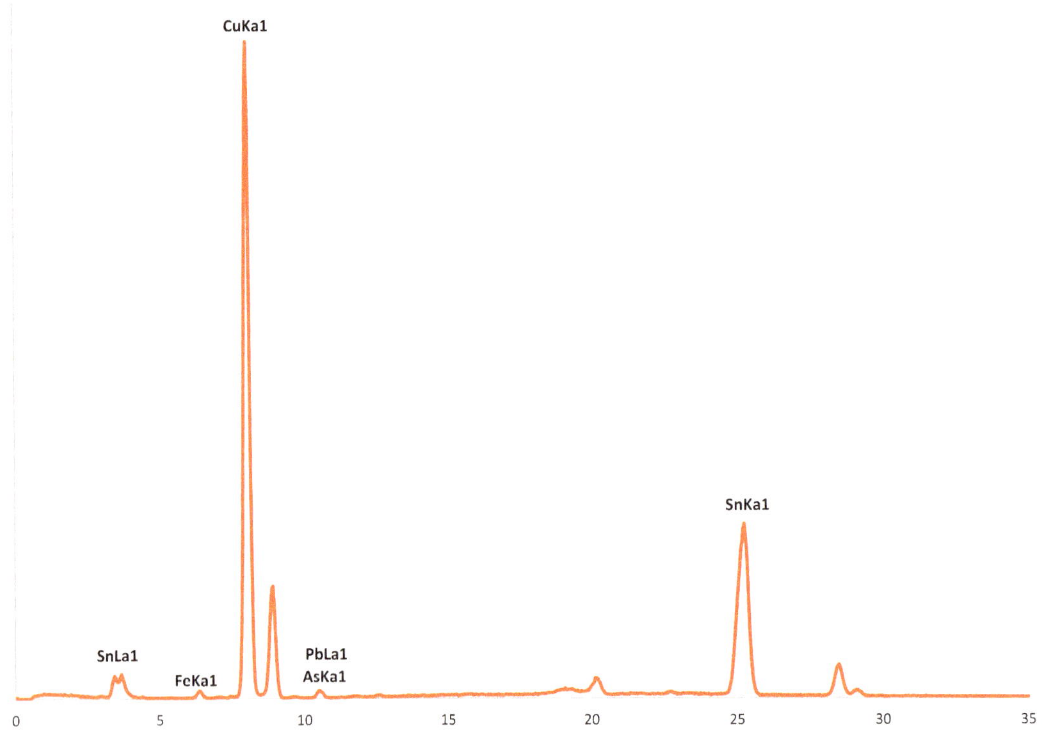

FIGURE 11.2 pXRF analysis of the pin fragment, **10/125** (SF 044), shows the most common composition found among the copper alloy artifacts analyzed from Lofkënd. The second most common alloy composition was similar to the spectrum shown, except for the lack of arsenic. Because the pXRF analysis was conducted on a corroded surface, it is not clear how much of an effect this had on the elements detected within the alloy and on the intensity of the peaks observed. Further analysis should be conducted using other techniques to determine better the types of alloys represented at Lofkënd.

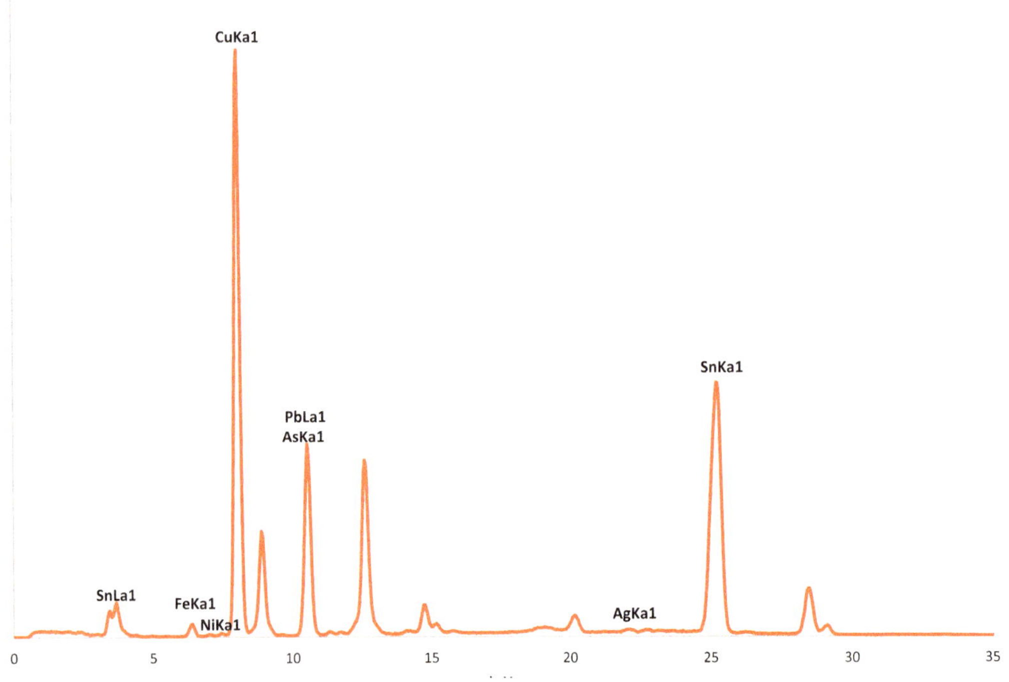

FIGURE 11.3 pXRF analysis of **10/57** (SF 300) showing the presence of minor amounts of arsenic, nickel, and silver. The presence of these elements in many of the copper alloy objects from Lofkënd suggests that the ore used may have come from a copper, arsenic, silver, and nickel mineralization belt.

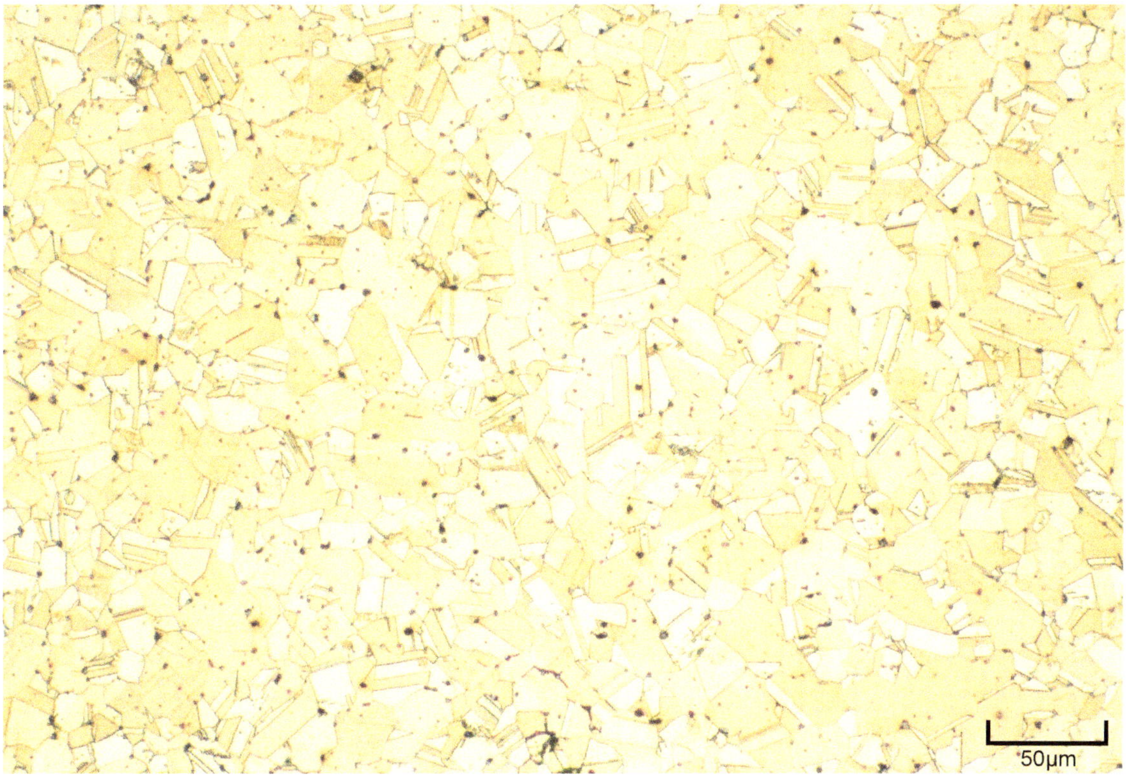

Figure 11.4 The etched sample from **10/17** (SF 434) reveals a structure composed of small twinned grains made through cold working and annealing. This structure is typical of many finds at Lofkënd. The small black holes seen in this sample indicate that the metal is porous as a result of casting.

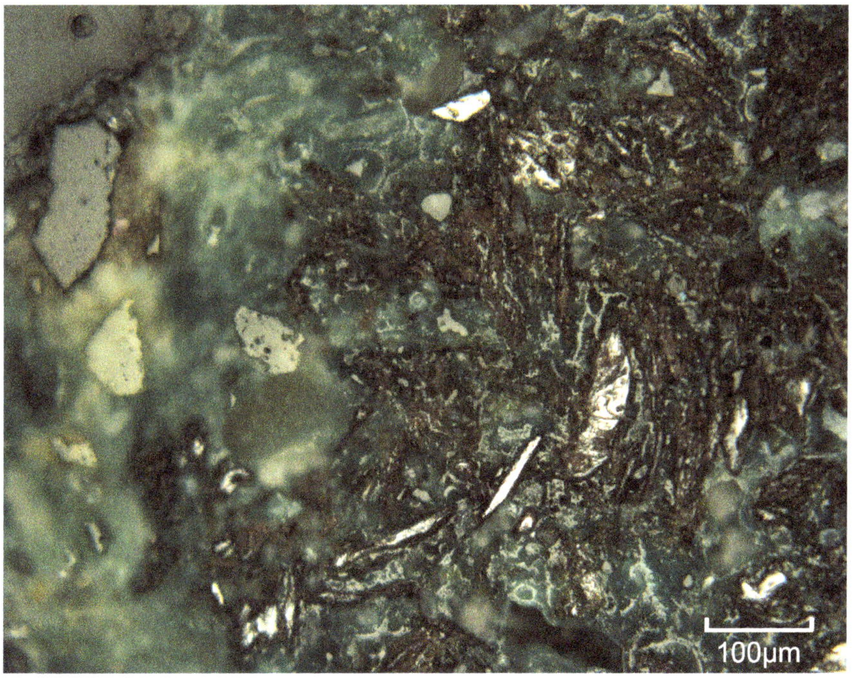

Figure 11.5 The unetched metallographic section of **10/30** (SF 105) reveals an unusual structure. There are no clear dendrites or grains. The structure consists of long copper inclusions in a corroded matrix. There are also large gray inclusions of lead (visible on the left side of the image). It is possible that this object is metal waste.

a

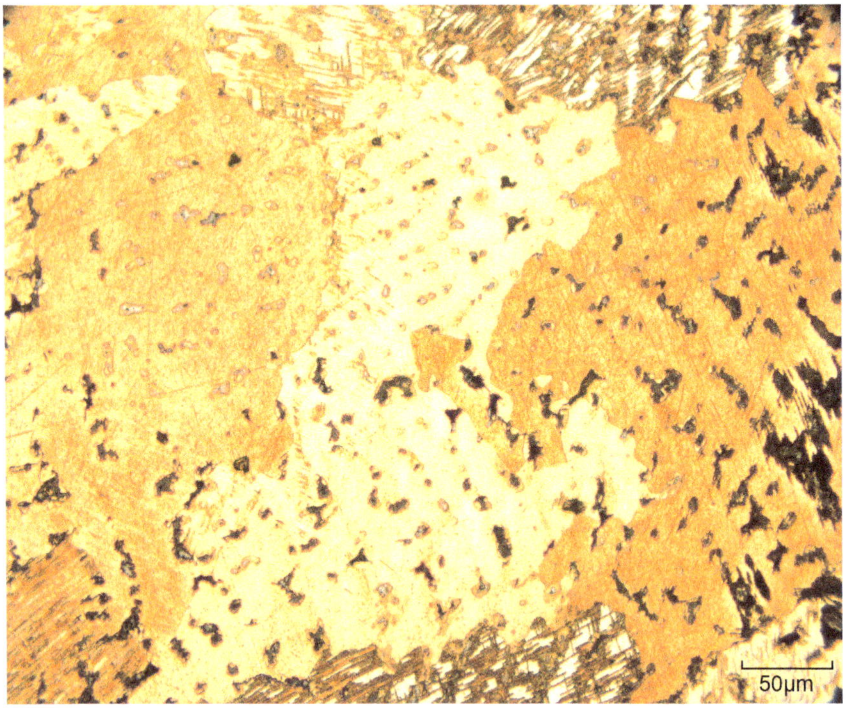

b

Figure 11.6 (a) The copper alloy head of pin **10/51** (SF 067) retains primarily a dendritic structure, suggesting that it was cast directly onto the iron pin. (b) At a higher magnification, the dendrites appear slightly deformed. This indicates that there was cold working of the piece after casting. Strain lines are also visible, formed due to heavy cold working after casting. It is possible that some final working of the copper alloy handle was required after casting to fit it onto the iron pin.

Figure 11.7 The light-colored inclusions within the sample taken from **10/59** (SF 336) appear flattened due to the hammering used to create the flat pendant. Other objects sampled from Lofkënd, such as **10/31** (SF 227), **10/61** (SF 163a–d), and **10/78** (SF 163h), also have inclusions that appear flattened due to cold working that took place during manufacture.

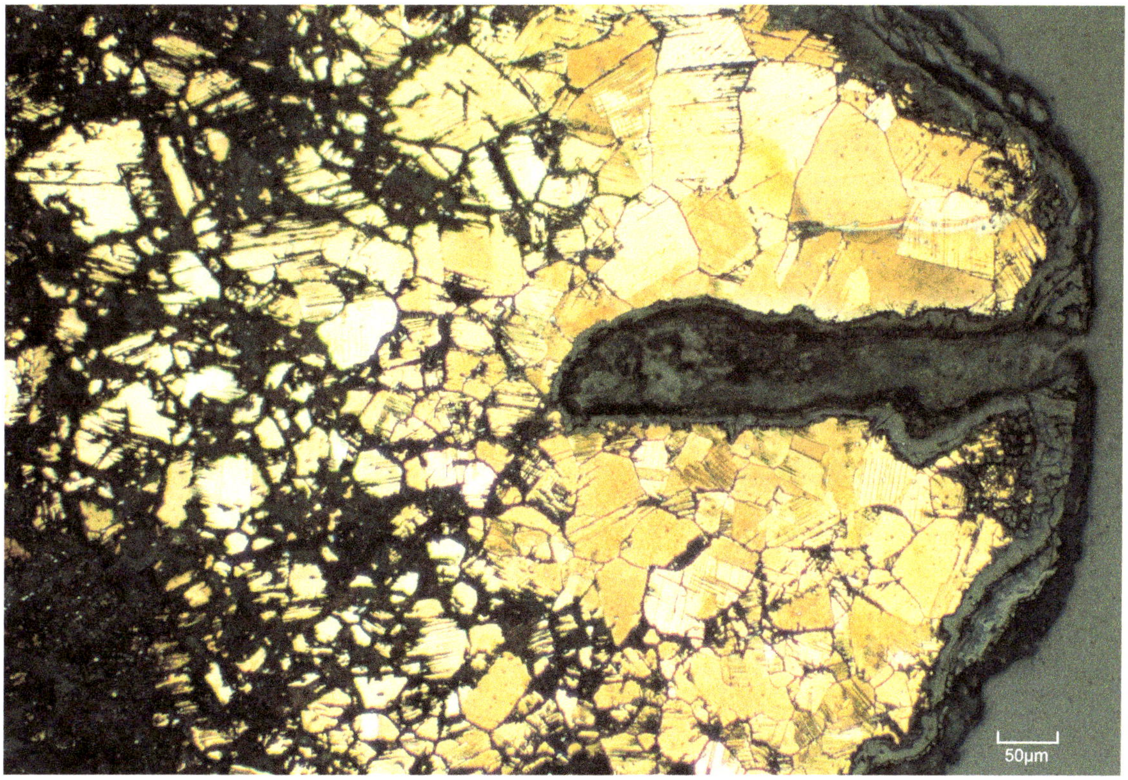

Figure 11.8 Examination of a sample from **10/77** (SF 163g) revealed that the object was made by folding a strip of metal that was then shaped into a ring. The ring was also heavily cold worked and annealed, indicated by twinned grains and strain lines. There is heavy corrosion on the left side of the sample, and corrosion is visible along the grains and the strain lines.

FIGURE 11.9 An unetched sample taken from **10/73** (SF 091d) shows that the structure is primarily dendritic, although one area (toward the lower section of the image) shows deformation of dendrites and formation of grains. This indicates cold working in at least one area of the object after casting. The strain lines indicate that it was heavily worked.

FIGURE 11.10 The metallographic structure of **10/75** (SF 163e) shows heavily worked grains. The grains toward the edge of the sample are smaller than those in the center and have more strain lines. This indicates heavier working at the edges of the ring.

Figure 11.11 This sample taken from **10/84** (SF 349), a headband, has a very corroded microstructure with a few small grains preserved. The appearance of the corrosion, where there is banding or alternating layers, suggests that it was formed due to the Liesegang phenomenon.

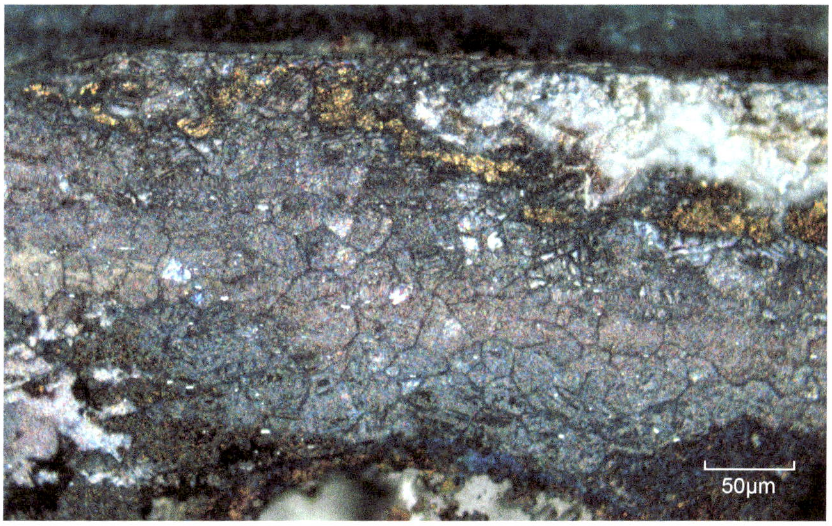

Figure 11.12 The headband **10/87** (SF 089 + 091) was completely corroded, but some of the metallographic structure was preserved in the corrosion layers. Grains are visible, indicating that the headband was cold worked. Strain lines are also visible caused by heavy cold working.

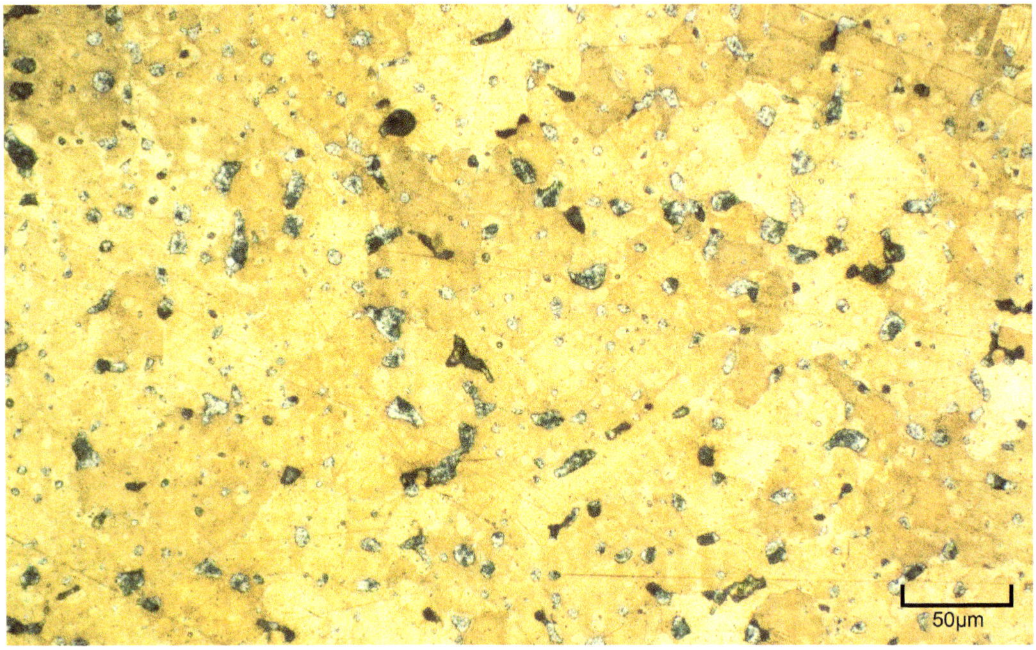

FIGURE 11.13 The sample taken from **10/135** (SF 333) shows a structure composed primarily of dendrites, with some grains at the edges. The twinning and strain lines indicate heavy cold working and annealing after casting. Lead in the alloy, appearing as gray areas with black specks, have some white inclusions within them. The white inclusions appear to be some material that has recrystallized within the lead. One suggestion is that it could be tin oxide, but further analysis would be required to determine this.

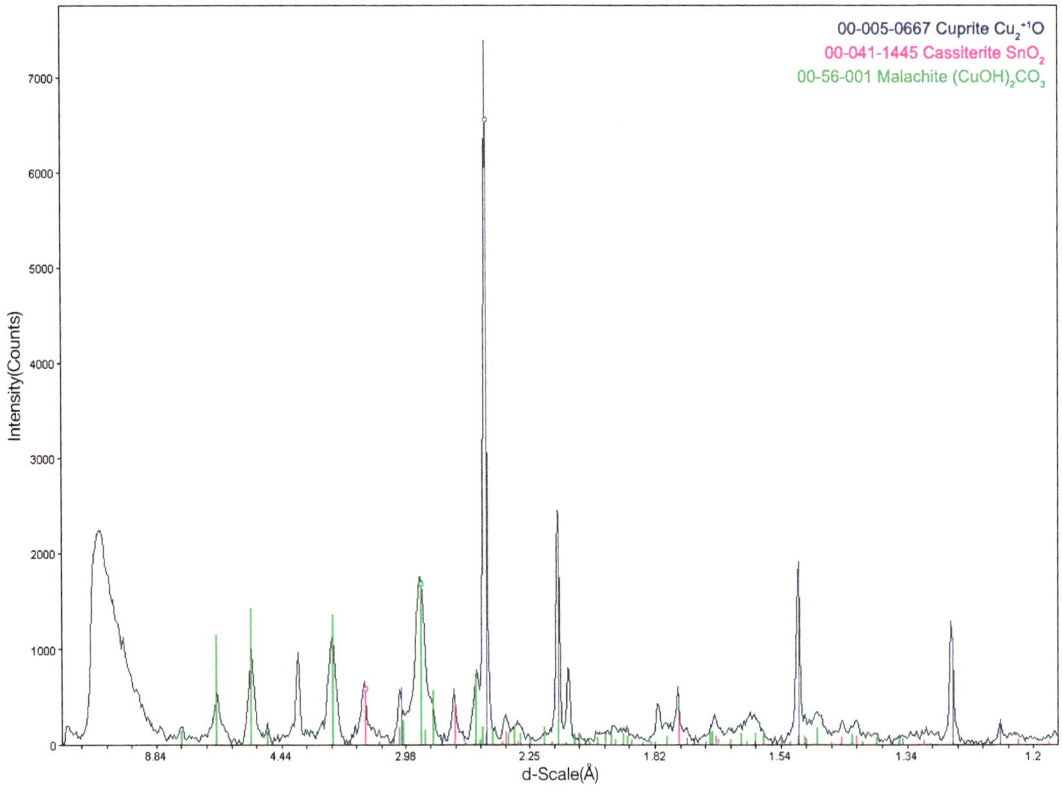

FIGURE 11.14 Corrosion from a spiral ornament or bead **10/66** (SF 313b) shows the presence of malachite (PDF #00-056-0001), cuprite (PDF#00-005-0667), and cassiterite (PDF #00-041-1445). Cassiterite was found on 6 of the 23 copper alloy objects sampled for corrosion studies.

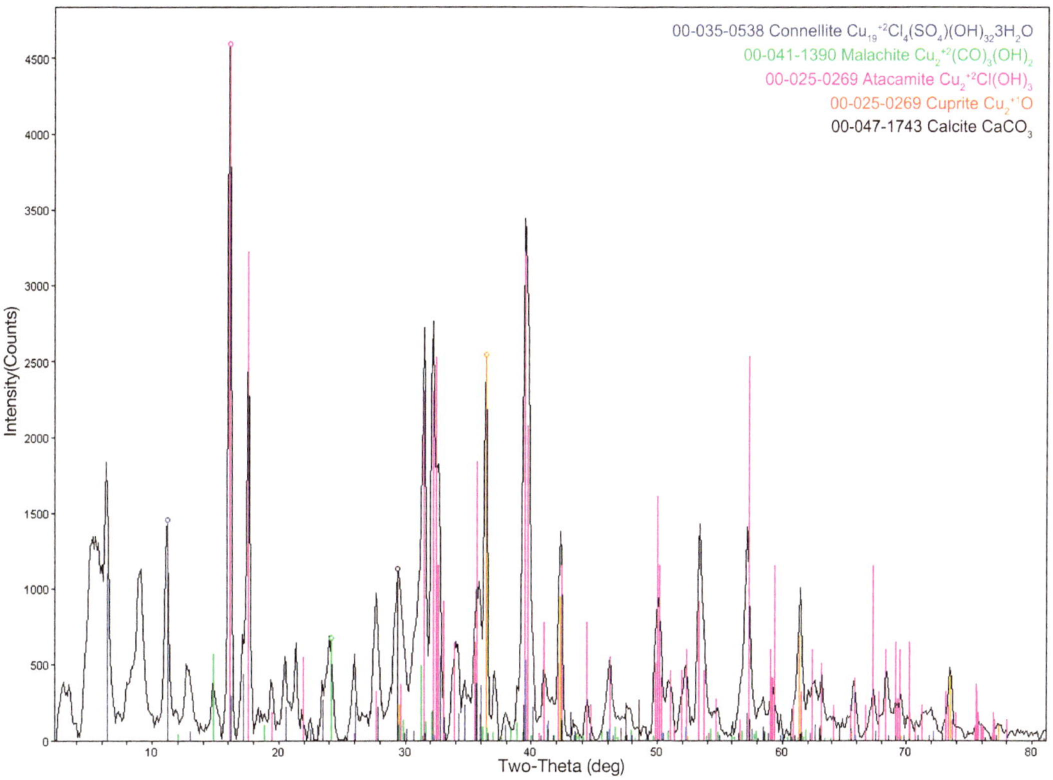

FIGURE 11.15 Connelite (PDF #00-035-0538), a copper chloride sulfate hydroxide hydrate, was found on a coiled wire or ring **10/83** (SF 413b) in a sample of blue-green corrosion. The corrosion sample also contained cuprite (PDF #00-005-0667), malachite (PDF #00-041-1390), atacamite (PDF #00-025-0269), and calcite (PDF #00-047-1743).

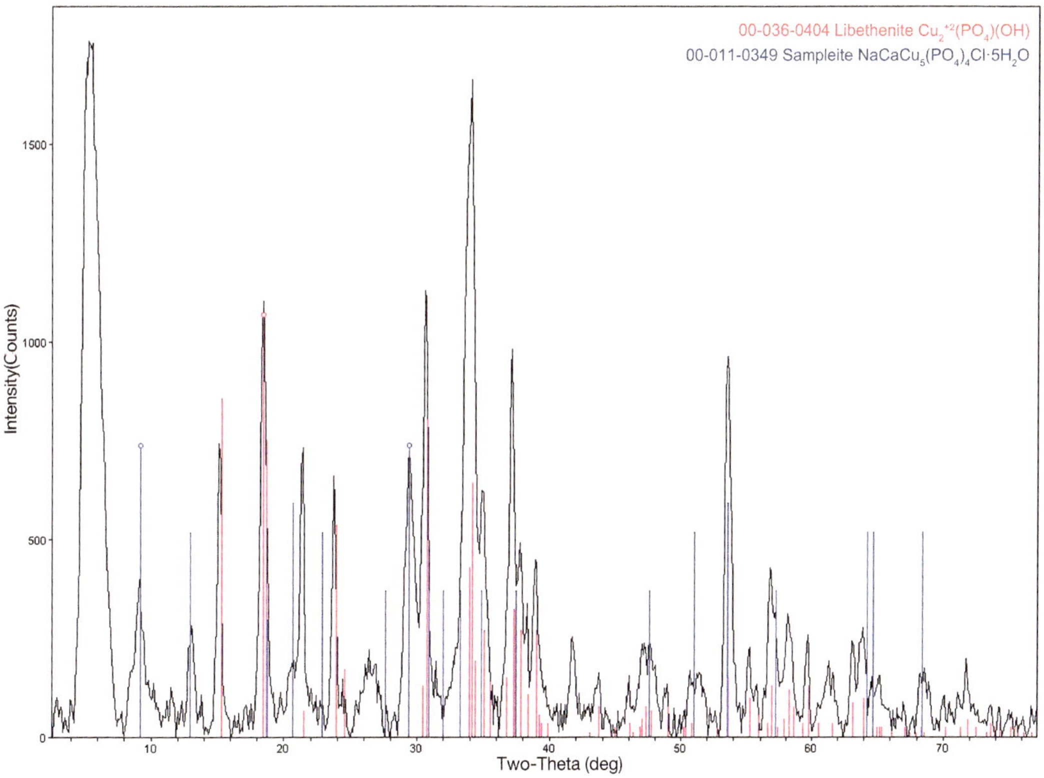

FIGURE 11.16 A sample of corrosion taken from the headband **10/86** (SF 255) contained both sampleite (PDF #00-011-0349) and libethenite (PDF #00-036-0404), two phosphate-containing copper corrosion products. The source of phosphorus for the formation of this corrosion product is from the dissolution or leaching of the bone with which the headband was in contact during burial.

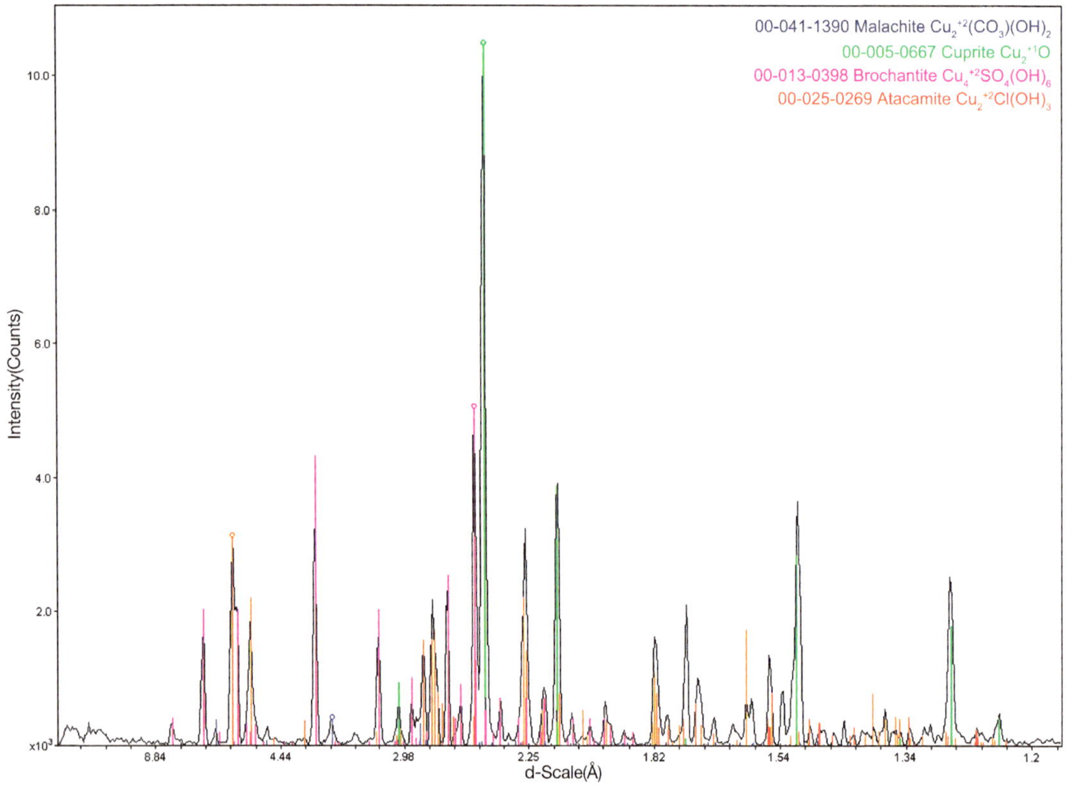

FIGURE 11.17 Samples of corrosion from five objects were found to contain brochantite (PDF #00-013-0398), a copper sulfate commonly found on outdoor sculpture in polluted environments. Brochantite was thought to have formed due to high levels of sulfur gases in the graves produced by decomposition of organic material and metabolic activity of microorganisms.

FIGURE 11.18 Three metallographic samples examined—**10/13** (SF 107), **10/18** (SF 157), and **10/75** (SF 163e)—did not show the presence of cuprite in the corrosion structure, as would be expected on copper alloy objects from a terrestrial environment. In the sample shown, taken from **10/13** (SF 107), the green corrosion appears directly over the metal core. The typical layer of cuprite expected between the metal core and the green corrosion layer is not present.

Chapter 12 Illustrations

FIGURE 12.1 Iron corrosion formed in somewhat linear pattern on fibula **10/24**, possibly following structure of textile

FIGURE 12.2 Detail of weave on pin **10/33** (SF 171) showing one element crossing over several others

FIGURE 12.3 Detail of possible cord or decorative strand of textile on pin **10/33**

FIGURE 12.4 The pattern preserved here on pin **10/35** seems to have elements crossing over and under each other as in plain weave, but has quite an open structure creating diamond or ovoid shapes when crossing

FIGURE 12.5 Detail of faced plain weave found on pin **10/37** (SF 046)

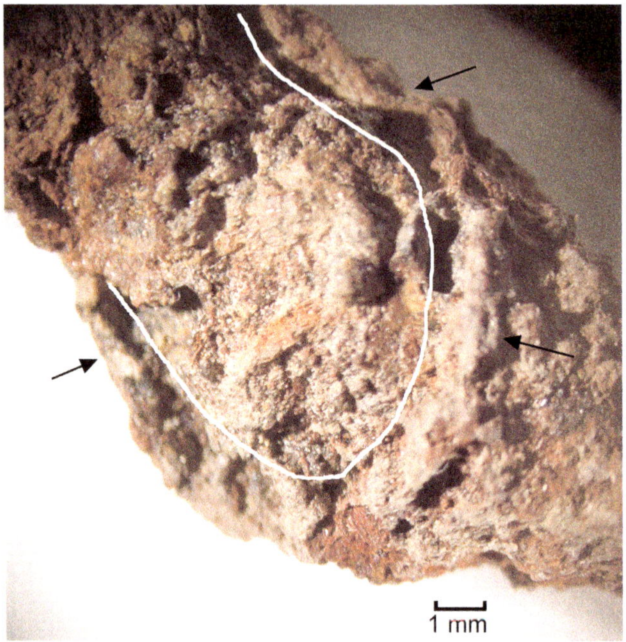

FIGURE 12.6 Detail of finished edge of textile on **10/37** (SF 046). The white line follows inside edge of textile and arrows indicate the outer surface of textile edge.

FIGURE 12.7 Detail of two-ply thread found on iron pin (**10/37**)

FIGURE 12.8 V-shaped elements found toward center of pin (**10/37**). It is not clear whether these represent a 2/1 twill weave. This weave pattern is different than what is found on other areas of the pin (**Fig. 12.5**) and may suggest that pseudomorphs from two different textiles are preserved or that one textile was made using different weaving techniques.

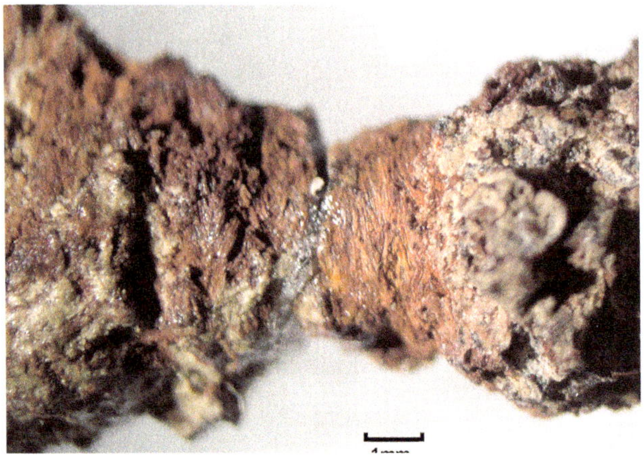

FIGURE 12.9 Fibers that appear to be wrapped around the central section of pin (**10/37**). This is also found on several other pins excavated from Lofkënd.

Figure 12.10 Photomicrograph of weave structure preserved near tip of pin (**10/42**). Pattern resembles plain weave with a selvedge preserved.

Figure 12.11 Ovoid- or triangular-shaped elements created by weave structure (**10/44**)

Figure 12.12 Area that appears to be broken or cut edge of textile showing elements in cross-section on iron pin **10/44**

Figure 12.13 Circular knob found on surface near head (left) and impression of diagonal yarns found on one area of knob (right) on pin **10/44**

Figure 12.14 Faced plain weave textile pseudomorph found on **10/45** (SF 226)

Figure 12.15 Triangular void left by crossing elements at tip of pin **10/45** either due to distortion during burial or different weave pattern/technique

Figure 12.16 Plain weave preserved on pin **10/46** (*left*) and image of selvedge (*right*)

Figure 12.17 Pseudomorph of 2/2 twill on iron pin **10/47** (SF 155)

Figure 12.18 V-shaped or plaited elements found on pin SF 155 (**10/47**)

Figure 12.19 Detail of faced plain weave found on bimetallic pin **10/50** (SF 111). The image on the right shows distortion of textile near pin head.

Figure 12.20 Photomicrograph of straight, parallel yarns with series of yarns running across them at 80° angle on iron bead **10/99**

Figure 12.21 Below the two areas of oblique angled yarns on bead 10/99 is what appears to be weave structure similar to that seen on **10/35** (SF 390)

Figure 12.22 Fibula **10/17** with fiber pseudomorphs wrapped around it

Figure 12.23 An example of the somewhat linear and globular corrosion, pictured here on spectacle fibula **10/14** (SF 170), found on some bronze objects. The corrosion pattern resembles a textile, though the exact weave pattern cannot be identified.

Figure 12.24 Image of Z-spun yarn on bronze disk **10/63**

FIGURE 12.25 Cord inside of the bead **10/99** (SF 124)

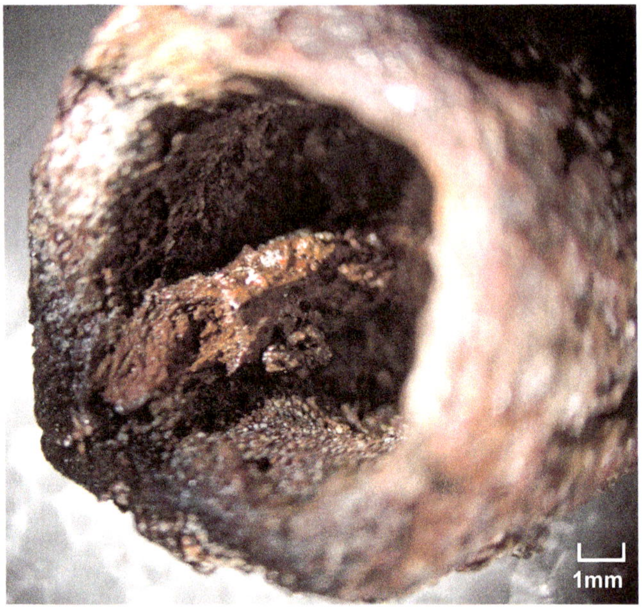

FIGURE 12.26 Cord inside iron bead **10/101** (SF 164a)

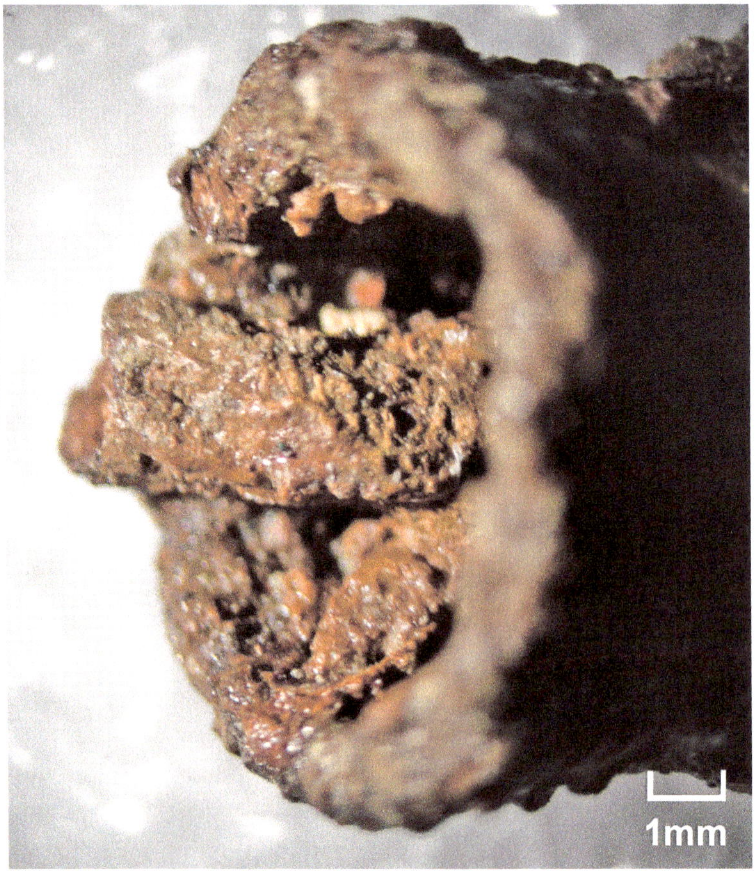

FIGURE 12.27 Cord inside of **10/89** (SF 164b) that extends slightly out of bead); (*right*) detail

Figure 12.28 Strap or cord found on front (*left*) and back (*right*) of headband **10/84**

Figure 12.29 Mass of fibers on edge of headband **10/84**. White, flat fibers are visible in image at right.

Figure 12.30 Wood pseudomorphs on tang or handle of blade **10/121** (SF 019)

Figure 12.31 Wood preserved in socket of spear shaft **10/118** (SF 263)

Figure 12.32 Organic pseudomorphs on **10/133** (SF 384) that resemble wood

Figure 12.33 Fragment containing cell-like structured pseudomorph as well as more striated, linear pseudomorphs

Chapter 13 Illustrations

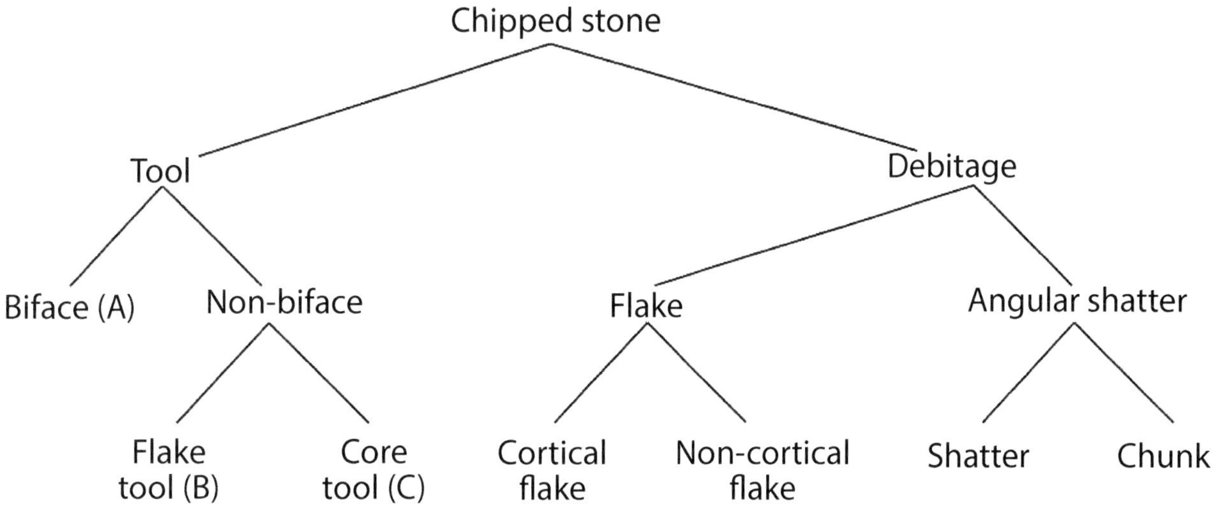

FIGURE 13.1 Schematic morphological lithic typology used in this analysis. Entries in the catalogue are labeled under the main headings of biface (A), flake tool (B), and core tool (C). An additional heading (D) was designated to separate a small collection of Middle Paleolithic flakes and flake tools. (Modified from Andrefsky 2005: fig. 4.7).

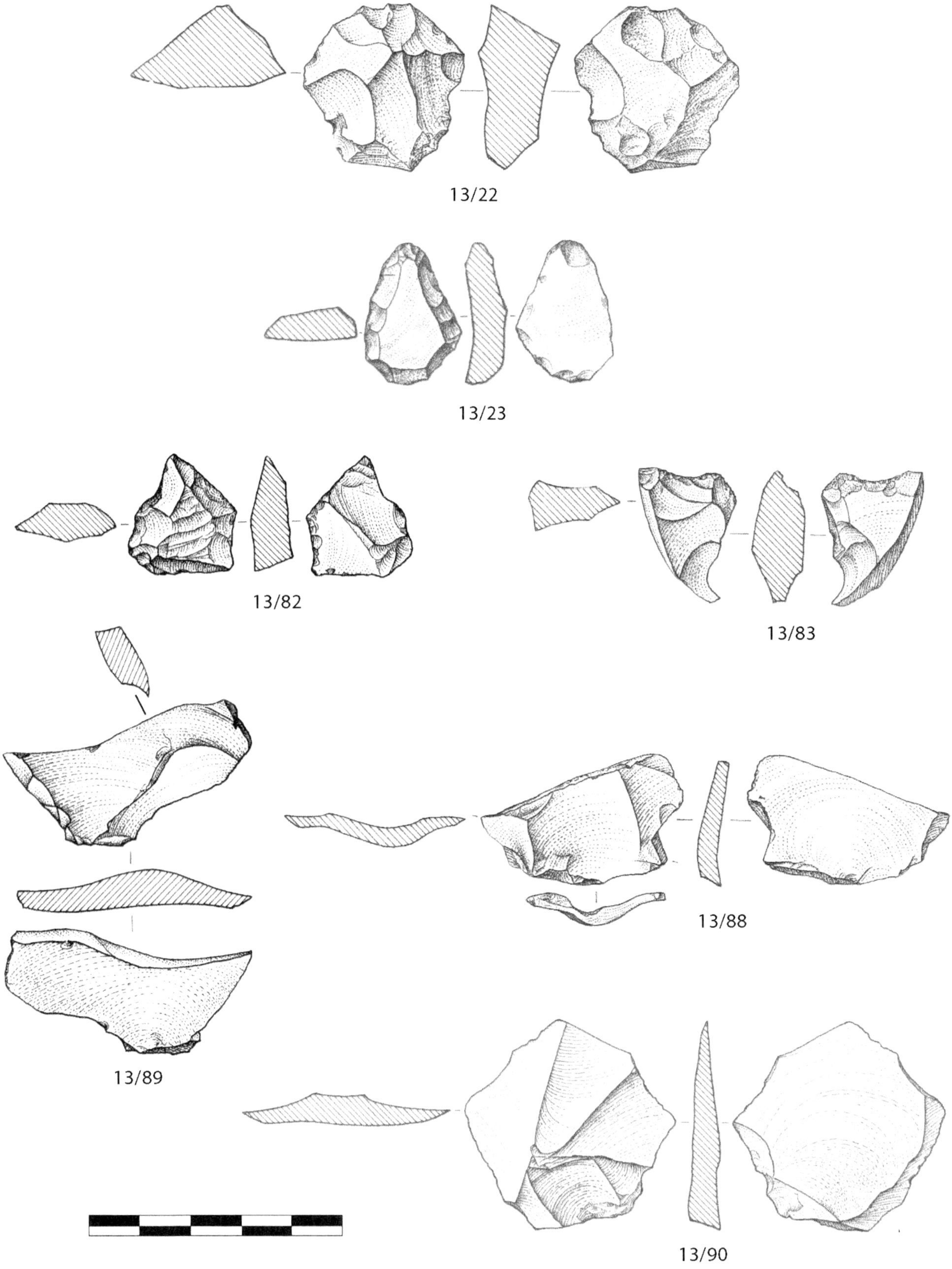

FIGURE 13.2 Middle Paleolithic, Upper Paleolithic, and Mesolithic artifacts (**13/22**, **13/23**, **13/82**, **13/83**, **13/88**, **13/89**, **13/90**)

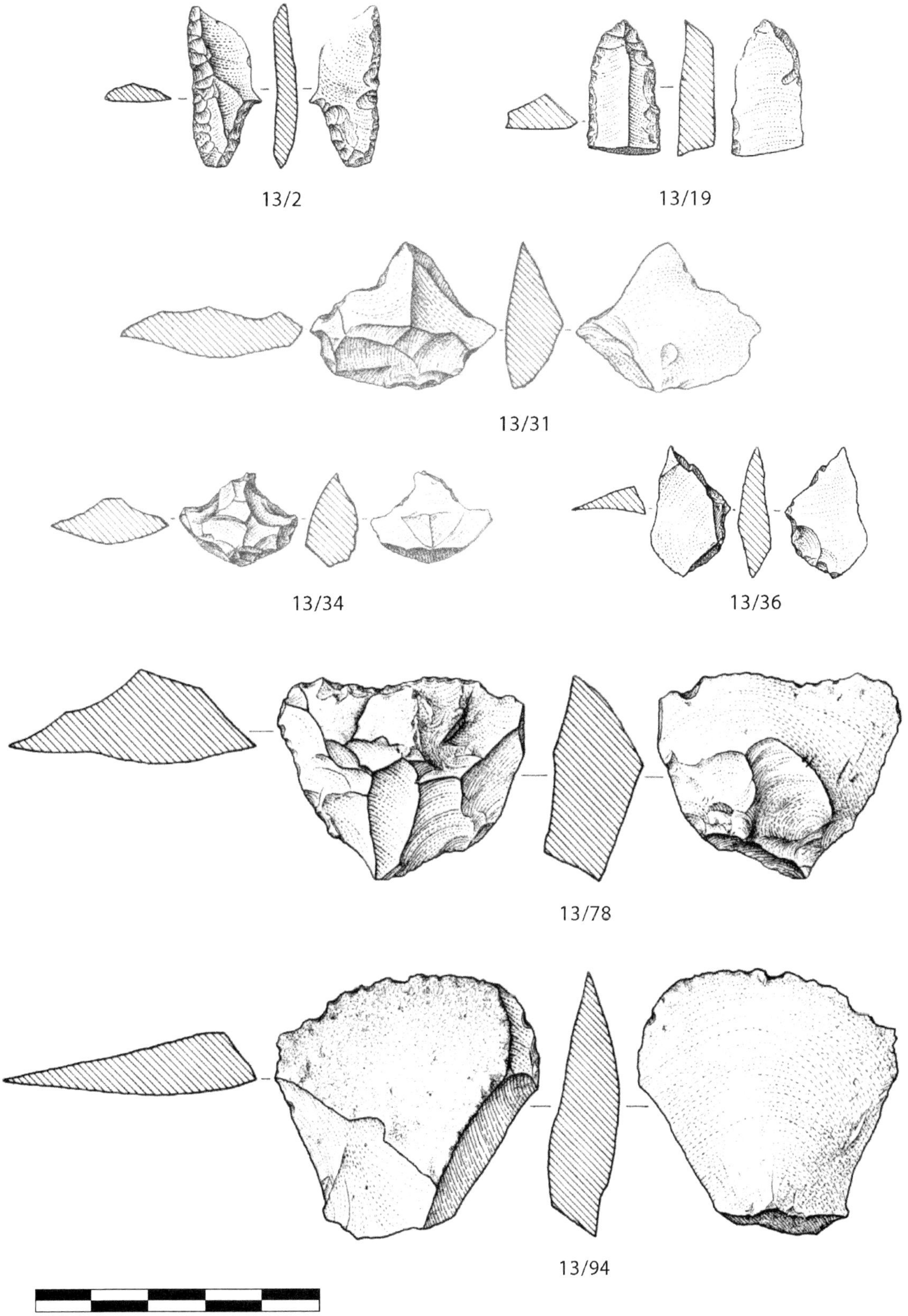

FIGURE 13.3 Middle Paleolithic, Upper Paleolithic, and Mesolithic artifacts (**13/2, 13/19, 13/31, 13/34, 13/36, 13/78, 13/94**)

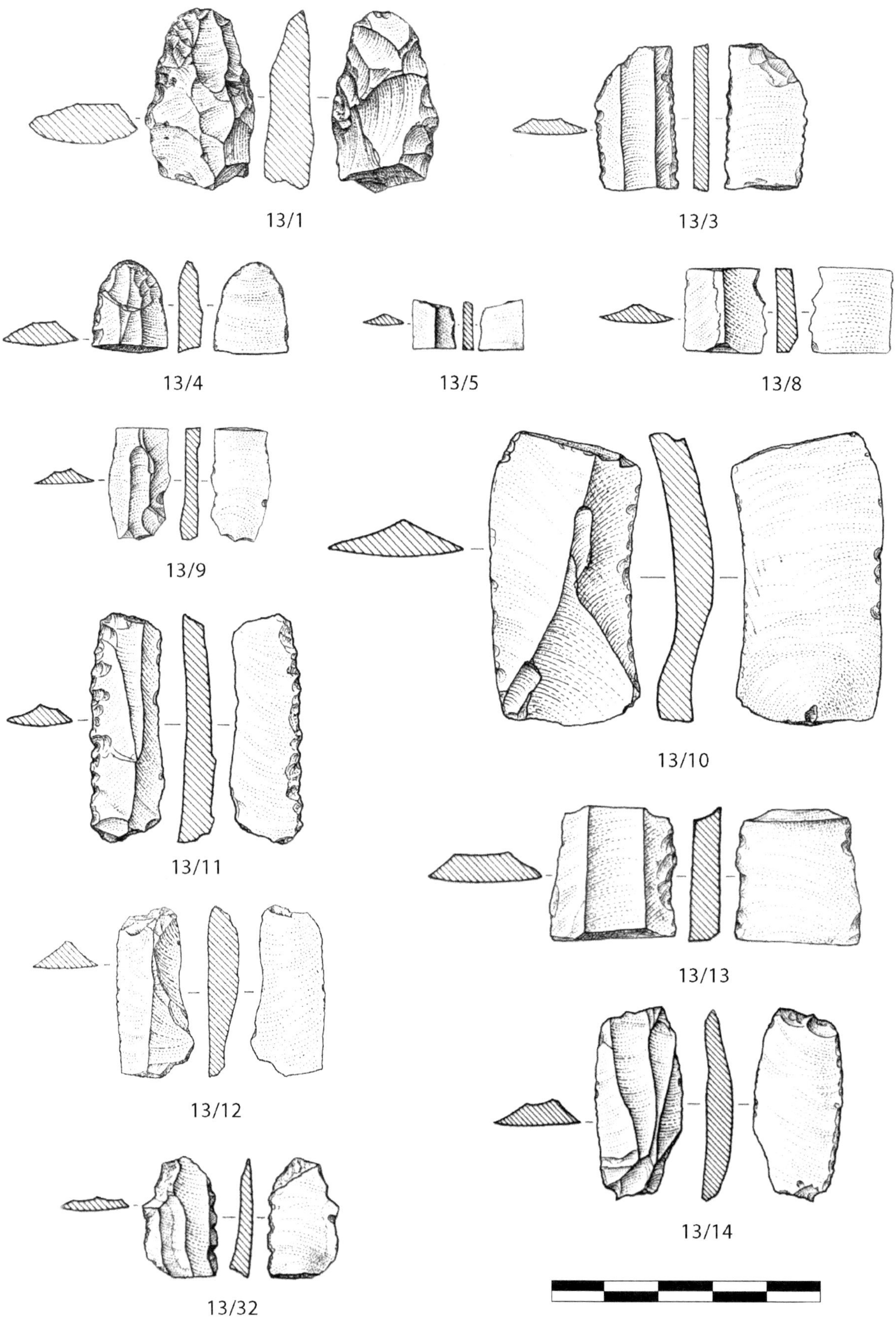

FIGURE 13.4 Neolithic/Bronze Age artifacts (**13/1**, **13/3**, **13/4**, **13/5**, **13/8**, **13/9**, **13/10**, **13/11**, **13/12**, **13/13**, **13/14**, **13/32**)

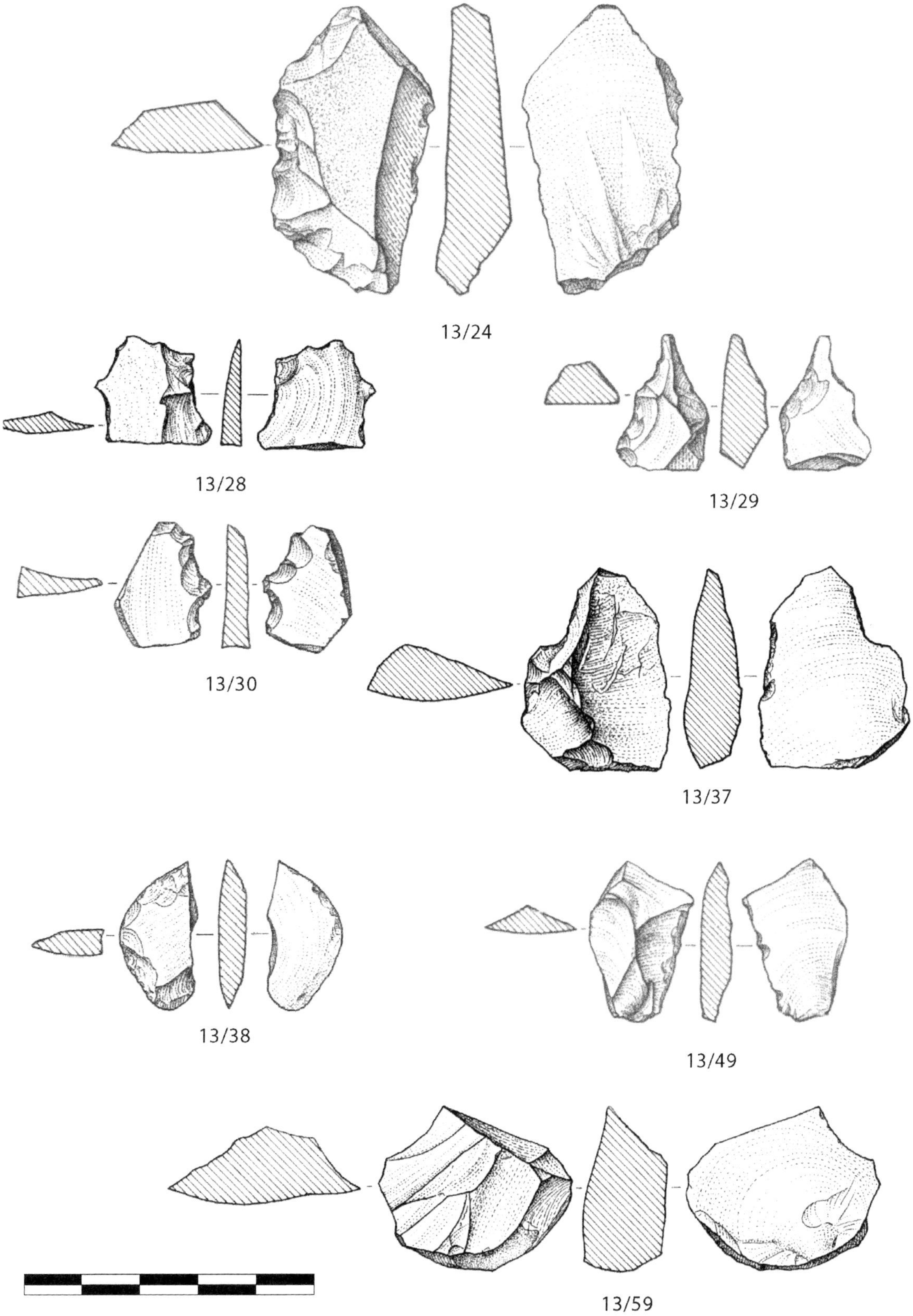

Figure 13.5 Non-diagnostic artifacts (**13/24**, **13/28**, **13/29**, **13/30**, **13/37**, **13/38**, **13/49**, **13/59**)

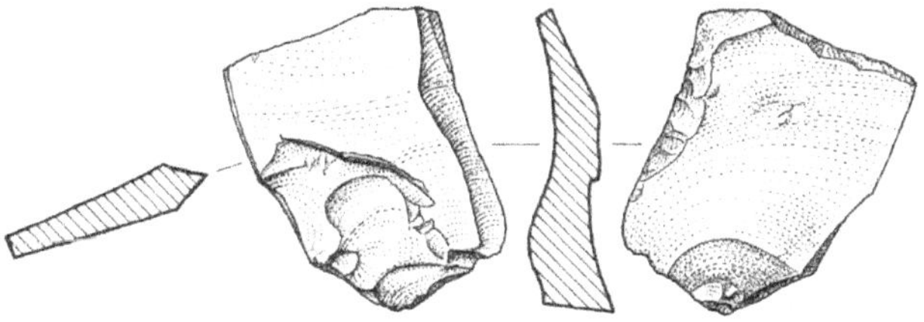

13/21

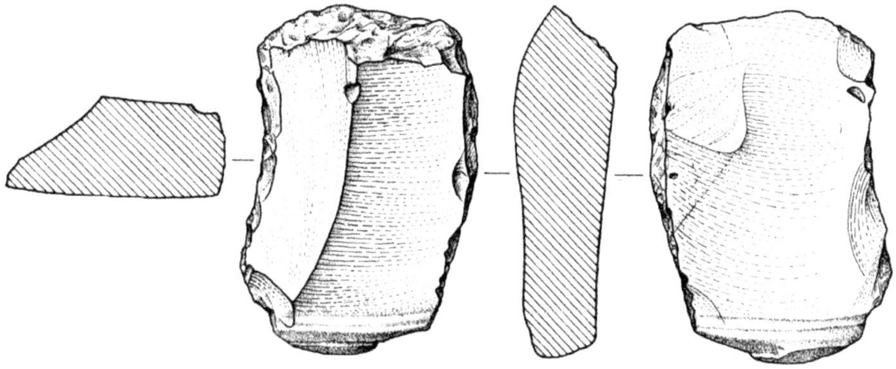

13/25

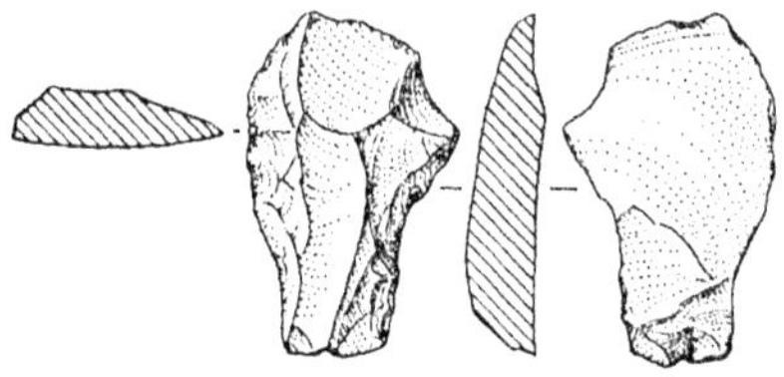

13/35

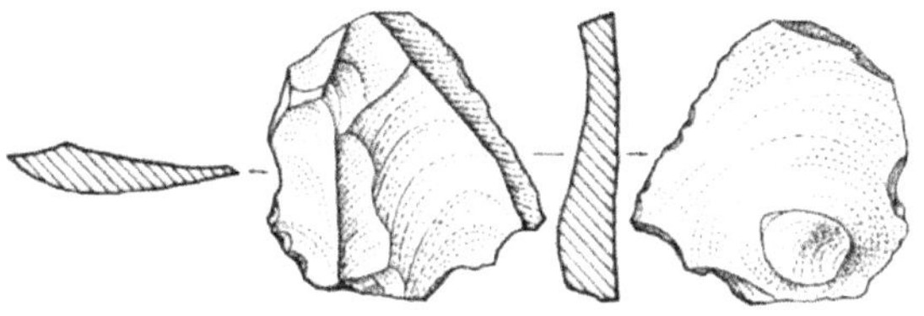

13/47

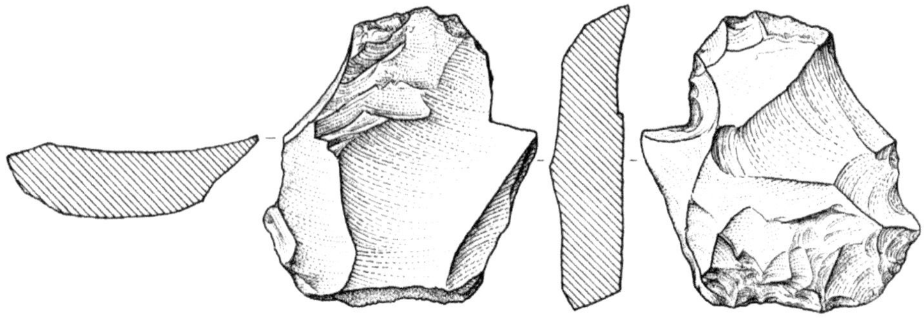

3/79

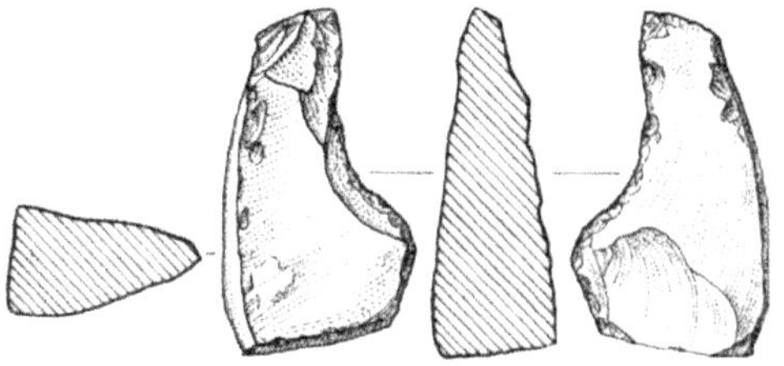

13/86

FIGURE 13.6 Non-diagnostic artifacts (**13/21**, **13/25**, **13/35**, **13/47**, **13/79**, **13/86**)

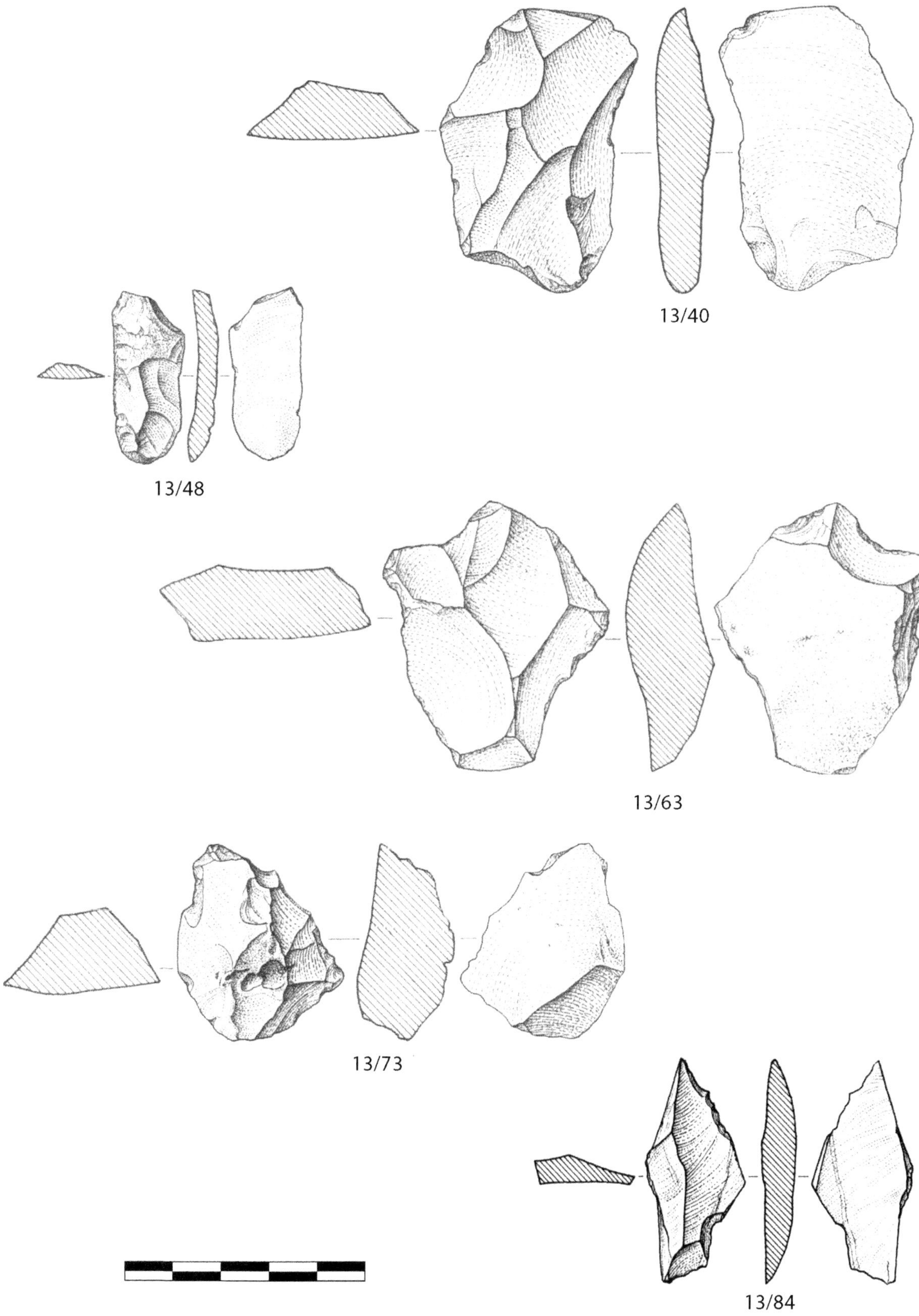

Figure 13.7 Non-diagnostic artifacts (**13/40**, **13/48**, **13/63**, **13/73**, **13/84**)

Figure 13.8 Paleolithic, Mesolithic, Neolithic, and Bronze Age artifacts (**13/1**, **13/3**, **13/6**, **13/13**, **13/14**, **13/22**, **13/31**, **13/32**, **13/34**, **13/92**)

FIGURE 13.9 Neolithic/Bronze Age and non-diagnostic artifacts (**13/25**, **13/29**, **13/30**, **13/51**, **13/52**, **13/73**, **13/74**)

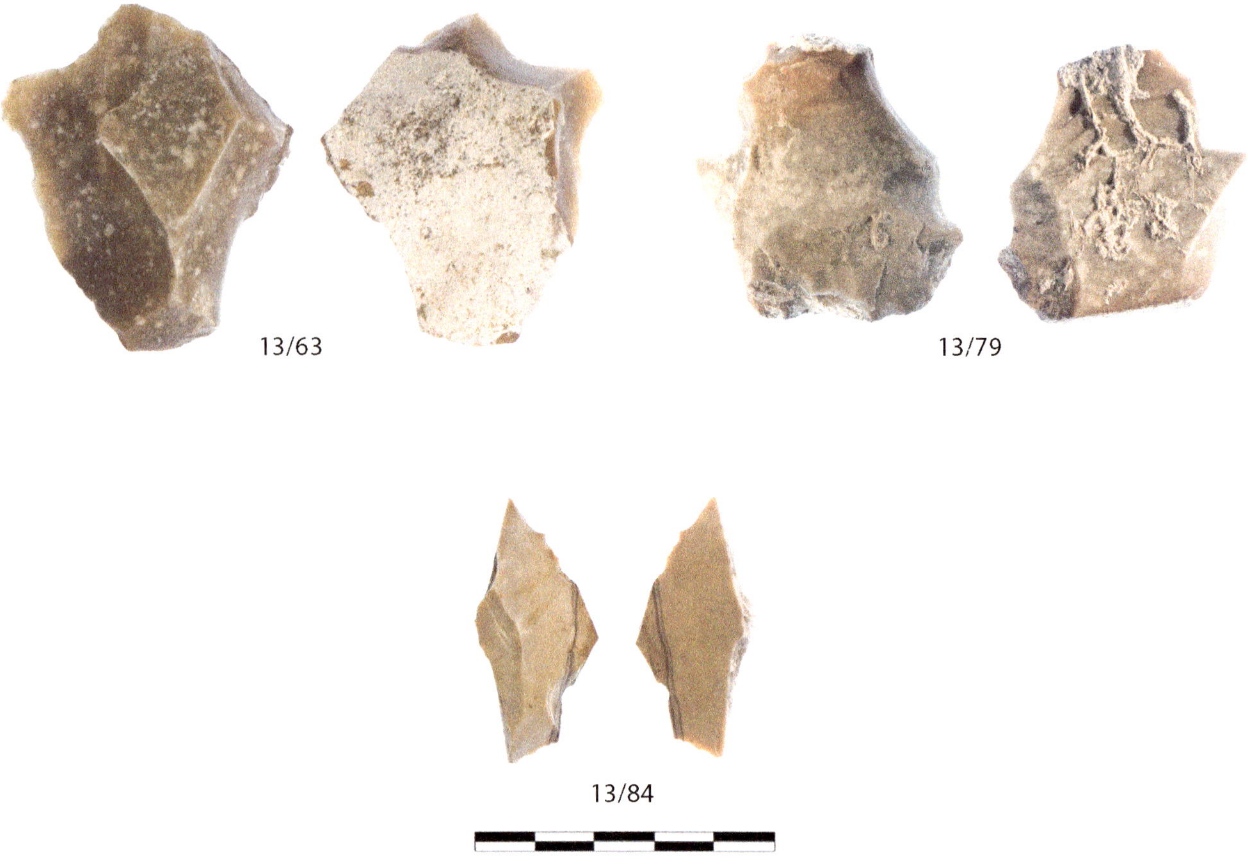

Figure 13.10 Non-diagnostic artifacts (**13/63**, **13/79**, **13/84**)

Chapter 14 Illustrations

FIGURE 14.1 Model of modern farmstead constructed from wattle and daub, Vlora Museum (photo Sarah Morris)

FIGURE 14.2 Ruin of modern wattle-and-daub house in the Lofkënd survey area (photo Jamie Aprile)

Figure 14.3 Modern wattle-and-daub hut in the Lofkënd survey area (photo Jamie Aprile)

a 

b 

Figure 14.4 (a) Haystack, Mallakastër Hills, near Lofkënd, with a capping of daub (photo Sarah Morris); (b) detail of haystack near Lofkënd tumulus with a capping of daub (photo Ian Coyle)

Figure 14.5 Pieces of daub from Neolithic Dimini, Greece, Athens, National Archaeological Museum (photo John Papadopoulos)

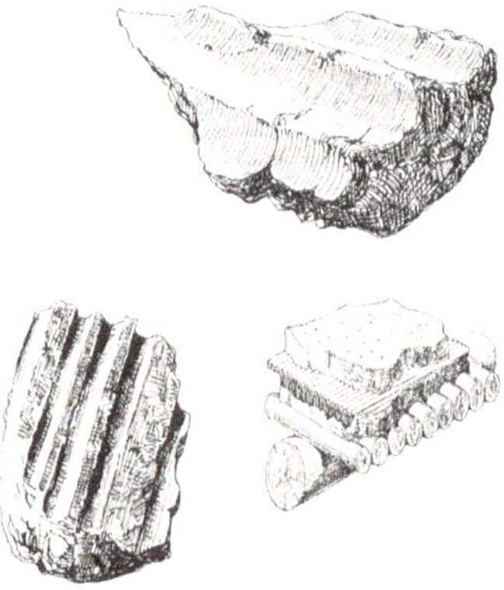

Figure 14.6 Pieces of daub and reconstruction of Middle Helladic roofing system from prehistoric Eutresis, Greece (after Goldman 1931:63, fig. 71)

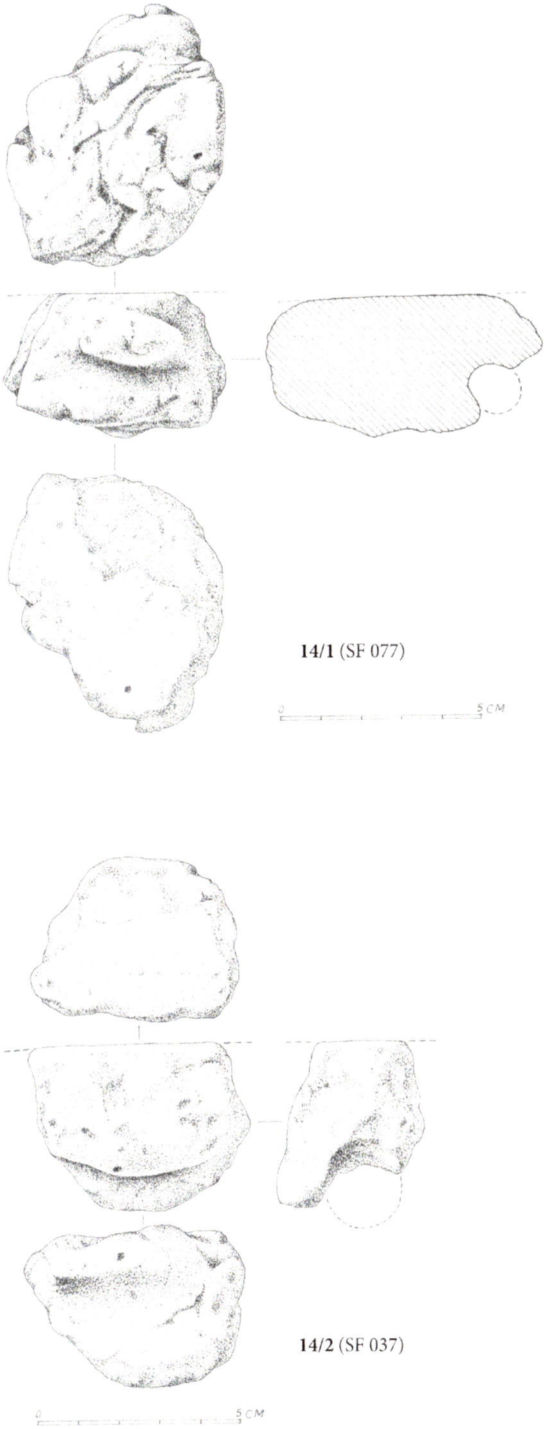

FIGURE 14.7 Drawings of daub from the Lofkënd tumulus (**14/1**, **14/2**)

FIGURE 14.8 Two views each of **14/1** and **14/2**

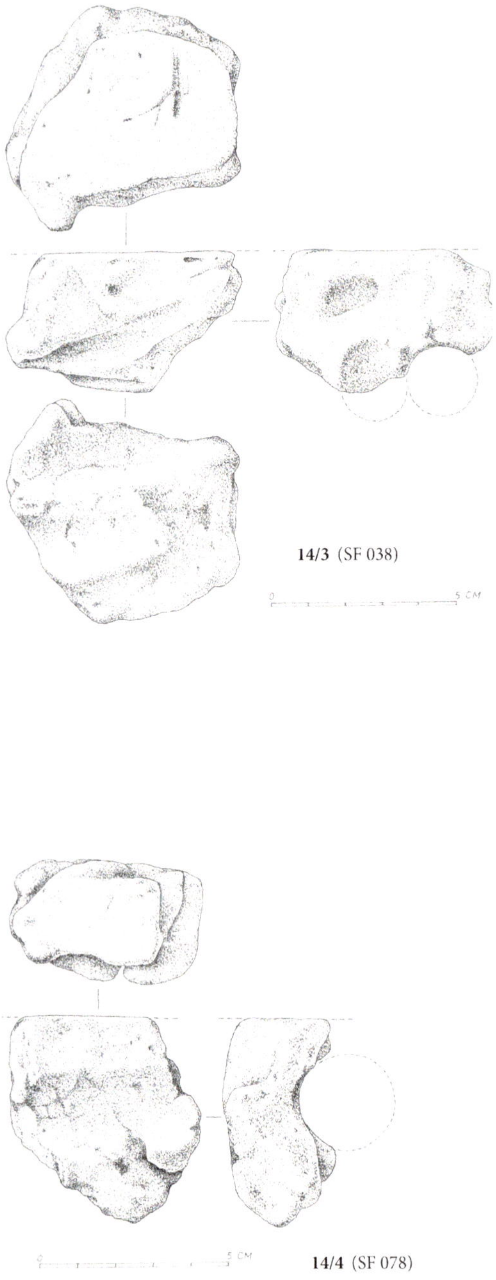

FIGURE 14.9 Multiple views of **14/3**, **14/4**

FIGURE 14.10 Multiple views of **14/3**, **14/4**

14/5 (SF 79) **14/6** (SF 128)

14/7 (SF 80)

FIGURE 14.11 Two views each of **14/5**, **14/6**, **14/7**

14/8 (SF 446) **14/9** (SF 445)

14/10 (SF 135) **14/11** (SF 39)

FIGURE 14.12 Two views each of **14/8**, **14/9**, **14/10**, **14/11**

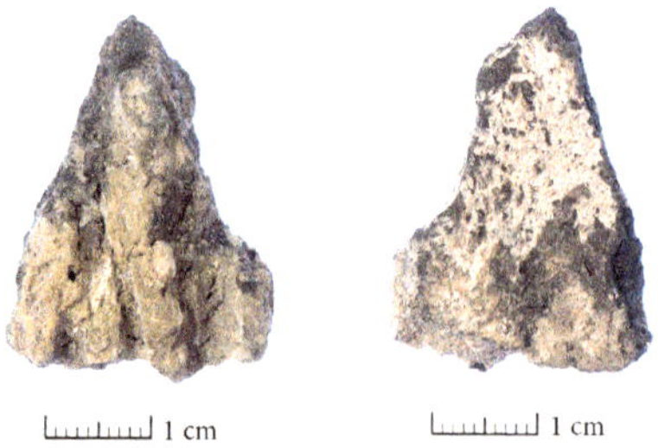

14/12 (SF 218)

14/13 (SF 127)

14/14 (SF 387)

14/15 (SF 310)

Figure 14.13 Two views each of **14/12, 14/13**, **14/14**, and **14/15**

14/16 (SF 356)

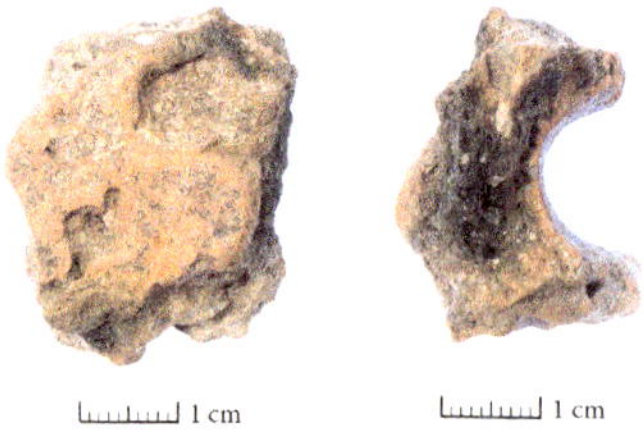

14/17 (SF 173)

Figure 14.14 Two views each of **14/16** and **14/17**

Chapter 15 Illustrations

FIGURE 15.1 Modern poster (1930s), Italian bitumen mining company (image courtesy Oliver Gilkes)

FIGURE 15.2 Roman silver coin of Apollonia (British Museum 1854, 1004.1AN1526788)

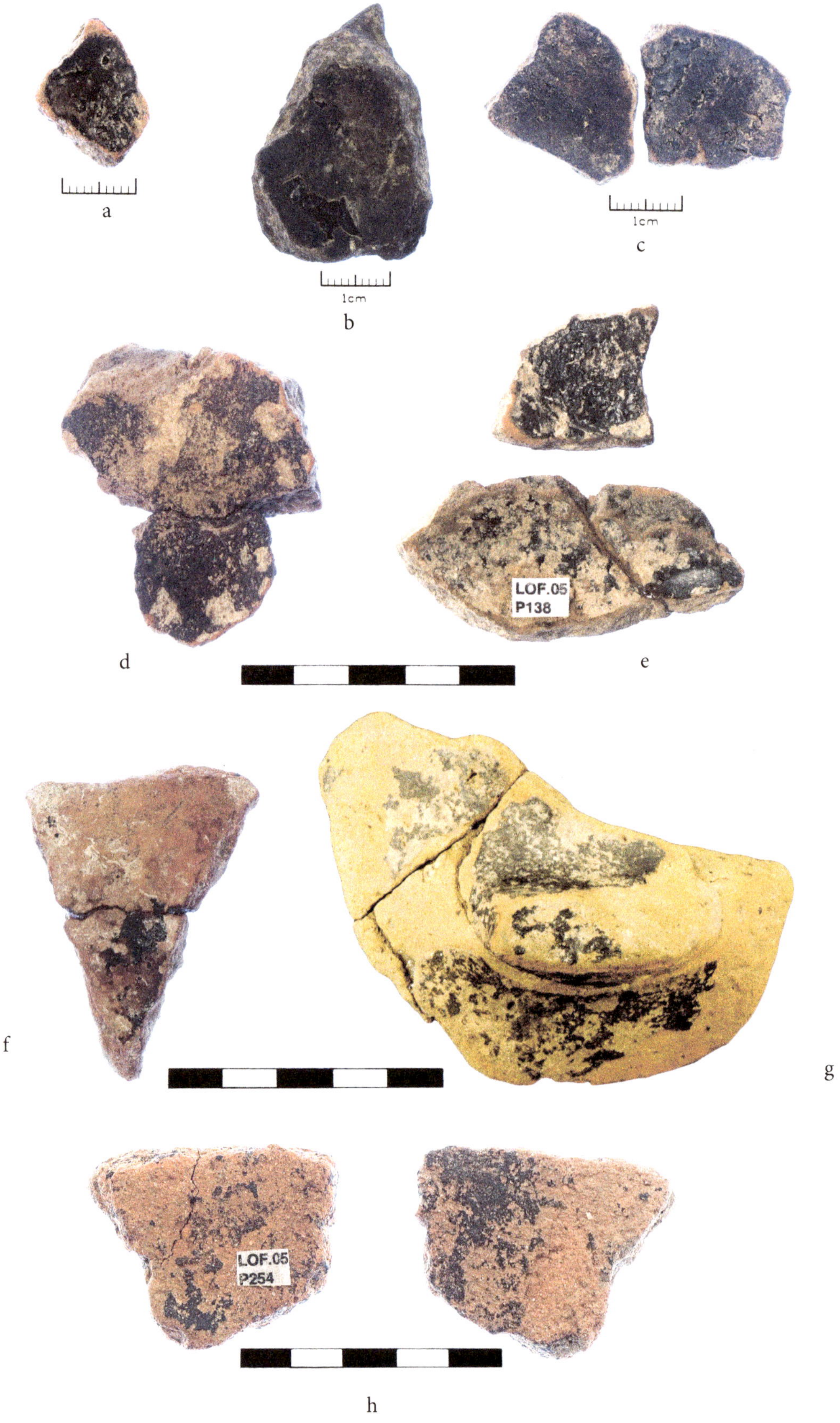

FIGURE 15.3 Sherds with bitumen: (a) **15/1**; (b) **15/2**; (c) **15/3**; (d) **15/6**; (e) **15/7**; (f) **15/8**; (g) **15/10**; (h) **15/11**

Figure 15.4 Large coarse vessel with bitumen, found in situ (**15/12** [SF 283])

FIGURE 15.5 Sherds with bitumen: (a) **15/13**; (b) **15/14**; (c) 15/**15**; (d) **15/16**; (e–f) **15/17**; (g) **15/18**; (h–i) **15/20**)

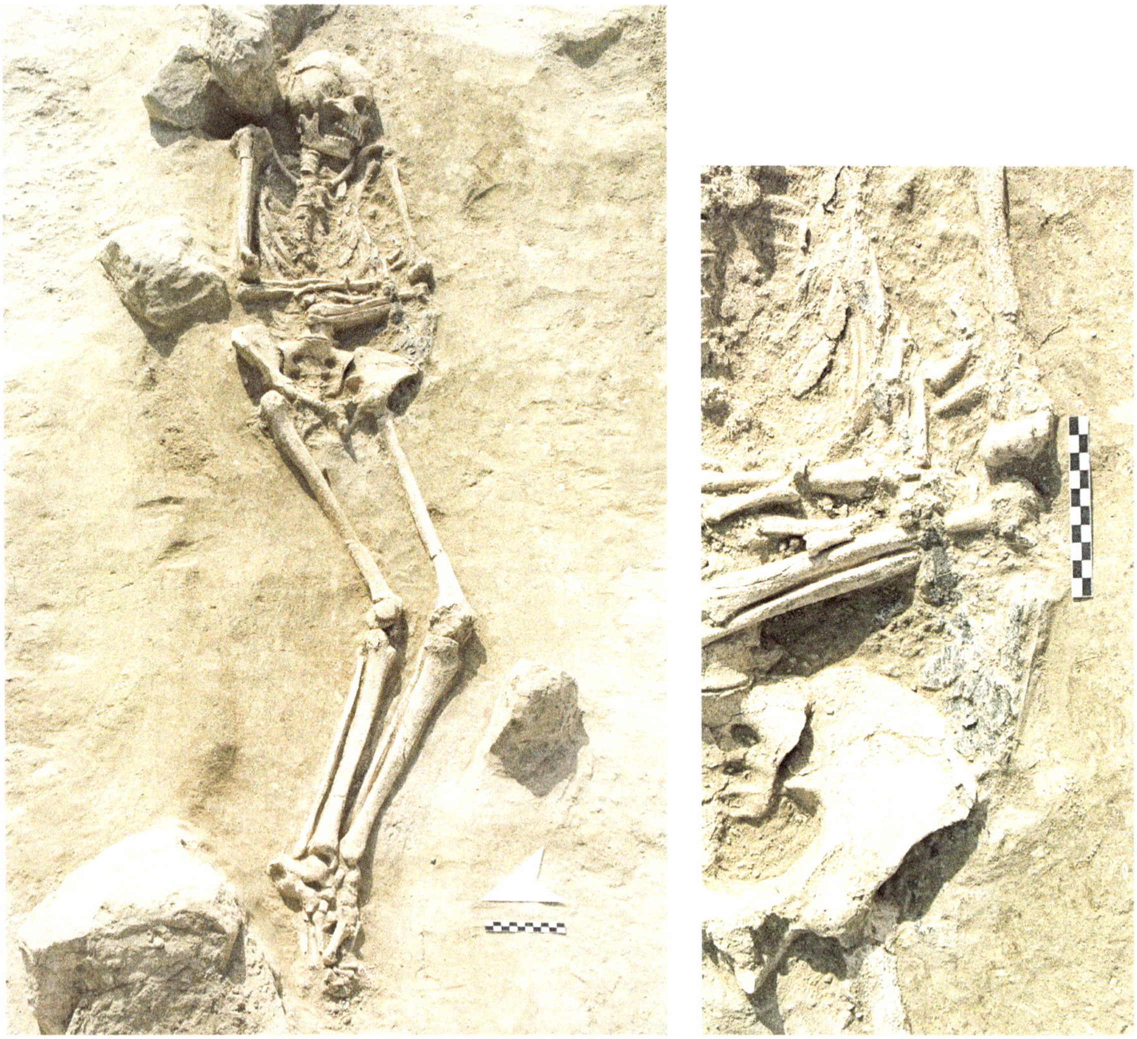

FIGURE 15.6 Tomb XX (50): (*left*) pelvis, ribs, and lower arms of adult female (20–25 years), overlaid with irregular strip of bitumen; (*right*) detail

Chapter 16 Illustrations

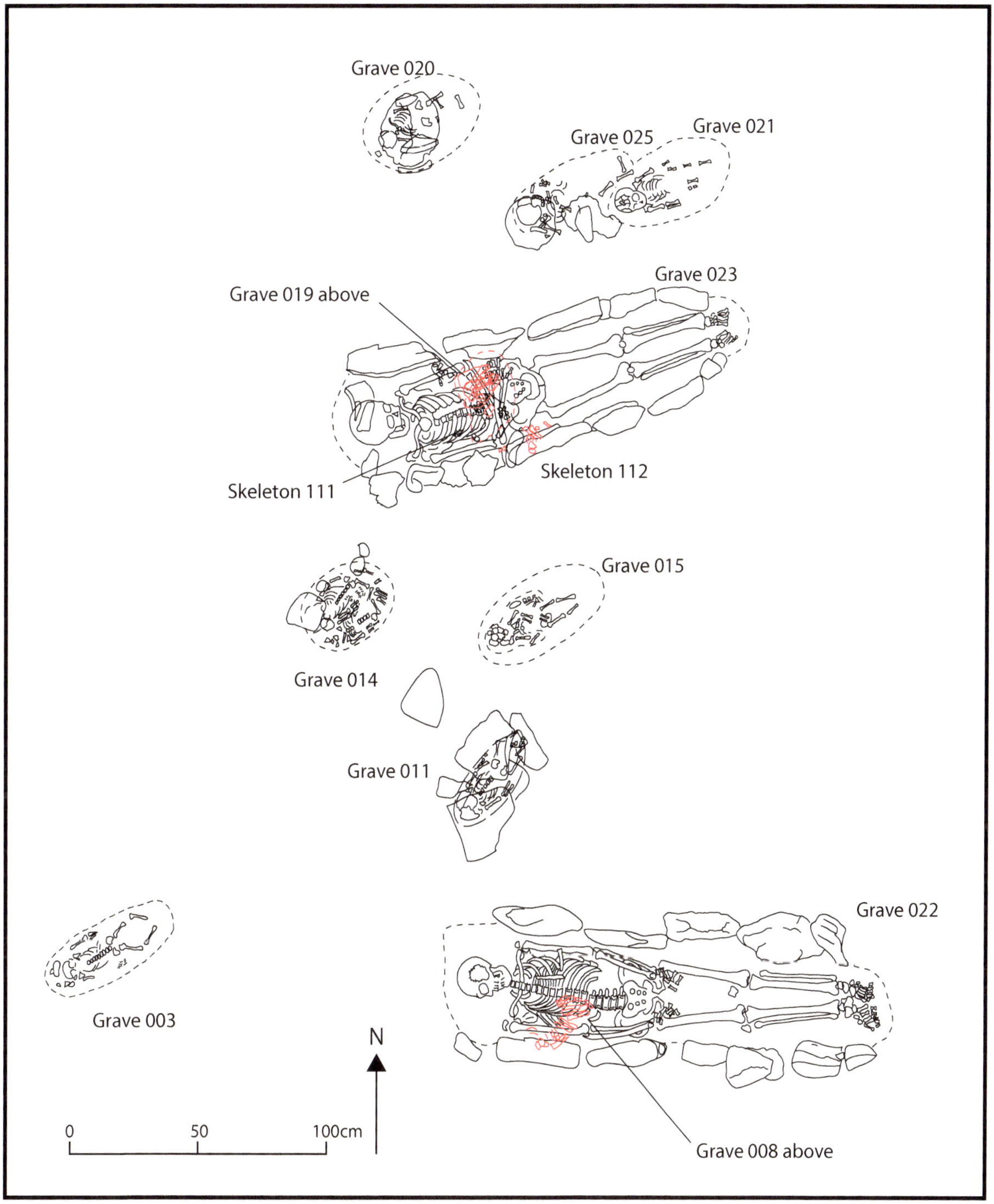

FIGURE 16.1 The modern burials in Trench 1, including stone cist graves (Tombs LXXXVI and XCII [Graves 22 and 23]) and associated animal burials, in red (Tombs LXXXVII and XCIII [Graves 8 and 19, Skeleton 112]). The remaining burials are human infants (figure modified from original illustrations by Max Farrar).

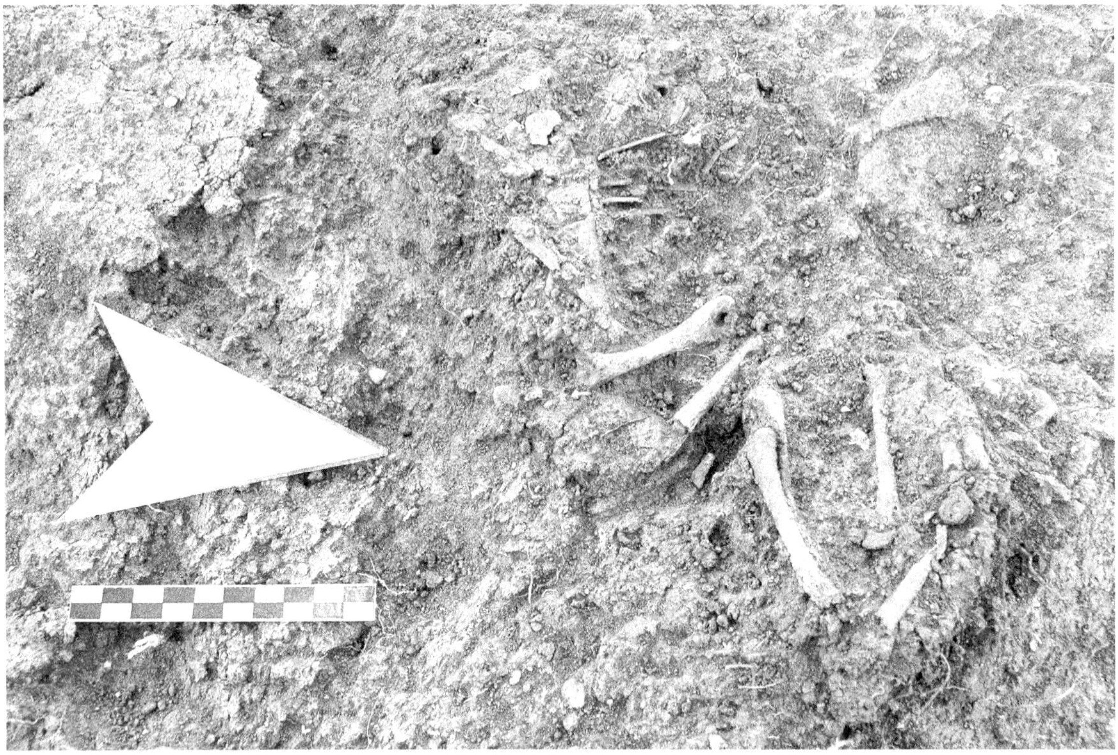

FIGURE 16.2 Tomb LXXXVII (Grave 8) as excavated

FIGURE 16.3 Skeleton 111 (human) and 112 (neonatal sheep) as excavated. Skeleton 112 is marked by the white circle; the remaining bones are human.

Figure 16.4 Tomb XCIII (Grave 19) as excavated

Figure 16.5 Tomb XCIII (Grave 19) cranium after removal of leg bones

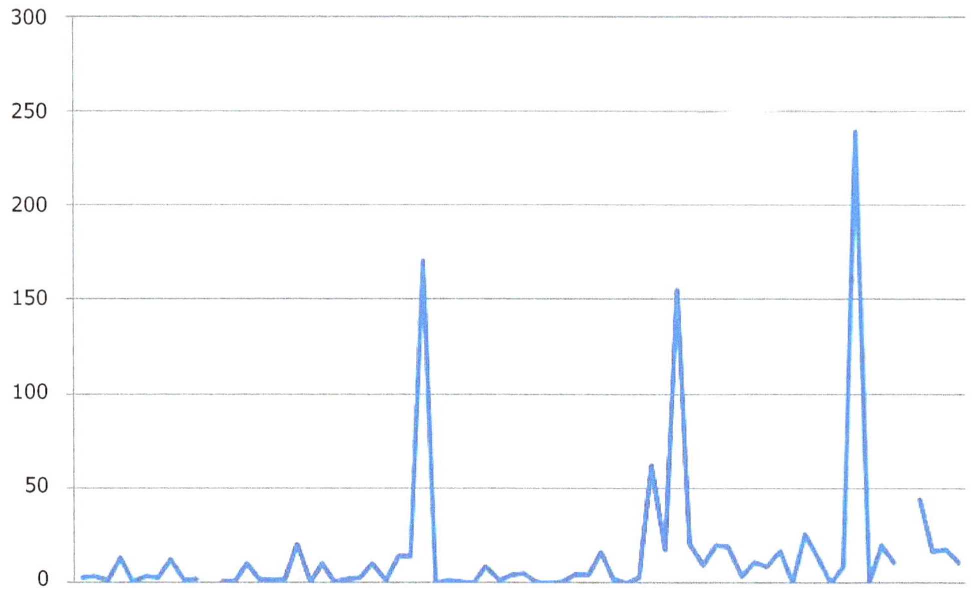

FIGURE 16.6 Distribution of *Helicella itala* in Tombs XCVII, XXXIX, XXXI. From left to right, the three main peaks correspond to Tombs XCVII (39), XXXIX (66), and XXXI (86). Left column = weight in grams. Most graves contained less than 50 g of *Helicella itala*.

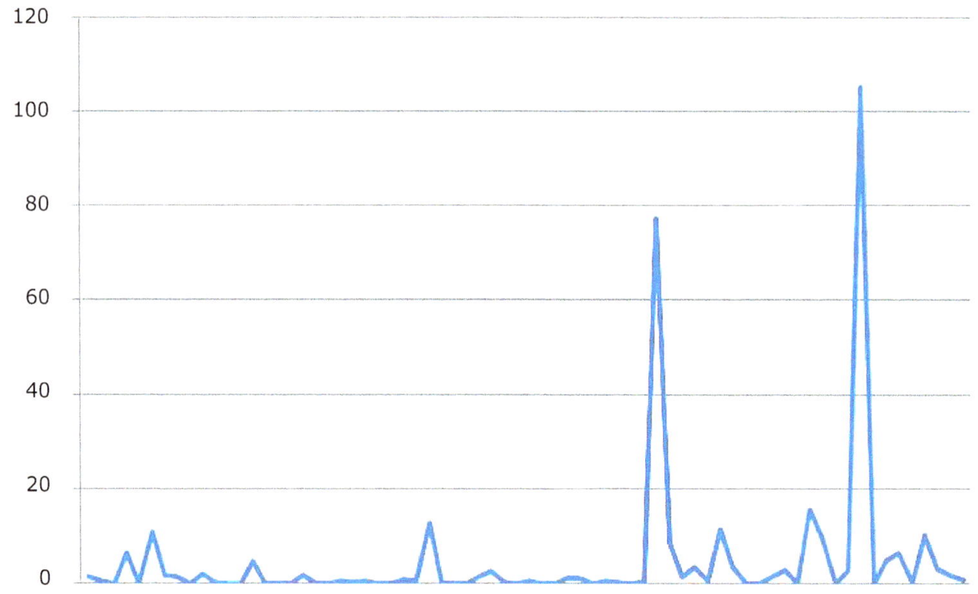

FIGURE 16.7 Distribution of *Pomatias elegans*. The two main peaks, from left to right, correspond to Tombs I (64) and XXXI (86). Most graves contained less than 20 g of *Pomatias elegans*.

FIGURE 16.8 *Pomatias elegans*, Trench 2, Unit 292

FIGURE 16.9 *Helicella itala*, Trench 5, Unit 548

FIGURE 16.10 *Monacha cartusiana*

FIGURE 16.11 *Trochoidea pyramidata*, Trench 4, Unit 201, tumulus edge

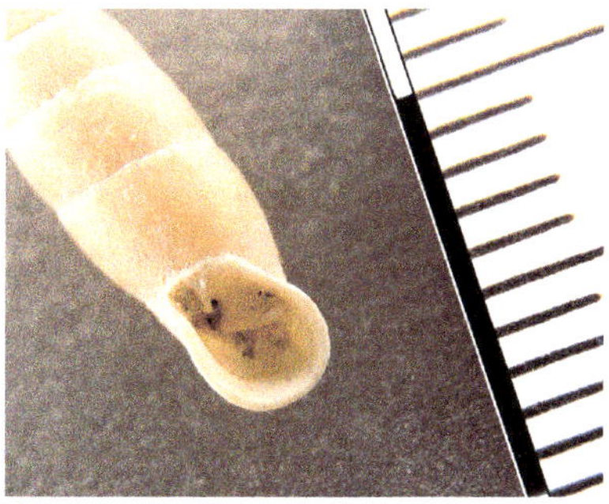

FIGURE 16.12 *Siciliaria stigmatica*, Trench 1, Unit 9

FIGURE 16.13 *Mastus pupa*, Trench 4, Unit 201, tumulus edge

Figure 16.14 *Poirctia delesserti*, Trench 1, Unit 9

Figure 16.15 *Cerastoderma edule*, Trench 4, Unit 115

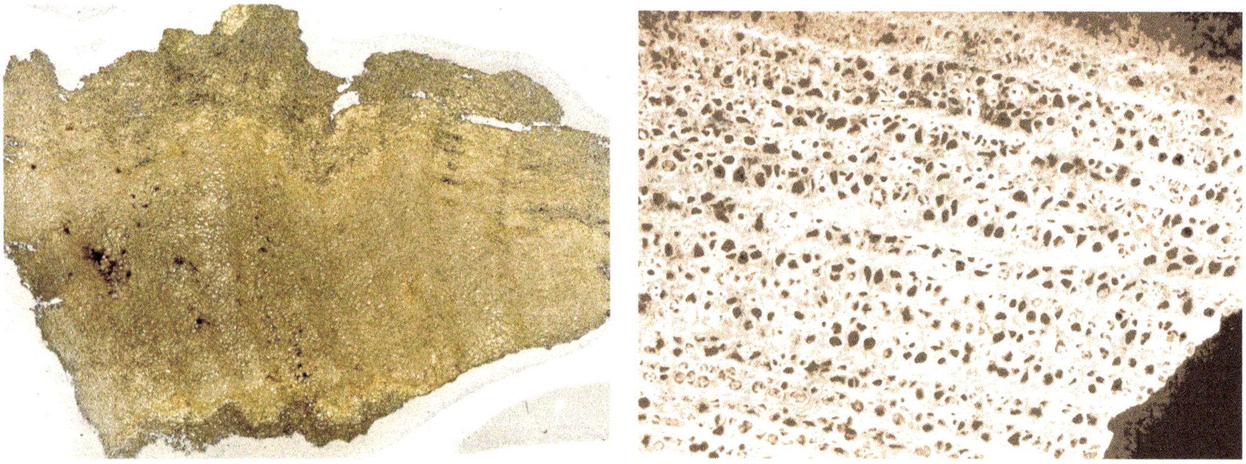

a b

Figure 16.16 Sections of fossil wood from Lofkënd

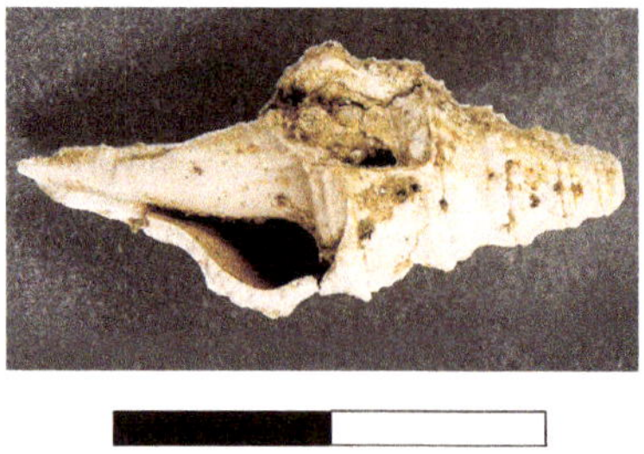

Figure 16.17 *Pleurotoma* sp. (Neogene)

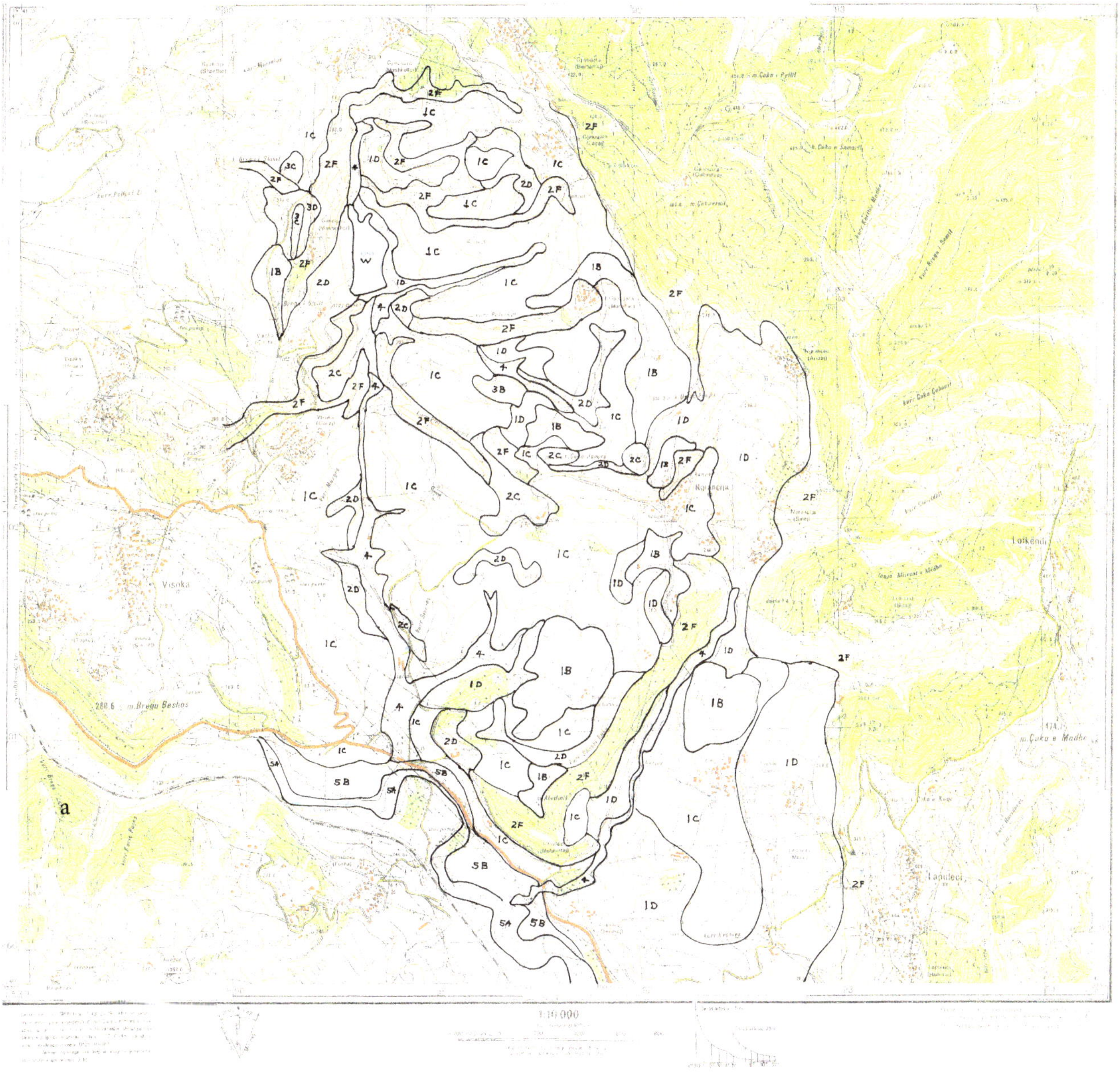

FIGURE 16.18 Regional soils map of the Lofkënd study area

Major units

1 Shale-derived soils occurring on a variety of slopes. Soils are well developed, clayey, and usually calcareous (has free lime). Cracking of surfaces during dry periods is very common. These soils are classified as Vertisols or at least in Vertic subgroup.
1B Gently sloping 3–8% **1C** Strongly sloping 8–16% **1D** Steeply sloping 16–30%

2 Sandstone-derived soils occurring on mostly steep and very steeply sloping areas. These soils are highly erodible, and in a number of locations the soils have very little vegetation because of the eroded condition and steep slopes. The sandstone bedrock will occasionally have lithified areas, but they are generally <0.5 m in thickness. Only the more stable slopes will show evidence of B horizon (subsoil) development. Soils have minimal productivity for agriculture.
2C Strongly sloping 8–16% **2D** Steeply sloping 16–30% **2F** Very steeply sloping 30–45%

3 Reddish soils developed from sandy-clayey deposits. Soils on the gently sloping areas have a thick argillic horizon (clayey Bt) with reddish brown colors (5 YR 4/6). The more steeply sloping areas were severely eroded and had shallow profiles.
3B Gently sloping 3–8% **3C** Strongly sloping 8–16% **3D** Steeply sloping 16–30%

4 Soils developed from localized alluvium in narrow upland valleys. These soils are generally nearly level to gently sloping.

5 Soils formed from alluvium along the Gjanicë River.
5A Recent alluvium, nearly level 0–3% **5B** Older alluvium with some horizonation, nearly level to gently sloping 0–8%

FIGURE 16.19 Landscape delineation of soil mapping unit 1B located north of the Lofkënd site

FIGURE 16.20 Profile of a shale-derived soil. Note the well-developed structure and slickenside structural units. The profile is located about 400 m southeast of the Lofkënd site.

FIGURE 16.21 Cracks occurring in the shale soil during dry season. Ruler is 15 cm in length.

FIGURE 16.22 Photograph of Unit 2 on the eastern side of the study area.
The mapping unit would be 2F, with very steeply sloping landscapes.

Figure 16.23 Photograph of Unit 2 that is mapped as 2D. This is a typical exposure of mapping unit 2D.

Figure 16.24 Landscape of Unit 1 showing the amount of erosion typically noted on this mapping unit. There is normally little vegetative cover associated with the mapping unit.

Figure 16.25 Reddish soils occurring on top of the hill, but with erosion around the edges, resulting in soil transport down slope

Figure 16.26 Isolated hill of reddish soils north of Visokë

FIGURE 16.27 Narrow valley with local alluvial soils

FIGURE 16.28 High terrace soil (soil mapping unit 5B) along the Gjanicë River

FIGURE 16.29 Profile S05AL1 at Lofkënd archaeological site. Profile extends from 0 to 130 cm.

FIGURE 16.30 Profile S05AL2 at Lofkënd archaeological site in Albania. Profile extends from 0 to 92 cm.

FIGURE 16.31 Landscape just south of Lofkënd showing unstable landscape with a thin layer of shale overlying sandstone. Arrow shows contact between sandstone and shale.

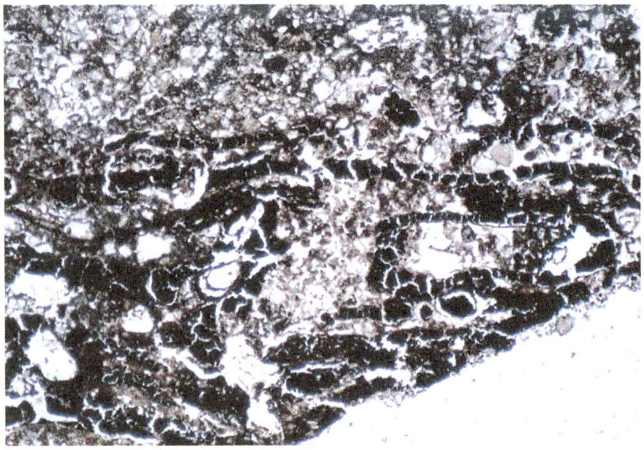

Figure 16.32 Sample 511-2. Scale: width of image 4.68 mm.

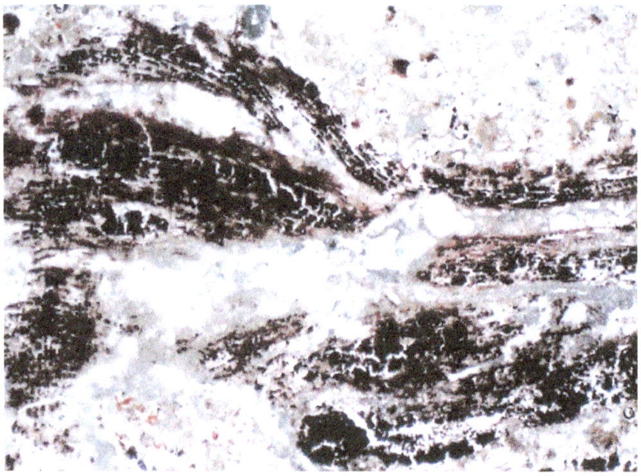

Figure 16.33 Sample 511-1. Scale: width of image 2.34 mm.

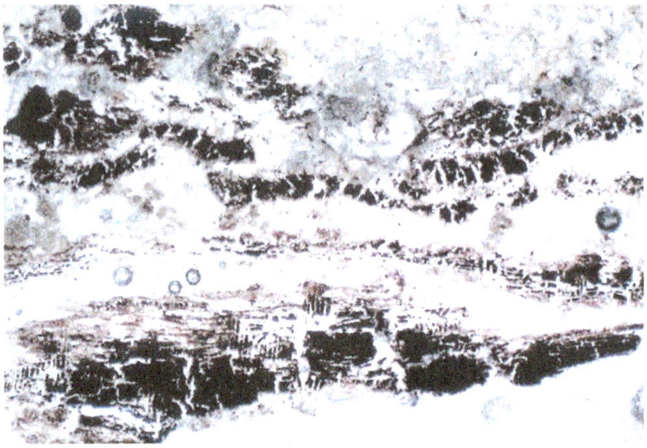

Figure 16.34 Sample 511-1. Scale: width of image 2.34 mm.

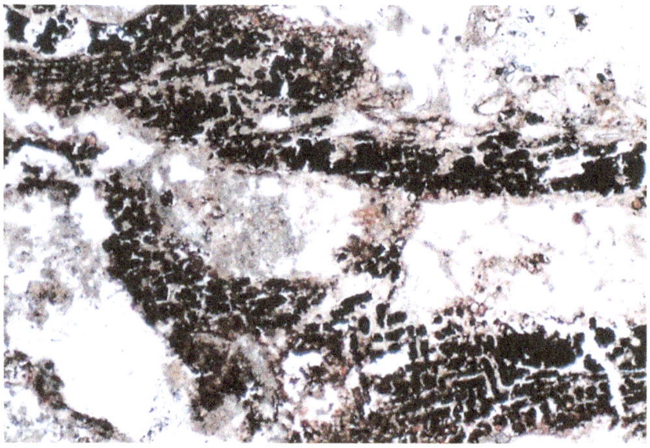

Figure 16.35 Sample 511-2. Scale: width of image 0.95 mm.

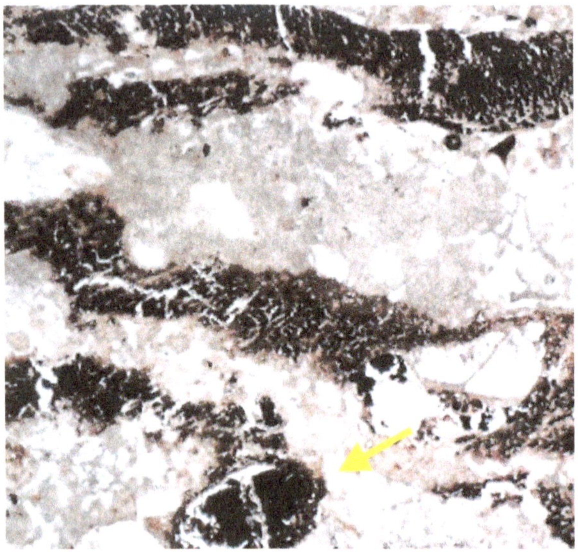

Figure 16.36 Sample 511-2. Scale: width of image 2.34 mm. Yellow arrow indicates probable "fruit" that should be identifiable based on shape and scale.

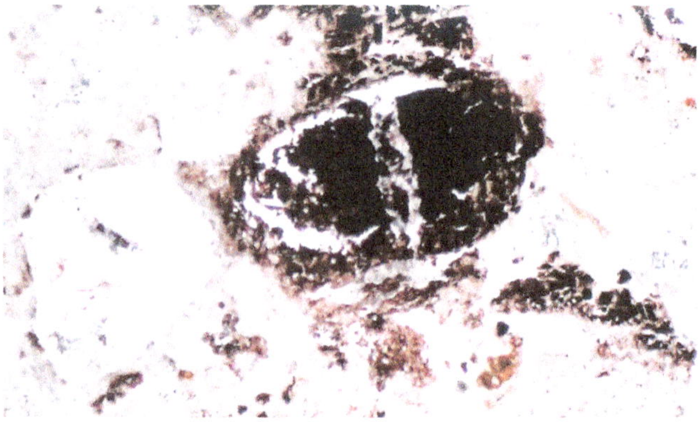

Figure 16.37 Sample 511-2. Scale: width of image 0.95 mm. Detail of probable fruit in **Figure 16.36.**

Chapter 17 Illustrations

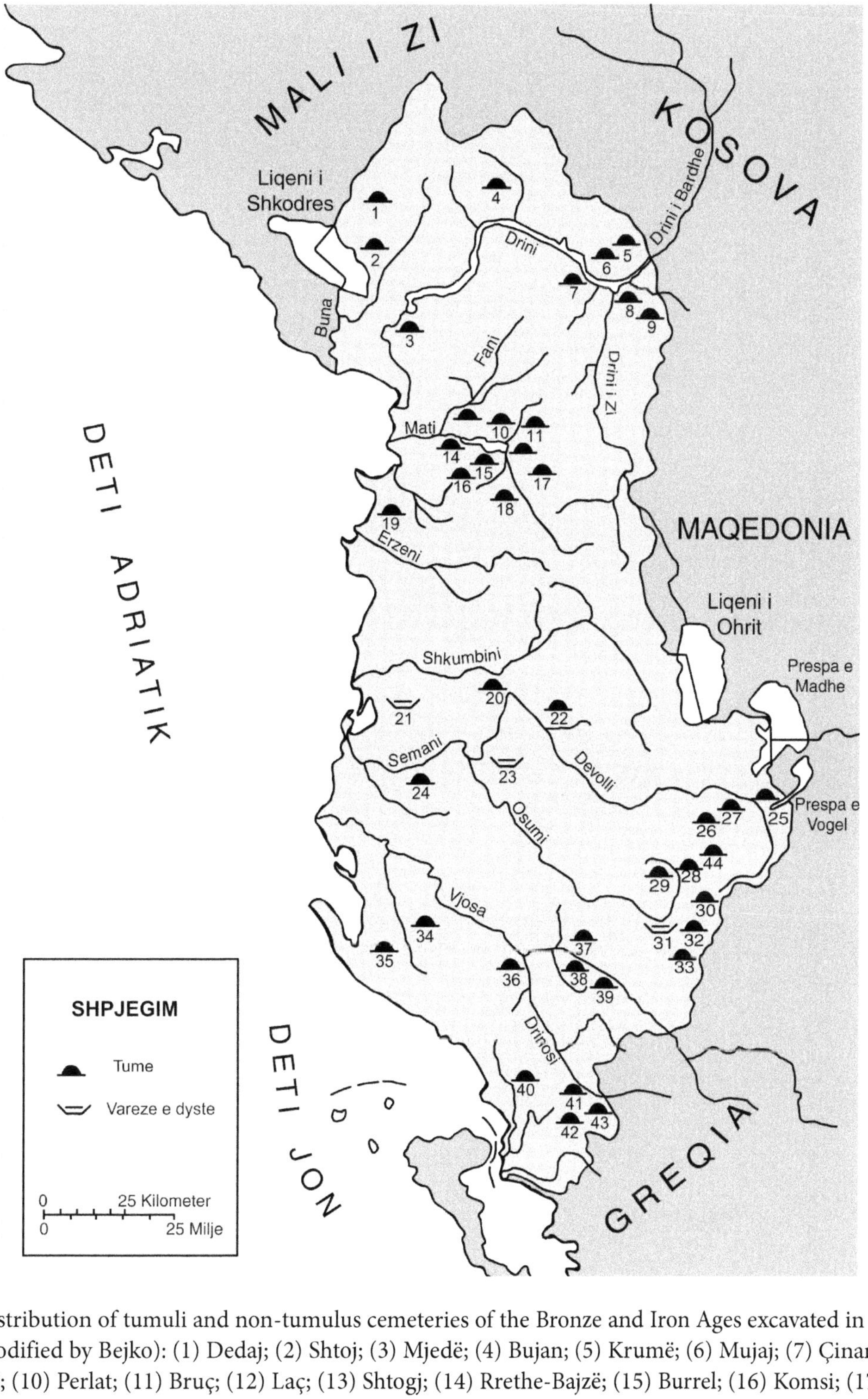

FIGURE 17.1 Distribution of tumuli and non-tumulus cemeteries of the Bronze and Iron Ages excavated in Albania (after Prendi 1988, modified by Bejko): (1) Dedaj; (2) Shtoj; (3) Mjedë; (4) Bujan; (5) Krumë; (6) Mujaj; (7) Çinamak; (8) Bardhoc; (9) Kënetë; (10) Perlat; (11) Bruç; (12) Laç; (13) Shtogj; (14) Rrethe-Bajzë; (15) Burrel; (16) Komsi; (17) Urakë; (18) Midhë; (19) Hamallaj; (20) Pazhok; (21) Gërmenj; (22) Cerujë; (23) Katundas; (24) Patos; (25) Shuec; (26) Barç; (27) Kuç i Zi; (28) Shtikë; (29) Luaras; (30) Psar; (31) Borovë; (32) Rehovë; (33) Prodan; (34) Vajzë; (35) Dukat; (36) Çepunë; (37) Piskovë; (38) Rrapckë; (39) Grabovë; (40) Bajkaj; (41) Vodhinë; (42) Bodrishtë; (43) Kakavijë; (44) Kamenicë

FIGURE 17.2 Final phases of excavation of Tumulus 1 at Barç (Korçë). Excavations by Zhaneta Andrea (1973).

FIGURE 17.3 Tumulus at Çinamak (Kukës). Excavations by Bep Jubani (1970).

Figure 17.4 View of Patos tumulus before excavations (1969) (courtesy Muzafer Korkuti)

Figure 17.5 View of graves in Patos tumulus (1969)

Figure 17.6 General view of the Kamenicë tumulus after excavations, conservation, and public presentation. Excavations by Lorenc Bejko, 2007 (photo Alket Islami).

FIGURE 17.7 General view of tumuli 9, 10, and 11 at Apollonia, after excavations by Maria Grazia Amore (2005); the double acropolis of Apollonia is visible in the upper left (photo Alket Islami)

FIGURE 17.8 Various burials at one level of Tumulus 9, Apollonia (2003) (courtesy Maria Grazia Amore)

Chapter 18 Illustrations

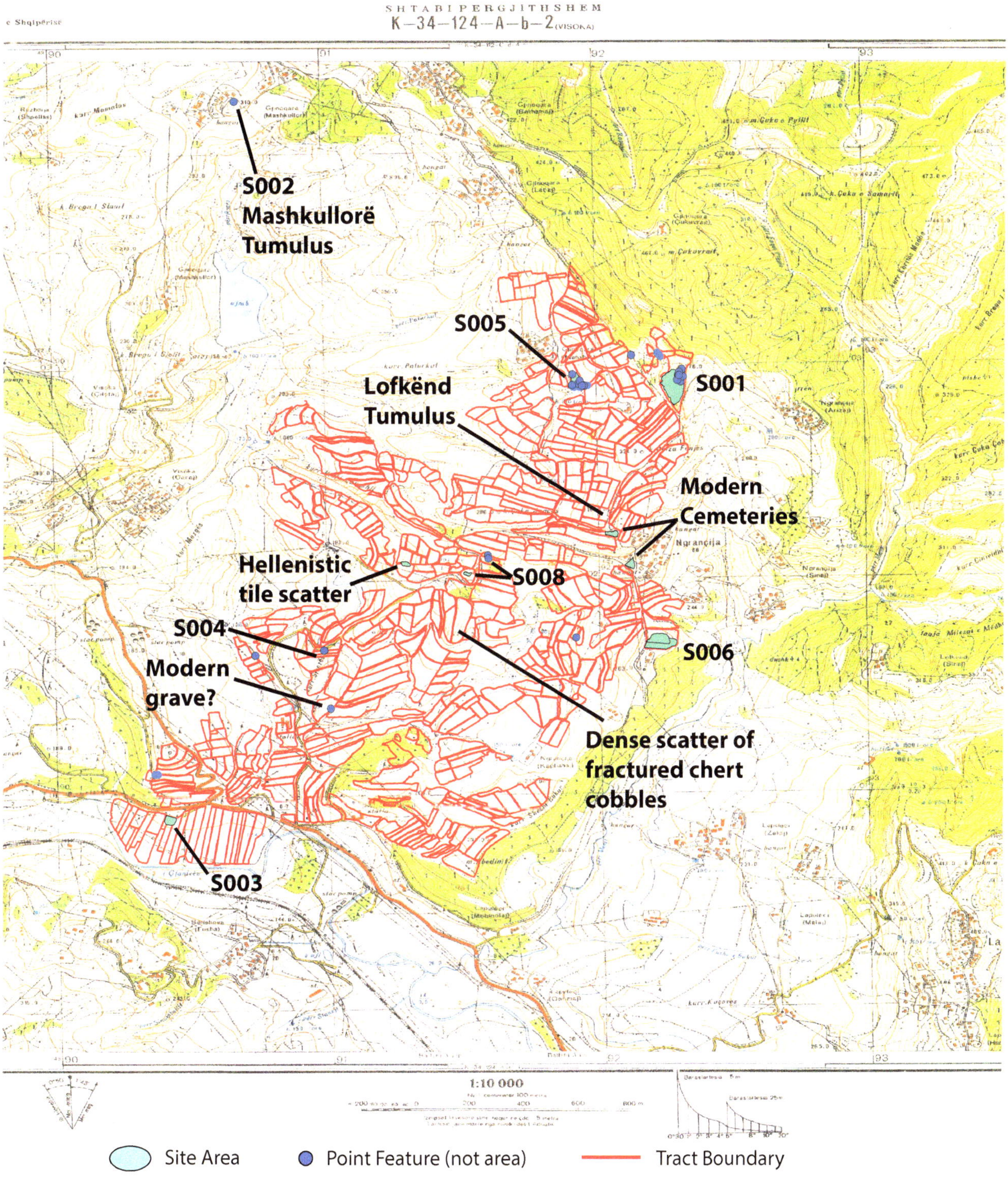

FIGURE 18.1 Plan of the survey area showing 2007 and 2008 tracts.
Sites and important features in the area are labeled (Site 007 not shown).

FIGURE 18.2 S001 from the south showing the highly eroded hilltop. The darker orange soil surrounded by olive trees near the top of the photo mark the highest point on the hilltop where the artifact concentration was greatest. Field walkers Shauna Kullman (foreground), Alison Adams, and Serena Vartazarian (background) shown collecting dog-leash units (photo Jamie Aprile).

FIGURE 18.3 Possible damaged grave in the central highest point of Site S001 (photo Jamie Aprile)

Figure 18.4 View of the Mashkullorë tumulus (S002) from the south showing modern graves and wall foundations. The large bushes mark pits in the surface (photo Jamie Aprile).

Figure 18.5 View of the terrace edge where Visokë A (S004) is eroding from the scarp (photo Jamie Aprile)

a

b

Figure 18.6 Views of reused architectural blocks in the abandoned school building in the modern village of Ngrançija (photos Jamie Aprile)

FIGURE 18.7 View of S008. The low mound in the center of the photo marks the location of the stable and teqe (tekke) ruins (photo Jamie Aprile)

FIGURE 18.8 Plan of all intensively surveyed tracts showing the distribution of all pottery recovered

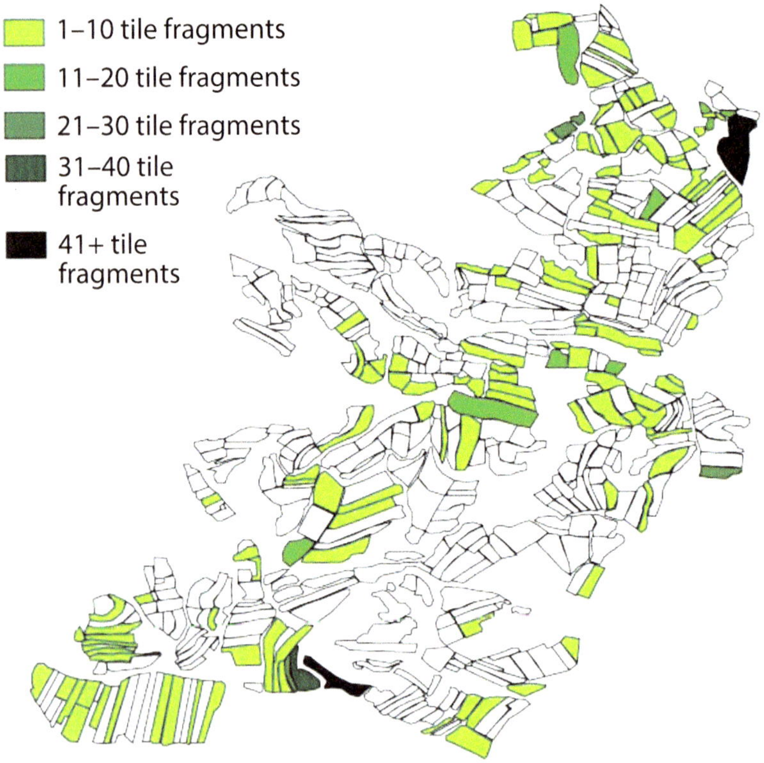

FIGURE 18.9 Plan of all intensively surveyed tracts showing the distribution of all tile collected

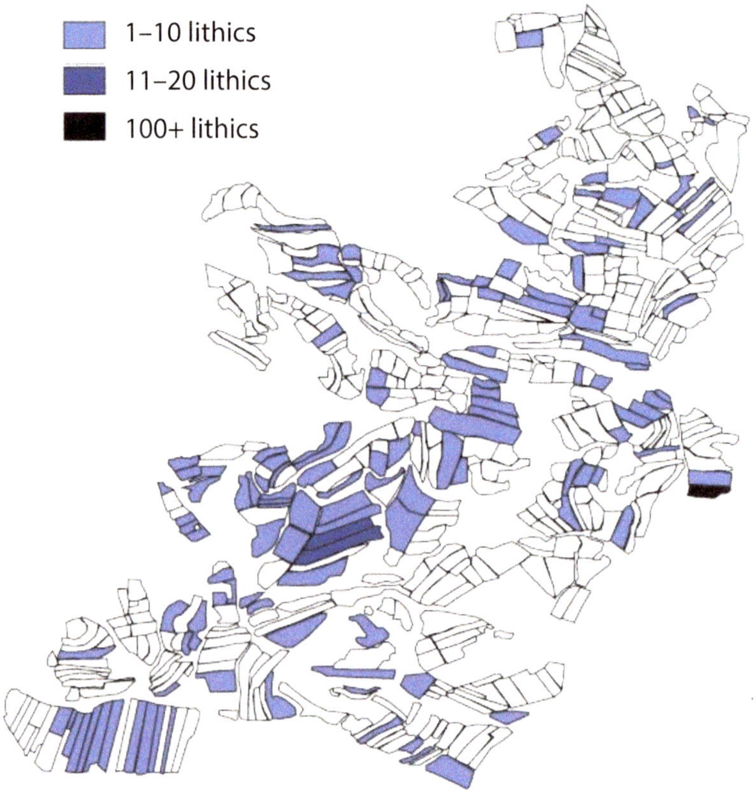

FIGURE 18.10 Plan of all intensively surveyed tracts showing the distribution of all lithic finds collected

Chapter 19 Illustrations

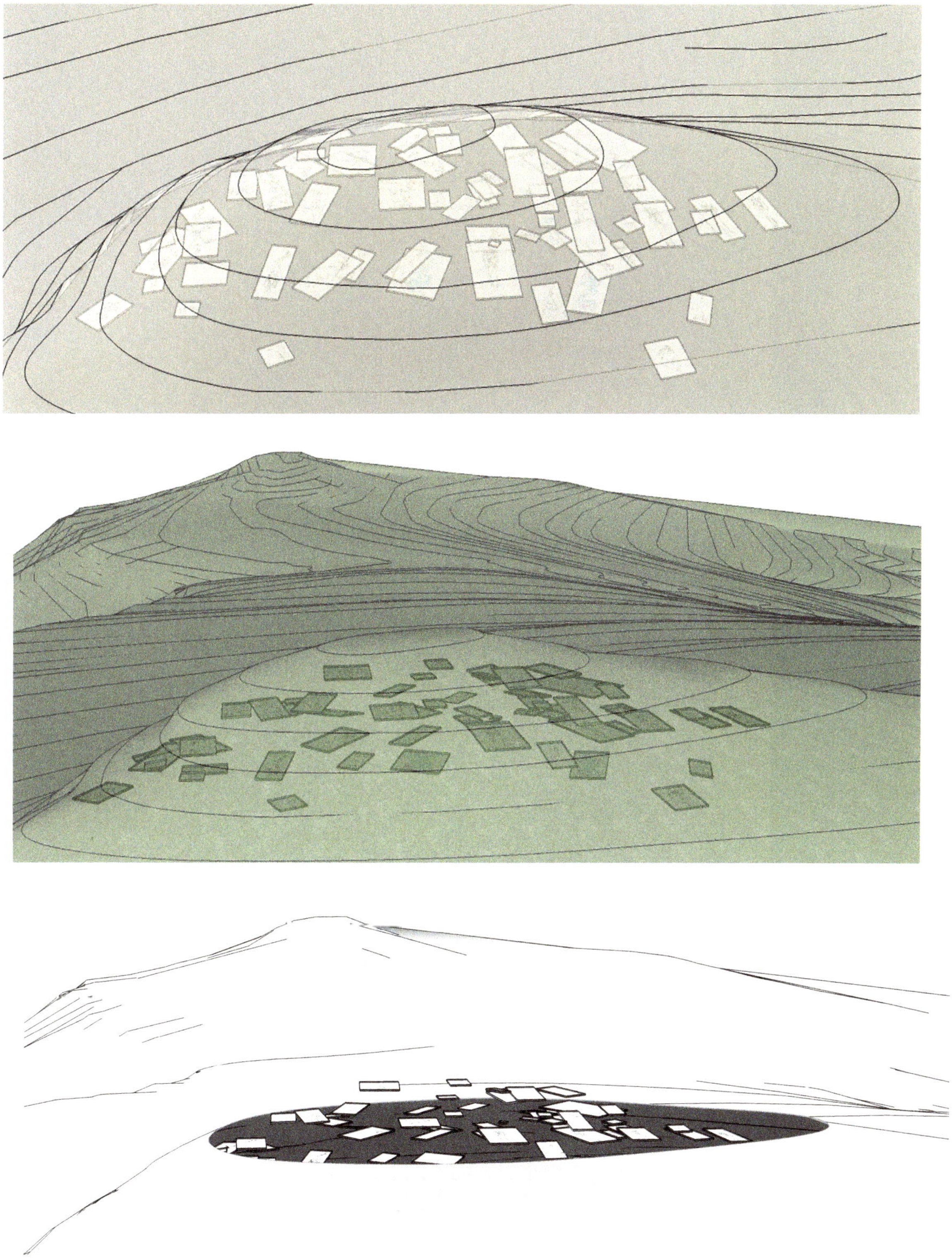

a

b

c

FIGURE 19.1 (a–c) Three preliminary renderings of the tumulus (looking west or west-northwest) from a three-dimensional model developed at the UCLA Experiential Technologies Center (Christopher Johanson and Itay Zaharovits)

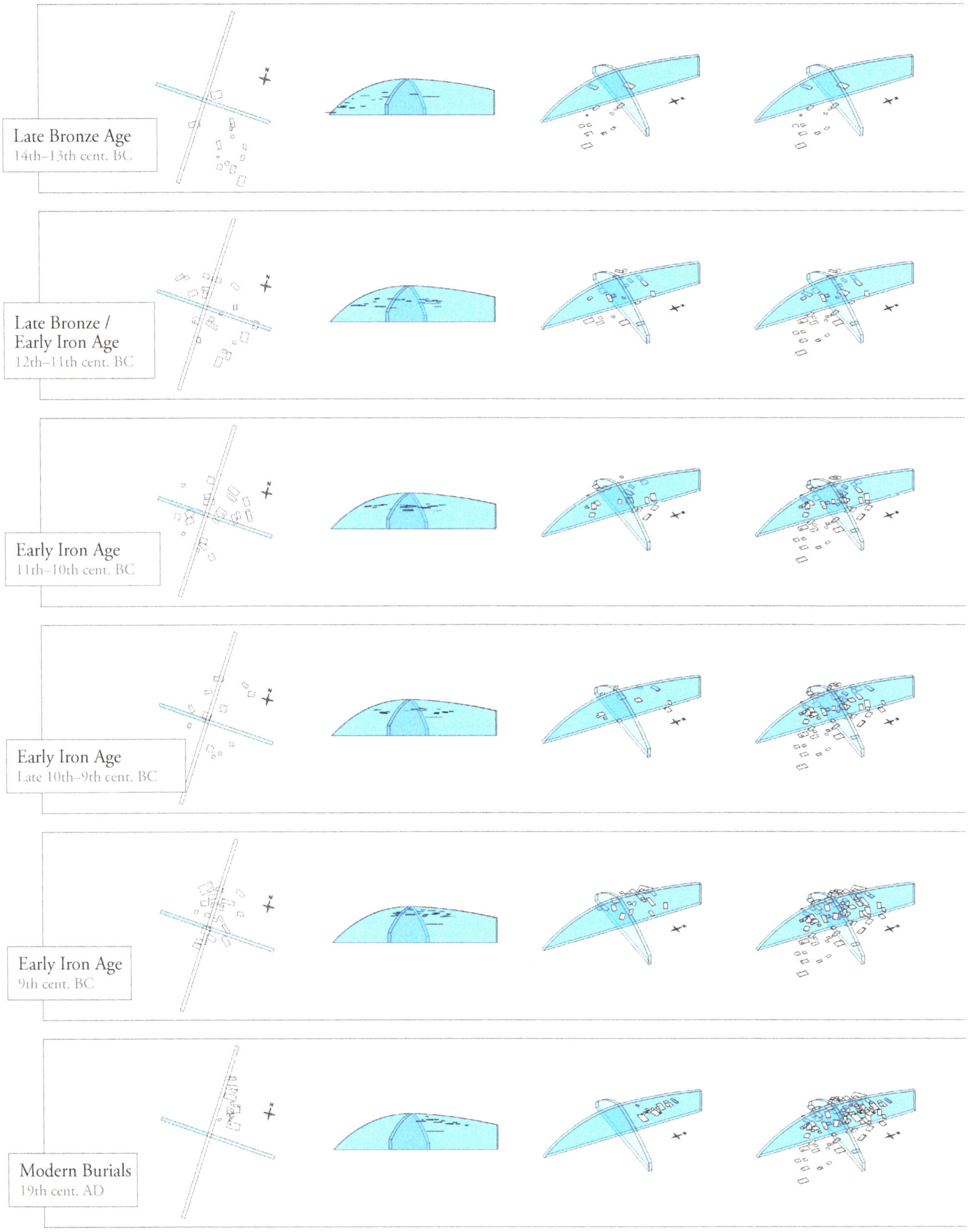

(*Facing Pages*) FIGURE 19.2 Phasing, chronology, and spatial arrangement of the burials. Information and graphic design by the UCLA Institute of Digital Research and Education (Jennifer Dillon) and the UCLA Experiential Technologies Center (Steven Hicks and Christopher Johanson). Three-dimensional digital model developed by the UCLA Experiential Technologies Center (Christopher Johanson, Steven Hicks, and Itay Zaharovits).

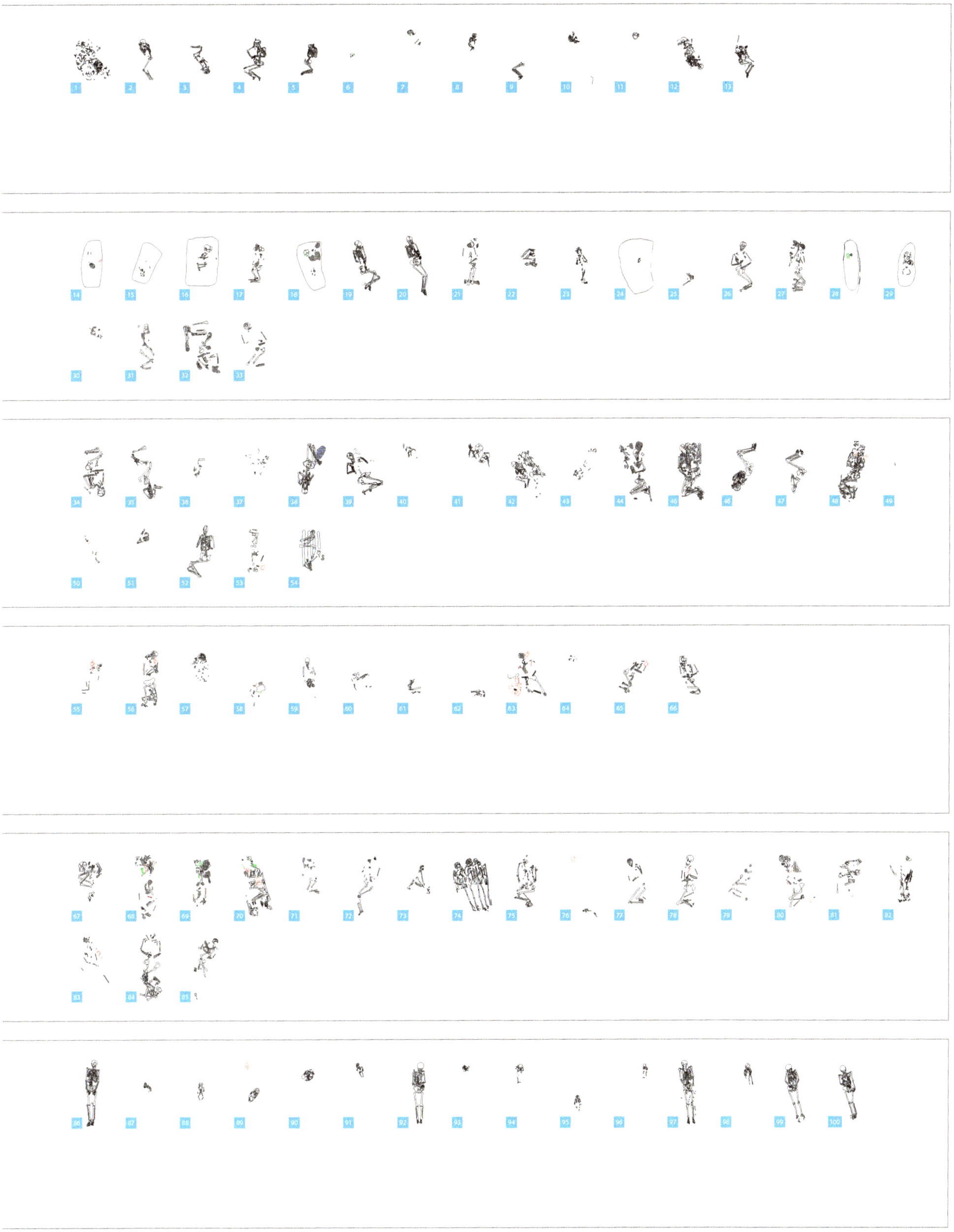

Figure 19.2 (*Continued from facing page*) Phasing, chronology, and spatial arrangement of the burials

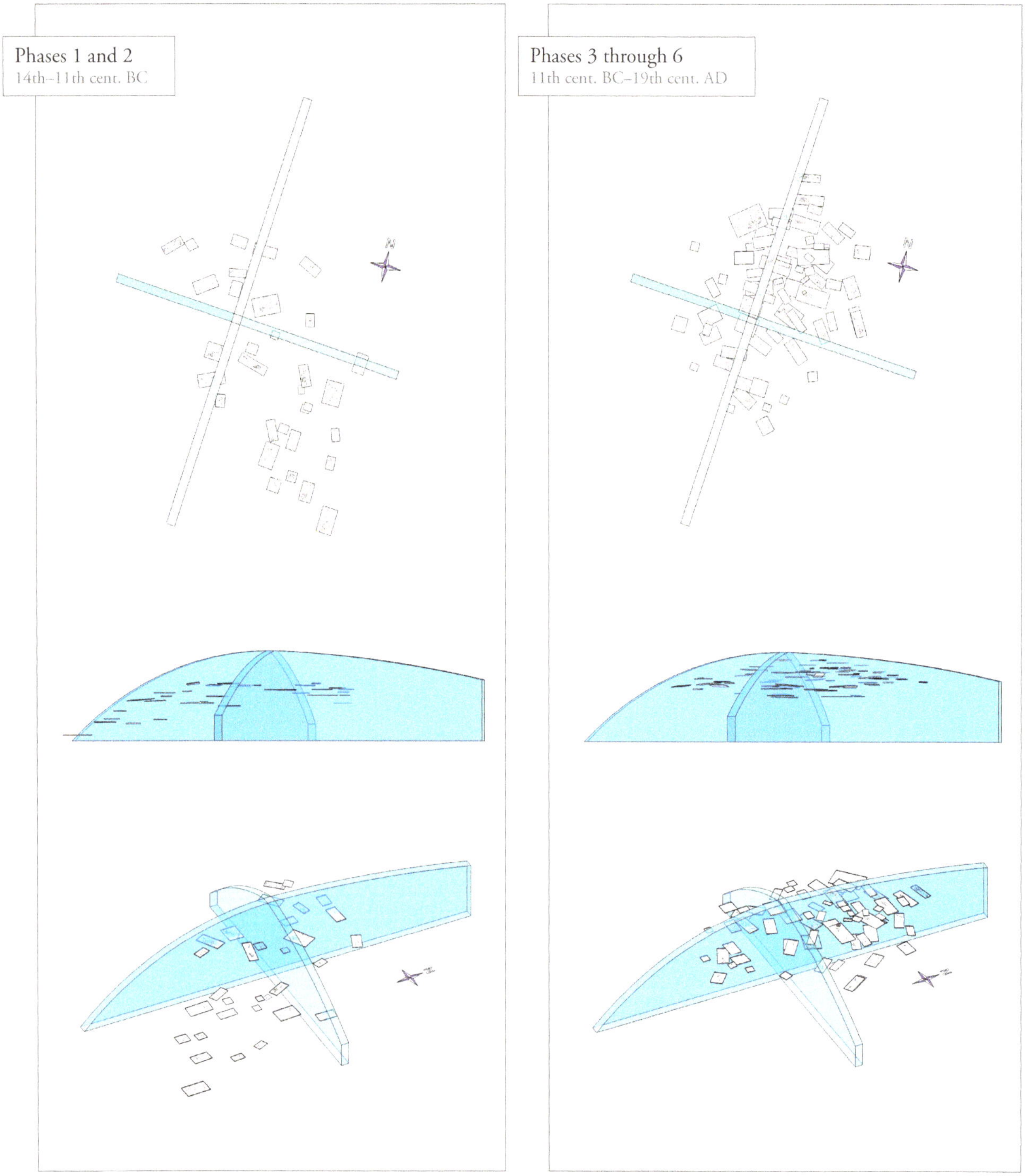

FIGURE 19.3 Two burial locations define the tumulus. One is centered on Tomb I. The other follows the southern slope of the original hill. Information and graphic design provided by the UCLA Institute of Digital Research and Education (Jennifer Dillon) and the UCLA Experiential Technology Center (Steven Hicks and Christopher Johanson). Three-dimensional digital model developed by the UCLA Experiential Technologies Center (Christopher Johanson, Steven Hicks, and Itay Zaharovits).

a

b

Figure 19.4 (a) Stylized rendering of pre-burial site of the tumulus; (b) comparative rendering of the tumulus as it might have appeared in Phase 6. Modeling and rendering developed by the UCLA Experiential Technologies Center (Marie Saldaña).

Chapter 20 Illustrations

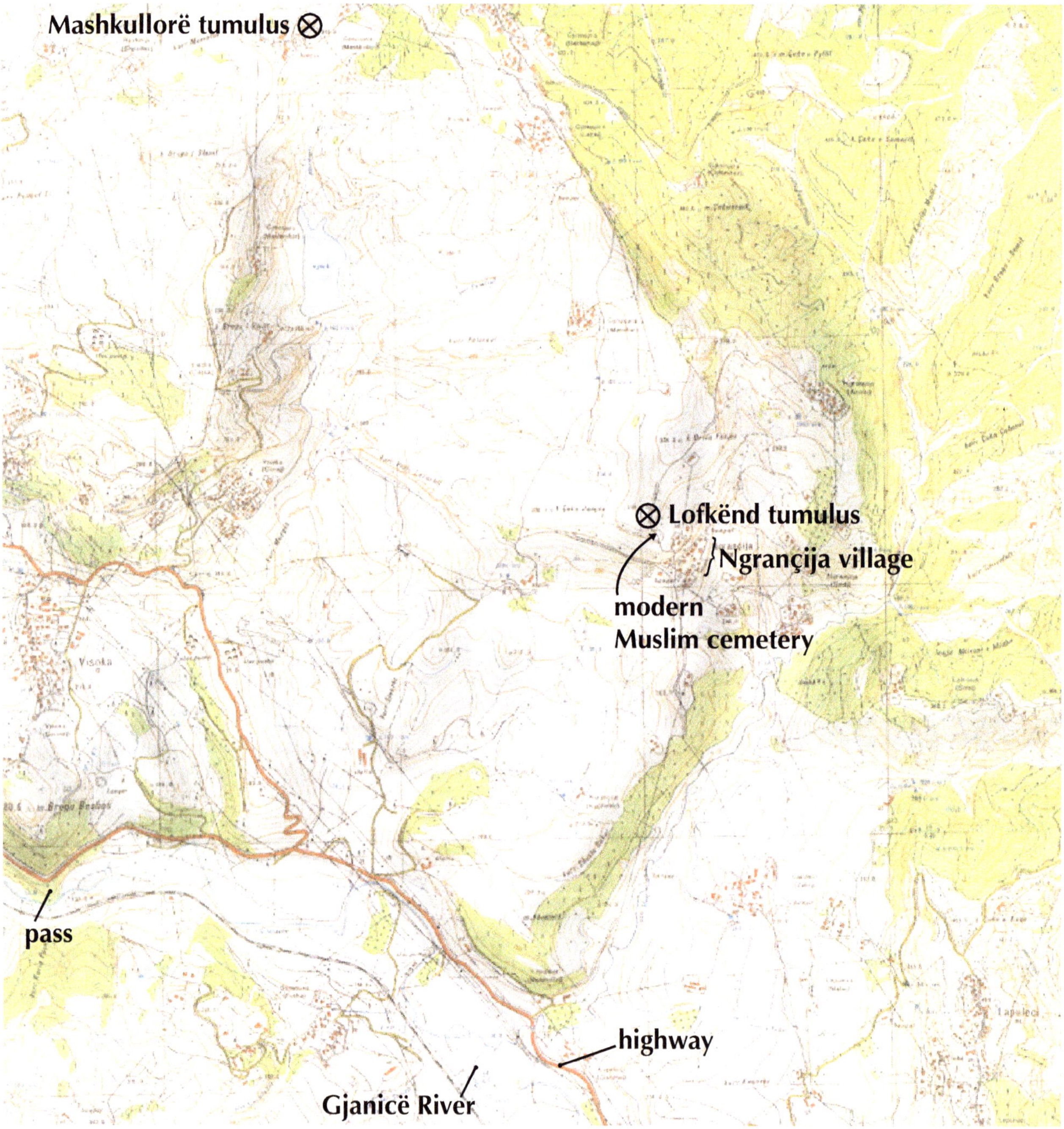

FIGURE 20.1 Map indicating the place of the Lofkënd and Mashkullorë tumuli, and other features in the landscape

FIGURE 20.2 View north-northeast from the Lofkënd tumulus, toward the Mallakastër ridge, which can been seen on the horizon (photo Samantha Martin-McAuliffe)

FIGURE 20.3 View south-southwest from the vicinity of the Lofkënd tumulus, toward the Gjanicë River

Figure 20.4 View in 2008 of the Lofkënd tumulus (rebuilt) from the village of Ngrançija to the immediate southeast (photo Samantha Martin-McAuliffe)

Figure 20.5 View in 2008 toward the Lofkënd tumulus (rebuilt) from the tumulus of Mashkullorë in the northwest (photo Samantha Martin-McAuliffe)

Figure 20.6 The site of Margëlliç, viewed from the tumulus of Lofkënd (photo Samantha Martin-McAuliffe)

a

b

FIGURE 20.7 (a) View south-southwest from the Lofkënd tumulus, toward the knobbly hill (seen at the center) that forms part of the narrow pass in the river valley; (b) basin of the Gjanicë River, looking toward the pass (in the center) (photo Samantha Martin-McAuliffe)

FIGURE 20.8 View south-southeast from the Lofkënd tumulus toward the ridge that forms the eastern edge of the crescent valley (photo Samantha Martin-McAuliffe)

FIGURE 20.9 View in 2008 toward the Lofkënd tumulus (rebuilt) from north of the highway (SH4) (photo Samantha Martin-McAuliffe)

FIGURE 20.10 View in 2008 toward the Lofkënd tumulus (rebuilt) from the river basin (photo Samantha Martin-McAuliffe)

FIGURE 20.11 Humphry Repton's trade card, engraved by Thomas Medland, depicting him as a landscape-gardener with his surveying tools

FIGURE 20.12 Close-up of tumulus (rebuilt) in 2008 showing the contrast of its contour with the surrounding hillside (photo Sarah Morris)

Figure 20.13 View from the south of the modern Muslim cemetery below the Lofkënd tumulus (photo Rich McDonald)

Figure 20.14 View of the modern Muslim cemetery (lower right) below the Mashkullorë tumulus (photo Samantha Martin-McAuliffe)

Chapter 21 Illustrations

FIGURE 21.1 Artist's conception of Lekë Dukagjini holding a scroll of the *Kanun* attributed to him

Figure 21.2 Shtjefën Konstantin Gjeçov (1874–1929)

Chapter 22 Illustrations

FIGURE 22.1 The tumulus with standing baulks at the end of the 2005 season, view from above northeast (photo Alket Islami)

FIGURE 22.2 The tumulus at the beginning of the 2006 season, as backfilled at the conclusion of the 2005 campaign (photo John Papadopoulos)

a

b

Figure 22.3 The bedrock of the tumulus at the end of excavations in 2007: (a) view from north (photo 1810); (b) view from south, with the pit for Tomb I in the center (photo 1824)

Figure 22.4 The bedrock of the tumulus at the end of excavations in 2007.
View from west, with the village of Lofkënd on the ridge in the distance (photo 3010).

Figure 22.5 Overview of the mud brick works adjacent to the site, looking northeast

a

b

Figure 22.6 Mixing the soil, straw, and water for the making of mud bricks: (a) mixing pits with water brought up to the site by donkey; (b) mixing the soil, straw, and water; (c, d) *on facing page*

c

d

FIGURE 22.6 (*continued*). Mixing the soil, straw, and water for the making of mud bricks: (c) treading of the mixture; (d) prepared mixture heaped into a pile

a

b

c

d

FIGURE 22.7 Making the mud bricks: (a) Baki Ymeri with double mold used to make all the bricks; (b) putting the mixture into the mold; (c) lifting the mold with the bricks to the drying area; (d) turning the bricks to promote drying

FIGURE 22.8 Dried mud bricks piled up, together with bricks at various stages of drying (in foreground)

FIGURE 22.9 Small stretch of the fortification wall of Bronze Age Hattuša (Turkey) as reconstructed (photo Charlie Steinmetz)

FIGURE 22.10 The mud bricks as they were stacked on either side of the baulks at the end of the 2006 season after having been exposed at the beginning of the 2007 season (photo 1664)

FIGURE 22.11 Mud bricks as built into a wall replicating the baulks, with openings to allow backfilling, at the end of the 2007 season (photo John Papadopoulos)

Figure 22.12 View from south of the reconstructed tumulus at the end of the 2007 season

Figure 22.13 View of the tumulus in the summer of 2008, a year after the reconstruction (photo 1883) (photo Ian Coyle)

Figure 22.14 The Patos tumulus in 1976, prior to excavations, view from west,with Skender Muçaj holding scale (photo Muzafer Korkuti)

FIGURE 22.15 The royal mounds at Gamla Uppsala, as shown in the *Suecia Antiqua et Hodierna* (ca. 1700)

FIGURE 22.16 Panoramic view of the Kamenicë tumulus in southeast Albania after the site was prepared as an open-air museum accessible to the public (photo Rich MacDonald)

UCLA Cotsen Institute of Archaeology Press

MONUMENTA ARCHAEOLOGICA